A PEARSON AUSTRALIA CUSTOM BOOK

Introduction to Mathematical Thinking

NMBR140

AF746502

This custom book is compiled for the Australian Catholic University from:

PEARSON MATHEMATICS NEW SOUTH WALES 7 STUDENT BOOK
MCLEOD ET AL.

PEARSON MATHEMATICS NEW SOUTH WALES 8 STUDENT BOOK
MCLEOD ET AL.

PEARSON MATHEMATICS NEW SOUTH WALES 9, STAGE 5.1, 5.2 & 5.3 STUDENT BOOK
MCLEOD ET AL.

PEARSON MATHEMATICS NEW SOUTH WALES 10, STAGE 5.1, 5.2 & 5.3 STUDENT BOOK
MCLEOD ET AL.

FOUNDATION MATHS
6TH EDITION
CROFT & DAVISON

Pearson Australia
707 Collins Street
Melbourne VIC 3008
Ph: 03 9811 2400

www.pearson.com.au

Project Management Team Leader:	Jill Gillies
Custom Portfolio Manager:	Lucie Bartonek
Portfolio Associate:	Eleanor Yeoell
Production Manager:	Candace Stuart
Production Controller:	Dom Harman

Printed and bound in Australia by The SOS + Print Media Group

ISBN: 978 1 4886 2611 1

Pearson

Table of Contents

About this Custom Book

Welcome to **Introduction to Mathematical Thinking NMBR140**.

The material included in this custom book has been specifically chosen to meet your course requirements. Please note that chapter, section and page numbers from the original source books still appear in this book.

The original page numbers do not run in consecutive order because only material relevant for your needs is in the book. So, to find your way around the book, please use the custom book page numbers.

The custom book page numbers are in the **Navigation Bar** at the top of each page and in the **Table of Contents**.

We hope you find this text easy to follow and enjoyable.

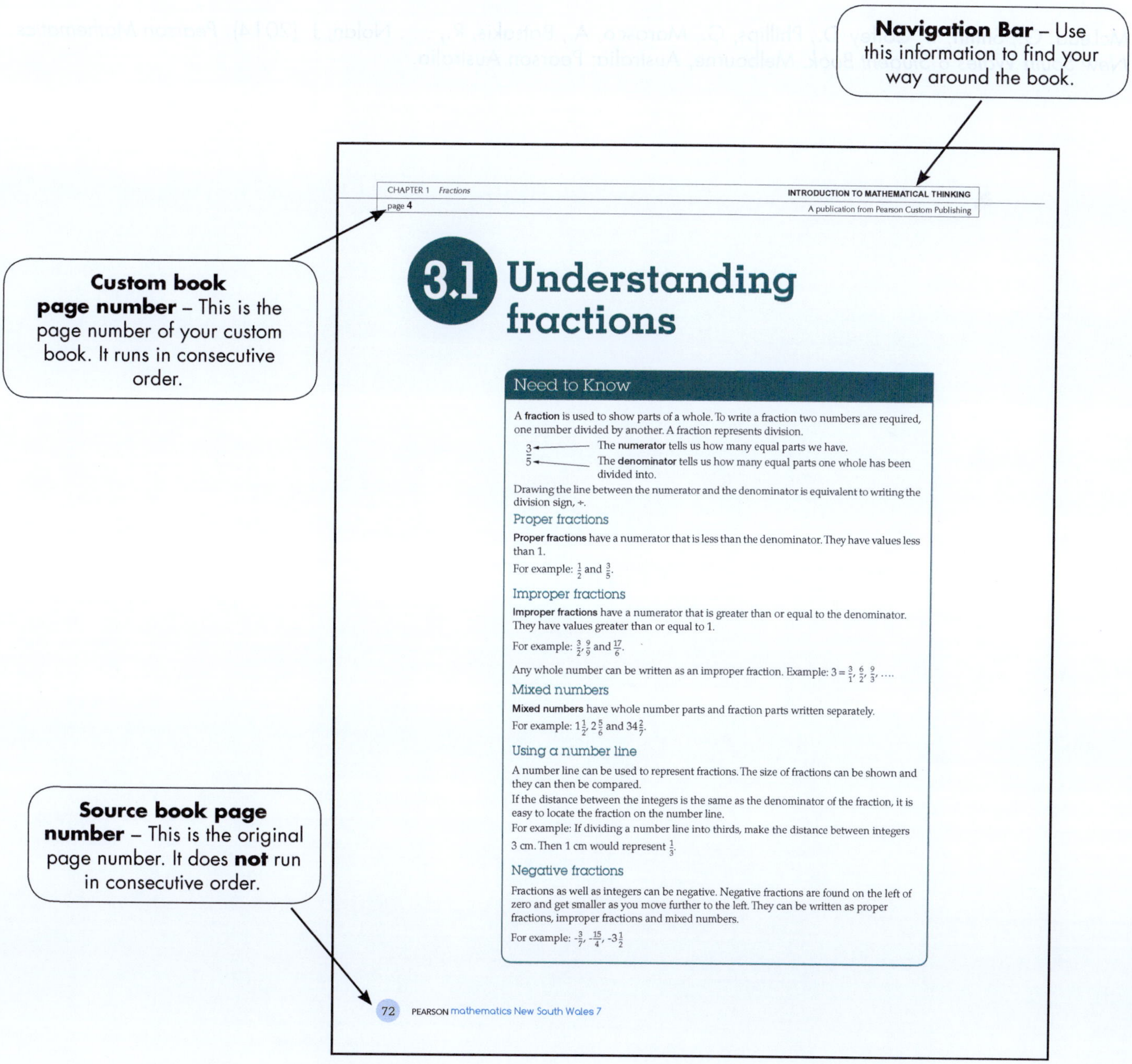

CHAPTER 1 *Fractions* INTRODUCTION TO MATHEMATICAL THINKING

page 4 A publication from Pearson Custom Publishing

3.1 Understanding fractions

Need to Know

A **fraction** is used to show parts of a whole. To write a fraction two numbers are required, one number divided by another. A fraction represents division.

$\frac{3}{5}$ — The **numerator** tells us how many equal parts we have. The **denominator** tells us how many equal parts one whole has been divided into.

Drawing the line between the numerator and the denominator is equivalent to writing the division sign, ÷.

Proper fractions

Proper fractions have a numerator that is less than the denominator. They have values less than 1.

For example: $\frac{1}{2}$ and $\frac{3}{5}$.

Improper fractions

Improper fractions have a numerator that is greater than or equal to the denominator. They have values greater than or equal to 1.

For example: $\frac{3}{2}, \frac{9}{9}$ and $\frac{17}{6}$.

Any whole number can be written as an improper fraction. Example: $3 = \frac{3}{1}, \frac{6}{2}, \frac{9}{3}, \ldots$

Mixed numbers

Mixed numbers have whole number parts and fraction parts written separately.

For example: $1\frac{1}{2}, 2\frac{5}{6}$ and $34\frac{2}{7}$.

Using a number line

A number line can be used to represent fractions. The size of fractions can be shown and they can then be compared.

If the distance between the integers is the same as the denominator of the fraction, it is easy to locate the fraction on the number line.

For example: If dividing a number line into thirds, make the distance between integers 3 cm. Then 1 cm would represent $\frac{1}{3}$.

Negative fractions

Fractions as well as integers can be negative. Negative fractions are found on the left of zero and get smaller as you move further to the left. They can be written as proper fractions, improper fractions and mixed numbers.

For example: $-\frac{3}{7}, -\frac{15}{4}, -3\frac{1}{2}$

72 PEARSON mathematics New South Wales 7

References

If you need to cite the material in this custom text, please find below the full references for each source text in APA style:

Croft, A., & Davison, R. (2016). *Foundation Maths* (6th ed.). Harlow, United Kingdom: Pearson Education Limited.

McLeod, G., Coffey, D., Marasco, A., Strasser, D., Aus, B., Byers, T., . . . Nolan, J. (2013). *Pearson Mathematics New South Wales 7 Student Book*. Melbourne, Australia: Pearson Australia.

McLeod, G., Coffey, D., Marasco, A., Strasser, D., Aus, B., Byers, T., . . . Phillips, G. (2013). *Pearson Mathematics New South Wales 9 Stages 5.1, 5.2 & 5.3 Student Book*. Melbourne, Australia: Pearson Australia.

McLeod, G., Coffey, D., Marasco, A., Strasser, D., Aus, B., Byers, T., . . . Phillips, G. (2014). *Pearson Mathematics New South Wales 10 Stages 5.1, 5.2 & 5.3 Student Book*. Melbourne, Australia: Pearson Australia.

McLeod, G., Smith, T., Coffey, D., Phillips, G., Marasco, A., Batsakis, R., . . . Nolan, J. (2014). *Pearson Mathematics New South Wales 8 Student Book*. Melbourne, Australia: Pearson Australia.

SECTION 1

Fractions, Decimals and Percentages

Fractions

3

Recall 3

Prepare for this chapter by attempting the following questions. If you have difficulty with a question, go to Pearson Places and download the Recall Worksheet.

1 Which of these have three-quarters of the whole shape shaded in?

(a) 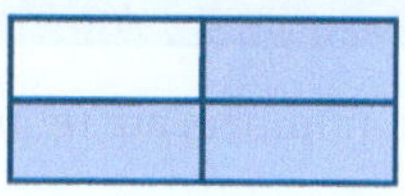
(b) 
(c) 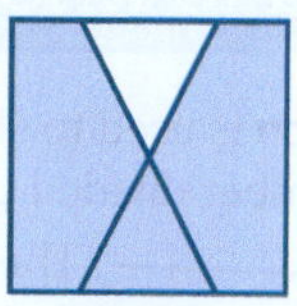
(d) 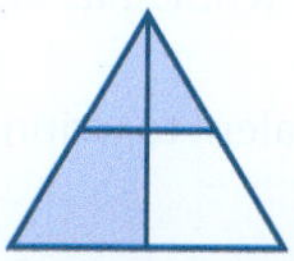
(e) 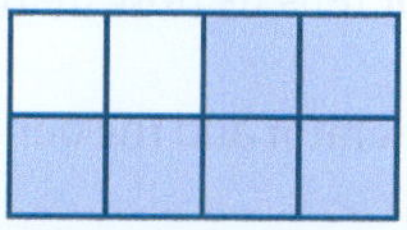
(f) 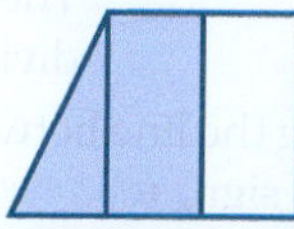

2 Write down a fraction to represent each of the following diagrams or descriptions.

(a) 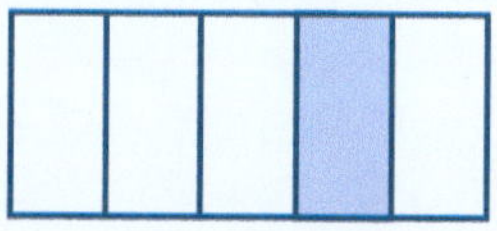
(b) 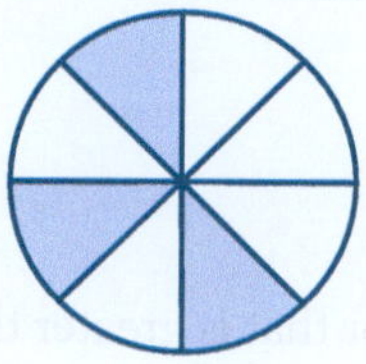
(c) 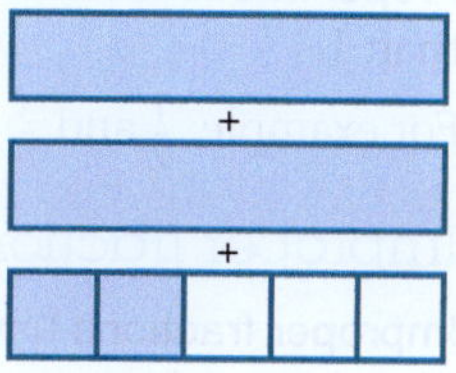

(d) One person's share of a birthday cake divided equally between 10 people.

(e) Joseph eats five out of eight equal-sized slices of a pizza.

(f) two-thirds (g) twelve-fifths (h) seven-sixteenths

3 (a) Write these in order from largest to smallest: $\frac{3}{4}, 1, 0, 2, 1\frac{1}{4}$

(b) Write these in order from smallest to largest: $\frac{3}{8}, \frac{1}{8}, 0, \frac{7}{8}, \frac{11}{8}, 1, \frac{9}{8}$

4 Calculate:

(a) $\frac{4}{7}+\frac{1}{7}$ (b) $\frac{8}{11}-\frac{2}{11}$ (c) $\frac{2}{3}+\frac{2}{3}$

5 (a) Write the first five multiples of 8. (b) Write the first five multiples of 12.

(c) Write the lowest common multiple (LCM) of 8 and 12.

(d) Write all the factors of 24. (e) Write all the factors of 36.

(f) Write the highest common factor (HCF) of 24 and 36.

Syllabus references

Number and Algebra: Fractions, Decimals and Percentages

✓ Compare fractions using equivalence; locate and represent positive and negative fractions and mixed numerals on a number line **(ACMNA152)**

✓ Solve problems involving addition and subtraction of fractions, including those with unrelated denominators **(ACMNA153)**

✓ Multiply and divide fractions and decimals using efficient written strategies and digital technologies **(ACMNA154)**

✓ Express one quantity as a fraction of another, with and without the use of digital technologies **(ACMNA155)**

3.1 Understanding fractions

Need to Know

A **fraction** is used to show parts of a whole. To write a fraction two numbers are required, one number divided by another. A fraction represents division.

$\frac{3}{5}$ — 3: The **numerator** tells us how many equal parts we have.
5: The **denominator** tells us how many equal parts one whole has been divided into.

Drawing the line between the numerator and the denominator is equivalent to writing the division sign, ÷.

Proper fractions

Proper fractions have a numerator that is less than the denominator. They have values less than 1.

For example: $\frac{1}{2}$ and $\frac{3}{5}$.

Improper fractions

Improper fractions have a numerator that is greater than or equal to the denominator. They have values greater than or equal to 1.

For example: $\frac{3}{2}, \frac{9}{9}$ and $\frac{17}{6}$.

Any whole number can be written as an improper fraction. Example: $3 = \frac{3}{1}, \frac{6}{2}, \frac{9}{3}, \ldots$

Mixed numbers

Mixed numbers have whole number parts and fraction parts written separately.

For example: $1\frac{1}{2}$, $2\frac{5}{6}$ and $34\frac{2}{7}$.

Using a number line

A number line can be used to represent fractions. The size of fractions can be shown and they can then be compared.

If the distance between the integers is the same as the denominator of the fraction, it is easy to locate the fraction on the number line.

For example: If dividing a number line into thirds, make the distance between integers 3 cm. Then 1 cm would represent $\frac{1}{3}$.

Negative fractions

Fractions as well as integers can be negative. Negative fractions are found on the left of zero and get smaller as you move further to the left. They can be written as proper fractions, improper fractions and mixed numbers.

For example: $-\frac{3}{7}, -\frac{15}{4}, -3\frac{1}{2}$

3.1

Worked Example 1

WE1

How many smileys are there in $\frac{4}{5}$ of this collection?

Working

$15 \div 5 = 3$, so $\frac{1}{5}$ of $15 = 3$

$\frac{4}{5}$ is 4 groups of 3 smileys,

so $\frac{4}{5}$ of $15 = 4 \times 3 = 12$

Worked Example 2

WE2

Copy this number line and show the positions of these fractions: $-\frac{3}{5}$, $\frac{4}{5}$ and $\frac{8}{5}$.

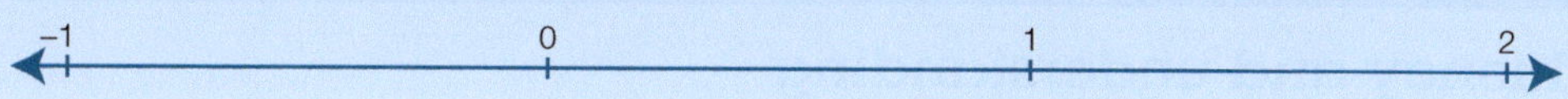

Working

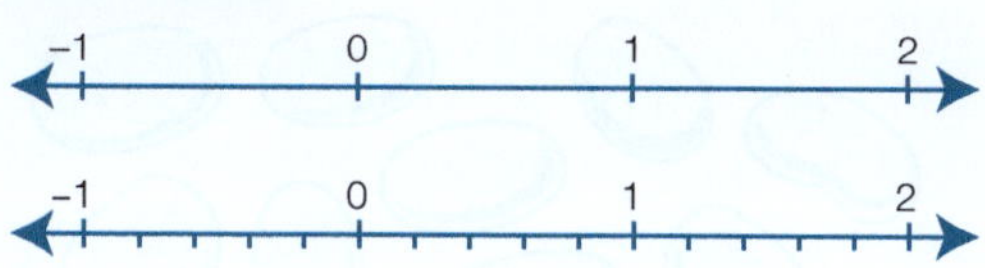

(Divide the number line into fifths.)

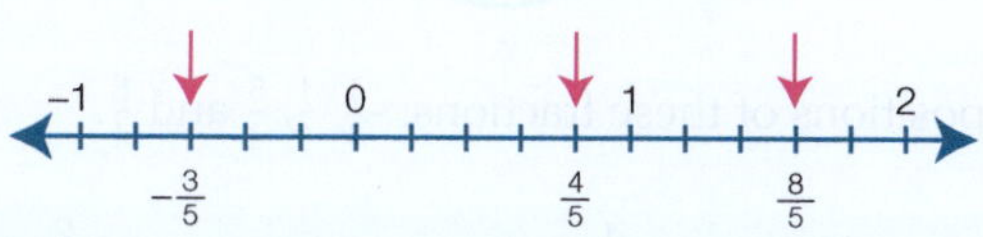

Worked Example 3

WE3

Write the whole number 9 as an improper fraction with a denominator of 4.

Working

$$9 = \frac{\square}{4}$$

$$= \frac{9}{1} \times \frac{4}{4}$$

$$= \frac{36}{4}$$

3.1

Worked Example 4

WE4

For the following, write the amount each student receives as a fraction (or as a mixed number if appropriate).

5 blocks of chocolate are shared equally among 8 students.

Working

$5 \div 8$

$= \frac{5}{8}$

Answers page 455

3.1 Understanding fractions

Navigator

Q1, Q2, Q3 (a), Q4, Q5, Q6, Q7 Columns 1 & 2, Q8, Q11 (a), Q13, Q16, Q17, Q19 (a), Q20 (b), Q22	Q1, Q2, Q3 (b), Q4, Q5, Q6, Q7 Columns 2 & 3, Q8 (d)–(f), Q9, Q10, Q11, Q13, Q14, Q16, Q17, Q18, Q19 (b), Q20 (b), Q22	Q2 (b), Q3 (b), Q4, Q6, Q7 Column 3, Q8 (d)–(f), Q9, Q10, Q11 (b), Q12, Q13, Q14, Q15, Q17, Q18, Q19 (a), Q20, Q21, Q22

Fluency and understanding

WE1

1 **(a)** How many smileys are there in $\frac{3}{4}$ of this collection?

(b) How many jelly beans are there in $\frac{2}{3}$ of this collection?

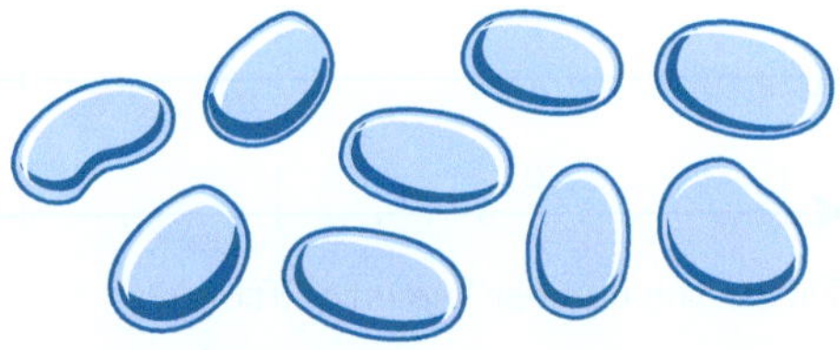

WE2

2 **(a)** Copy this number line and show the positions of these fractions: $-\frac{1}{4}, \frac{1}{4}, \frac{5}{4}$ and $\frac{8}{4}$.

(b) Copy this number line and show the positions of these fractions: $-\frac{5}{6}, \frac{2}{6}, -\frac{1}{6}$ and $\frac{8}{6}$.

WE3

3 **(a)** Write the whole number 5 as an improper fraction with a denominator of:

(i) 2 **(ii)** 7 **(iii)** 11 **(iv)** 5 **(v)** 1

(b) Write the whole number 13 as an improper fraction with a denominator of:

(i) 2 **(ii)** 5 **(iii)** 8 **(iv)** 13 **(v)** 1

3.1

4 (a) For each of the following, write the amount each student receives as a fraction (or as a mixed number if appropriate). WE4

(i) 1 pizza is shared equally between 2 students.

(ii) 2 apples are shared equally between 3 students.

(iii) 6 packets of lollies are shared equally between 5 students.

(iv) 10 packets of biscuits are shared equally between 7 students.

(b) In which of the above situations does a student receive more than one whole?

5 Here are 16 lollies. How many will you eat if you eat these fractions of the total?

(a) $\frac{1}{4}$

(b) $\frac{3}{4}$

(c) $\frac{1}{8}$

(d) $\frac{5}{8}$

6 (a) What fraction of the bananas in this bunch are rotten?

(b) What fraction of the flowers in this vase are red?

7 Write the value of the fraction indicated by the arrow on each of the number lines below.

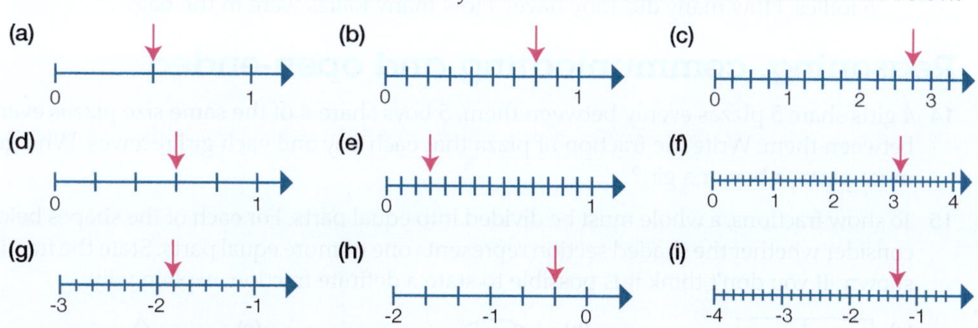

8 Write a fraction or a mixed number to show each of these:

(a) a numerator of 8 and a denominator of 17

(b) denominator of 4 and a numerator of 15

(c) nine chocolate biscuits in a packet of 20

(d) two wholes and two thirds

(e) 4 complete pairs of socks and one odd sock

(f) 3 whole 24-piece blocks of chocolate, with 7 extra pieces.

3.1

9 Write a fraction to show each of these:

(a) the weekend days as a fraction of a whole week

(b) 1 hour out of a whole day

(c) 1 second out of a whole minute

(d) 17 minutes out of a whole hour

(e) 157 mL of cola drunk from a 375 mL can

(f) 421 L of water in a 500 L rainwater tank.

10 What fraction of this collection of shapes are:

(a) stars **(b)** stars or hearts **(c)** not hearts?

11 **(a)** Draw a diagram to show that if 5 pizzas are shared equally between 6 students, then each student receives $\frac{5}{6}$ of a pizza.

(b) Draw a diagram to show that if 5 blocks of chocolate are shared equally between 4 students, then each student receives 1 full block and $\frac{1}{4}$ of a second block.

12 Tim is 149 cm tall. His dad is 185 cm tall.

(a) Write Tim's height as a fraction of his dad's height.

(b) Write his dad's height as a fraction of Tim's height.

13 **(a)** After driving 8 km, the Johnson family are $\frac{1}{4}$ of the way to the zoo. How far away from the zoo do the Johnsons live?

(b) Toby and his mum shared a bag of lollies: $\frac{2}{3}$ for Toby and $\frac{1}{3}$ for his mum. His mum had 6 lollies. How many did Toby have? How many lollies were in the bag?

Reasoning, communicating and open-ended

14 4 girls share 5 pizzas evenly between them. 5 boys share 4 of the same size pizzas evenly between them. Write the fraction of pizza that each boy and each girl receives. Who gets more pizza, a boy or a girl?

15 To show fractions, a whole must be divided into equal parts. For each of the shapes below, consider whether the shaded section represents one or more equal parts. State the fraction shown. If you don't think it is possible to state a definite fraction, explain why.

(a)

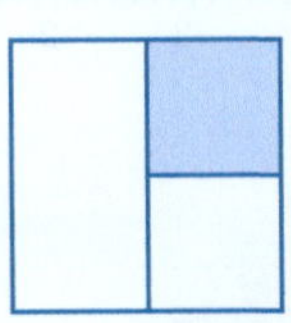

(b)

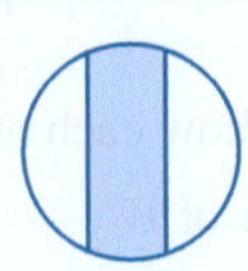

(c)

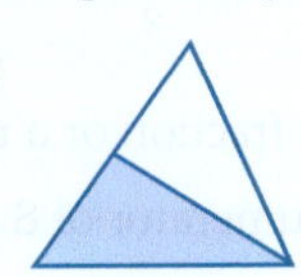

(d)

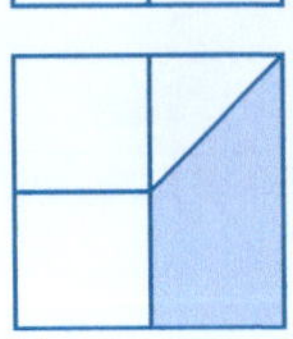

(e)

(f)

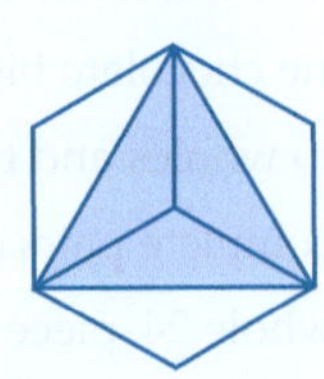

16 The fraction of this circle that is shaded is:

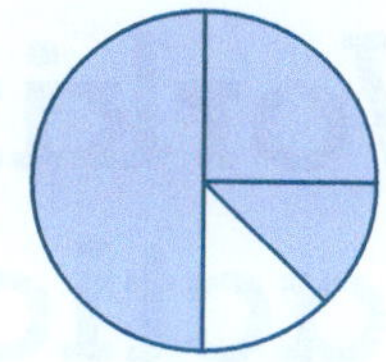

A $\frac{3}{4}$ **B** $\frac{4}{5}$ **C** $\frac{5}{6}$ **D** $\frac{7}{8}$

17 The fraction of this circle that is shaded is:

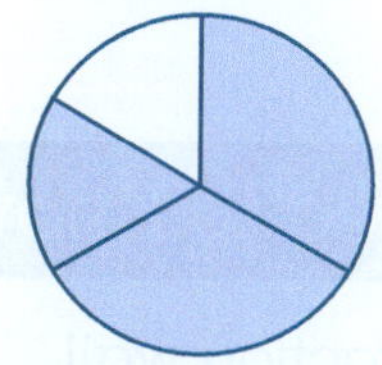

A $\frac{3}{4}$ **B** $\frac{4}{5}$ **C** $\frac{5}{6}$ **D** $\frac{7}{8}$

18 How many whole numbers would appear on the section of the number line between $\frac{1}{2}$ and $\frac{11}{2}$?

19 Draw a diagram that shows:

(a) three-eighths of an object **(b)** $\frac{5}{6}$ of a collection of objects **(c)** $\frac{9}{3}$

20 **(a)** This shape  represents $\frac{1}{5}$ of a whole. Draw two examples of what one whole could look like.

(b) This shape 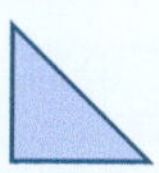 represents $\frac{1}{4}$ of a whole. Draw two examples of what two wholes could look like.

21 Draw a diagram to help you explain why the fraction of this tile that is shaded is $\frac{1}{2}$.

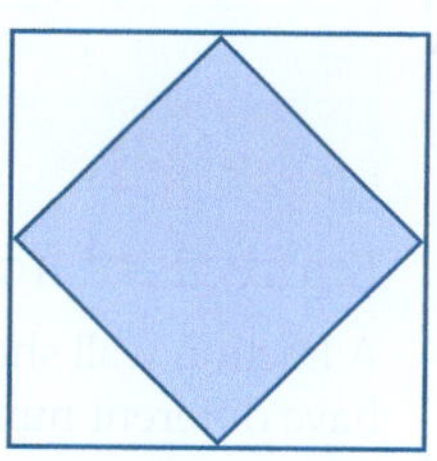

22

Oliver's diagram is supposed to show which of the fractions $\frac{1}{2}$ or $\frac{1}{3}$ is the larger fraction.

(a) Explain to Oliver what is wrong with his diagram.

(b) Draw a more accurate diagram to help you explain where he has gone wrong.

3.2 Working with fractions

Need to Know

A fraction wall

Equivalent fractions

A fraction wall shows **equivalent fractions**; fractions that represent the same amount but have different numerators and denominators. For example: $\frac{1}{2} = \frac{2}{4} = \frac{3}{6} = \frac{4}{8} = \frac{5}{10} = \frac{6}{12}$.

(These fractions are all shaded in green on the fraction wall. The dark blue sections are equal to $\frac{1}{3}$.)

Equivalent fractions are found either by multiplying both the numerator and the denominator by the same number, or by dividing both the numerator and the denominator by the same number.

Simplifying fractions

To write a fraction in its **simplest form**, or to **simplify** it, we divide the numerator and the denominator by their highest common factor (HCF). The process of simplifying fractions is often called '**cancelling**'.

Answers to fraction questions should always be written in simplest form.

Mixed numbers and improper fractions

To write an improper fraction as a mixed number, divide the numerator by the denominator to get the whole number part. We then write the remainder as the numerator of the fraction part.

To write a mixed number as an improper fraction, write the whole number part as an improper fraction with the same denominator. Then, add the proper fraction part.

3.2

Worked Example 5

WE 5

Write pairs of equivalent fractions by copying and completing the following.

(a) $\frac{7}{10} = \frac{\square}{40}$ (b) $\frac{27}{36} = \frac{3}{\square}$

Working

(a) $\frac{7}{10} = \frac{7}{10} \times \frac{4}{4}$

$= \frac{28}{40}$

(b) $\frac{27}{36} = \frac{3 \times \cancel{9}^1}{4 \times \cancel{9}^1}$

$= \frac{3}{4}$

Worked Example 6

WE 6

Write the following fraction and mixed number in their simplest form.

(a) $\frac{24}{60}$ (b) $3\frac{12}{28}$

Working

Method 1: Find the HCF

(a) HCF of 24 and 60 = 12

$\frac{^2\cancel{24}}{_5\cancel{60}}$

$= \frac{2}{5}$

(b) HCF of 12 and 28 = 4

$3\frac{^3\cancel{12}}{_7\cancel{28}}$

$= 3\frac{3}{7}$

Working

Method 2: Do several small divisions

(a) $\frac{^{12}\cancel{24}}{_{30}\cancel{60}}$

$= \frac{^6\cancel{12}}{_{15}\cancel{30}}$

$= \frac{^2\cancel{6}}{_5\cancel{15}}$

$= \frac{2}{5}$

(b) $3\frac{^6\cancel{12}}{_{14}\cancel{28}}$

$= 3\frac{^3\cancel{6}}{_7\cancel{14}}$

$= 3\frac{3}{7}$

Worked Example 7

WE 7

Write the improper fraction $\frac{34}{7}$ as a mixed number.

Working

$34 \div 7 = 4 \text{ rem } 6$

$\frac{34}{7} = 4\frac{6}{7}$

3.2

Worked Example 8

WE8

Write the mixed number $7\frac{3}{5}$ as an improper fraction.

Working

$$7\frac{3}{5} = 7 + \frac{3}{5}$$
$$= \frac{35}{5} + \frac{3}{5}$$
$$= \frac{38}{5}$$

Answers page 456

3.2 Working with fractions

Navigator

Q1 Columns 1–3, Q2 Columns 1–3, Q3 Columns 1-3, Q4 Columns 1–3, Q5 (a) & (b), Q6, Q7, Q8, Q9, Q10, Q11, Q15, Q16, Q17, Q18 (a)–(c), Q19, Q22	Q1 Columns 2 & 3, Q2 Columns 2 & 3, Q3 Columns 2 & 3, Q4 Columns 2 & 3, Q5, Q6, Q7, Q8, Q9, Q10, Q12, Q13, Q15, Q16, Q17, Q18, Q19, Q21, Q22	Q1 Column 4, Q2 Column 4, Q3 Column 4, Q4 Column 4, Q5, Q6, Q7, Q8, Q9, Q12, Q13, Q14, Q15, Q16, Q17, Q20, Q21, Q22

Fluency and understanding

WE5

1 Write pairs of equivalent fractions by copying and completing the following.

(a) $\frac{3}{4} = \frac{\square}{20}$ (b) $\frac{1}{3} = \frac{7}{\square}$ (c) $\frac{1}{2} = \frac{\square}{52}$ (d) $\frac{2}{5} = \frac{16}{\square}$

(e) $\frac{8}{100} = \frac{\square}{25}$ (f) $\frac{15}{100} = \frac{\square}{20}$ (g) $\frac{42}{70} = \frac{6}{\square}$ (h) $\frac{7}{11} = \frac{\square}{99}$

(i) $\frac{4}{3} = \frac{\square}{27}$ (j) $\frac{28}{21} = \frac{4}{\square}$ (k) $\frac{24}{16} = \frac{\square}{4}$ (l) $\frac{72}{36} = \frac{\square}{9}$

(m) $\frac{8}{\square} = \frac{88}{55}$ (n) $\frac{\square}{5} = \frac{42}{35}$ (o) $\frac{54}{\square} = \frac{9}{6}$ (p) $\frac{\square}{36} = \frac{4}{9}$

WE6

2 Write the following fractions and mixed numbers in their simplest form.

(a) $\frac{5}{10}$ (b) $\frac{3}{9}$ (c) $\frac{3}{12}$ (d) $\frac{7}{21}$

(e) $\frac{4}{10}$ (f) $\frac{8}{22}$ (g) $\frac{10}{16}$ (h) $\frac{36}{24}$

(i) $2\frac{14}{21}$ (j) $3\frac{5}{30}$ (k) $1\frac{6}{20}$ (l) $12\frac{24}{30}$

(m) $6\frac{20}{45}$ (n) $3\frac{25}{100}$ (o) $2\frac{35}{42}$ (p) $100\frac{12}{84}$

Hint

If you need to revise how to find the HCF, go to pages 34 and 36.

3 Write these improper fractions as mixed numbers. WE7

(a) $\frac{7}{5}$ (b) $\frac{13}{6}$ (c) $\frac{23}{4}$ (d) $\frac{15}{7}$

(e) $\frac{37}{10}$ (f) $\frac{48}{5}$ (g) $\frac{44}{7}$ (h) $\frac{59}{10}$

(i) $\frac{77}{9}$ (j) $\frac{91}{12}$ (k) $\frac{107}{100}$ (l) $\frac{97}{40}$

4 Write these mixed numbers as improper fractions. WE8

(a) $1\frac{3}{4}$ (b) $3\frac{1}{5}$ (c) $6\frac{2}{3}$ (d) $5\frac{9}{10}$

(e) $4\frac{3}{8}$ (f) $4\frac{7}{11}$ (g) $10\frac{2}{7}$ (h) $6\frac{8}{11}$

(i) $6\frac{5}{12}$ (j) $3\frac{9}{100}$ (k) $6\frac{8}{9}$ (l) $8\frac{7}{9}$

5 (a) Which one of the following is the simplest form of $3\frac{35}{45}$?

A $3\frac{5}{9}$ B $3\frac{3}{4}$ C $3\frac{7}{9}$ D $3\frac{70}{90}$

(b) Which of the following fractions is not equivalent to $\frac{10}{15}$?

A $\frac{2}{3}$ B $\frac{30}{45}$ C $\frac{40}{60}$ D $\frac{20}{25}$

(c) As a mixed number, $\frac{33}{4}$ equals:

A $3\frac{3}{4}$ B $7\frac{4}{5}$ C $8\frac{1}{4}$ D $8\frac{1}{3}$

(d) As an improper fraction, $2\frac{5}{8}$ equals:

A $\frac{7}{8}$ B $\frac{21}{8}$ C $\frac{25}{8}$ D 21

6 Use the fraction wall on page 78 to identify the fraction or fractions that are equivalent to:

(a) $\frac{3}{5}$ (b) $\frac{2}{8}$ (c) $\frac{5}{6}$ (d) $\frac{9}{12}$

7 (a) Copy each of these figures. Write down the fraction that is shaded.

(b) In each figure, rule two diagonal lines joining the opposite vertices (corners). Name the equivalent fraction of the shaded part that you have created.

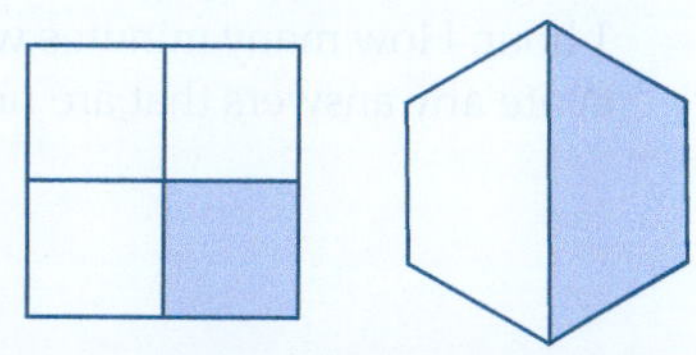

8 **(a)** The top side of this number line is divided into quarters, and the bottom side into halves.

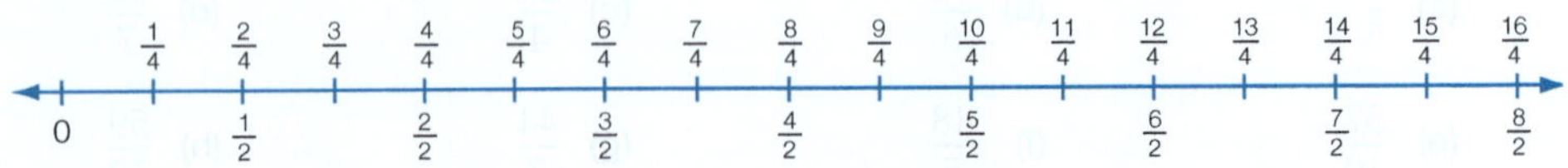

What do you observe about the positions of the equivalent pairs of fractions $\frac{1}{2}$ and $\frac{2}{4}$, $\frac{2}{2}$ and $\frac{4}{4}$, $\frac{3}{2}$ and $\frac{6}{4}$, and $\frac{4}{2}$ and $\frac{8}{4}$?

(b) Copy the number line below. Divide the top side into thirds and the bottom side into ninths. Use your observation from **(a)** to identify pairs of equivalent fractions.

9 Monica, Michelle and Travis are sharing an apple pie that has been cut into 12 equal slices. Monica and Michelle decide to have $\frac{1}{4}$ of the apple pie each and give Travis the remainder. How many slices of pie did each person receive? Draw a diagram that shows this information.

10 Six pizzas were each sliced into 8 equal pieces. If $4\frac{3}{8}$ of the pizzas were eaten, how many slices is that?

11 A block of chocolate has 24 individual squares. How many squares of the chocolate will be gone if $\frac{7}{12}$ of the chocolate is eaten?

12 Talia shaded a fraction of this long rectangle.

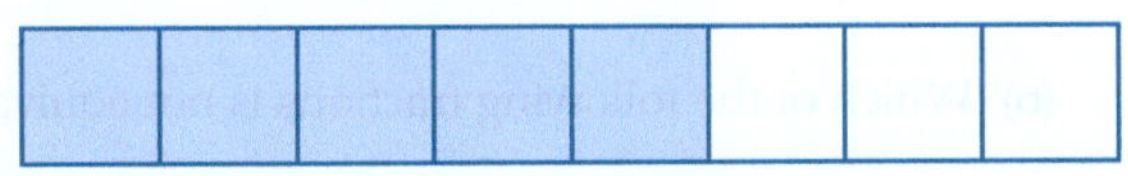

How many squares on this grid should Talia shade to show a fraction equivalent to the fraction she shaded in the rectangle above?

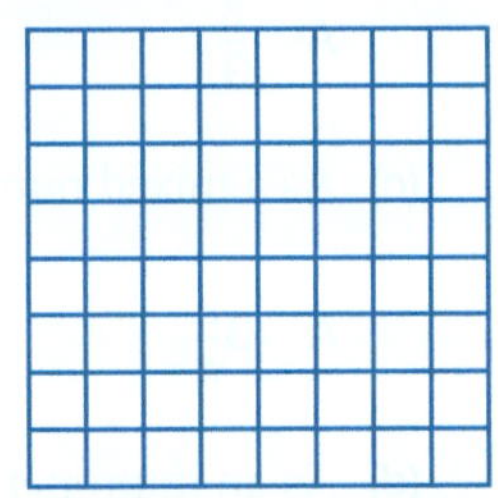

13 Write your answers in simplest form.

(a) What fraction of an hour has passed from 3:49 p.m. to 3:51 p.m.?

(b) What fraction of an hour has passed from 11:03 a.m. to 11:23 a.m.?

(c) What fraction of an hour has passed from 8:13 p.m. to 8:25 p.m.?

(d) What fraction of an hour has passed from 2:36 a.m. to 2:45 a.m.?

14 Imagine that the whole rectangle on the top of the fraction wall on page 78 represents 1 hour. How many minutes would each of the individual fractions on the wall represent? Write any answers that are not whole numbers as mixed numbers.

3.2

Reasoning, communicating and open-ended

15 How many whole numbers would appear on the section of the number line between $\frac{4}{3}$ and $\frac{51}{9}$? (Hint: Convert these improper fractions to mixed numbers.)

16 Which of the following fractions could not be the simplified form of an original fraction that had a denominator of 40?

A $\frac{3}{8}$ **B** $\frac{2}{5}$ **C** $\frac{3}{7}$ **D** $\frac{9}{10}$

17 Which rows of fractions in the fraction wall on page 78 have no equivalent fractions on the wall? Explain why.

18 For each of the following problems, it may be useful to draw and label a number line with the known fractions. Then, locate the position of the unknown fraction and identify it. Equivalent fractions can help you do this.

Which fraction lies exactly in-between the following fractions on the number line?

(a) $\frac{2}{5}$ and $\frac{3}{5}$ **(b)** $\frac{3}{7}$ and $\frac{4}{7}$ **(c)** $\frac{1}{4}$ and $\frac{2}{4}$

(d) $\frac{6}{10}$ and $\frac{7}{10}$ **(e)** $\frac{1}{9}$ and $\frac{6}{9}$ **(f)** $\frac{2}{11}$ and $\frac{7}{11}$

19 Write three fractions that are equivalent to $\frac{4}{5}$.

20 Write three improper fractions that simplify to $1\frac{2}{3}$.

21 For each of these pairs of fractions, draw a diagram that shows they are equivalent.

(a) $\frac{1}{3}, \frac{5}{15}$ **(b)** $\frac{3}{4}, \frac{12}{16}$

22 From the fraction wall on page 78 choose three fractions with three different denominators that have no equivalent fractions in the wall.

3.3 Estimating and comparing fractions

Need to Know

The **Lowest Common Denominator (LCD)** is the lowest common multiple of the denominators of fractions being compared.

For example, the LCD of $\frac{1}{3}$ and $\frac{2}{5}$ is 15.

Comparing fractions using the LCD

- Find the lowest common multiple of the denominators.
- Rewrite each fraction as an equivalent fraction with this multiple as the new denominator.
- Write '>' (greater than), '<' (less than) or '=' (equal to) between the fractions.

For example, to compare $\frac{1}{3}$ and $\frac{2}{5}$, write each with the same LCD. $\frac{1}{3} = \frac{5}{15}$ and $\frac{2}{5} = \frac{6}{15}$, so $\frac{2}{5} > \frac{1}{3}$.

If working with mixed numbers, write them as improper fractions first.

Comparing the denominators

If two fractions have the same numerator but different denominators, the *larger fraction* is the one with the *smaller number as the denominator.*

For example, $\frac{5}{9} > \frac{5}{12}$, $\frac{3}{7} > \frac{3}{8}$.

Comparing to a common fraction or integer

Visualise what a common number such as $\frac{1}{2}$, $\frac{1}{4}$, 1 or 2 'looks like' by imagining a shape, such as a circle or a rectangle, with that fraction shaded in. How would your fraction look in comparison? Would it be bigger or smaller?

For example, $\frac{2}{5} < \frac{1}{2}$ and $\frac{4}{7} > \frac{1}{2}$, so $\frac{1}{3} < \frac{4}{7}$.

Comparing using a number line

A number line can be used by dividing each side into different parts equal to the denominators of each fraction.

For example, to compare $\frac{4}{5}$ and $\frac{2}{3}$, divide one side of the number line into fifths and the other side into thirds. This shows that $\frac{4}{5}$ lies to the right of $\frac{2}{3}$, so $\frac{4}{5} > \frac{2}{3}$.

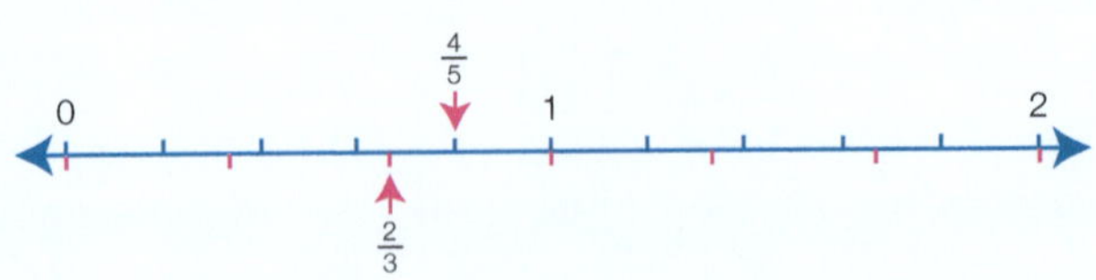

One number is larger than another if it lies to the right of it on the number line.

3.3

Worked Example 9

WE 9

By finding the LCD, determine which fraction in each of the following pairs is larger.

(a) $\frac{3}{5}$ or $\frac{4}{7}$ (b) $\frac{15}{6}$ or $2\frac{7}{12}$

Working

(a) $\frac{3}{5}$ or $\frac{4}{7}$

Multiples of 5: 5, 10, 15, 20, 25, 30, 35, 40, ...

Multiples of 7: 7, 14, 21, 28, 35, 42, 49, ...

LCD = 35

$\frac{3}{5}\times\frac{7}{7}=\frac{21}{35}$ $\frac{4}{7}\times\frac{5}{5}=\frac{20}{35}$

$\frac{21}{35}>\frac{20}{35}$

So, $\frac{3}{5}>\frac{4}{7}$

(b) $2\frac{7}{12}=\frac{31}{12}$

$\frac{15}{6}$ or $\frac{31}{12}$

LCD = 12

$\frac{15}{6}\times\frac{2}{2}=\frac{30}{12}$

$\frac{30}{12}<\frac{31}{12}$

So, $\frac{15}{6}<2\frac{7}{12}$

3.3 Estimating and comparing fractions

Answers page 457

Navigator

Q1 Columns 1–3, Q2, Q3 Columns 1 & 2, Q5, Q7, Q8, Q9, Q10, Q12, Q13, Q15, Q17	Q1 Columns 2 & 3, Q2, Q3 Columns 1–3, Q4, Q5, Q6, Q7, Q8, Q9, Q10, Q12, Q13, Q14, Q15, Q17	Q1 Column 4, Q2, Q3, Q4, Q5, Q6, Q7, Q8, Q9, Q10, Q11, Q12, Q13, Q14, Q16, Q17

Fluency and understanding

1 By finding the LCD, determine which fraction in each of the following pairs is larger.

WE 9

(a) $\frac{3}{5}, \frac{7}{15}$ (b) $\frac{13}{12}, \frac{5}{4}$ (c) $\frac{7}{8}, \frac{5}{6}$ (d) $\frac{5}{9}, \frac{7}{12}$

(e) $\frac{2}{3}, \frac{5}{7}$ (f) $\frac{1}{6}, \frac{2}{9}$ (g) $\frac{3}{8}, \frac{5}{12}$ (h) $\frac{7}{4}, \frac{11}{6}$

(i) $1\frac{3}{4}, \frac{11}{8}$ (j) $\frac{5}{2}, 2\frac{1}{4}$ (k) $\frac{16}{7}, 2\frac{1}{5}$ (l) $2\frac{2}{7}, 2\frac{4}{9}$

Hint

Can't remember how to find the LCD? Turn to pages 34 and 35 to revise LCM first.

3.3

2 Use the fraction wall on page 78 to determine which fraction in each of the following pairs is larger.

(a) $\frac{3}{7}$ or $\frac{3}{10}$ **(b)** $\frac{7}{9}$ or $\frac{7}{8}$ **(c)** $\frac{4}{5}$ or $\frac{5}{6}$ **(d)** $\frac{1}{3}$ or $\frac{2}{5}$

(e) $\frac{9}{12}$ or $\frac{3}{4}$ **(f)** $\frac{5}{11}$ or $\frac{4}{9}$ **(g)** $\frac{6}{7}$ or $\frac{7}{8}$ **(h)** $\frac{3}{5}$ or $\frac{7}{11}$

3 Copy the following fraction pairs. Use any appropriate strategy to decide which fraction is larger, and place < or > between them. If the fractions are equivalent, write = between them.

(a) $\frac{2}{7}$ ______ $\frac{5}{7}$ **(b)** $\frac{4}{5}$ ______ $\frac{4}{7}$ **(c)** $\frac{1}{2}$ ______ $\frac{5}{8}$ **(d)** $\frac{8}{11}$ ______ $\frac{9}{10}$

(e) $\frac{2}{4}$ ______ $\frac{3}{6}$ **(f)** $\frac{6}{12}$ ______ $\frac{7}{10}$ **(g)** $\frac{3}{7}$ ______ $\frac{5}{8}$ **(h)** $\frac{5}{9}$ ______ $\frac{2}{5}$

(i) $\frac{5}{6}$ ______ $\frac{7}{8}$ **(j)** $\frac{7}{11}$ ______ $\frac{8}{10}$ **(k)** $\frac{6}{10}$ ______ $\frac{3}{5}$ **(l)** $\frac{4}{5}$ ______ $\frac{6}{9}$

(m) $\frac{2}{4}$ ______ $\frac{4}{2}$ **(n)** $\frac{24}{6}$ ______ $\frac{8}{2}$ **(o)** $\frac{18}{11}$ ______ $\frac{9}{5}$ **(p)** $\frac{9}{7}$ ______ $\frac{11}{9}$

4 **(a)** Draw a number line from 0 to 2. Indicate the approximate positions of the following fractions on the number line, then use the number line to write the fractions in ascending order (smallest to largest).

$\frac{3}{4}, \frac{2}{3}, \frac{6}{4}, 1\frac{1}{3}$

(b) Draw a number line from 0 to 3. Indicate the approximate positions of the following fractions on the number line, then use the number line to write the fractions in descending order (largest to smallest).

$\frac{5}{2}, \frac{1}{2}, \frac{4}{10}, 1\frac{7}{10}$

5 Ali gets 7 out of 8 shots at goal in the basket. Joseph gets 10 out of 12 shots in. By comparing fractions, determine who is the more accurate shooter.

6 In a box containing 16 chocolates, 5 of them contain nuts. A larger box with 64 of the same brand of chocolates contains 23 nut chocolates. Which box contains the greater fraction of chocolates with nuts?

7 Use the fraction wall on page 78 to determine which fraction or fractions on the wall lie in-between the following (include all equivalent fractions).

(a) $\frac{9}{10}$ and $\frac{11}{12}$ **(b)** $\frac{5}{8}$ and $\frac{7}{10}$ **(c)** $\frac{3}{8}$ and $\frac{4}{9}$ **(d)** $\frac{2}{11}$ and $\frac{3}{11}$

8 **(a)** Which arrow is pointing closest to the location of $\frac{9}{10}$ on this number line?

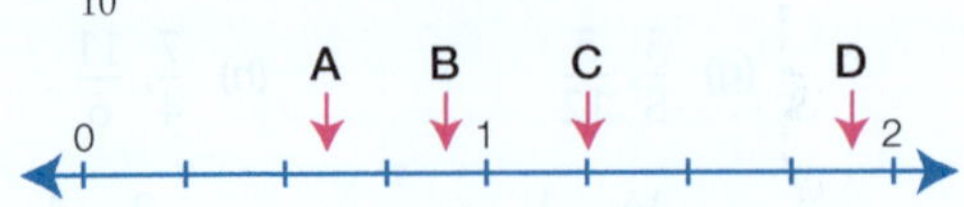

(b) Which arrow is pointing closest to the location of $1\frac{1}{3}$?

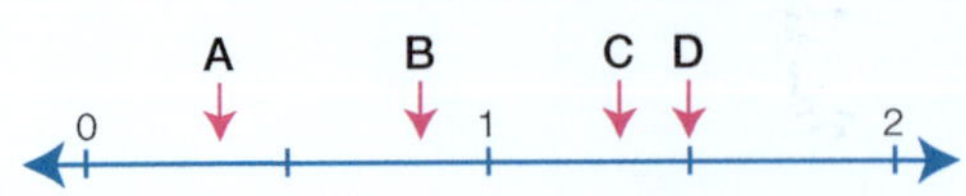

9 **(a)** Which one of the following is closest to $\frac{5}{8}$?

A $\frac{2}{9}$ **B** $\frac{1}{3}$ **C** $\frac{1}{2}$ **D** $\frac{7}{8}$

(b) Which one of the following is closest to 2?

A $\frac{11}{12}$ **B** $\frac{3}{2}$ **C** $\frac{7}{4}$ **D** $\frac{7}{3}$

Reasoning, communicating and open-ended

10 Write each of these lists in ascending order (from smallest to largest). Explain using a series of brief points, or steps, how you decided the order.

(a) $\frac{2}{3}, \frac{5}{6}, \frac{7}{8}, \frac{1}{2}, \frac{3}{4}, \frac{10}{11}$ **(b)** $\frac{4}{7}, 1, \frac{7}{4}, \frac{6}{7}, \frac{7}{8}, 0$

11 Write each of these lists in descending order (from largest to smallest). Explain briefly how you decided the order.

(a) $\frac{4}{7}, \frac{2}{9}, \frac{2}{3}, 1, \frac{11}{21}, \frac{45}{63}$ **(b)** $\frac{21}{16}, 1\frac{3}{80}, \frac{5}{4}, 1\frac{7}{20}, \frac{39}{40}, 1\frac{3}{10}$

12 **(a)** Estimate the fraction of the Australian flag that is blue.

(b) Estimate the fraction of Australia that is taken up by the state of Queensland.

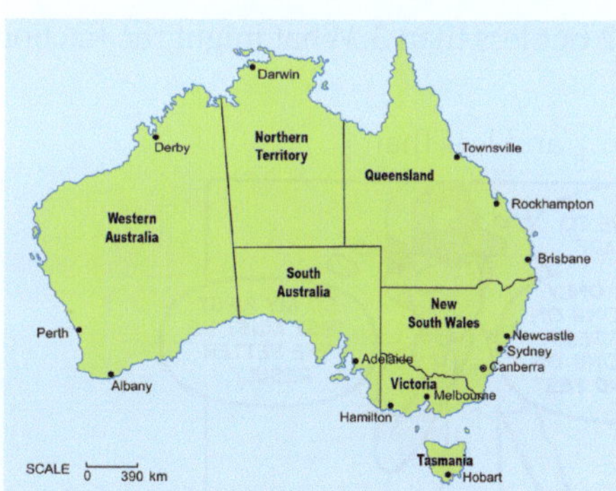

(c) Estimate the fraction of this jug that has cordial in it.

13 For each of the following pairs of identical shapes, state whether the fraction of the shape that is shaded is larger in **A** or in **B**. If the fraction shaded is the same in **A** and in **B**, write 'same'.

(a) **A** 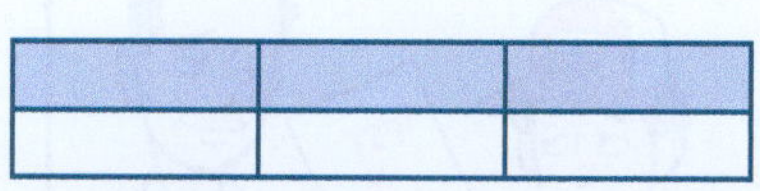 **B** 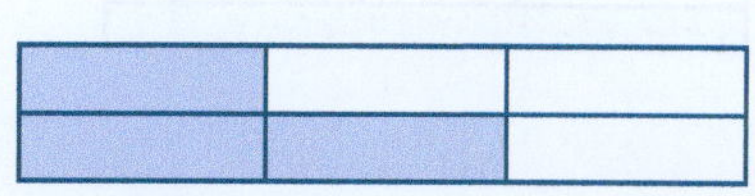

(b) **A** 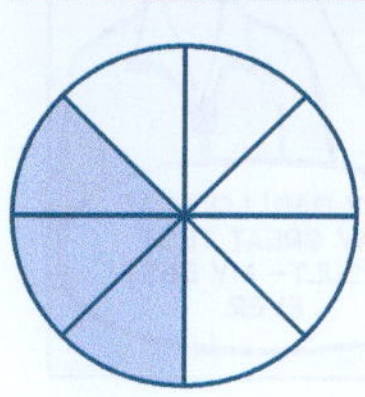 **B** 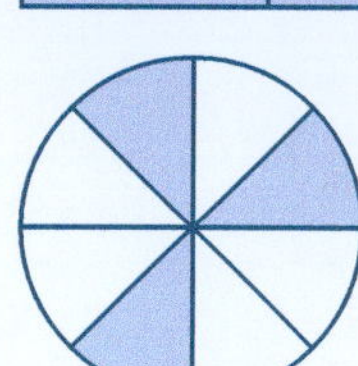

3.3

(c) A 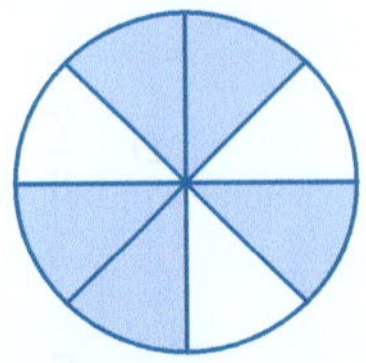 B 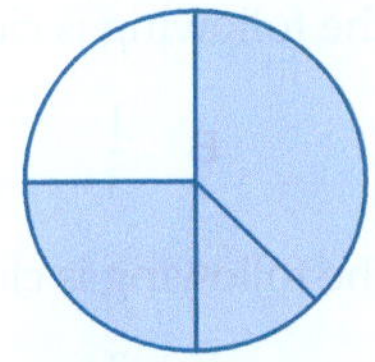

(d) A 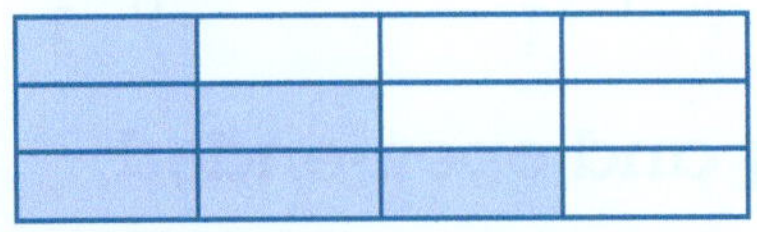 B 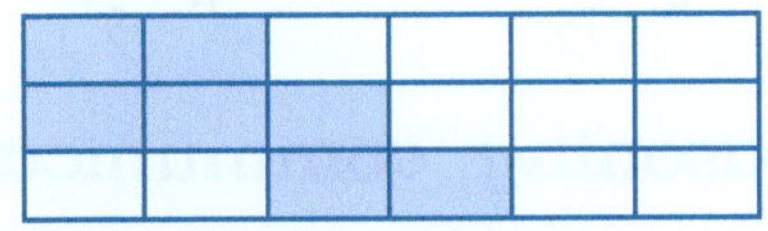

(e) A 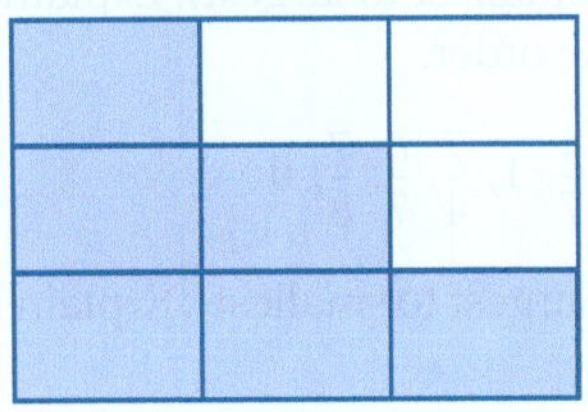 B 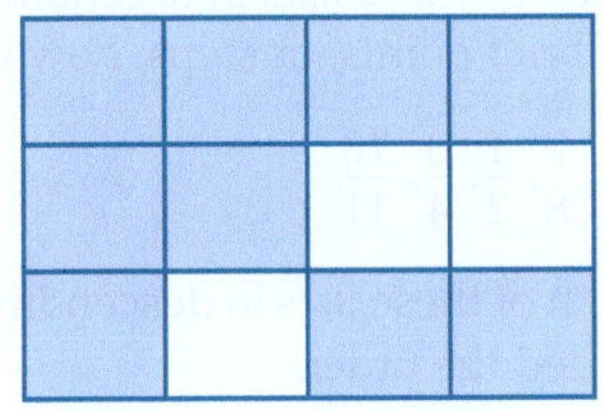

(f) A 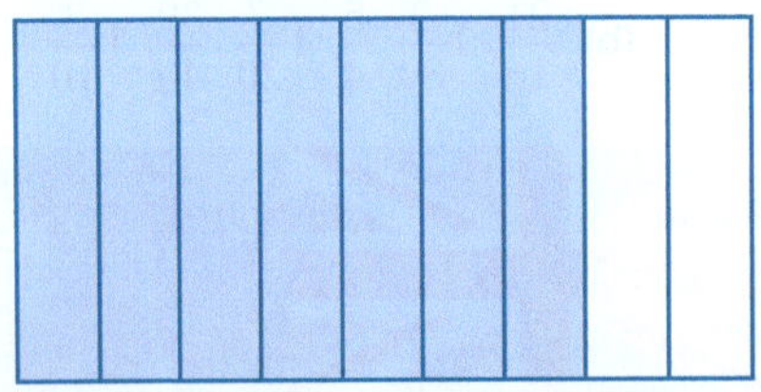 B 

14 Angus has 100 mL left of his 500 mL bottle of cola. Gianni has 50 mL left in his 300 mL bottle.

(a) Who has drunk more cola?

(b) Who has drunk a greater fraction of the cola in their bottle?

15 (a) A proper fraction is greater than $\frac{3}{4}$ but less than 1. What might the fraction be? Give two possibilities.

(b) An improper fraction is greater than 2 but less than 3. What might the fraction be? Give two possibilities.

16 Write three fractions that are greater than $\frac{3}{5}$ and less than $\frac{4}{5}$.

17

Is Jason's thinking correct? Which is his better test result? Explain your choice.

Adding and subtracting fractions

3.4

Need to Know

Adding fractions using the LCD

To add or subtract fractions, we must have fractions with the same denominator. If the denominators are the same, the size of the fraction parts is the same, and we can simply count how many of each part we have.

For example, $\frac{3}{7} + \frac{2}{7} = \frac{5}{7}$ $\frac{3}{7} - \frac{2}{7} = \frac{1}{7}$.

If the denominators are different, each fraction must be written as an equivalent fraction with the same denominator, the LCD (Lowest Common Denominator).

For example, $\frac{1}{3} + \frac{2}{5} = \frac{5}{15} + \frac{6}{15}$ and $\frac{2}{5} - \frac{1}{3} = \frac{6}{15} - \frac{5}{15}$.

$= \frac{11}{15}$ $= \frac{1}{15}$

Estimating answers to fraction problems

To estimate an answer to adding or subtracting fractions, compare each fraction to another fraction. For example, $\frac{3}{5} + \frac{3}{7} \approx 1$ as $\frac{3}{5}$ is slightly $> \frac{1}{2}$ and $\frac{3}{7}$ is slightly $< \frac{1}{2}$. The actual answer is $1\frac{1}{35}$.

Adding fractions using the grid method

1 Draw two identical squares or rectangles.
2 Divide one horizontally using the denominator of the first fraction.
3 Divide the other vertically using the denominator of the second fraction.
4 Lay one square or rectangle on top of the other to form a grid.
5 Where the grid squares are shaded twice due to the overlap, shade an equal number of blank grid squares to remove the overlap.
6 The resultant grid represents the fraction sum.

Adding and subtracting mixed numbers

To add or subtract mixed numbers, we can either:

- add or subtract the whole number parts and fraction parts separately

or

- convert the mixed numbers to improper fractions, then add or subtract them in the usual way.

Adding and subtracting negative fractions

Negative fractions can be added and subtracted using the same processes as used for integers and positive fractions.

For example, $\frac{1}{3} + -\frac{2}{5} = \frac{1}{3} - \frac{2}{5}$

$= \frac{5}{15} - \frac{6}{15}$

$= -\frac{1}{15}$

Technology Exploration: Calculator

Working with fractions on a calculator

The key that can be used for entering fractions into your calculator will often look something like this [$\frac{\square}{\square}$]. To enter a fraction, press [$\frac{\square}{\square}$], then key in the numerator and the denominator, using the cursor keys to move between them. To enter a mixed number, press [SHIFT] [$\frac{\square}{\square}$].

On some calculators, the key looks like this: [$a^b/_c$], or this [$^a/_b$]. To enter a fraction, such as $\frac{5}{6}$, type [5] [$a^b/_c$] [6]. You may need to press [=] for the fraction to be displayed on the screen. To enter a mixed number, such as $4\frac{3}{8}$, press [4] [$a^b/_c$] [3] [$a^b/_c$] [8].

The screen will use symbols such as ⌐, ⌙ or _ to separate the whole numbers, numerator and denominator.

Check that you can work with fractions correctly on your calculator by entering some additions and subtractions for which you already know the answers. For example, if you enter $\frac{3}{8} + 1\frac{2}{3}$ correctly, the answer of $2\frac{1}{24}$ will appear on the screen.

If you want to convert a mixed number that you have already entered to an improper fraction, press [SHIFT] and the [$a^b/_c$] key, or [SHIFT] and the [S⇔D] key. (You might need to press [=] to see the result.)

You will also notice that pressing the [S⇔D] key, or pressing the [$a^b/_c$] key twice, converts a fraction to its decimal form.

Worked Example 10

WE10

Calculate the following, giving your answers in simplest form. Use estimation to judge whether your answers are reasonable.

(a) $\frac{2}{3} + \frac{3}{4}$ (b) $\frac{7}{10} - \frac{1}{5}$

Working

(a) Estimate: between 1 and 2

$\frac{2}{3} + \frac{3}{4}$ LCD = 12

$= \frac{8}{12} + \frac{9}{12}$

$= \frac{17}{12}$

$= 1\frac{5}{12}$

Reasonable

(b) Estimate: about $\frac{1}{2}$

$\frac{7}{10} - \frac{1}{5}$ LCD = 10

$= \frac{7}{10} - \frac{2}{10}$

$= \frac{5}{10}$

$= \frac{1}{2}$

Exact

3.4

Worked Example 11

WE11

Add $\frac{1}{3}$ and $\frac{2}{5}$ by using the grid method.

Working

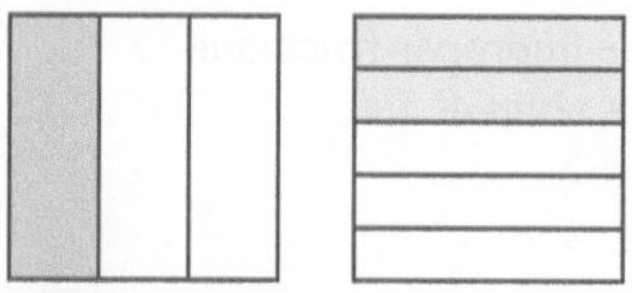

$\frac{1}{3}+\frac{2}{5}$

(Draw two identical squares or rectangles. Divide them up and shade in each fraction.)

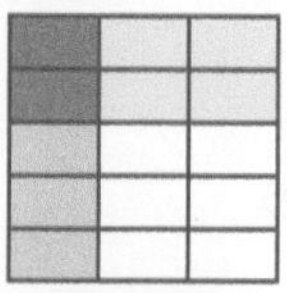

(Lay one square or rectangle on top of the other to form a grid.)

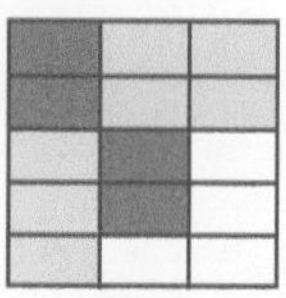

(Where the grid squares are shaded twice due to overlap, shade an equal number of blank grid squares to 'remove' the overlap. The resultant shaded grid represents the fraction sum.)

$\frac{11}{15}$

Worked Example 12

WE12

Calculate $4\frac{5}{6}+2\frac{3}{4}$. Write your answer as a mixed number in simplest form.

Working

Method 1: Add wholes and fractions separately

Estimate: between 7 and 8

$4\frac{5}{6}+2\frac{3}{4}$

$=4+2+\frac{5}{6}+\frac{3}{4}$

$=6+\frac{5}{6}+\frac{3}{4}$

$=6+\frac{10}{12}+\frac{9}{12}$

$=6+\frac{19}{12}$

$=6+1\frac{7}{12}$

$=7\frac{7}{12}$

Reasonable

Working

Method 2: Use improper fractions

Estimate: between 7 and 8

$4\frac{5}{6}+2\frac{3}{4}$

$=\frac{29}{6}+\frac{11}{4}$

$=\frac{58}{12}+\frac{33}{12}$

$=\frac{91}{12}$

$=7\frac{7}{12}$

Reasonable

3.4

Worked Example 13 (WE13)

Calculate $6\frac{5}{8} - 2\frac{3}{4}$. Write your answer as a mixed number in simplest form.

Working

Method 1: Subtract wholes and fractions separately. Estimate: less than 4

$$6\frac{5}{8} - 2\frac{3}{4}$$
$$= 6 - 2 + \frac{5}{8} - \frac{3}{4}$$
$$= 4 + \frac{5}{8} - \frac{3}{4}$$
$$= 4 + \frac{5}{8} - \frac{6}{8}$$
$$= 3 + \frac{8}{8} + \frac{5}{8} - \frac{6}{8}$$
$$= 3 + \frac{13}{8} - \frac{6}{8}$$
$$= 3\frac{7}{8}$$

Working

Method 2: Use improper fractions Estimate: less than 4

$$6\frac{5}{8} - 2\frac{3}{4}$$
$$= \frac{53}{8} - \frac{11}{4}$$
$$= \frac{53}{8} - \frac{22}{8} \quad \text{LCD} = 8$$
$$= \frac{31}{8}$$
$$= 3\frac{7}{8}$$

Answers page 458

3.4 Adding and subtracting fractions

Navigator

Q1 Columns 1 & 2, Q2, Q3 Columns 1 & 2, Q4 Columns 1 & 2, Q5, Q6 Column 1, Q7, Q8 Column 1, Q9, Q10, Q11, Q13, Q16 Column 1, Q17, Q19, Q20	Q1 Columns 2 & 3, Q2, Q3 Columns 2 & 3, Q4 Columns 2 & 3, Q5, Q6 Column 2, Q7, Q8 Column 2, Q11, Q12, Q13, Q15, Q16, Q17, Q18, Q19, Q20	Q1 Column 4, Q3 Columns 3 & 4, Q4 Columns 3 & 4, Q5, Q6 Column 3, Q7, Q8 Column 3, Q10, Q12, Q13, Q14, Q15, Q16, Q17, Q18, Q19, Q20

Fluency and understanding

WE10

Hint

Make sure you don't add or subtract fractions until you've rewritten them with the same denominator.

1 Calculate the following, giving your answers in simplest form. Use estimation to judge whether your answers are reasonable.

(a) $\frac{5}{14} + \frac{1}{7}$ (b) $\frac{2}{5} - \frac{7}{20}$ (c) $\frac{5}{8} - \frac{7}{40}$ (d) $\frac{4}{55} + \frac{2}{11}$

(e) $\frac{5}{6} + \frac{1}{3}$ (f) $\frac{15}{49} + \frac{5}{7}$ (g) $\frac{3}{5} + \frac{13}{20}$ (h) $\frac{3}{4} + \frac{1}{5}$

(i) $\frac{1}{4} + \left(-\frac{1}{6}\right)$ (j) $\frac{5}{9} - \left(-\frac{3}{4}\right)$ (k) $\frac{1}{6} - \left(-\frac{3}{7}\right)$ (l) $\frac{11}{15} - \left(-\frac{7}{20}\right)$

(m) $\frac{11}{12} - \left(-\frac{3}{10}\right)$ (n) $\frac{23}{30} + \left(-\frac{7}{20}\right)$ (o) $\frac{3}{5} - \left(-\frac{9}{14}\right)$ (p) $\frac{41}{45} + \left(-\frac{11}{18}\right)$

3.4

2 Add the following fractions by using the grid method. WE11

(a) $\frac{2}{3}+\frac{1}{2}$ (b) $\frac{3}{4}+\frac{1}{7}$

3 Calculate the following. Write your answers as mixed numbers in simplest form. WE12

(a) $3\frac{1}{7}+4\frac{2}{7}$ (b) $8\frac{1}{9}+5\frac{5}{9}$ (c) $2\frac{3}{8}+7\frac{1}{4}$ (d) $6\frac{7}{10}+1\frac{1}{5}$

(e) $1\frac{3}{4}+1\frac{1}{2}$ (f) $3\frac{1}{5}+2\frac{9}{10}$ (g) $4\frac{1}{2}+2\frac{3}{4}$ (h) $1\frac{7}{12}+7\frac{2}{3}$

(i) $6\frac{4}{5}+5\frac{1}{2}$ (j) $3\frac{8}{9}+7\frac{1}{6}$ (k) $1\frac{4}{11}+3\frac{2}{5}$ (l) $1\frac{4}{25}+2\frac{3}{20}$

4 Calculate the following. Write your answers as mixed numbers in simplest form. WE13

(a) $3\frac{5}{7}-2\frac{1}{7}$ (b) $3\frac{2}{9}-2\frac{5}{9}$ (c) $5\frac{7}{9}-1\frac{1}{3}$ (d) $7\frac{11}{12}-6\frac{1}{4}$

(e) $3\frac{1}{12}-1\frac{5}{6}$ (f) $4\frac{7}{10}-2\frac{1}{5}$ (g) $4\frac{1}{10}-2\frac{4}{5}$ (h) $6\frac{1}{18}-4\frac{1}{6}$

(i) $2\frac{1}{9}-1\frac{1}{6}$ (j) $2\frac{1}{4}-1\frac{2}{5}$ (k) $2\frac{1}{8}-1\frac{1}{6}$ (l) $5\frac{1}{8}-2\frac{2}{3}$

5 (a) $6-\frac{3}{8}$ (b) $4-\frac{2}{3}$ (c) $8-\frac{5}{9}$

(d) $3-\frac{4}{5}$ (e) $2-\frac{5}{6}$ (f) $5-\frac{4}{7}$

6 Find the following. You might like to use your calculator to check your answers.

(a) $\frac{3}{8}-\frac{1}{6}-\frac{1}{3}$ (b) $\frac{1}{2}-\frac{3}{4}-\frac{1}{3}$ (c) $\frac{7}{10}-\frac{4}{5}-\frac{1}{2}$

(d) $1\frac{1}{2}+\frac{2}{5}+2\frac{3}{10}$ (e) $2\frac{1}{3}+5\frac{3}{4}-1\frac{2}{5}$ (f) $3\frac{1}{4}-2\frac{1}{5}-3\frac{1}{3}$

7 (a) $\frac{5}{8}-\frac{1}{3}=$

A $\frac{4}{24}$ B $\frac{7}{24}$ C $\frac{4}{5}$ D $\frac{23}{24}$

(b) $1\frac{2}{3}+\frac{3}{20}=$

A $1\frac{1}{17}$ B $1\frac{5}{23}$ C $1\frac{5}{20}$ D $1\frac{49}{60}$

8 (a) $-\frac{3}{4}+\frac{4}{5}$ (b) $-\frac{3}{8}+\frac{2}{7}$ (c) $-\frac{4}{9}+\frac{2}{5}$

(d) $-\frac{2}{3}-\frac{1}{2}$ (e) $-\frac{6}{7}-\frac{3}{4}$ (f) $-\frac{6}{7}-\frac{2}{9}$

(g) $-2\frac{1}{2}+1\frac{1}{4}$ (h) $-7\frac{1}{3}+5\frac{1}{4}$ (i) $-6\frac{2}{3}+4\frac{1}{5}$

(j) $-1\frac{1}{3}-2\frac{1}{5}$ (k) $-4\frac{1}{2}-3\frac{3}{7}$ (l) $-8\frac{1}{9}-2\frac{4}{7}$

3.4

9 When Eric got in his car and started driving, his fuel gauge looked like this:

Eric stopped to get petrol when his fuel gauge looked like this:

After putting some petrol in his car, Eric's fuel gauge looked like this:

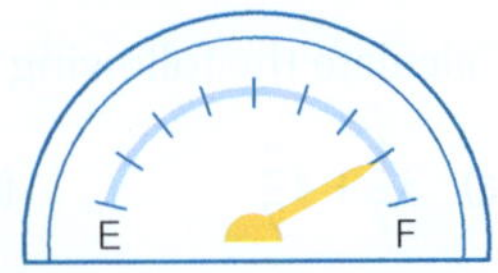

Write a fraction addition or a subtraction to answer the following.

(a) What fraction of the tank of fuel did Eric use between the first two gauge readings? Write your answer in simplest form.

(b) What fraction of the tank of petrol did Eric add when he was at the petrol station? Write your answer in simplest form.

10 What fraction of a large pizza was eaten if Steven had $\frac{5}{12}$ of the pizza and Lara had $\frac{1}{3}$ of it? Write your answer in simplest form.

11 Jaydeep, Corey and Isaac were sharing the driving on a trip along the coast. If Jaydeep drove $\frac{3}{5}$ of the distance and Isaac drove $\frac{3}{20}$, what fraction of the total distance did Corey drive?

12 Belinda worked part-time at a fast food restaurant. Her hours for the three days she worked one week were $2\frac{1}{4}$, $3\frac{1}{2}$ and $4\frac{1}{3}$. Calculate the total number of hours she worked for the week.

13 Five large packets of cereal were bought for a Year 7 camp. On the first day $1\frac{1}{3}$ packets were eaten, on the second day $\frac{3}{4}$ of a packet was eaten and on the third day $2\frac{1}{6}$ packets were eaten. How much cereal was available for the fourth day?

Reasoning, communicating and open-ended

14 A fraction is added to $\frac{3}{5}$. The lowest common denominator for the two fractions is 40. Which of the following fractions was added?

A $\frac{2}{6}$ **B** $\frac{3}{8}$ **C** $\frac{13}{20}$ **D** $\frac{11}{15}$

15 A fraction is added to $\frac{3}{10}$. Which of the following would give an improper fraction as the result?

A $\frac{7}{15}$ **B** $\frac{7}{12}$ **C** $\frac{3}{5}$ **D** $\frac{18}{25}$

16 **(a)** $11 - 2\frac{1}{3}$ **(b)** $8 - 4\frac{1}{4}$ **(c)** $11 - 5\frac{1}{8}$

(d) $5 - 1\frac{2}{7}$ **(e)** $9 - 3\frac{2}{9}$ **(f)** $15 - 7\frac{5}{6}$

17 This question appeared on a Year 7 maths test: 'What whole number is $\frac{7}{8} + \frac{12}{13}$ approximately equal to?' Several students answered 19 or 21, both of which are incorrect. How might the students have arrived at their incorrect answers? Describe how you could use estimation to arrive at the correct answer.

18 Find two fractions that add to $\frac{3}{4}$ and do not have a denominator of 4.

19 Use a diagram to help you explain why $\frac{3}{10} + \frac{4}{10} = \frac{7}{10}$, not $\frac{7}{20}$. Why do we add the numerators, but not the denominators?

20

Who do you think is right, Jake, his teacher, or both? Explain why.

Problem solving

Fill in the blanks

1 Fill in the blanks with positive whole numbers to make mathematically true statements. Do not use the same number twice within a statement. See if you can come up with at least two different combinations.

(a) $\frac{\blacklozenge}{4} + \frac{1}{\blacklozenge} = \frac{\blacklozenge}{20}$

(b) $\frac{\bigstar}{\bigstar} - \frac{\bigstar}{6} = \frac{\bigstar}{12}$

2 Use four of the digits 1, 3, 4, 5, 6, 7 to make two fractions whose sum is close to but less than 1.

$\frac{\blacksquare}{\blacksquare} + \frac{\blacksquare}{\blacksquare} < 1$

Try to come up with at least three combinations. Which combination is the closest to 1?

Strategy options

- Guess and check.
- Test all possible combinations.

Half-time 3

Ex. 3.1

1 There are 15 cookies in a bag.

(a) If I eat 7 cookies, what fraction of the total cookies is this?

(b) If I eat $\frac{2}{5}$ of the cookies in the bag, how many cookies do I eat?

Ex. 3.2

2 Simplify the following.

(a) $\frac{18}{48}$ (b) $\frac{9}{36}$ (c) $3\frac{9}{12}$ (d) $7\frac{35}{50}$

Ex. 3.2

3 Convert the following improper fractions to whole numbers or mixed numbers.

(a) $\frac{27}{4}$ (b) $\frac{40}{8}$ (c) $\frac{50}{10}$ (d) $\frac{37}{7}$

Ex. 3.3

4 Determine which fraction in each of these pairs is the larger number, and write a < or > sign between them. If they are equivalent, write an '=' sign between them.

(a) $\frac{5}{8}$ ____ $\frac{2}{3}$ (b) $\frac{4}{9}$ ____ $\frac{15}{27}$ (c) $\frac{48}{56}$ ____ $\frac{6}{7}$ (d) $\frac{11}{5}$ ____ $\frac{9}{4}$

Ex. 3.1, 3.2, 3.3

5 Draw a number line from 0 to 2. Divide the bottom side of the number line into quarters and the top side into eighths.

(a) Show the positions of these fractions with a labelled arrow: $\frac{3}{4}, \frac{5}{8}, \frac{5}{4}, \frac{9}{8}, \frac{7}{4}, \frac{14}{8}$.

(b) Identify five pairs of equivalent fractions that are shown on the number line.

(c) Use the number line to determine which fraction out of the following pairs is larger.

(i) $\frac{5}{4}$ or $\frac{9}{8}$ (ii) $\frac{11}{8}$ or $\frac{7}{4}$

Ex. 3.4

6 Calculate the following, writing your answers in simplest form.

(a) $\frac{3}{8}+\frac{3}{4}$ (b) $\frac{7}{15}-\frac{3}{10}$ (c) $\frac{7}{9}+\frac{5}{6}$ (d) $\frac{11}{18}-\frac{5}{12}$

Ex. 3.3

7 Place the following in ascending order by first locating them on a number line.

$\frac{1}{3}, \frac{6}{3}, \frac{9}{4}, \frac{3}{4}, 2\frac{2}{3}$

Ex. 3.2

8 7 pizzas were each sliced into 6 equal pieces.

(a) If $4\frac{5}{6}$ of the pizzas were eaten, how many slices was that?

(b) If $3\frac{1}{2}$ of the pizzas were eaten, how many slices were left over?

Ex. 3.4

9 Calculate the following.

(a) $2\frac{7}{12}+2\frac{1}{10}$ (b) $1\frac{16}{25}+2\frac{13}{20}$ (c) $2\frac{13}{20}-1\frac{3}{100}$ (d) $3\frac{9}{10}-2\frac{8}{25}$

Ex. 3.1, 3.2

10 Dave is leading a group of 12 people on a bushwalk. He has brought 30 L of water to supply the group.

Ex. 3.4

(a) How many litres does each group member receive? Write your answer as both an improper fraction and as a mixed number in simplest form.

(b) One of the group members has drunk $\frac{3}{4}$ of a litre by lunchtime. How many litres do they have left?

Multiplying fractions 3.5

Need to Know

The multiplication sign, ×, means *'lots of'*, or simply *'of'*.

For example, 8 lots of \$5 = 8 × \$5 = \$40.

A **unit fraction** is a fraction with a numerator of 1.

Finding a fraction of a whole number

Either:

- find the unit fraction (by dividing by the denominator), then multiply by the numerator

or

- write the whole number as an improper fraction and multiply the numerators together and the denominators together. Cancel any common factors between numerators and denominators before multiplying (this avoids having to work with large numbers).

Finding a fraction of a fraction

$\frac{4}{5} \times 15 = 12$

and $\frac{3}{4} \times 12 = 9$

so $\frac{3}{4} \times \frac{4}{5} \times 15 = 9$

However, $\frac{3}{5} \times 15 = 9$

so $\frac{3}{4} \times \frac{4}{5} = \frac{3}{5}$.

Multiplying fractions

- Write mixed numbers as improper fractions before multiplying

  For example, $\frac{3}{16}$ of $5\frac{3}{5} = \frac{3}{16}$ of $\frac{28}{5}$

- Replace 'of' with a '×' symbol

  $\frac{3}{16}$ of $\frac{28}{5} = \frac{3}{16} \times \frac{28}{5}$

- Cancel any common factors

  $\frac{3}{\cancel{16}_4} \times \frac{\cancel{28}^7}{5} = \frac{3}{4} \times \frac{7}{5}$

- Multiply the numerators together and the denominators together

  $\frac{3}{4} \times \frac{7}{5} = \frac{21}{20}$

- Simplify the answer further if necessary.

  $\frac{21}{20} = 1\frac{1}{20}$

3.5

Worked Example 14

WE 14

Find $\frac{2}{5}$ of 40.

Working

Method 1: Find the unit fraction, then multiply

$\frac{1}{5}$ of $40 = 8$

$\frac{2}{5}$ of $40 = 2 \times 8$

$= 16$

Working

Method 2: Multiply two fractions

$\frac{2}{5}$ of 40

$= \frac{2}{{}_1\cancel{5}} \times \frac{\cancel{40}^8}{1}$

$= \frac{16}{1}$

$= 16$

Worked Example 15

WE 15

Find $\frac{3}{4}$ of $\frac{4}{5}$.

Working

$\frac{3}{4}$ of $\frac{4}{5}$

$= \frac{3}{{}_1\cancel{4}} \times \frac{\cancel{4}^1}{5}$

$= \frac{3}{5}$

3.5

Worked Example 16

WE16

Find $4 \times 2\frac{3}{20}$. Write your answer as a mixed number, if appropriate.

Working

$4 \times 2\frac{3}{20}$

$= \frac{4}{1} \times \frac{43}{20}$

$= \frac{\cancel{4}^{1}}{1} \times \frac{43}{\cancel{20}_{5}}$

$= \frac{43}{5}$

$= 8\frac{3}{5}$

3.5 Multiplying fractions

Answers page 459

Navigator

Q1 Columns 1–3, Q2 Columns 1–3, Q3 Columns 1 & 2, Q4 Column 1, Q5, Q6, Q8, Q10, Q11, Q13, Q14, Q16, Q19	Q1 Columns 3 & 4, Q2 Columns 3 & 4, Q3 Columns 3 & 4, Q4 Column 2, Q5, Q7, Q8, Q9, Q10, Q11, Q13, Q14, Q15, Q19	Q1 Column 4, Q2 Column 4, Q3 Column 4, Q4 Column 3, Q5, Q7, Q8, Q10, Q11, Q12, Q13, Q15, Q16, Q17, Q18, Q19

Fluency and understanding

WE14

1 Find the following.

(a) $\frac{1}{3}$ of 18 (b) $\frac{1}{4}$ of 12 (c) $\frac{1}{6}$ of 18 (d) $\frac{1}{9}$ of 36

(e) $\frac{3}{4}$ of 28 (f) $\frac{4}{5}$ of 30 (g) $\frac{6}{7}$ of 56 (h) $\frac{3}{8}$ of 72

(i) $\frac{3}{2}$ of 20 (j) $\frac{6}{5}$ of 100 (k) $\frac{11}{10}$ of 50 (l) $\frac{5}{4}$ of 60

WE15

2 Find the following.

(a) $\frac{1}{2}$ of $\frac{2}{13}$ (b) $\frac{2}{9} \times \frac{1}{2}$ (c) $\frac{4}{5} \times \frac{5}{9}$ (d) $\frac{3}{7} \times \frac{7}{13}$

(e) $\frac{5}{7} \times \frac{1}{10}$ (f) $\frac{3}{5}$ of $\frac{10}{11}$ (g) $\frac{5}{12}$ of $\frac{6}{7}$ (h) $\frac{5}{6}$ of $\frac{1}{11}$

(i) $\frac{6}{7} \times \frac{3}{8}$ (j) $\frac{9}{10} \times \frac{5}{6}$ (k) $\frac{4}{7}$ of $\frac{2}{3}$ (l) $\frac{3}{10} \times \frac{9}{25}$

3.5

WE16

Hint

When cancelling common factors, don't cancel out two things on the same line—cancel something on the bottom with something on the top.

3 Find the following. Write your answers as mixed numbers, if appropriate.

(a) $3 \times \frac{2}{9}$ (b) $\frac{3}{4} \times 8$ (c) $\frac{5}{8} \times 2$ (d) $3 \times \frac{5}{12}$

(e) $5 \times 2\frac{3}{10}$ (f) $8 \times 1\frac{3}{4}$ (g) $1\frac{4}{5} \times 15$ (h) $5\frac{1}{12} \times 8$

(i) $5\frac{1}{3} \times \frac{3}{8}$ (j) $3\frac{1}{2} \times 3\frac{1}{5}$ (k) $4\frac{1}{5} \times 4\frac{2}{7}$ (l) $1\frac{1}{8} \times 3\frac{5}{9}$

4 Calculate the following, giving your answers in simplest form.

(a) $\frac{3}{11} \times \frac{11}{5} \times \frac{1}{4}$ (b) $\frac{2}{7} \times \frac{7}{9} \times \frac{1}{5}$ (c) $\frac{6}{7} \times \frac{5}{8} \times \frac{2}{3}$

(d) $1\frac{2}{3} \times \frac{12}{13} \times \frac{1}{2}$ (e) $3\frac{2}{5} \times \frac{5}{6} \times \frac{7}{17}$ (f) $\frac{4}{9} \times 3\frac{1}{2} \times \frac{6}{7}$

5 (a) $\frac{2}{5}$ of $2\frac{1}{2}$ cups of sugar is:

A $\frac{1}{5}$ cup **B** $\frac{4}{10}$ cup **C** 1 cup **D** $2\frac{1}{5}$ cups

(b) $\frac{1}{5}$ of $\frac{1}{3}$ of 15 is:

A 1 **B** 2 **C** 3 **D** 5

6 A group of 7 workers shared third prize in a local raffle. The prize was worth $2170.

(a) What fraction of the prize money does each worker receive?

(b) How much money does each worker get?

7 The proceeds of a $48 000 inheritance were to be shared between four cousins. Each cousin was to receive a specific fraction of the inheritance. Michael was to receive $\frac{5}{24}$, Marcus $\frac{1}{12}$, Lily $\frac{3}{8}$ and Julie $\frac{1}{3}$. How much money did each cousin receive?

8 A chocolate bar was made up of 45 equal pieces. Ling had to share it equally with her brother and sister, Gao and Chen.

(a) How many pieces did each person receive? Copy this chocolate block into your book and divide it up to show each person's share.

(b) Ling then decided to divide her share of chocolate equally between herself and four of her friends.

(i) How many pieces did each person receive?

(ii) What fraction of the original block is this?

(iii) On your drawing of the chocolate bar, show the fraction that Ling and her friends receive.

9 Annika cut one-quarter of an apple pie in halves to share with her brother Arjun. What fraction of the original whole pie do they each now have?

10 Draw a rectangle that is 5 cm long and 4 cm wide. Divide it lengthways into fifths. Shade in $\frac{1}{5}$. Now, take the shaded section and divide it into quarters. Shade in $\frac{3}{4}$ of this section.

(a) What fraction of the original rectangle have you shaded twice?

(b) Write the fraction multiplication that is shown by your diagram.

11 A petrol tank was filled to its capacity of 52 litres but now has the petrol gauge reading shown.

(a) How much petrol has been used?

(b) How much petrol is still in the tank?

12 Melissa surveyed 75 Year 7 students. $\frac{2}{3}$ said they enjoyed studying mathematics. $\frac{4}{5}$ of those who enjoyed mathematics also said they enjoyed studying science.

(a) How many students said they enjoyed mathematics?

(b) How many students said they enjoyed mathematics and science?

13 Angelina is making Custard Cream cookies. Below are the ingredients that make 24 cookies.

180 g butter $\frac{1}{2}$ cup icing sugar $1\frac{1}{4}$ cups self-raising flour $\frac{1}{3}$ cup custard powder

Rewrite the ingredients to make 60 cookies. (Hint: What do you need to multiply 24 by to get 60?)

Reasoning, communicating and open-ended

14 Without doing the calculation, which is larger: $1\frac{1}{2} \times 12$ or $\frac{1}{3} \times 9$?

15 Without doing any calculations, say which of the following would give the biggest answer.

A $\frac{1}{2} \times \frac{3}{4}$ **B** $\frac{2}{5} \times \frac{2}{3}$ **C** $\frac{3}{4} \times \frac{6}{5}$ **D** $1\frac{1}{2} \times 1\frac{2}{5}$

16 The missing number in the statement $\frac{5}{12} \times \frac{\square}{10} = \frac{5}{8}$ is:

A 4 **B** 5 **C** 10 **D** 15

17 Write two fractions that multiply to give $\frac{1}{6}$ as the simplified answer.

18 $\frac{1}{\square} \times 3\square = 1\square$

Fill each of the boxes with a single digit (0–9) to make the statement correct. There are six possible combinations. Try to find at least three of them.

19 Asher was asked to find $\frac{3}{5}$ of 60. His answer was 4. Here is his working:

$$\frac{\cancel{3}^{1}}{5} \times \frac{\cancel{60}^{20}}{1} = \frac{20}{5} = 4$$

Asher knows his answer is not correct, because he knows that $\frac{3}{5}$ is bigger than one half, and 4 is nowhere near half of 60! Explain where Asher has gone wrong. Include the correct working and answer to the problem.

3.6 Dividing fractions

Need to Know

Whole numbers divided by fractions

If 3 whole apples are cut into quarters, how many apple quarters will there be?

'How many quarters in 3' can be written as $3 \div \frac{1}{4} = ?$

There are 4 quarters in each apple, so there are 12 quarters in 3 apples: $3 \div \frac{1}{4} = 12$.

If the apples are cut into fifths, there would be fifteen fifths: $3 \div \frac{1}{5} = 15$.

If the apples are cut into sixths, there would be eighteen sixths: $3 \div \frac{1}{6} = 18$.

The answer is found by multiplying the whole number by the denominator of the fraction.

For example:

$3 \div \frac{1}{4}$	$3 \div \frac{1}{5}$	$3 \div \frac{1}{6}$
$= 3 \times \frac{4}{1}$	$= 3 \times \frac{5}{1}$	$= 3 \times \frac{6}{1}$
$= 12$	$= 15$	$= 18$

The fractions have been turned upside down or 'inverted', and the division sign changed to multiplication.

Inverting a fraction (turning it upside down) is also known as 'finding the **inverse**'.

The inverse of $\frac{1}{4}$ is $\frac{4}{1}$ or simply 4.

To divide a whole number by a fraction, multiply by the inverse of the fraction.

Fractions divided by fractions

The 'invert the fraction you are dividing by and multiply' method also applies when dividing one fraction by another.

For example, $\frac{1}{5} \div \frac{1}{3} = \frac{1}{5} \times \frac{3}{1}$

$= \frac{3}{5}$

Using equivalent fractions

Another method of dividing fractions is to convert both numbers to equivalent fractions with the same denominator (LCD). Then perform division on the numerators.

Convert mixed numbers to improper fractions before dividing.

3.6

Worked Example 17

WE17

Calculate $9 \div \frac{3}{8}$.

Working

Method 1: Multiply by the inverse

$\frac{9}{1} \div \frac{3}{8}$

$= \frac{9}{1} \times \frac{8}{3}$

$= \frac{\cancel{9}^{3}}{1} \times \frac{8}{\cancel{3}_{1}}$

$= \frac{3 \times 8}{1 \times 1} = \frac{24}{1}$

$= 24$

Working

Method 2: Use equivalent fractions with the same denominator

LCD = 8

$9 = \frac{72}{8}$

$9 \div \frac{3}{8}$

$= \frac{72}{8} \div \frac{3}{8}$

$= 72 \div 3$

$= 24$

Worked Example 18

WE18

Calculate $1\frac{1}{6} \div \frac{2}{3}$.

Working

Method 1: Multiply by the inverse

$1\frac{1}{6} = \frac{7}{6}$

$\frac{7}{6} \div \frac{2}{3}$

$= \frac{7}{6} \times \frac{3}{2}$

$= \frac{7}{\cancel{6}_{2}} \times \frac{\cancel{3}^{1}}{2}$

$= \frac{7 \times 1}{2 \times 2}$

$= \frac{7}{4}$

$= 1\frac{3}{4}$

Working

Method 2: Use equivalent fractions with the same denominator

LCD = 6

$1\frac{1}{6} = \frac{7}{6}$

$\frac{2}{3} = \frac{4}{6}$

$1\frac{1}{6} \div \frac{2}{3}$

$= \frac{7}{6} \div \frac{4}{6}$

$= 7 \div 4$

$= 1\frac{3}{4}$

3.6

Answers page 460

3.6 Dividing fractions

Navigator

Q1 Columns 1–3, Q2 Columns 1 & 2, Q4, Q5, Q6, Q8, Q9, Q12, Q15	Q1 Columns 3 & 4, Q2 Columns 3 & 4, Q3, Q4, Q5, Q6, Q7, Q8, Q9, Q11, Q12, Q15	Q1 Column 4, Q2 Column 4, Q3, Q4, Q5, Q7, Q8, Q9, Q10, Q11, Q12, Q13, Q14, Q15

Fluency and understanding

WE17

1 Calculate the following.

(a) $6 \div \frac{1}{4}$ (b) $2 \div \frac{1}{6}$ (c) $3 \div \frac{1}{7}$ (d) $4 \div \frac{1}{9}$

(e) $4 \div \frac{2}{9}$ (f) $4 \div \frac{4}{5}$ (g) $5 \div \frac{3}{7}$ (h) $7 \div \frac{4}{9}$

(i) $6 \div 1\frac{1}{8}$ (j) $8 \div 2\frac{2}{3}$ (k) $6 \div 3\frac{1}{5}$ (l) $3 \div 2\frac{1}{4}$

WE18

2 Calculate the following.

(a) $\frac{1}{2} \div \frac{1}{4}$ (b) $\frac{1}{4} \div \frac{1}{6}$ (c) $\frac{3}{4} \div \frac{2}{3}$ (d) $\frac{7}{10} \div \frac{1}{2}$

(e) $\frac{1}{3} \div 2$ (f) $\frac{1}{5} \div 4$ (g) $\frac{3}{4} \div 3$ (h) $\frac{9}{10} \div 6$

(i) $\frac{9}{10} \div 1\frac{3}{5}$ (j) $\frac{7}{6} \div 2\frac{1}{12}$ (k) $\frac{4}{5} \div 1\frac{8}{9}$ (l) $\frac{9}{10} \div 2\frac{2}{3}$

(m) $2\frac{2}{3} \div 1\frac{1}{3}$ (n) $5\frac{1}{2} \div 5\frac{3}{8}$ (o) $4\frac{1}{2} \div 1\frac{1}{5}$ (p) $4\frac{5}{7} \div 2\frac{3}{4}$

3 (a) $5 \div \frac{15}{8} =$

A $\frac{8}{75}$ B $\frac{3}{8}$ C $2\frac{2}{3}$ D $\frac{75}{8}$

(b) $\frac{16}{7} \div \frac{20}{49} =$

A $\frac{5}{28}$ B $\frac{320}{343}$ C $\frac{343}{320}$ D $5\frac{3}{5}$

4 Copy and complete the following. Do all working in your head.

(a) There are ______ halves in one, so $1 \div \frac{1}{2} =$ ______.

(b) There are ______ halves in three, so $3 \div \frac{1}{2} =$ ______.

(c) There are ______ quarters in one, so $1 \div \frac{1}{4} =$ ______.

(d) There are ______ quarters in two, so $2 \div \frac{1}{4} =$ ______.

3.6

5 The inverse of $\frac{1}{3}$ is:

A $\frac{1}{6}$ B $\frac{2}{3}$ C 1 D 3

6 There is $\frac{5}{8}$ of a metre of ribbon in the textiles class cupboard. Students require $\frac{1}{10}$ of a metre each for their projects. How many students will be able to use the ribbon?

7 Scientists in a medical laboratory work in shifts of $3\frac{1}{2}$ hours each. How many whole shifts are to be worked during a period of 42 hours?

8 Hannah is a baker whose specialty is scones. She has a sack containing 80 cups of flour. Her scone recipe uses $2\frac{2}{3}$ of a cup of flour for each batch of scones. How many batches of scones will Hannah get from her sack of flour?

9 On a 15 km fun run course, drinks stations are placed every $2\frac{1}{2}$ km from the start. How many drinks stations will there be on the course? (There are no drinks stations at the start or finish lines.)

10 The hit movie 'Exterminator 5' runs for $1\frac{3}{4}$ hours. A local cinema complex plans to screen it back to back (alternating the screenings between 2 cinemas). The first screening will start at 1:30 p.m. The last screening must finish by 11 p.m. How many times can the movie be screened within this time period?

Reasoning, communicating and open-ended

11 The missing number from the statement $\frac{3}{4} \div \frac{\square}{8} = 2$ is:

A 1 B 2 C 3 D 4

12 Without doing a calculation, state which will give the larger number as a result:
$\frac{4}{5} \div \frac{1}{3}$, or $\frac{4}{5} \div \frac{1}{3}$?

13 Consider this sequence: $5 \div \frac{3}{4} = 6\frac{2}{3}$, $5 \div \frac{3}{5} = 8\frac{1}{3}$, $5 \div \frac{3}{6} = 10$, $5 \div \frac{3}{7} = 11\frac{2}{3}$.

(a) Why does the quotient (the answer to the division) increase as the denominator of the fractions in red is increased?

(b) Predict the answer to the next term in the sequence; i.e. $5 \div \frac{3}{8}$ will be ...

14 Use digits from 0 to 9 to fill in the boxes to make a correct statement. Find at least two different combinations.

$4 \div \frac{2}{\square} = 1\square$

15 Here is Gabi's working on a fraction division problem:

$3 \div \frac{1}{4} = 3 \div 4 = \frac{3}{4}$

Her friend Natalie has worked it out differently. This is what Natalie wrote:

$3 \div \frac{1}{4} = \frac{3}{1} \times \frac{4}{1} = 12$

'That can't be right,' said Gabi, '12 is way too big to be the answer!'

Which of the girls' working is correct? Explain what the other has done wrong.

3.7 Mixed fraction problems

Need to Know

The order of operations rules that apply to whole numbers also apply to fractions.

1 Simplify brackets first.
2 Perform any multiplication or division working from left to right.
3 Perform any addition or subtraction working from left to right.

Worked Example 19

WE19

Simplify the following using the correct order of operations.

(a) $3+\frac{2}{5}\times\frac{1}{2}$

(b) $\frac{6}{7}\times\left(1\frac{2}{3}+\frac{1}{4}\right)-\frac{3}{7}$

Working

(a) $3+\frac{2}{5}\times\frac{1}{2}$

$=3+\frac{\cancel{2}^{1}\times 1}{5\times\cancel{2}^{1}}$

$=3+\frac{1}{5}$

$=3\frac{1}{5}$

(b) $\frac{6}{7}\times\left(1\frac{2}{3}+\frac{1}{4}\right)-\frac{3}{7}$

$=\frac{6}{7}\times\left(\frac{5}{3}+\frac{1}{4}\right)-\frac{3}{7}$

$=\frac{6}{7}\times\left(\frac{20}{12}+\frac{3}{12}\right)-\frac{3}{7}$

$=\frac{6}{7}\times\frac{23}{12}-\frac{3}{7}$

$=\frac{\cancel{6}^{1}\times 23}{7\times{}^{2}\cancel{12}}-\frac{3}{7}$

$=\frac{23}{14}-\frac{3}{7}$

$=\frac{23}{14}-\frac{6}{14}$

$=\frac{17}{14}$

$=1\frac{3}{14}$

3.7

Technology Exploration: Calculator

Entering fractions on a calculator

Enter the required fractions using the fraction keys on your calculator and the operations as they occur in the question. Do not press enter until you have entered the entire question. Your calculator will evaluate using the correct order of operations.

For example, $\frac{3}{4}-\frac{3}{5}\div 1\frac{1}{2}+1\frac{8}{9}\times 2\frac{1}{4}=4\frac{3}{5}$

3.7 Mixed fraction problems

Answers page 460

Equipment required: a calculator for Questions **5** and **6**

Navigator

Q1 Column 1, Q2 (a), Q3 (a) & (b), Q4, Q5, Q6, Q7, Q8, Q9, Q14, Q17	Q1 Column 2, Q2, Q3, Q4, Q5, Q6, Q7, Q8, Q9, Q10, Q12, Q14, Q15, Q16, Q18	Q1 Columns 2 & 3, Q2, Q3, Q5, Q6, Q7, Q8, Q10, Q11, Q12, Q13, Q15, Q16, Q17, Q18

Fluency and understanding

WE 19

1 Simplify the following using the correct order of operations.

(a) $2+\frac{3}{4}\times\frac{1}{3}$ (b) $5-\frac{7}{8}\times\frac{1}{7}$ (c) $\frac{5}{11}\div(5+4)$

(d) $\frac{3}{4}\times\left(\frac{1}{3}+\frac{1}{6}\right)\times\frac{1}{2}$ (e) $\frac{7}{8}\times\left(\frac{9}{14}-\frac{2}{7}\right)$ (f) $\frac{5}{6}\times\left(\frac{2}{5}+\frac{1}{3}\right)$

(g) $1\frac{5}{8}+\left(1\frac{3}{4}-1\frac{1}{8}\right)\times 5$ (h) $2\frac{3}{4}+\left(4\frac{1}{2}-3\frac{1}{4}\right)\times\frac{1}{2}$ (i) $1\frac{3}{5}+\left(3\frac{3}{10}-1\frac{1}{5}\right)\div 2$

(j) $1\frac{1}{4}-\left(3\frac{1}{3}-2\frac{5}{6}\right)\div 2$ (k) $3\frac{3}{8}+\left(8\frac{1}{4}-6\frac{3}{8}\right)\div 3$ (l) $\frac{7}{9}\div\left(1\frac{5}{6}-1\frac{3}{5}\right)\times\frac{1}{5}$

2 (a) $\frac{3}{7}\times\frac{1}{3}+5=$

A $5\frac{1}{7}$ B $5\frac{1}{3}$ C $5\frac{5}{21}$ D $5\frac{7}{9}$

(b) $6-\frac{9}{13}\div\frac{3}{4}=$

A $5\frac{1}{13}$ B $5\frac{25}{52}$ C $5\frac{12}{13}$ D $7\frac{1}{13}$

3 Write out the following questions using fractions, the operations symbols (×, ÷, +, −) and brackets where necessary. Then, evaluate using the correct order of operations.

(a) Add $\frac{7}{8}$ and $\frac{5}{6}$, and then multiply the total by 2.

(b) Find the product of $\frac{4}{5}$ and $\frac{1}{3}$, then add $1\frac{1}{2}$.

(c) Find the difference between $2\frac{4}{5}$ and $1\frac{2}{3}$, then multiply by 3.

(d) Find out how many $\frac{2}{3}$s in 5, then multiply by 4.

3.7

4 A survey was taken of 120 people as they left Wally's Sandwich Bar. Write your answers to the following in simplest form.

(a) If 80 were male, what fraction were male?

(b) What fraction were female?

(c) If 45 were under 21 years of age, what fraction were under 21?

(d) What fraction were 21 or over?

(e) If $\frac{3}{4}$ had bought a drink, how many had bought a drink?

(f) Of those who bought a drink, $\frac{2}{5}$ had bought a salad roll. How many people had a drink and a salad roll?

5 Use a calculator to evaluate the following.

(a) $6\frac{7}{9} \times 5\frac{2}{11}$ **(b)** $14\frac{1}{7} \times 3\frac{1}{3} \times 5\frac{2}{5}$ **(c)** $11\frac{1}{4} \times 8\frac{4}{5} \div 6\frac{1}{9}$

(d) $39\frac{3}{4} \div 6\frac{2}{3}$ **(e)** $7\frac{3}{8} \times 4\frac{1}{9} \div 2\frac{1}{3}$ **(f)** $37\frac{4}{5} \div 12\frac{1}{4} \times 8\frac{3}{10}$

6 For each of the following:

(i) estimate your answer

(ii) calculate the answer

(iii) use a calculator to check your answer

(iv) compare your estimate to the exact answer.

(a) $\left(4\frac{1}{2} + 3\frac{2}{5}\right) \times 8\frac{1}{4}$

(b) $11\frac{2}{3} \div 2\frac{7}{8} \times \left(5\frac{1}{4} - 2\frac{1}{8}\right)$

(c) $6\frac{1}{4} - 3\frac{1}{3} \div 5\frac{1}{7}$

7 Evaluate:

(a) $\frac{3}{11}$ of $4\frac{5}{7}$ **(b)** $\frac{7}{8}$ of $11\frac{1}{5}$ **(c)** $\frac{5}{7}$ of $8\frac{1}{6}$

8 Grant is training in his backyard pool for the 1500 m freestyle at the next Olympics. Unfortunately, the pool is only 20 m in length.

(a) What fraction of the total 1500 m race would he have swum after completing just 1 lap?

(b) How many laps would he have to complete to swim 1500 m?

(c) What fraction of the total 1500 m race would he have swum after completing 6 laps?

(d) How far would he have swum if he had completed $\frac{1}{10}$ of the race distance?

(e) How far would he have swum if he had completed $\frac{3}{4}$ of the race distance?

9 Valerie is running laps of her school oval to train for Rigby High's Mini Marathon. She ran $4\frac{1}{2}$ laps before having to rest. After her rest, she ran another $3\frac{1}{4}$ laps before stopping again. Then, she struggled through another $\frac{1}{3}$ of a lap. How many laps did she complete altogether?

3.7

10 One weekend, Louisa and Andrew walk from Ferndale to Greenhill and back, a total of $2\frac{3}{4}$ km. The next weekend they walk from Ferndale through Greenhill to Highvale, which is $5\frac{1}{3}$ km.

How far (in km) is it from:

(a) Ferndale to Greenhill

(b) Greenhill to Highvale

(c) Ferndale to Highvale and back again?

11 Jeremy worked at an amusement park for 4 hours at the normal hourly rate, then another $2\frac{1}{2}$ hours at $1\frac{1}{2}$ times the normal rate. What was the equivalent number of normal hours he was to be paid?

Reasoning, communicating and open-ended

12 Theresa, Isabelle and Bryce worked on a mathematical problem and came up with the answers $8\frac{1}{2}$, $8\frac{2}{3}$ and $8\frac{3}{5}$, respectively. The correct answer was $8\frac{7}{12}$. Which of the three students was closest to the correct answer?

13 Despina's department store advertises a winter sale in which they claim everything is between $\frac{1}{3}$ and $\frac{1}{2}$ off. For which of the following items is this incorrect?

Item	Normal price	Sale price
Bath towels	$22	$12
Camera	$900	$600
Cutlery set	$250	$120
Microwave oven	$540	$350
Jeans	$60	$35

14 The Happy Valley Farming Cooperative has divided its square block of land in the way shown in the diagram. Each section of land is exactly half of the section next to it.

The May family has been given the shaded region to farm. If the total block of land is 1, what fraction of the total *doesn't* the May family farm?

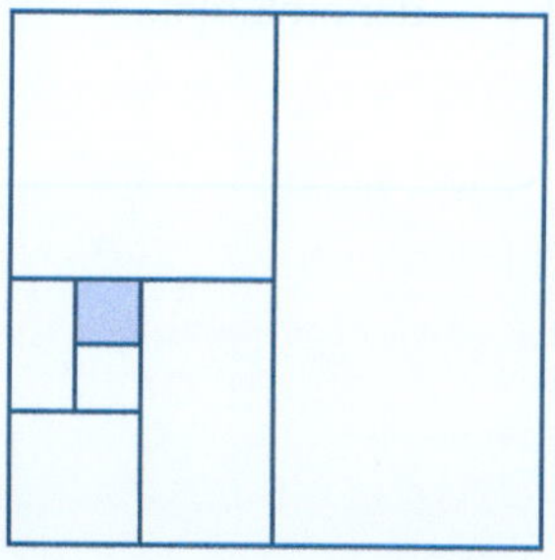

15 Prashanth has agreed to sell badges for a charity. He receives a large box and places them, with permission, at the school reception. After several days, Prashanth took an estimate of the number of badges he had left. He emptied out the box, divided the pile of badges roughly into halves, and put half back in the box. He then halved the pile he had left, and returned half to the box. He did this one more time, after which he counted the number of badges in the pile, and found he had 13.

(a) What fraction of the total number of badges does 13 approximately represent?

(b) How many badges did Prashanth have in the box, approximately?

Hint

Draw a diagram or flowchart to show this process.

3.7

16 Write down three numbers in fraction form that are less than $\frac{7}{8}$ but greater than $\frac{1}{2}$.

17 A and B are two different numbers selected from the digits 0, 1, 2, 3, 4, 5, 6, 7, 8, 9, where A is always greater than B.

(a) What values can $\dfrac{A+B}{A-B}$ have? Give three possibilities.

(b) What is the largest value this fraction can have?

18 'Ideal fractions' are pairs of fractions whose sum and product are the same. (They give the same number when added as when multiplied.)

(a) Show that $\frac{7}{3}$ and $\frac{7}{4}$ are ideal fractions.

(b) Show that $\frac{5}{3}$ and $\frac{5}{2}$ are ideal fractions.

(c) Look carefully at the numerators and denominators of the fraction pairs in **(a)** and **(b)**. Try to find a pattern.

(d) Write down another pair of ideal fractions.

Problem solving

Mega-equivalents

1 Use four of the digits 1, 2, 3, 4, 5, 6, 7, 8, 9, 0 to make a fraction equivalent to $\frac{1}{2}$ that has two digits in the numerator and in the denominator (e.g. $\frac{34}{68}$).

2 Now, use six digits to make another fraction equivalent to $\frac{1}{2}$ that has three digits in the numerator and denominator.

3 See if you can form another fraction equivalent to $\frac{1}{2}$ using eight digits (four in numerator and denominator).

4 Can you make a fraction equivalent to $\frac{1}{2}$ using all 10 digits?

Strategy options

- Guess and check.
- Test all possible combinations.

Challenge 3

1 If $\frac{1}{3}$ of a number is 16, then $\frac{3}{8}$ of this number is:

A 8 **B** 16 **C** 18 **D** 24

2 Kerry bought a new water tank for her garden. The tank is initially empty, but fills to $\frac{2}{3}$ full of capacity after rain. The tank now holds 480 litres of water. How many litres does the tank hold when full?

3 If $\frac{\nabla}{16}$ lies between $\frac{1}{4}$ and $\frac{3}{8}$, and ∇ is a whole number, then ∇ equals:

A 5 **B** 6 **C** 7 **D** 8

4 Two proper fractions (in simplest form) are subtracted, the first with a denominator of 10, and the second with a denominator of 6. If the answer is given in simplest form, the largest possible numerator that it could have is:

A 8 **B** 9 **C** 11 **D** 22

5 The number halfway between $\frac{1}{5}$ and $\frac{1}{20}$ is:

A $\frac{1}{10}$ **B** $\frac{1}{8}$ **C** $\frac{7}{40}$ **D** $\frac{1}{4}$

6 Two frogs live together in a backyard pond. One frog croaks every 4 minutes, the other frog croaks every $2\frac{1}{2}$ minutes. If they both croak together at 11 a.m., what is the first time after 1 p.m. that they will croak together?

7 $2 + \dfrac{6}{1 + \frac{1}{2}}$ equals:

A 2 **B** 4 **C** 6 **D** 8

8 Express this fraction in its simplest form:

$$\cfrac{1}{3 + \cfrac{1}{3 + \cfrac{1}{3 + \cfrac{1}{3}}}}$$

9 If 60 is added to one-third of a number, the resulting value is double the number. What is the number?

10 Five numbers are put in a row from smallest to largest. The difference between each adjacent number (a number and the number next to it) is the same. If the first number is $\frac{1}{8}$ and the last is $\frac{1}{6}$, what are the three numbers in-between, in simplest form?

11 Tom was born into a strange mathematical family. His grandmother sent him a money voucher for his birthday. The card said:

Dear Tom,
I am giving you the following amount to spend as you wish:
$\frac{1}{2}$ of $\frac{2}{3}$ of $\frac{3}{4}$ of $\frac{4}{5}$ of $\frac{5}{6}$ of $\frac{6}{7}$ of $\frac{7}{8}$ of $\frac{8}{9}$ of $\frac{9}{10}$ of \$1000.
Love, Grandma

How much birthday money did Tom receive?

Chapter review 3

D.I.Y. Summary

Key Words

cancelling	improper fraction	mixed number	simplest form
denominator	inverse	numerator	simplify
equivalent fractions	Lowest Common Denominator (LCD)	proper fraction	unit fraction
fraction			

Copy and complete the following using the words and phrases from this list, where appropriate, to write a summary for this chapter. A word or phrase may be used more than once.

1 In a fraction, the __________ tells us how many parts we have, and the __________ tells us the size of each part.

2 A __________ has a value between 0 and 1.

3 To __________ a fraction, divide the numerator and denominator by a common factor. This process is known as __________.

4 Dividing by a fraction is the same as multiplying by the __________ of the fraction.

5 An __________ has a numerator that is greater than or equal to the denominator.

6 __________ are located at exactly the same position on a number line.

7 A __________ is a fraction with a numerator of 1.

Fluency and understanding

1 Here are 3 dozen (36) eggs. **Ex. 3.1, 3.2**

(a) If $\frac{4}{9}$ of the eggs are slightly cracked, how many eggs are damaged?

(b) The farmer's 3 best hens laid 15 of the eggs pictured. What fraction of the total is this? Write your answer in simplest form.

2 Write the value of the fractions shown by the arrows on this number line: **Ex. 3.1**

3 Write **(a)** 6 and **(b)** 11 as improper fractions with denominators of: **Ex. 3.1**

(i) 3 **(ii)** 10 **(iii)** 1

4 **(a)** 3 pizzas are shared between 5 students. Write the amount of pizza each student gets as a fraction. **Ex. 3.1**

(b) 7 packets of lollies are shared between 4 students. Write the number of packets each student receives as a mixed number.

5 Find equivalent fractions by copying and completing the following. Ex 3.2

(a) $\frac{2}{7} = \frac{12}{\square}$ **(b)** $\frac{12}{18} = \frac{\square}{6}$ **(c)** $\frac{15}{30} = \frac{45}{\square}$ **(d)** $\frac{\square}{24} = \frac{5}{4}$

6 Write the following fractions in their simplest form. Ex 3.2

(a) $\frac{12}{15}$ **(b)** $\frac{48}{20}$ **(c)** $3\frac{60}{72}$ **(d)** $1\frac{8}{40}$

7 (a) Write $3\frac{2}{7}$ as an improper fraction. Ex 3.2

(b) Write $\frac{50}{9}$ as a mixed number.

8 Copy and complete the following by inserting <, > or =. Ex 3.3

(a) $\frac{5}{13}$ ______ $\frac{1}{2}$ **(b)** $\frac{6}{16}$ ______ $\frac{30}{80}$ **(c)** $\frac{5}{12}$ ______ $\frac{3}{8}$ **(d)** $\frac{8}{9}$ ______ $\frac{9}{10}$

9 Evaluate the following. Estimate first to check that your answers are reasonable. Ex 3.4

(a) $\frac{5}{12} + \frac{7}{8}$ **(b)** $\frac{7}{18} - \frac{2}{9}$ **(c)** $\frac{2}{3} - \frac{2}{7}$ **(d)** $\frac{1}{6} + \frac{7}{10}$

10 Evaluate the following. Write your answers as mixed numbers in simplest form. Ex 3.4

(a) $3\frac{2}{5} - 1\frac{3}{4}$ **(b)** $2\frac{1}{4} + 5\frac{11}{12}$ **(c)** $8 - 2\frac{5}{7}$ **(d)** $4\frac{5}{6} - 4\frac{2}{3}$

11 Calculate the following. Ex 3.5

(a) $\frac{3}{7}$ of \$28 **(b)** $\frac{2}{9}$ of $\frac{3}{4}$

12 Evaluate the following. Ex 3.5

(a) $\frac{6}{11} \times \frac{55}{18}$ **(b)** $2\frac{4}{9} \times 5$ **(c)** $\frac{2}{3} \times \frac{9}{16} \times 1\frac{1}{7}$

13 Evaluate the following. Ex 3.6

(a) $9 \div \frac{3}{4}$ **(b)** $21 \div 3\frac{2}{7}$ **(c)** $\frac{6}{5} \div \frac{32}{15}$ **(d)** $3\frac{1}{4} \div 3\frac{1}{3}$

14 Simplify by using the correct order of operations. Ex 3.7

(a) $4 + \frac{5}{6} \times 2\frac{1}{3}$ **(b)** $7\frac{2}{5} - \left(3\frac{1}{10} - \frac{3}{5}\right)$ **(c)** $\frac{7}{8} \times \left(\frac{3}{4} + 4 \times \frac{3}{4}\right)$

15 (a) What fraction sum is represented by the following two grids?

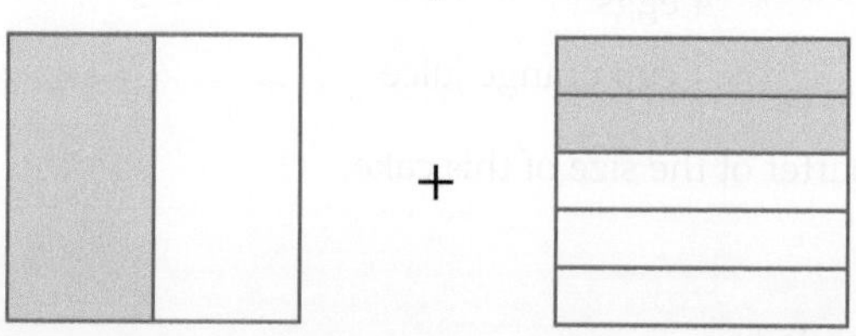

(b) Calculate the fraction sum using the grid method or any other method.

16 Write a fraction or a mixed number in simplest form to show each of these:

(a) eight chocolate biscuits in a packet of 24

(b) two complete '6-packs' of soft drink cans, with one can left over

(c) one packet of twelve colour pencils and four pencils in a second packet

(d) nine hours sleep in one day.

17 There are 12 boys and 15 girls in class 7A. 6 of the students have blond hair, and 2 have red hair. Write the following as fractions of the whole class, in simplest form.

(a) the number of girls

(b) the number of blond-haired students

(c) the number of students with neither blond nor red hair

18 Write each set of numbers in order from smallest to largest, by using strategies for comparing fractions.

(a) $\frac{3}{4}, \frac{1}{2}, 1, \frac{1}{3}, \frac{3}{5}$

(b) $2, \frac{13}{5}, \frac{5}{4}, \frac{30}{50}, 1\frac{2}{5}$

19 Jamal gets 7 out of 9 shots at goal in the netball ring, whereas Kayla gets 9 out of 12 shots in. Determine who is the more accurate shooter by comparing fractions.

20 Elise and Reece each have an identical block of chocolate. Elise eats $\frac{2}{5}$ and Reece eats $\frac{2}{3}$.

(a) How much more has Reece eaten, as a fraction of a block?

(b) Together, have Elise and Reece eaten more or less than a whole block of chocolate? Estimate, then calculate your answer.

21 Gary had $1\frac{1}{2}$ bags of cement in his shed. He used $\frac{2}{3}$ of a bag to mix up some concrete. What fraction of a full bag does he have left?

22 Jarrod owns a petrol station that has three main fuel tanks. There is a dip stick in each tank that shows the level of fuel remaining in each tank. How many litres are left in each tank?

(a) Full tank = 48 000 L

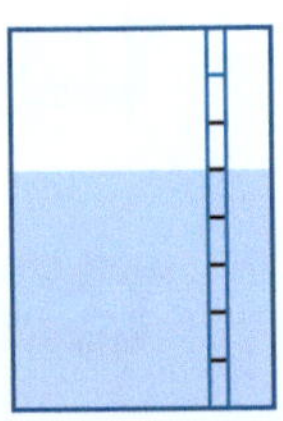

(b) Full tank = 60 000 L

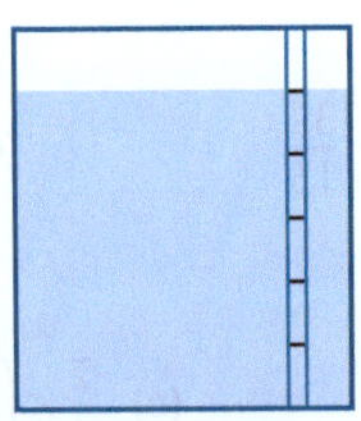

(c) Full tank = 30 000 L

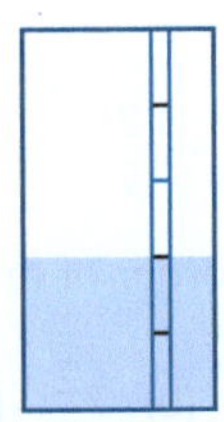

23 Here are the ingredients for a choc-orange cake:

180 g butter	$\frac{2}{3}$ cup castor sugar	4 eggs
$1\frac{3}{4}$ cups self-raising flour	2 tablespoons cocoa	$\frac{1}{2}$ cup orange juice

Rewrite the ingredients to make a cake that is one-quarter of the size of this cake.

Reasoning and communicating

24 State the fraction of the area of these shapes that is shaded.

(a)

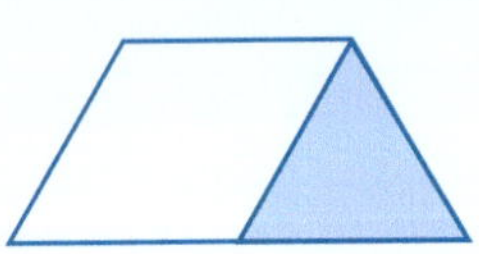

(b)

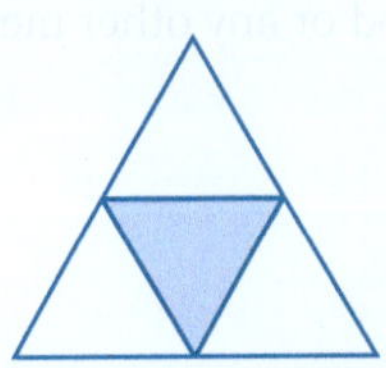

(c)

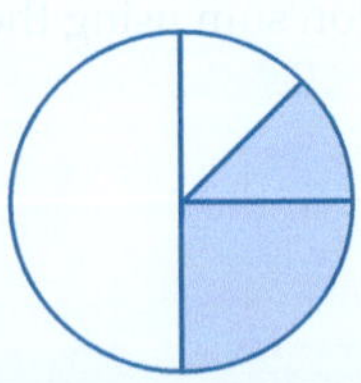

25 45 minutes into a fun run, Katya had completed 4 km of the 10 km course, whereas Mabok had completed 7 km of the 15 km course.

(a) Who had the least distance still to run?

(b) Who had completed a bigger fraction of their particular course?

26 Which of the following will give an answer of 1?

A $\frac{1}{4} \div \frac{1}{4}$ **B** $\frac{1}{2} \div \frac{1}{4}$ **C** $1 \div \frac{1}{4}$ **D** $4 \div \frac{1}{4}$

NAPLAN practice 3

Numeracy: Non-calculator

1 Which diagram does not have exactly $\frac{1}{2}$ of the area shaded?

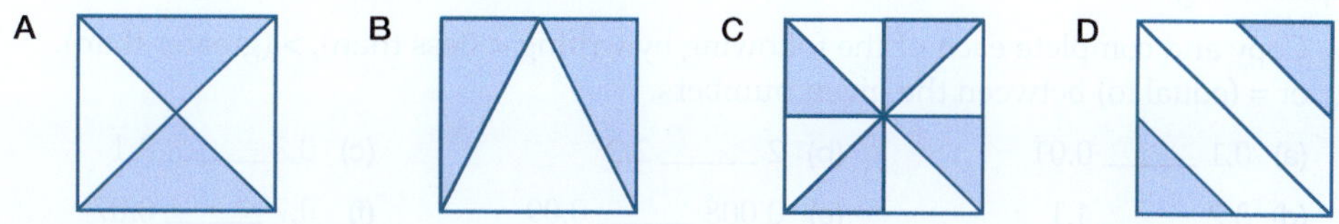

2 Which arrow is pointing closest to the location of $\frac{11}{8}$ on this number line?

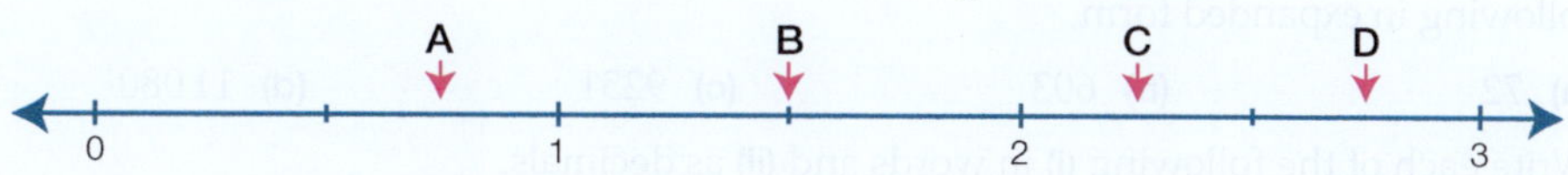

3 Kevin is mixing sand, gravel and cement to make concrete. Sand makes up $\frac{1}{2}$ of the mixture, and gravel is $\frac{1}{3}$. What fraction of the mixture is cement?

A $\frac{1}{6}$ **B** $\frac{1}{5}$ **C** $\frac{1}{4}$ **D** $\frac{1}{3}$

4 Esther is making vanilla ice-cream using these ingredients that make 6 servings:

5 eggs
6 cups milk
1 cup sugar
$1\frac{1}{2}$ teaspoons vanilla
2 cans condensed milk

How many teaspoons of vanilla will Esther need to make 30 servings?

A $5\frac{1}{2}$ **B** $7\frac{1}{2}$ **C** $30\frac{1}{2}$ **D** 45

Numeracy: Calculator allowed

5 A school has 200 students. 48 of the students are involved in the school production. The fraction of students who are involved in the school production is closest to:

A one-fifth **B** one-quarter **C** one-third **D** one-half

6 Which of the following fractions is greater than 2 but less than 3?

A $\frac{3}{4}$ **B** $\frac{12}{9}$ **C** $\frac{17}{7}$ **D** $\frac{15}{5}$

7 Imran ran $\frac{1}{6}$ of a cross-country course. He ran 3 km. How long is the course?

A 6 km **B** 9 km **C** 12 km **D** 18 km

8 Which fraction lies exactly halfway between $\frac{3}{8}$ and $\frac{4}{8}$ on the number line?

A $\frac{3}{16}$ **B** $\frac{1}{4}$ **C** $\frac{1}{3}$ **D** $\frac{7}{16}$

4 Decimals, percentage and ratio

Recall 4

Prepare for this chapter by attempting the following questions. If you have difficulty with a question, go to Pearson Places and download the Recall Worksheet.

1 Copy and complete each of the following by writing < (less than), > (greater than), or = (equal to) between the given numbers.

(a) 0.1 ______ 0.01 **(b)** 2 ______ 2.2 **(c)** 0.3 ______ 0.1
(d) 3.2 ______ 4.1 **(e)** 0.008 ______ 0.09 **(f)** 0.7 ______ 0.07

2 The number 43 can be written in expanded form as 'four tens and three ones'. Write the following in expanded form.

(a) 72 **(b)** 603 **(c)** 9251 **(d)** 11 080

3 Write each of the following **(i)** in words and **(ii)** as decimals.

(a) $\frac{7}{10}$ **(b)** $\frac{8}{1000}$ **(c)** $\frac{3}{100}$ **(d)** $1\frac{9}{10}$

4 Calculate:

(a) $\begin{array}{r} 34 \\ +\ 76 \\ \hline \end{array}$ **(b)** $\begin{array}{r} 925 \\ +\ 610 \\ \hline \end{array}$ **(c)** $\begin{array}{r} 67 \\ 409 \\ +\ 3 \\ \hline \end{array}$ **(d)** $\begin{array}{r} 459 \\ 6013 \\ +\ 27 \\ \hline \end{array}$

5 Calculate:

(a) $\begin{array}{r} 74 \\ -\ 25 \\ \hline \end{array}$ **(b)** $\begin{array}{r} 823 \\ -\ 376 \\ \hline \end{array}$ **(c)** $\begin{array}{r} 8289 \\ -\ 384 \\ \hline \end{array}$ **(d)** $\begin{array}{r} 2000 \\ -\ 352 \\ \hline \end{array}$

6 Calculate:

(a) 2×17 **(b)** 25×96 **(c)** 3×48 **(d)** 51×800

7 Perform the following divisions.

(a) $362 \div 2$ **(b)** $9459 \div 9$ **(c)** $5600 \div 200$ **(d)** $4500 \div 30$

8 Calculate:

(a) 70×100 **(b)** $12 \times 10\,000$ **(c)** 9200×1000
(d) $80 \div 10$ **(e)** $3600 \div 100$ **(f)** $250\,000 \div 1000$
(g) 1.2×10 **(h)** 6.3×100 **(i)** 0.0427×1000
(j) $58 \div 10$ **(k)** $901 \div 100$ **(l)** $76.2 \div 1000$

9 Write these percentages as fractions in simplest form.

(a) 50% **(b)** 10% **(c)** 25% **(d)** 20% **(e)** 1% **(f)** 60%

Syllabus references

Number and Algebra: Fractions, Decimals and Percentages

- ✓ Multiply and divide fractions and decimals using efficient written strategies and digital technologies **(ACMNA154)**
- ✓ Round decimals to a specified number of decimal places **(ACMNA156)**
- ✓ Find percentages of quantities and express one quantity as a percentage of another, with and without the use of digital technologies **(ACMNA158)**
- ✓ Investigate terminating and recurring decimals **(ACMNA184)**
- ✓ Connect fractions, decimals and percentages and carry out simple conversions **(ACMNA157)**

Number and Algebra: Ratio and Rates

- ✓ Recognise and solve problems involving simple ratios **(ACMNA173)**

Number and Algebra: Financial Mathematics

- ✓ Investigate and calculate 'best buys', with and without digital technologies **(ACMNA174)**

Place value and comparing decimals

4.1

Need to Know

Decimal place value

Like fractions, decimals are used to write numbers that have a part that is less than one. The **decimal point** is used to separate the whole number part from the part less than one.

258.237 195

whole number part decimal point part less than one (decimal part)

The number of **digits** after the **decimal point** of a number show how many **decimal places** that number has.

Because the number above has six digits after the decimal point, we say it has six decimal places.

Place value table

Hundreds 100	Tens 10	Ones 1	•	Tenths $\frac{1}{10}$ (0.1)	Hundredths $\frac{1}{100}$ (0.01)	Thousandths $\frac{1}{1000}$ (0.001)	Ten-thousandths $\frac{1}{10000}$ (0.0001)	Hundred-thousandths $\frac{1}{10000}$ (0.00001)	Millionths $\frac{1}{100000}$ (0.000001)
2	5	8	•	2	3	7	1	9	5

If there is no digit for a particular **place value** in a decimal number, we use a zero to show this. For example, five and two tenths and three thousandths is written as 5.203. It is important to include the zero between the 2 and the 3. Leaving it out would mean 5.203 becomes 5.23, which is a different number (that is, five and two tenths and three hundredths).

Expressing decimals in different forms

Decimal numbers can be written in several ways:

- Decimal form: 258.237 195
- Expanded fractional form: $258 + \frac{2}{10} + \frac{3}{100} + \frac{7}{1000} + \frac{1}{10\,000} + \frac{9}{100\,000} + \frac{5}{1\,000\,000}$
- Expanded word form: *two* hundreds, *five* tens, *eight* ones, *two* tenths, *three* hundredths, *seven* thousandths, *one* ten-thousandth, *nine* hundred-thousandths and *five* millionths.

Comparing decimals using a number line

If a unit on a number line is divided into 10, each division represents one tenth, $\frac{1}{10}$ or 0.1. If the one-tenth interval is divided into 10 again, each division represents one hundredth, $\frac{1}{100}$ or 0.01. Dividing one-hundredth into 10 gives divisions of one thousandth, $\frac{1}{1000}$ or 0.001.

The positions of the decimal numbers 0.04, 0.2, 0.565 and 0.85 are shown with arrows.

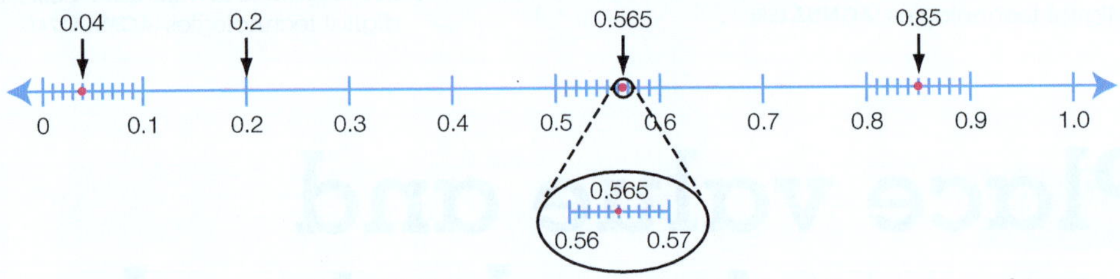

The interval between 0.56 and 0.57 has been magnified to show the position of 0.565 more clearly. It shows 1 hundredth divided up into 10 parts, each equal to 1 thousandth.

$\frac{1}{100} + \frac{100}{1000}$

The decimal number with the most digits is not necessarily the largest; e.g. $0.565 < 0.85$.

A number's size depends on the value and position of the digits; e.g. $0.2 > 0.04$.

Worked Example 1

WE1

(a) Write $4 + \frac{3}{10} + \frac{6}{1000} + \frac{5}{10\,000}$ as a decimal.

(b) Write 6.2807 in expanded fractional form.

Working

(a) $4 + \frac{3}{10} + \frac{6}{1000} + \frac{5}{10\,000}$

$= 4.3065$

(b) 6.2807

$= 6 + \frac{2}{10} + \frac{8}{100} + \frac{7}{10\,000}$

Worked Example 2

WE2

(a) Write seven units, three tenths, six hundredths, seven thousandths and four hundred-thousandths as a decimal.

(b) Write 28.0045 in expanded word form.

Working

(a) 7.367 04

(b) 28.0045

= Two tens, eight ones, four thousandths and five ten-thousandths

Worked Example 3

WE 3

(a) Write the value of the 6 in 4.368 as a fraction.

(b) Write the value of the 9 in 0.0109 in words.

Working

(a) $\frac{6}{100}$

(b) nine ten-thousandths

Worked Example 4

WE 4

Write < or > between each of the following pairs of decimals to make a true statement.

(a) 5.173 _____ 5.0731 (b) 0.472 39 _____ 0.4731 (c) 3.5 _____ 3.51

Working

(a) $5.173 > 5.0731$ (b) $0.47239 < 0.4731$ (c) $3.5 < 3.51$

4.1 Place value and comparing decimals

Answers page 462

Navigator

Q1, Q2, Q3, Q4 Column 1, Q5 Column 1, Q7, Q8 (a) & (b), Q9 Column 1, Q10, Q11, Q13, Q14, Q15, Q17, Q19	Q1, Q2, Q3, Q4 Column 2, Q5 Column 2, Q6, Q8 (c) & (d), Q9 Column 2, Q10, Q11, Q12, Q14, Q15, Q16, Q18, Q19	Q1 (b), Q2 (b), Q3, Q4 Column 2, Q5 Column 2, Q6, Q7, Q8 (c) & (d), Q9 Column 2, Q11, Q12, Q13, Q14, Q15, Q16, Q17, Q18, Q19

Fluency and understanding

WE 1

1 (a) Write each of the following as a decimal.

(i) $45 + \frac{4}{10} + \frac{6}{100} + \frac{2}{1000}$

(ii) $12 + \frac{5}{10} + \frac{1}{100} + \frac{9}{1000} + \frac{3}{10\,000} + \frac{7}{100\,000} + \frac{2}{1\,000\,000}$

(iii) $3 + \frac{7}{10} + \frac{9}{100} + \frac{8}{1000} + \frac{5}{10\,000}$

(iv) $1 + \frac{8}{1000} + \frac{3}{10\,000} + \frac{3}{100\,000} + \frac{2}{1\,000\,000}$

(v) $\frac{7}{10} + \frac{8}{100} + \frac{6}{1000} + \frac{6}{100\,000}$

(vi) $7 + \frac{3}{100} + \frac{3}{1000} + \frac{4}{100\,000} + \frac{7}{1\,000\,000}$

Hint

Don't forget to write a zero if you have no digits to go in a place value column.

4.1

(b) Write each of the following in expanded fractional form.

(i) 6.63	**(ii)** 0.921	**(iii)** 0.7345
(iv) 7.826	**(v)** 23.913 04	**(vi)** 45.004 589

WE2

2 **(a)** Write each of the following as a decimal.

(i) six ones and five tenths

(ii) nine tenths and seven hundredths

(iii) two tenths, seven hundredths and three thousandths

(iv) three tens, seven ones, four tenths, two hundredths and one ten-thousandth

(v) one ten, four ones, nine tenths, five hundredths, seven thousandths, six ten-thousandths, two hundred-thousandths and three millionths

(vi) seven hundreds, four thousandths, five ten-thousandths and nine hundred-thousandths

(b) Write each of the following in expanded word form.

(i) 5.2	**(ii)** 4.9	**(iii)** 34.17	**(iv)** 0.61
(v) 2.794	**(vi)** 7.5092	**(vii)** 35.865 43	**(viii)** 0.820 027

WE3

3 **(a)** Write the value of the 2 in each of the following as a fraction.

(i) 6.012	**(ii)** 0.004 52	**(iii)** 3.287	**(iv)** 2.034

(b) Write the value of the 7 in each of the following in words.

(i) 5.734	**(ii)** 0.0076	**(iii)** 1.2037	**(iv)** 8.130 037

WE4

4 Write $<$ or $>$ between each of the following pairs of decimals to make a true statement.

(a) 2.4 _____ 0.42	**(b)** 2.32 _____ 1.955
(c) 0.65 _____ 0.57	**(d)** 0.3003 _____ 0.333
(e) 4.7038 _____ 4.7312	**(f)** 8.251 _____ 8.2501
(g) 7.02 _____ 7.002	**(h)** 4.7367 _____ 4.7376
(i) 0.927 _____ 0.927 34	**(j)** 6.013 _____ 6.01
(k) 3.406 _____ 3.4063	**(l)** 0.9995 _____ 0.9986

5 Write TRUE or FALSE for each of the following.

(a) $7.5 < 5.77$	**(b)** $4.1 > 4.12$
(c) $6.08 > 6.8$	**(d)** $67.54 < 67.504$
(e) $3.023 < 3.203$	**(f)** $0.547 > 0.547\,08$
(g) $2.000\,012 < 2.0001$	**(h)** $4.145\,29 > 4.200\,01$

6 53.017 expressed in expanded word form is:

A five tens, three ones, one tenth and seven hundredths

B five tens, three ones, one hundredth and seven thousandths

C five tenths, three hundredths, one ten-thousandth and seven hundred-thousandths

D five tens, three ones, one tenth and seven thousandths

7 Nine hundredths, four thousandths and three ten-thousandths is equal to:

A 0.090 43	**B** 0.0943	**C** 0.943	**D** 9.43

8 For each of the following sets of numbers, copy the number line shown, then mark the positions of the numbers with a labelled arrow. (You may have to estimate the position of some.)

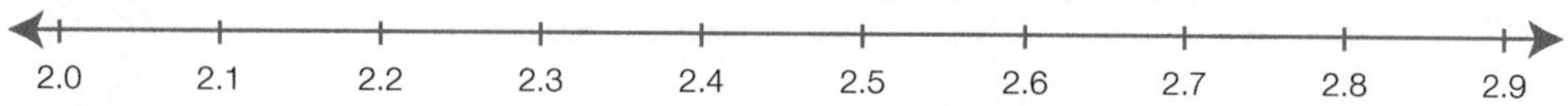

(a) 2.05, 2.09, 2.6, 2.12, 2.59
(b) 2.0, 2.8, 2.88, 2.9, 2.805
(c) 2.7, 2.4, 2.07, 2.04, 2.407
(d) 2.2, 2.4, 2.85, 2.35, 2.05

9 Write each set of decimals in order from smallest to largest.

(a) 2.3, 2.03, 2.13
(b) 8.7, 8.007, 8.67
(c) 6.646, 6.6403, 6.64
(d) 0.0095, 0.0905, 0.0509
(e) 5.3281, 5.38, 5.003821
(f) 3.616, 3.116, 3.661
(g) 0.92, 0.29, 0.092
(h) 0.85, 0.815, 0.086

10 Joel wrote down the five-digit number as he saw it on his electronic stopwatch: 37901. However, he forgot to put in the decimal point. Anna knew that the race was timed to a thousandth of a second and was able to put the decimal point in the correct place. Write the number with the decimal point in the correct place.

11 The difference in times between two skiers in a downhill race was 0.437 of a second. Write this:

(a) in expanded fraction form
(b) as a single fraction.

12 For each of the following sets of decimals, draw a section of the number line, marked in either tenths or hundredths, and indicate the position of each number in the set.

(a) 0.4, 0.72, 1.01
(b) 1.7, 1.25, 0.95
(c) 0.06, 0.045, 0.038
(d) 2.9, 3.2, 2.75, 3.06

13 Jane records her three best practice times for the 100 m sprint. They are as follows: 13.95 seconds, 13.08 seconds and 13.69 seconds.

(a) Which was Jane's fastest time?
(b) Which was Jane's slowest time?

14 Egor throws a shot-put the following distances during a competition: 25.6 m, 25.56 m and 25.081 m. Which distance was Egor's longest throw?

Reasoning, communicating and open-ended

15 Write the value of the decimals indicated by the arrows on the following scales.

(a)

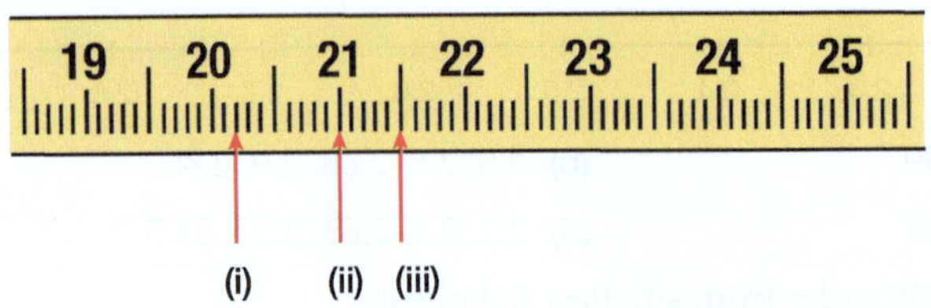

(b)

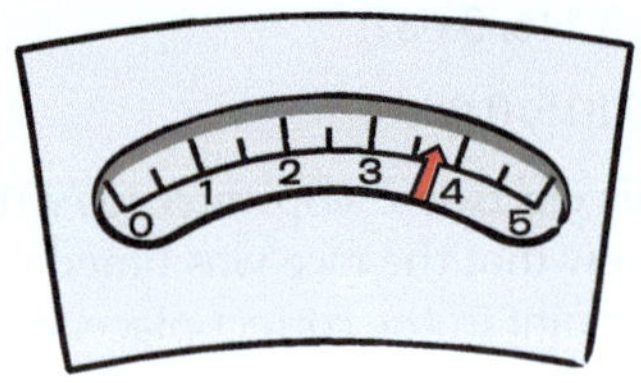

(c)

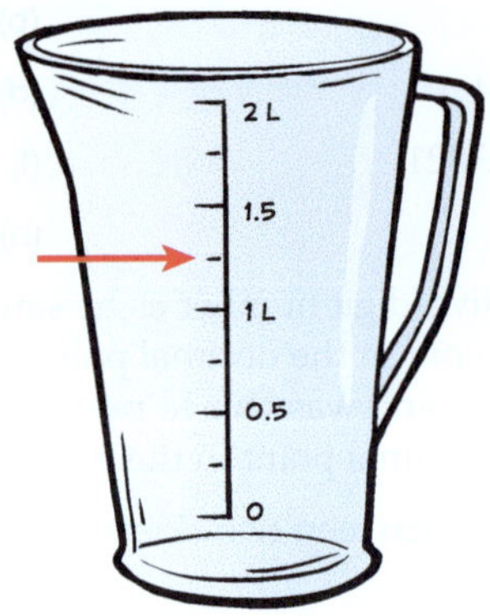

(d)

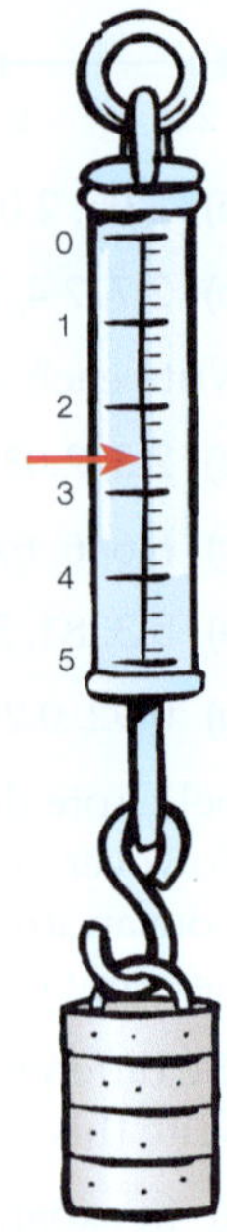

16 State the mistakes that the following students have made.

(a) Minh states that the decimal 34.162 has a 2 in the hundredths column.

(b) Al writes four hundreds, nine ones, six tenths and seven hundredths as 409.067.

(c) Max writes seven ones, eight hundredths and nine ten-thousandths as 0.7809.

(d) Polly writes 0.9056 as $\frac{9}{10} + \frac{5}{100} + \frac{6}{1000}$.

17 **(a)** Write down at least six numbers between 4.5 and 4.7.

(b) Put your numbers from **(a)** in ascending order.

18 **(a)** Use any digit from 0 to 9 in each box to make a correct statement. Digits can be repeated. Find at least three different answers.

□.□□ < 5.□

(b) Explain any limitations on the other numbers that can be used when certain numbers are used in some boxes.

19 Donna has cut out the digits 2, 5 and 7, as well as a decimal point, from coloured paper to stick on a poster to show a decimal number. If she uses all three digits:

(a) how many different decimal numbers with two decimal places could she make

(b) how many decimal numbers greater than 5.4 could she make?

Rounding decimals

4.2

Need to Know

Rounding decimal numbers

Step 1 Determine the number of decimal places that you are required to **round** to. The digit in this place will either remain as it is, or will increase by one.

Step 2 Look at the next digit to the right of this digit. If this digit is:

- 0, 1, 2, 3 or 4, leave the digit considered in **Step 1** and delete all digits after it
- 5, 6, 7, 8 or 9, increase the digit considered in **Step 1** by one and delete all digits after it.

If the number in the place value you are rounding to is 9 make the 9 a 0, and add one to the digit in front of it. For example, 1.499 rounded to the nearest hundredth is 1.50. Write a 0 in the hundredths column (even though it makes no difference to the value of the number).

For example:

The number line below shows how the number 1.364 is rounded:

- To the nearest tenth (1 decimal place): 1.364 lies between the tenths values of 1.3 and 1.4.

  The 6 in the hundredths place value column places it closer to 1.4.

  1.364 rounded to the nearest tenth is *rounded up* to 1.4.

- To the nearest hundredth (2 decimal places): 1.364 lies between the hundredths values of 1.36 and 1.37.

  The 4 in the thousandths column places it closer to 1.36.

  1.364 rounded to the nearest hundredth is *rounded down* to 1.36.

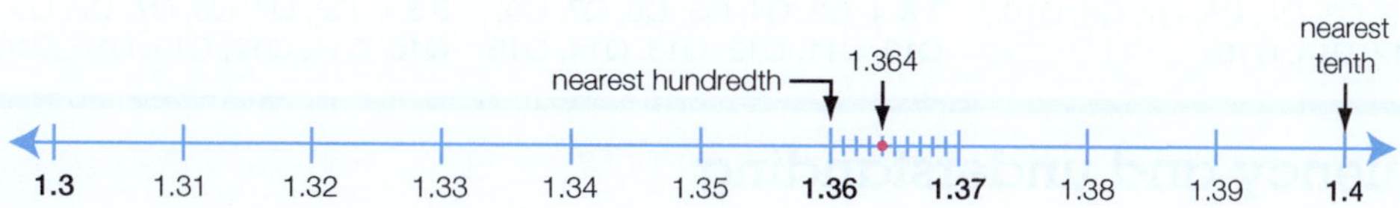

Rounding money

In Australia, the 5-cent coin is the smallest coin in use so, if money is being paid in cash, the following rules apply.

If the amount ends in…	*then it is rounded…*	*the last digit becomes…*
1 or 2 cents	down	0
3 or 4 cents	up	5
6 or 7 cents	down	5
8 or 9 cents	up	0

If the amount ends in 0 or 5 cents, then the exact amount is paid.

Rounding does not apply to electronic transactions.

4.2

Worked Example 5

WE5

Round the following decimals to the number of decimal places shown in brackets.

(a) 3.785 (2) (b) 0.958 34 (3) (c) 6.214 96 (4)

Working

(a) 3.785
3.78(5)
3.79

(b) 0.958 34
0.958(3)4
0.958

(c) 6.214 96
6.214 9(6)
6.2150

Worked Example 6

WE6

Round the following amounts of money to the nearest 5 cents.

(a) \$8.53 (b) \$7.02 (c) \$4.99

Working

(a) \$8.55 (Here, we round the 3 up to the nearest 5 cents.)

(b) \$7.00 (Here, we round the 2 down to the nearest multiple of 10 cents, which is 0.)

(c) \$5.00 (Here, we round the 9 up to the nearest multiple of 10 cents, which is 100.)

Answers page 463

4.2 Rounding decimals

Navigator

Q1 Columns 1 & 2, Q2 Columns 1–3, Q3, Q4, Q5, Q7, Q8, Q10, Q12, Q14, Q16	Q1 Columns 2 & 3, Q2 Columns 3 & 4, Q3, Q4, Q5, Q6, Q7, Q9, Q10, Q11, Q12, Q13, Q14, Q16	Q1 Columns 2 & 3, Q2 Columns 3 & 4, Q3, Q4, Q6, Q7, Q8, Q9, Q10, Q11, Q12, Q13, Q15, Q16

Fluency and understanding

WE5

Hint
Don't just cut off the digits, check the next digit first.

1 Round the following decimals to the number of decimal places shown in brackets.

(a) 4.88 (1) (b) 6.72 (1) (c) 7.635 (2)
(d) 4.552 (2) (e) 0.6416 (3) (f) 3.2772 (2)
(g) 0.314 42 (4) (h) 11.828 55 (2) (i) 2.917 96 (2)
(j) 18.499 95 (3) (k) 90.899 99 (4) (l) 18.999 999 (3)
(m) 23.4097 (2) (n) 57.0804 (3) (o) 259.899 99 (4)

WE6

2 Round the following amounts of money to the nearest 5 cents.

(a) \$4.52 (b) \$2.76 (c) \$11.03 (d) \$23.88
(e) \$176.35 (f) \$542.06 (g) \$789.54 (h) \$56.75
(i) \$6774.99 (j) \$1149.97 (k) \$8989.99 (l) \$9999.99

3 0.837 256 rounded to the nearest ten-thousandth is:

A 0.837 B 0.8372 C 0.837 26 D 0.8373

4 At the 2009 World Athletics Championships, Usain Bolt set a new 100 m record time of 9.58 seconds. Round this time to the nearest tenth of a second.

5 The total shown on the receipt for Russell's grocery shopping is $53.67. If Russell is paying in cash, how much money will he have to hand over?

6 Sir Donald Bradman's cricket batting average was 99.943 (rounded to the nearest thousandth). What would be his average rounded to the nearest:

(a) tenth **(b)** hundredth?

7 The top temperature recorded one day in Albury was 41.6°C. What figure should the weather presenter read out in his weather report as the day's maximum if he rounds the temperature reading to the nearest whole number (i.e. no decimal places)?

8 The price of a box of chocolates is $11.97. If Beau buys the chocolates and pays with cash, how much will he pay?

9 A physics student calculates the speed of sound to be 352.153 metres per second. In case her equipment is not accurate enough for this result to be reliable, she decides to round it to the nearest tenth. What value does she write in her report, assuming she rounds off correctly?

Reasoning, communicating and open-ended

10 Ruth calculates the length of shelf required for a wall unit to be 1.266 66 m. As her tape measure is only accurate enough to measure in centimetres (cm), she needs to round off this figure to the nearest cm. If 1 cm = 0.01 m, what length should she measure?

11 In a cycling sprint race at a velodrome, the following results were recorded: 12.149 s, 12.143 s, 12.138 s, 12.152 s and 12.157 s.

(a) Write these in order from fastest to slowest.

(b) It was found that there was a small error in the timing equipment and the results were to be recorded to the nearest hundredth. What were the new times?

(c) Does this affect the race result?

12 **(a)** The total cost of Rani's shopping was $20.34. Round this amount to the nearest 5 cents.

The individual costs of the items Rani bought were:

Cereal	$6.93
Yoghurt	$4.78
Bananas	$2.51
Juice	$3.63
Pasta	$2.49

(b) How much more would Rani have paid if the individual prices of the items were each rounded to the nearest 5 cents?

13 Write at least three different decimal numbers that, when rounded, would give the number 3.79.

14 Write down two prices that would give $100 dollars when rounded to:

(a) the nearest 5 cents **(b)** the nearest dollar.

15 Taking 2.548 46 and rounding it to four decimal places gives 2.5485, then rounding this to three decimal places gives 2.549.

What answer do you get when rounding 2.548 46 to three decimal places?

Explain why this is different, and what is wrong with doing it the first way.

16 Why do you think 1 cent and 2 cent coins were taken out of circulation? List two or three possible reasons why a government might make this decision.

4.3 Decimals and fractions

Need to Know

Writing decimals as fractions

A decimal can be written in expanded fraction form which can be simplified to a single fraction or mixed number.

For example: $3.72 = 3 + \frac{7}{10} + \frac{2}{100}$

$= 3 + \frac{70}{100} + \frac{2}{100}$

$= 3 + \frac{72}{100}$

$= 3\frac{72}{100}$ or $\frac{372}{100}$

$= 3\frac{18}{25}$ or $\frac{93}{25}$

Writing fractions as decimals

There are two methods that can be used.

Method 1: Write the fraction as an equivalent fraction with a denominator of 10, 100, 1000, etc.

The digits in the numerator of this new fraction are the decimal values.

For example: $\frac{4}{5} = \frac{8}{10} = 0.8$ and $\frac{13}{20} = \frac{65}{100} = 0.65$

This method is only useful for fractions with denominators that have factors of 2, 5 or 10.

Method 2: Add a decimal point and as many zeroes as required to the number in the numerator and divide by the number in the denominator.

For example: $\frac{7}{8} = 7.000 \div 8 = 0.8$ and $\frac{37}{40} = 37 \div 40 = 3.70000 \div 4 = 0.925$

A number that has a finite number of decimals is called a **terminating decimal**.

For example: 1.4 or 0.375

Recurring decimals

When the division gives a repeating pattern of decimal digits, it is called a **recurring decimal**. If a recurring pattern appears, stop the division and write the answer with a dot above the repeating digit or pattern.

For example: $\frac{2}{3} = 0.\dot{6}$, $\frac{1}{12} = 0.08\dot{3}$, $\frac{4}{7} = 0.\overline{571428}$ or $0.\dot{5}7142\dot{8}$

4.3

Technology Exploration: Calculator

Fractions and decimals on a calculator

On a scientific calculator, the a b/c or S⇔D key can be used to convert between the fraction and the decimal forms of a number. Pressing the key again will convert it from decimal form back to the fraction.

Recurring decimals are displayed as approximations with the last digit rounded up or down as appropriate.

For example: $\frac{2}{3}$ is displayed as 0.666 666 667 when the display is fixed to 9 decimal places. The last decimal place has been rounded up.

Worked Example 7

WE7

Convert each of the following decimals to fractions, giving your answers in simplest form.

(a) 6.28 **(b)** 0.3125

Working

(a) 6.28

$= 6\frac{28}{100}$ or $\frac{628}{100}$

$= 6\frac{7}{25}$ or $\frac{157}{25}$

(b) 0.3125

$= \frac{3125}{10\,000}$

$= \frac{625}{2000}$

$= \frac{25}{80}$

$= \frac{5}{16}$

Worked Example 8

WE8

Convert each of the following fractions to decimals.

(a) $\frac{7}{5}$ (b) $\frac{3}{8}$ (c) $\frac{4}{3}$

Working

(a) $\frac{7}{5}$

$= \frac{14}{10}$

$= 1\frac{4}{10}$

$= 1.4$

$\frac{7}{5} = 1.4$

(b) $\frac{3}{8}$

$= \frac{3.0}{8}$

$8\overline{)3.0}$ gives $0.$

$= \frac{3.000}{8}$

$8\overline{)3.0^60^40}$ gives 0.375

$= 0.375$

(c) $\frac{4}{3}$

$= 4.000 \div 3$

$3\overline{)4.^10^10^10}$ gives 1.333

$= 1.\dot{3}$

Answers page 464

4.3 Decimals and fractions

Navigator

Q1 Columns 1–3, Q2 Columns 1–3, Q3, Q4, Q5, Q7, Q8, Q9, Q10, Q12 (a)	Q1 Columns 2–4, Q2 Columns 2–4, Q3, Q4, Q6, Q7, Q8, Q9, Q10, Q11, Q12	Q1 Columns 3 & 4, Q2 Columns 3 & 4, Q6, Q7, Q8, Q9, Q10, Q11, Q12 (b), Q13

Equipment required: Calculator may be used for Question **2 (i)–(p)**

Fluency and understanding

WE7

1 Convert each of the following decimals to fractions, giving your answers in simplest form.

(a) 4.1 (b) 6.5 (c) 8.6 (d) 9.4

(e) 4.71 (f) 2.37 (g) 0.35 (h) 2.48

(i) 5.009 (j) 0.884 (k) 6.128 (l) 3.172

(m) 7.045 (n) 0.088 (o) 6.0015 (p) 1.0075

Hint

Make sure you put the correct number of zeroes in the denominator.

4.3

WE8

2 Convert each of the following fractions to decimals.

(a) $\frac{7}{10}$ (b) $\frac{9}{100}$ (c) $\frac{17}{100}$ (d) $\frac{123}{1000}$

(e) $\frac{1}{5}$ (f) $\frac{3}{25}$ (g) $\frac{13}{50}$ (h) $\frac{7}{4}$

(i) $\frac{7}{8}$ (j) $\frac{11}{16}$ (k) $\frac{23}{40}$ (l) $\frac{47}{60}$

(m) $\frac{1}{9}$ (n) $\frac{7}{12}$ (o) $\frac{17}{6}$ (p) $\frac{16}{15}$

3 0.55 expressed as a fraction in simplest form is:

A $\frac{55}{1000}$ B $\frac{11}{200}$ C $\frac{11}{25}$ D $\frac{11}{20}$

4 2.000 47 is equal to:

A $2\frac{47}{100}$ B $2\frac{47}{1000}$ C $2\frac{47}{10\,000}$ D $2\frac{47}{100\,000}$

5 As part of a science experiment, Galina's reaction time was measured to be 0.48 seconds. What fraction of a second is this?

6 One inch is approximately equal to 2.54 centimetres. What is this when expressed as a fraction in simplest form?

7 Hair grows at an average rate of 1.35 cm per month. Write this as a mixed number in simplest form.

8 A virus particle has a diameter of 0.000024 mm. Write this as a fraction in simplest form.

9 A red blood cell's diameter is $\frac{1}{200\,000}$ of a metre. Write this measurement in decimal form.

Reasoning, communicating and open-ended

10 A bank offers its customers an interest rate of 5.85 per cent. Write this rate in single fraction form. Is this rate higher than the $5\frac{3}{4}$ per cent offered by another bank?

11 Jeremy has five different containers and is trying to place them in ascending order of size. The volumes of all five containers are measured in litres but some are given using decimals and others using fractions. The tubs have the following volumes: $4\frac{1}{5}$, 4.375, 4.865, $4\frac{9}{10}$, $4\frac{1}{2}$. Help Jeremy put them in order from the smallest to the largest.

12 (a) Write three different fractions that have decimal values between 3.4 and 3.5 and have a denominator of 100.

(b) Write three different fractions that have decimal values between 3.4 and 3.5 where each fraction has a different denominator that is not a power of 10 (such as 100 or 1000).

13 Two numbers are rounded to one decimal place to give 5.6. What could those two numbers be? Give your answers in fraction form.

4.4 Decimal addition and subtraction

Need to Know

To add or subtract decimals, line up the decimal points so that digits with the same place value are underneath each other.

If the numbers have different numbers of decimal places, zeroes can be written in the 'empty' place value columns to help them line up correctly.

Worked Example 9

WE9

Calculate: 12.45 + 6 + 0.3678

Working

```
   12.¹4500
    6.0000
+   0.3678
----------
   18.8178
```

Worked Example 10

WE10

Calculate: 15 – 2.147

Working

```
   1⁴5.⁹0⁹0¹0
–  02.147
---------
   12.853
```

4.4 Decimal addition and subtraction

Answers page 464

Navigator

Q1 Columns 1 & 2, Q2 Columns 1 & 2, Q3, Q4, Q5, Q7, Q8, Q9, Q10, Q13, Q14	Q1 Column 2, Q2 Column 2, Q3, Q4, Q5, Q6, Q8, Q10, Q11, Q12, Q13, Q15, Q17	Q1 Column 3, Q2 Column 3, Q4, Q5, Q6, Q8, Q11, Q12, Q13, Q15, Q16, Q17

Fluency and understanding

1 Calculate: WE9

(a) $8.3 + 7.9$ (b) $4.85 + 8.09$ (c) $0.237 + 0.677$

(d) $25.61 + 0.038$ (e) $5.098 + 21.32$ (f) $0.025 + 39.786$

(g) $9.703 + 5.624 + 7.5$ (h) $7.35 + 0.609 + 2.3$ (i) $0.648 + 7.31 + 0.9$

(j) $6 + 5.017 + 12.9$ (k) $2.59 + 15 + 0.005$ (l) $0.0004 + 3.583 + 8$

2 Calculate: WE10

(a) $5.7 - 3.8$ (b) $7.05 - 2.93$ (c) $23.982 - 11.735$

(d) $9.663 - 2.7$ (e) $4.984 - 1.8$ (f) $7.238 - 3.4$

(g) $9.5 - 2.24$ (h) $16.2 - 8.75$ (i) $3.7 - 0.931$

(j) $8 - 7.44$ (k) $3 - 2.92$ (l) $93 - 0.698$

3 Which of the following shows the correct way to set out the addition of 0.56, 15.092 and 2.7?

A	B	C	D
0.5 6 1 5.092 + 2.7 ——— ———	0.56 15.092 + 2.7 ——— ———	0.56 15.092 + 2.7 ——— ———	0.56 15.092 + 2.7 ——— ———

4 Daily rainfall totals for three days over a long weekend were 3.78, 2.5 and 6.42 millimetres. What was the total rainfall over this three-day period?

5 John loads his shopping trolley with several items priced as shown.

1 frozen lasagna	\$5.71
2 L orange juice	\$4.25
1 tub of yoghurt	\$3.77
1 packet of cereal	\$6.14

(a) Find the total cost of John's purchases.

(b) Round the total to the nearest 5 cents.

6 Cristina's bank account balance was \$335.96 just before she withdrew \$40.45 to pay a bill. How much did she have left in her account after the withdrawal?

7 In a gymnastics competition, Adelia scores the following from the five judges: 7.5, 8.5, 7.9, 8 and 8.6. Find her total score.

8 A family drives to a holiday resort and records the car's odometer readings as shown.

Departure: 234.8 km Arrival: 502.7 km

How far did they travel to get to the resort?

4.4

9 During a car tour of Tasmania, Helena travelled distances of 25.64 km, 165.35 km and 5.97 km all in one day. How many kilometres did she travel altogether that day?

10 Georgina pays for \$36.35 worth of groceries with a \$100 note. How much change should she receive?

11 Nico pours 1.625 litres of milk from a full 2-litre container. How much milk is left in the container?

Reasoning, communicating and open-ended

12 Amanthi's credit card bill shows the following purchase amounts: \$23.56, \$40.15, \$7.89, \$18.48, \$45.50, \$21.73 and \$8.59. The total at the bottom of her statement is \$184.38, which she is sure is wrong.

(a) What should the total be? **(b)** How do you think the error was made?

13 Tina sells flowers at a market stall. One day, she takes in \$473.50 from the sale of flowers. If the flowers cost Tina \$128.35, and she had to pay a fee of \$14.50 to rent the stall for one day, what profit did she make for the day?

14 **(a)** Write two numbers that add up to 2.871.

(b) Write three numbers, each with three decimal places, that add up to 3.86.

(c) Explain how you chose the last digit of the numbers you used in **(b)**.

15

$$\begin{array}{r} 4.\square \\ +\ \square.8\square \\ \hline 7.\square5 \\ \hline \end{array}$$

(a) Use one digit from 0 to 9 to fill each box and make a correct sum. Find at least three different combinations.

(b) Was your choice of digits restricted? Explain how.

16 **(a)** Write two decimal numbers, each with three decimal places, that have a difference of 8.712.

(b) Write two decimal numbers, each with three decimal places, that have a difference of 8.71.

(c) Explain how you chose the last digit of the numbers you used in **(b)**.

17 Two Australian swimmers finish first and second in the women's 100 m freestyle at the Olympics. The difference in their times is 0.04 seconds. The Olympic record was 53.52 seconds. Assuming that they both break this record by less than $\frac{2}{10}$ of a second, what could their times be?

Decimal multiplication 4.5

Need to Know

Multiplying a decimal by a whole number

Multiply as though both values are whole numbers. Then, place a decimal point in the answer so that it has the same number of decimal places as the decimal number in the question.

To check whether your decimal point is in the right place, estimate your answer before calculating it.

Multiplying a decimal by a power of 10

Multiplying a number by 10 makes each digit in the number ten times its previous value. For example, $43.25 \times 10 = 432.5$.

To multiply by a power of 10, move the decimal point to the *right* the same number of place values as the zeroes in the power of 10. Write a zero in any empty place values.

$$\begin{aligned} 0.45 \times 10^1 &= 0.45 \times 10 \\ &= 4.5 \end{aligned}$$

$$\begin{aligned} 0.45 \times 10^2 &= 0.45 \times 100 \\ &= 45 \end{aligned}$$

$$\begin{aligned} 0.45 \times 10^3 &= 0.45 \times 1000 \\ &= 450 \end{aligned}$$

Dividing a decimal by a power of 10

Dividing a number by 10 makes each digit in the number one-tenth of its previous value.

For example, $65.1 \div 10 = 6.51$.

To divide by a power of 10, move the decimal point to the *left* the same number of place values as the zeroes in the power of 10. Write a zero in any empty place values.

$$\begin{aligned} 78.9 \div 10^1 &= 78.9 \div 10 \\ &= 7.89 \end{aligned}$$

$$\begin{aligned} 78.9 \div 10^2 &= 78.9 \div 100 \\ &= 0.789 \end{aligned}$$

$$\begin{aligned} 78.9 \div 10^3 &= 78.9 \div 1000 \\ &= 0.0789 \end{aligned}$$

Multiplying one decimal by another

Step 1 Write both numbers as the division of a whole number and a power of 10.

Step 2 Multiply as though both numbers are whole numbers.

Step 3 Divide the product by each of the powers of 10 identified in **Step 1**. Show this by moving the decimal place to the left.

Remember, while 'move the decimal point' is used, it is the digits themselves that actually move as their values increase or decrease.

4.5

Worked Example 11

WE11

Calculate:

(a) 4.172×3 (b) 52.63×14 (c) 0.0003×2

Working

(a) $4.172 \times 3 \approx 4 \times 3$
≈ 12 (estimate)

$$\begin{array}{r} 4.172 \\ \times \quad {}^{2}\;\;3 \\ \hline 12.516 \\ \hline \end{array}$$

$4.172 \times 3 = 12.516$

(b) $52.63 \times 14 \approx 50 \times 14$
$\approx 14 \times 5 \times 10$
≈ 700 (estimate)

$$\begin{array}{r} 52.63 \\ {}^{1\,2\,1} \\ \times \quad 14 \\ \hline 21052 \\ 52630 \\ \hline 736.82 \\ \hline \end{array}$$

$52.63 \times 14 = 736.82$

(c)
$$\begin{array}{r} 0.0003 \\ \times \quad 2 \\ \hline 0.0006 \\ \hline \end{array}$$

$0.0003 \times 2 = 0.0006$

Worked Example 12

WE12

Calculate:

(a) 0.6295×7000 (b) $9.81 \times 340\,000$

Working

(a) 0.6295×7000
$= 0.6295 \times 7 \times 1000$
4.4065×1000
$= 4406.5$

$$\begin{array}{r} 0.6295 \\ {}^{4\;2\,6\,3} \\ \times \quad 7 \\ \hline 4.4065 \\ \hline \end{array}$$

(b) $9.81 \times 340\,000$
$= 9.81 \times 34 \times 10\,000$
$333.54 \times 10\,000$
$= 3\,335\,400$

$$\begin{array}{r} 9.81 \\ \times \quad {}^{2\,3}\;34 \\ \hline 3924 \\ {}^{1}2{}^{1}9430 \\ \hline 333.54 \\ \hline \end{array}$$

4.5

Worked Example 13

WE13

Calculate:

(a) 0.63×0.4 (b) 0.014×0.002

Working

(a) 0.63×0.4
$= 63 \div 100 \times 4 \div 10$
$= 252 \div 1000$
$= 0.252$

$$\begin{array}{r} 0.63 \\ \times \quad 0.4 \\ \hline 0.252 \end{array}$$

(b) 0.014×0.002
$= 14 \div 1000 \times 2 \div 1000$
$= 28 \div 1\,000\,000$
$= 0.000\,028$

$$\begin{array}{r} 0.014 \\ \times \quad 0.002 \\ \hline 0.000\,028 \end{array}$$

4.5 Decimal multiplication

Answers page 464

Navigator

Q1 Columns 1 & 2, Q2 Columns 1 & 2, Q3 Columns 1 & 2, Q4, Q5, Q7, Q9, Q11, Q12, Q13, Q14, Q17, Q21, Q23, Q24	Q1 Columns 2 & 3, Q2 Columns 2 & 3, Q3 Columns 2 & 3, Q4, Q5, Q6, Q7, Q9, Q10, Q11, Q13, Q14, Q16, Q17, Q19, Q21, Q22, Q23, Q24, Q25	Q1 Column 3, Q2 Column 3, Q3 Column 3, Q4, Q6, Q8, Q9, Q10, Q11, Q13, Q14, Q15, Q16, Q18, Q19, Q20, Q22, Q23, Q24, Q25

Fluency and understanding

1 Calculate: WE11

(a) 7.6×4 (b) 5.3×7 (c) 6.5×9
(d) 5×4.62 (e) 0.54×4 (f) 6×12.33
(g) 8.047×8 (h) 0.619×7 (i) 3×18.309
(j) 22.08×13 (k) 48.76×25 (l) 37.09×32
(m) 0.0004×2 (n) 0.0005×7 (o) 1.0006×3

Hint

Remember, the number of decimal places in your answer should be the same as the number in the decimal being multiplied.

2 Calculate: WE12

(a) 2.4×60 (b) 3.27×70 (c) 2.991×40
(d) 0.57×200 (e) 4.34×500 (f) 6.625×400
(g) 4.591×6000 (h) 6.38×4300 (i) 3.14×9300
(j) $0.571 \times 68\,000$ (k) $0.9465 \times 71\,000$ (l) $7.4832 \times 55\,000$

3 Calculate: WE13

(a) 0.6×0.8 (b) 0.4×0.6 (c) 0.9×0.4
(d) 9.32×0.7 (e) 4.12×0.3 (f) 5.26×0.8
(g) 7.74×0.31 (h) 9.46×0.22 (i) 3.76×0.18
(j) 0.004×0.6 (k) 0.0308×0.3 (l) 0.072×0.05

4.5

4 A reasonable estimate for the answer to 1.8×15 is:

A 3 B 16.8 C 30 D 300

5 How many decimal places will there be in the answer to 12.789×41?

A 1 B 2 C 3 D 4

6 How many decimal places will there be in the answer to 0.36×1.04?

A 2 B 3 C 4 D 5

7 Renée buys 52 components from an electronics shop for \$0.35 each. What is the total cost of the components?

8 A pill bottle contains 37 pills, each of mass 0.287 grams. What is the total mass of the bottle's contents?

9 Write the following as whole numbers.

(a) 3.5 million (b) 4.26 billion (c) 1.3 trillion (1 trillion = 1000 billion)

(d) 57.08 million (e) 10.2 trillion (f) 20.045 billion

10 A snail crawls 0.041 metres in an hour. If it kept up this pace, how far could it move in a day?

11 Amy orders 27.5 metres of timber priced at \$4.29 per metre from her local hardware store. How much will the timber cost her in total? (Round your answer to the nearest cent.)

12 Gavin sells 300 punnets of berries to a supermarket. If he is paid \$1.72 for each punnet, how much money does he receive?

13 In 2010, Mexican businessman Carlos Slim was declared the world's richest man, with a fortune worth 53.5 billion US dollars. Write this amount as a whole number.

14 (a) Marita fills her car's fuel tank with petrol at a cost of \$1.35 per litre. If she adds 70 litres, how much must she pay?

(b) Ali pumps 9.5 litres of petrol into his car's tank. How much did the fuel cost him if he paid the same price per litre as Marita? Round your answer to the nearest cent.

15 The instructions on a bag of fertiliser say to spread 0.75 kilograms per square metre of area to be fertilised. Raisa wants to fertilise 2600 square metres of property using this product. How many kilograms of fertiliser should she use?

16 Bryan has calculated that he needs 3.6 square metres of tiles for his bathroom. What will he have to pay if the tiles he wants cost \$22.39 per square metre? (Round your answer to the nearest cent.)

Reasoning, communicating and open-ended

17 Kim works in an ice-cream parlour, and one hot day sells 500 scoops of vanilla. If there is 0.185 litres of ice-cream in each scoop and the day started with 130 litres of vanilla in the freezer, how many litres of vanilla ice-cream are left at the end of the day?

18 Yoghurt contains 167 kilojoules per 100 grams. How many kilojoules would a 175 gram tub contain?

19 In April 2010, the Australian Bureau of Statistics calculated that the population of Australia was 22 296 411.

(a) Write this number as a decimal number, with the number of millions as the whole number part, and the rest of the number rounded to one decimal place.

(b) Why would your answer to **(a)** be the number you would expect to see reported in the news, instead of the 'exact' value?

20 **(a)** A woodwork teacher needs 50 pieces of pine, each 0.135 metres long, to give to students to make a particular model. What total length (in metres) of timber does the teacher need? (Ignore any lost length due to saw cuts.)

(b) If he can buy the pine in 3.5 m lengths, how many pieces does he need, and how much does he have left over?

21 A multi-bag of mini chocolate bars costs $3.95 and contains 12 bars. Each mini chocolate bar can also be purchased individually for 40 cents.

(a) How much would 12 mini chocolate bars cost if they were purchased individually?

(b) How much do you save by purchasing the multi-bag?

(c) Charlotte wants to buy 30 mini bars for a party. Should she buy them individually, or in multi-bags? Which way is cheaper?

22 Wayne is following his grandfather's recipe for tomato sauce, which requires 3.5 pounds of tomatoes. (A pound is an old imperial unit of mass.) Because he only has scales in kilograms, Wayne decides to convert the mass to kilograms. A website tells him that 1 pound equals 0.454 kilograms. What answer should he get when he converts the mass of the tomatoes to kilograms? Round your answer to one decimal place.

23 Francesca read her electricity meter one Saturday, and again at the same time the next Saturday. She recorded the following readings.

First Saturday: 7562.3 kilowatt-hours
Second Saturday: 7638.6 kilowatt-hours

(a) How many kilowatt-hours of electricity did Francesca's household use in the week?

(b) If electricity costs $0.13 per kilowatt-hour, find the total cost of the electricity supplied in the week.

24 Jayapal has been asked to work out the length if every student in Year 7 lay end to end. He has been told that there are 89 students in Year 7 and that their average height is 1.63 m.

This is his working out:

$89 \times 1 = 89$
$89 \times 6 = 534$
$89 \times 3 = 267$

$$\begin{array}{r} 89 \\ 534 \\ +\,267 \\ \hline =880 \end{array}$$

Answer: 8.80 m.

(a) Where has Jayapal made his error? Explain how he could use estimation to tell that he was incorrect.

(b) Explain how he could modify his method to get the correct answer.

(c) Show how you would complete this problem.

25 Two numbers are multiplied together to give 68.4.

(a) Find at least two possible pairs of numbers with one number being a multiple of 10.

(b) Find at least two possible pairs of numbers with neither being a multiple of 10.

Half-time 4

Ex 4.3 1 Convert the following decimals into fractions (or mixed numbers where appropriate) in simplest form.

(a) 0.25 (b) 0.3 (c) 3.54 (d) 9.345

Ex 4.1 2 Put the following decimals in order from smallest to largest.

9.01, 19.12, 0.03, 0.321, 9.003 24

Ex 4.4 3 Calculate:

(a) $12.5 + 7.9$ (b) $124.54 + 8.379$ (c) $0.32 + 2.5$

(d) $15 - 8.98$ (e) $127.061 - 43.2$ (f) $6 - 4.9$

Ex 4.2 4 Round the following numbers to three decimal places.

(a) 4.8765 (b) 13.0037 (c) 2.3498 (d) 23.032 49

Ex 4.5 5 Calculate the following.

(a) 7.6×4 (b) 28.3×5 (c) 1.37×9

(d) 45.2×30 (e) 0.26×2000 (f) 6.78×500

(g) 4.3×1.7 (h) 39.82×15.4 (i) 1.2×0.0005

Ex 4.4 6 Franji is saving up to purchase a games console. She has saved the following amounts over the last four weeks: \$25, \$47.50, \$18.70, \$9.35.

(a) How much has she saved in total?

(b) If the console is \$149, how much more does Franji have to save?

Ex 4.1 7 Write the following in decimal form.

(a) $23 + \frac{4}{10} + \frac{9}{100} + \frac{7}{10000}$ (b) $14 + \frac{235}{1000}$

(c) 2 hundreds, 9 tens, 3 ones, 4 tenths, 1 hundredth and 5 thousandths

(d) 9 tens, 0 ones, 6 tenths, 2 hundredths and 7 ten-thousandths

Ex 4.2 8 Round the following amounts of money to the nearest 5 cents.

(a) \$1.93 (b) \$35.08 (c) \$57.44 (d) \$189.98

Ex 4.3 9 Write the following fractions as decimals, using the correct notation for recurring decimals if necessary.

(a) $\frac{4}{5}$ (b) $\frac{17}{20}$ (c) $\frac{2}{3}$ (d) $\frac{7}{12}$

Ex 4.1 10 The times recorded by the finalists in a track cycling sprint race were: 59.023 s, 59.105 s, 59.23 s, 59.125 s. Write the times in order from fastest to slowest.

Ex 4.4, 4.5 11 Melinda is out buying some ingredients for a cooking class. She buys:

12 packets of pasta at \$2.49 each
18 tins of tomatoes at \$1.35 each
6 kg of mince at \$8.99 per kg
3 packets of cheese at \$4.78 per packet

Find the total cost of Melinda's shopping.

Decimal division

4.6

Need to Know

Dividing a decimal by a whole number

Place a decimal point in the quotient in line with the decimal point in the dividend, then perform the division.

For example:

$$\begin{array}{r} 8.14 \\ 7\overline{)\,56.98} \end{array}$$

quotient → 8.14
divisor → 7
dividend ← 56.98

Dividing a decimal by a multiple of 10

Write the divisor as a whole number multiplied by a power of 10. Perform two separate divisions. Divide by the whole number. Then divide by the power of 10 by moving the decimal point to the left the same number of places as zeroes in the power.

For example:

$$\begin{aligned} 23.4 \div 9000 &= 23.4 \div (9 \times 1000) \\ &= 23.4 \div 9 \div 1000 \\ &= 2.6 \div 1000 \\ &= 0.0026 \end{aligned}$$

Dividing a decimal by another decimal

Multiply both numbers by a power of 10 or another number so that the divisor is now a whole number. Perform the division.

For example:

$$\begin{aligned} &13.52 \div 1.2 \\ &= 135.2 \div 12 \text{ (multiplying both numbers by 10)} \\ &= 11.27 \end{aligned}$$

$$\begin{aligned} &0.75 \div 2.5 \\ &= 1.5 \div 5 \text{ (multiplying by 2)} \\ &= 0.3 \end{aligned}$$

Note: While the words 'moving the decimal point' are often used, it is the digits that actually move.

4.6

Worked Example 14

WE14

Calculate the following. Round your answers to three decimal places if necessary.

(a) $5.28 \div 3$ (b) $0.41 \div 7$

Working

(a) $\begin{array}{r} 1.\ 7\ 6 \\ 3\overline{)\,5.{}^{2}2{}^{1}8} \end{array}$

(b) $\begin{array}{r} 0.0\ 5\ 8\ 5\ldots \\ 7\overline{)\,0.4{}^{4}1{}^{6}0{}^{4}0} \end{array}$

0.059 (to 3 d.p.)

Worked Example 15

WE15

Calculate the following. Round your answers to three decimal places where necessary.

(a) $9.76 \div 400$ (b) $14.7 \div 80$

Working

(a) $9.76 \div 400$
$= 9.76 \div 4 \div 100$
$= 2.44 \div 100$
$= 0.0244$
$= 0.024$ (to 3 d.p.)

$\begin{array}{r} 2.44 \\ 4\overline{)\,9.{}^{1}7{}^{1}6} \end{array}$

Working

(b) $14.7 \div 80$
$= 14.7 \div 8 \div 10$
$= 1.8375 \div 10$
$= 0.183\,75$
$= 0.184$ (to 3 d.p.)

$\begin{array}{r} 1.\ 8\ 3\ 7\ 5 \\ 8\overline{)\,14.{}^{6}7{}^{3}0{}^{6}0{}^{4}0} \end{array}$

Worked Example 16

WE16

Calculate the following. Round your answer to three decimal places if necessary.

(a) $3.35 \div 0.005$ (b) $10.52 \div 0.9$

Working

(a) $3.35 \div 0.005$
$= (3.35 \times 1000) \div (0.005 \times 1000)$
$= 3350 \div 5$
$= 670$

$\begin{array}{r} 6\ 70 \\ 5\overline{)\,33{}^{3}50} \end{array}$

Working

(b) $10.52 \div 0.9$
$= (10.52 \times 10) \div (0.9 \times 10)$
$= 105.2 \div 9$
$= 11.689$ (to 3 d.p.)

$\begin{array}{r} 1\ 1.\ 6\ 8\ 8\ 8 \\ 9\overline{)\,10{}^{1}5.{}^{6}2{}^{8}0{}^{8}0{}^{8}0} \end{array}$

4.6 Decimal division

Answers page 465

Navigator

Q1 Columns 1 & 2, Q2 Columns 1 & 2, Q3 Column 1, Q4, Q5, Q6, Q7, Q8, Q9, Q10, Q11, Q12, Q13, Q14, Q16, Q17, Q19, Q23, Q24	Q1 Columns 2 & 3, Q2 Column 2, Q3 Columns 2 & 3, Q4, Q5, Q7, Q8, Q9, Q10, Q11, Q12, Q13, Q14, Q15, Q17, Q19, Q20, Q23, Q24	Q1 Column 3, Q2 Column 3, Q3 Column 3, Q4, Q5, Q7, Q8, Q9, Q10, Q11, Q12, Q13, Q14, Q15, Q17, Q18, Q19, Q20, Q21, Q22, Q23, Q24

Fluency and understanding

1 Calculate the following. Round your answers to three decimal places if necessary. WE14

(a) 8.48 ÷ 4 **(b)** 12.96 ÷ 3 **(c)** 14.22 ÷ 9
(d) 3.3 ÷ 6 **(e)** 2.4 ÷ 7 **(f)** 5.1 ÷ 9
(g) 39.42 ÷ 6 **(h)** 10.6 ÷ 7 **(i)** 0.47 ÷ 9
(j) 0.42 ÷ 6 **(k)** 1.22 ÷ 3 **(l)** 18.46 ÷ 7
(m) 154.33 ÷ 5 **(n)** 84.365 ÷ 8 **(o)** 347.507 ÷ 7

Hint

Don't forget to write a zero in the quotient if at any stage the divisor 'won't go'.

2 Calculate the following. Round your answers to three decimal places if necessary. WE15

(a) 3.6 ÷ 40 **(b)** 8.1 ÷ 90 **(c)** 36.216 ÷ 60
(d) 96.58 ÷ 200 **(e)** 2.296 ÷ 400 **(f)** 23.4 ÷ 600
(g) 233.4 ÷ 3000 **(h)** 40.6 ÷ 8000 **(i)** 18.9 ÷ 7000
(j) 68.4 ÷ 120 **(k)** 3.19 ÷ 110 **(l)** 8.97 ÷ 1300

3 Calculate the following. Round your answers to three decimal places if necessary. WE16

(a) 5.14 ÷ 0.2 **(b)** 4.12 ÷ 0.4 **(c)** 5.1 ÷ 0.6
(d) 15.48 ÷ 0.04 **(e)** 9.018 ÷ 0.09 **(f)** 1.736 ÷ 0.07
(g) 2.382 ÷ 0.006 **(h)** 0.3996 ÷ 0.009 **(i)** 0.196 ÷ 0.008
(j) 7.7 ÷ 0.011 **(k)** 3.78 ÷ 0.012 **(l)** 1.16 ÷ 0.008
(m) 1.74 ÷ 0.0002 **(n)** 8.05 ÷ 0.0007 **(o)** 13.36 ÷ 0.0008

4 How many decimal places will the answer to 298.12 ÷ 100 have?

A 1 **B** 2 **C** 3 **D** 4

5 Evaluating which of the following will give the answer to 3.2 ÷ 0.005?

A 0.0032 ÷ 5 **B** 320.0 ÷ 5 **C** 3200.0 ÷ 5 **D** 32 000.0 ÷ 5

6 A block of metal of mass 2.312 kg is to be divided equally into 8 pieces. What will the mass of each piece be?

7 A company makes a profit of 7.86 million dollars. If the profit is to be divided equally among the four company owners, how much would each owner get? (Answer in millions of dollars.)

8 The total bill for a wedding reception for 100 guests is $9398.55. How much does this work out to per guest? (Round your answer appropriately.)

9 Anthony needs to mix 0.07 litres of environmentally friendly pest repellent with water each time he fills the tank of his garden sprayer. If he has 0.84 litres of repellent left, how many sprayer tankfuls can he mix?

4.6

10 A jeweller calculated that a popular style of necklace contains 0.009 g of gold. How many necklaces could she make using the 2.332 g of gold in her workshop?

11 Eleanor's car travels 37.29 km on three litres of fuel. How far would it travel on one litre of fuel?

12 A 4000 millilitre container of orange juice costs \$8.90. Determine the cost per 100 mL. (Round your answer appropriately.)

13 Simon buys 7 kg of spicy sausages for \$32.95. How much is this per kilogram?

14 Regina has a tub containing 5.8 kilograms of worms which she uses for fishing. If the average worm has a mass of 0.003 kilograms, how many are in her tub?

15 Hai is making a long-distance call on his mobile. The call costs \$0.90 per minute. For how many minutes can he continue his call if he has \$5.85 credit left on his phone?

16 A 1.2 kilometre taxi trip costs Joan \$5.70. How much is this per kilometre?

Reasoning, communicating and open-ended

17 The captains of two netball teams were discussing which of their teams was taller. The heights of Angie's team were 1.49 m, 1.54 m, 1.65 m, 1.67 m, 1.58 m, 1.73 m and 1.56 m. The heights of Jessica's team were 1.73 m, 1.58 m, 1.59 m, 1.62 m, 1.45 m, 1.51 m and 1.62 m. Add the seven heights and divide the result by seven to find the average (mean) height of each team. Which of the two teams was the taller, on average? Answer correct to two decimal places.

18 Monica and Celeste were trying to determine the best bargain of three different jars of coffee. The 200 g jar of BestCafe cost \$5.95, the 500 g jar of Maxine House was \$14.85 and the 300 g tin of Brazilian Roast was \$6.75. Which is the best buy in terms of price per 100 grams?

19 Karen was counting the coins in her cash register drawer and found that she had \$1.65 in 5-cent coins, \$2.30 in 10-cent coins, \$5.20 in 20-cent coins and \$7.50 in 50-cent coins. How many of each type of coin did she have?

20 Jacob received a bill from a rent-a-car firm that gave the total cost as \$310. This included the daily cost of \$24.80 and \$1.20 per kilometre travelled. If Jacob had hired the car for 2 days, how many kilometres did he travel?

21 Write three numbers, each with three decimal places, that when divided by 4 give an answer between 0.5 and 1.

22 For each of the following, write two equivalent divisions by multiplying the divisor by a number that is not a power of 10.

(a) $10.05 \div 0.5$ **(b)** $36.04 \div 0.02$

23 The area of a rectangle is calculated by multiplying the length and the width. If the area of a rectangle is 8.45 m^2, what might be the length and width of the sides? Give two possible values for the side dimensions. For each of the side dimensions, work out the related perimeter (distance along the boundary of the rectangle).

24 Tran divided 5.25 by 5 and wrote down an answer of 1.5.

(a) What error did Tran make?

(b) Explain to Tran what he needs to do to calculate decimal divisions accurately.

Percentages, fractions and decimals 4.7

Need to Know

The words **'per cent'** literally mean 'for every hundred' or 'out of one hundred'. The percentage symbol is %. Percentages can be thought of and written as fractions with a denominator of 100. Percentages can also be thought of as division by 100.

100% is equivalent to 1 whole, $\frac{1}{1}$ and 1.0.

Writing a percentage as a fraction

- Write the percentage value as the numerator of a fraction with a denominator of 100.
- Simplify the fraction if possible.

For example: $25\% = \frac{25}{100}$

$= \frac{1}{4}$

Writing a percentage as a decimal

- Divide the percentage value by 100.

For example: $25\% = 25 \div 100$

$= 0.25$

Writing a fraction as a percentage

We can use one of two methods.

Method 1: Write the fraction as an equivalent fraction with a denominator of 100. Use the numerator of this new fraction as the percentage.

For example: $\frac{1}{4} = \frac{25}{100}$

$= 25\%$

Method 2: Multiply the fraction by 100%.

For example: $\frac{1}{4} = \frac{1}{4} \times 100\%$

$= 25\%$

Writing decimals as percentages

Multiply the decimal by 100%.

For example: $0.25 = 0.25 \times 100\%$

$= 25\%$

Rational numbers

All fractions, terminating decimals, recurring decimals and percentages are known as **rational numbers** as they can all be written in the form of $\frac{a}{b}$, where a and b are integers and $b \neq 0$.

Worked Example 17

WE 17

Write 12% as a fraction in simplest form.

Working

$$12\% = \frac{12}{100}$$
$$= \frac{3}{25}$$

Worked Example 18

WE 18

Write 45% as a decimal.

Working

$$45\% = 45 \div 100$$
$$= 0.45$$

Worked Example 19

WE 19

Write each of the following fractions as percentages.

(a) $\frac{4}{5}$ (b) $\frac{27}{40}$

Working

(a) $\frac{4}{5} = \frac{4}{5} \times \frac{20}{20}$

$= \frac{80}{100}$

$= 80\%$

(b) $\frac{27}{40} = \frac{27}{40} \times \frac{100}{1}\%$

$= \frac{27}{{}_{2}\cancel{40}} \times \frac{{}^{5}\cancel{100}}{1}\%$

$= \frac{135}{2}\%$

$= 67.5\%$

Worked Example 20

WE 20

Write each of the following decimals as percentages.

(a) 0.04 (b) 0.125

Working

(a) $0.04 = 0.04 \times 100\%$
$= 4\%$

(b) $0.125 = 0.125 \times 100\%$
$= 12.5\%$

4.7

4.7 Percentages, fractions and decimals

Answers page 465

Navigator

Q1, Q2, Q3 Columns 1 & 2, Q4, Q5, Q6, Q7, Q9, Q11, Q12, Q15, Q16, Q17	Q1, Q2, Q3 Columns 2 & 3, Q4, Q5, Q7, Q8, Q9, Q11, Q12, Q13, Q14, Q15, Q16, Q17	Q1, Q2, Q3 Columns 3 & 4, Q4, Q5, Q8, Q9, Q10, Q11, Q12, Q13, Q14, Q16, Q17

Fluency and understanding

1 Write each of the following percentages as fractions in simplest form. WE17

(a) 15% (b) 24% (c) 18% (d) 64%

(e) 110% (f) 122% (g) 140% (h) 198%

2 Write each of the following percentages as decimals. WE18

(a) 34% (b) 13% (c) 78% (d) 92%

(e) 120% (f) 123% (g) 254% (h) 320%

3 Write each of the following fractions as percentages. WE19

(a) $\frac{2}{5}$ (b) $\frac{3}{4}$ (c) $\frac{3}{10}$ (d) $\frac{7}{20}$

(e) $\frac{11}{10}$ (f) $\frac{3}{2}$ (g) $\frac{28}{25}$ (h) $\frac{6}{5}$

(i) $\frac{3}{8}$ (j) $\frac{1}{3}$ (k) $\frac{5}{6}$ (l) $\frac{4}{9}$

(m) $\frac{13}{15}$ (n) $\frac{19}{30}$ (o) $\frac{33}{40}$ (p) $\frac{44}{60}$

4 Write each of the following decimals as percentages. WE20

(a) 0.87 (b) 0.34 (c) 0.65 (d) 0.96

(e) 1.34 (f) 3.58 (g) 5.32 (h) 1.21

5 Complete the following table.

Percentage	Fraction (simplest form)	Decimal
40%		
	$\frac{7}{10}$	
		0.05
	$\frac{1}{50}$	
75%		
	$\frac{5}{4}$	

6 $\frac{1}{4}$, $\frac{1}{2}$, $\frac{1}{10}$, $\frac{1}{5}$ are all common fractions.

(a) Change each of these fractions into decimals.

(b) Change each of these decimals into percentages.

4.7

7 95% of all the animal species in the world are insects. Write this as a fraction in simplest form.

8 At Valley High School, $\frac{5}{8}$ of the students are studying Chinese. Write this fraction as a percentage.

9 Place the following in ascending order (smallest to largest).

(a) 82%, 0.03, $\frac{7}{50}$, $\frac{3}{5}$, 0.8, 57%

(b) $\frac{1}{2}$, 73%, 1.05, $\frac{3}{10}$, 43%, 0.05

10 The Petronas Towers in Kuala Lumpur are approximately 0.6 times the size of the tallest building in the world, the Burj Khalifa in Dubai.

(a) Write the height of the Petronas Towers as a fraction of the height of the Burj Khalifa.

(b) Now, write the equivalent percentage.

Reasoning, communicating and open-ended

11 A rugby stadium contains 43 500 people. 37% of the crowd supports England and the rest supports Australia. What percentage of the crowd supports Australia?

12 Allison is a journalist. She has surveyed 150 people and found that 6 believe that aliens exist.

(a) Write the survey result:

(i) as a fraction in simplest form

(ii) as a percentage

(iii) as a decimal.

(b) Which of the three forms of the result would you recommend that Allison uses as the headline for the article she is writing? Give a reason for your answer.

(c) Use your answer from **(b)** to write a potential headline.

13 $\frac{3}{4}$ is 3 lots of $\frac{1}{4}$, or $3 \times \frac{1}{4}$. How is this fact reflected in the percentage values for $\frac{1}{4}$ and $\frac{3}{4}$?

14 13% of a class have red hair, 62% of the class have black hair.

(a) What is the total percentage of the class that has red or black hair?

(b) What is this total as a decimal?

(c) What percentage of the class has a hair colour that is not red or black?

(d) Write your answer to **(c)** as a decimal.

15 Write down at least three fractions that have equivalent percentage values between 10 and 40 per cent.

16 Find at least three percentages that have decimal values between 0.34 and 0.35.

17 Estimate the percentage of one weekday that you would spend:

(a) sleeping (b) eating (c) studying (d) socialising.

Using percentages

4.8

Need to Know

Finding the percentage of a quantity

Two methods can be used.

Method 1

Step 1 Break the percentage into smaller amounts that are easy to calculate.

Step 2 Find each percentage.

Step 3 Add the parts together.

For example: $12\% \text{ of } 84 = (10 + 2)\% \text{ of } 84$

$= 10\% \text{ of } 84 + 2\% \text{ of } 84$

$= 8.4 + 1.68$

$= 10.08$

Method 2

Step 1 Write the % as a fraction of 100 (simplifying the fraction can often help) or as a decimal.

Step 2 Multiply the amount by the fraction or decimal.

For example: $12\% \text{ of } 84 = \frac{12}{100} \times \frac{84}{1}$ or 0.12×84

$= \frac{1008}{100}$

$= 10.08$

Writing one amount as a percentage of another

Step 1 Write the amount as the numerator of a fraction, with the other amount as the denominator.

Step 2 Convert this fraction to a percentage. If the percentage is not a whole number, then rounding to one decimal place gives an accurate enough answer in most cases.

Technology Exploration: Calculator

Percentages on a calculator

The [%] key on a calculator can be used to find the percentage of a quantity.

For example: To calculate 12% of 84, press [8] [4] [×] [1] [2] then press the [%] key. 10.08 will be displayed. You may need to use [SHIFT] to access the % function. Notice that the quantity is input first.

The [%] key can also be used to write one quantity as a percentage of another.

For example: To write 13 as a percentage of 65, press [1] [3] [÷] [6] [5], then press the [%] key. The display will give you the percentage of 20%.

4.8

Worked Example 21

WE 21

Calculate the following percentage: 15% of 70.

Working

Method 1: Use basic percentages

(a) $15\% = 10\% + 5\%$

10% of 70

$= \frac{\cancel{10}^1}{\cancel{100}^1} \times 70$

$= 7$

5% of 70

$= (10\% \text{ of } 70) \div 2$

$= 7 \div 2$

$= 3.5$

$15\% \text{ of } 70 = 7 + 3.5$

$= 10.5$

Working

Method 2: Fraction or decimal multiplication

(b) 15% of 70

$= 15\% \times 70$

$= \frac{15}{100} \times 70$ or 0.15×70

$= \frac{{}^3\cancel{15}}{{}_2\cancel{100}} \times \frac{\cancel{70}}{1}$ or 1.5×7

$= \frac{3 \times 7}{2 \times 1}$ or 10.5

$= \frac{21}{2}$

$= 10.5$

Worked Example 22

WE 22

Write the following as percentages. Round your answers to one decimal place, if necessary.

(a) Melissa got 21 out of 25 questions on her grammar test correct.

(b) 13 of the 24 students in class 7C have brown hair.

Working

(a) $\frac{21}{25} = \frac{21}{25} \times \frac{4}{4}$

$= \frac{84}{100}$

$= 84\%$

Melissa got 84% of the test questions correct.

Working

(b) $\frac{13}{24}$

$= \frac{13}{{}_6\cancel{24}} \times \frac{{}^{25}\cancel{100}}{1}\%$

$= \frac{325}{6}\%$

$= 54.2\%$

54.2% of the students in 7C have brown hair.

4.8

4.8 Using percentages

Answers page 466

Navigator

Q1, Q2, Q3 Columns 1 & 2, Q4, Q6, Q7, Q9, Q10, Q11, Q13, Q14, Q16, Q18, Q20	Q1, Q2, Q3, Q4, Q5, Q7, Q8, Q9, Q11, Q12, Q13, Q14, Q15, Q16, Q18, Q20	Q1 Columns 2 & 3, Q2 (d)–(i), Q3, Q4, Q5, Q8, Q9, Q11, Q12, Q13, Q15, Q16, Q17, Q18, Q19, Q20

Equipment required: Calculator may be used for Questions **2, 7–12, 15–17**

Fluency and understanding

WE 21

1 Calculate each of the following percentages.

(a) 50% of $36 (b) 10% of 450 m (c) 20% of 840 L
(d) 80% of 120 m (e) 90% of 500 kg (f) 2% of $8400
(g) 13% of 72 kg (h) 27% of $354 (i) 21% of 876 m
(j) 28% of 231.25 m (k) 34% of 210.4 L (l) 4% of $32.65

WE 22

2 Write the following as percentages. Round your answers to one decimal place, if necessary.

(a) 17 out of 20 on a spelling quiz.
(b) 5 out of 22 students in the class have blond hair.
(c) There are 100 kittens at the animal shelter, 40 of which are female.
(d) 4 out of 5 students surveyed own an MP3 player.
(e) 50 pairs of shoes were sold, of which 38 pairs were black.
(f) 54 out of 70 cars counted on a freeway had only 1 person in them.
(g) 15 of the 27 cars in the staff carpark are white or silver.
(h) 3 out of every 4 people on the bus said they were travelling to work.
(i) 28 out of 34 people surveyed said they were non-smokers.

3 Calculate the following using the 'basic percentages method' or any other mental strategy.

(a) 50% of $25 (b) 10% of 75 kg (c) 25% of 60 km
(d) 20% of 40 L (e) 1% of 360 m (f) 5% of $120
(g) 60% of 350 kg (h) 75% of 800 mL (i) 35% of $400

4 Mehmet successfully shot 13 basketball goals out of 18 attempts. The calculation that converts this result to a percentage would be set out as follows.

A $\frac{13}{100} \times \frac{18}{1}$ B $\frac{13}{18} \times \frac{100}{1}$ C $\frac{18}{13} \times \frac{100}{1}$ D $\frac{100}{13} \times \frac{18}{1}$

5 20% of $420 is:

A $8.40 B $20 C $84 D $200

6 40% of the 25 students in class 7A have black hair. How many students is this?

7 16 out of the 28 biscuits in a packet have cream centres. What percentage of the biscuits in the packet are cream biscuits?

8 A 2 litre bottle of orange drink contains 35% orange juice. Work out how many litres of orange juice is contained in the bottle. (It may help to work in mL.)

9 A lion can sleep for up to 20 hours a day. What percentage of a whole day is this?

4.8

10 There are 11 boys and 15 girls in class 7B. Write the number of boys and the number of girls as percentages of the total number of students in the class.

11 Damian kicked 7 goals from 10 attempts. Franco kicked 11 goals from 16 attempts.

(a) Express the boys' results as percentages.

(b) Who is the more accurate goalkicker?

12 Australia scored 381 runs in a cricket match against India. If Shane Watson scored 126 of those runs, what percentage of the team's total did he score?

13 Shops often uses percentages to advertise savings during sales. The percentage is used to calculate the amount that is removed from the price of the goods. This amount is known as the 'discount'.

(a) Calculate the dollar value of the discount on the following.

(i) \$40 shirt, 20% discount

(ii) \$125 jacket, 25% discount

(iii) \$65 jeans, 50% discount

(iv) \$239 suitcase, 10% discount

(v) \$799 camera, 40% discount

(vi) \$1299 sofa, 5% discount

(b) Now, find the sale price of the above items by subtracting the discount amount from the original price.

Reasoning, communicating and open-ended

14 8% of the apples in a crate were rotten. What percentage of the apples were not rotten?

15 The 180 students at St John's College all completed an online survey. The results showed that:

22% of the students had blue eyes

34% spoke two or more languages fluently

7% of them were left handed.

(a) Calculate the number of students in each category using the above three percentages.

(b) Considering that we are dealing with numbers of students, what is strange about your answer to **(a)**?

(c) What might be the actual numbers of students represented by each of the percentages?

16 Chelsea and Sonia are netballers. In one game, Chelsea scored 17 goals from 25 attempts, and Sonia scored 15 goals from 20 attempts. In the next game, Chelsea scored 19 goals from 27 attempts, and Sonia scored 17 goals from 23 attempts. Use percentages to show which one of the girls improved the accuracy of her goal shooting. Round your figures to one decimal place, if necessary.

17 The energy in food is measured in kilojoules (kJ). The National Heart Foundation of Australia recommends that no more than 30% of our daily energy intake should come from eating fats.

(a) A healthy, active woman should eat up to 9000 kJ of food a day. Up to how many kJ in the form of fats can she eat in 1 day?

(b) 1 gram of fat gives the body 37 kJ of energy. How many grams of fat would provide a woman with the maximum daily amount of kilojoules allowed from fat? Round your answer to the nearest whole number.

18 Suggest some reasons as to why some stores use percentages to advertise savings, rather than dollar amounts.

19 Mark saved \$5.00 when a store had a sale. Determine at least three different percentage discounts and original prices for the item that Mark purchased.

20 Fedora and Tanya were asked to show their working to the following question. 'Calculate 15% of 80'.

Here is Tanya's working:

$$\frac{15}{8\not{0}} \times \frac{10\not{0}}{1}$$
$$= \frac{150}{8}$$
$$= 18.75$$

Here is Fedora's working:

15% of 80
$= 0.15 \times 80$
$= 12$

Which of the girls' working is correct? Give a piece of advice to the other girl to help her avoid similar mistakes in the future.

Problem solving

The great half-price sale

Tricia and Carly were out shopping. They headed towards their favourite clothes shop. On the shop window was a sign that read: 'Take 50% off all marked prices!' Stuck over the top of that sign was another sign, saying: 'Take another 50% off reduced prices!'

'Eh?' said Carly. '50% is one half. If I take half off the price, then take off another half, there won't be any price left! That means everything in the shop is free!'

'I don't think so', said Tricia. 'How can you run a business like that?'

'Well', said Carly, 'If I pick up a pair of $100 jeans and take two lots of 50% off the price, what do they cost now?'

'Err...', said Tricia.

Can you work it out? How much would the jeans cost?

Strategy options

- Draw a diagram.
- Break problem into manageable parts.

4.9 Ratio

Need to Know

Ratios

A **ratio** is a comparison of two or more amounts of the same type of quantity, written using a ':' symbol.

For example: 2 lengths can be compared.

The order in which a ratio is stated must be made clear. If the ratio of teachers to students in a class is 1 : 20, then the ratio of students to teachers is 20 : 1.

A **part : part ratio** compares separate parts of a whole.

A **part : whole ratio** compares a part (or parts) to a whole.

Ratios are written without any units.

Ratios, fractions and percentages

Ratios can be written as fractions or percentages.

For example, if 11 : 20 is a part : whole ratio, it can be written as $\frac{11}{20}$ or 55%.

Equivalent ratios

Equivalent ratios are created by multiplying or dividing each part in the ratio by the same number.

For example, 1 : 4 = 50 : 200.

Simplifying ratios

Ratios can be simplified by dividing each part in the ratio by a common factor. Dividing by the highest common factor (HCF) will give the ratio in simplest form.

For example, 60 : 300 = 1 : 5.

Worked Example 23

WE 23

Write a ratio for each of the following.

(a) the number of windows in the middle building compared to the number of windows in the smallest building

(b) the number of floors in the buildings, from left to right

Working

(a) Windows in middle building: 12
Windows in smallest building: 7
12 : 7

(b) 5 floors to 3 floors to 2 floors
5 : 3 : 2

Worked Example 24

WE 24

A large enclosure at the zoo contains 5 giraffes and 12 zebras.

(a) Write a ratio comparing the number of zebras to the number of giraffes.

(b) Write a ratio comparing the number of giraffes to the total number of animals.

(c) Write the number of giraffes as a fraction of the total number of animals.

(d) Write the number of zebras as a percentage of the total number of animals, correct to the nearest whole number.

Working

(a) zebras : giraffes
= 12 : 5

(b) $5 + 12 = 17$
giraffes : total animals
= 5 : 17

(c) $\frac{5}{17}$

(d) zebras : total animals
= 12 : 17

$\frac{12}{17} \times \frac{100}{1}\%$
= 71% (to nearest whole number)

4.9 Ratio

Answers page 466

Navigator

Q1, Q2, Q3 Columns 1 & 2, Q4 Columns 1 & 2, Q5, Q6, Q7, Q9, Q11, Q14, Q15, Q17, Q18, Q19	Q1 Column 2, Q2, Q3 Columns 2 & 3, Q4 Columns 2 & 3, Q5, Q7, Q8, Q9, Q10, Q11, Q13, Q14, Q15, Q17, Q18, Q19	Q1 Column 2, Q2, Q3 Column 4, Q4 Column 4, Q5, Q7, Q8, Q9, Q10, Q11, Q12, Q13, Q15, Q16, Q17, Q18, Q19

Fluency and understanding

WE 23

1 Write a ratio for each of the following.

(a) circles to squares

(b) blue circles to red circles

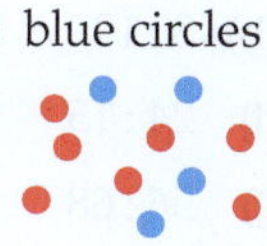

(c) happy faces to sad faces

(d) ducks to swans

4.9

(e) spider to flies to butterflies

(f) rulers to pencils to erasers

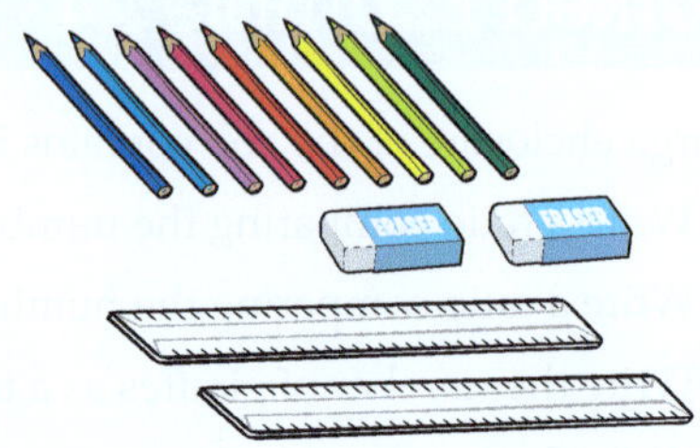

(g) bees to flowers to leaves

(h) large mass to medium mass to small mass

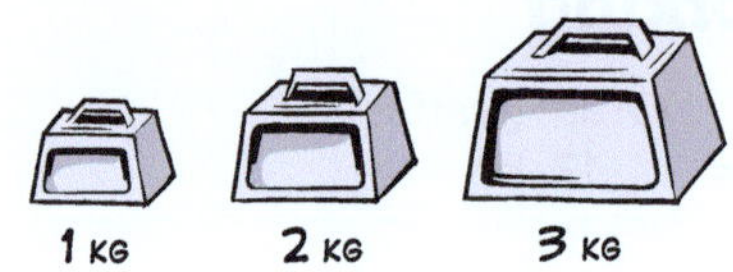

WE24

2 Sumi has 3 canaries and 2 mice as pets.
   (a) Write a ratio comparing the number of mice to the number of canaries.
   (b) Write a ratio comparing the number of canaries to the total number of pets.
   (c) Write the number of canaries as a fraction of the total number of pets.
   (d) Write the number of mice as a percentage of the total number of pets.

3 Create pairs of equivalent ratios by multiplying or dividing each part in the ratio by the number given in brackets beside it.

(a) $1:3\ (\times 2)$	(b) $1:8\ (\times 2)$	(c) $1:11\ (\times 2)$	(d) $1:33\ (\times 2)$
(e) $4:5\ (\times 3)$	(f) $3:7\ (\times 4)$	(g) $6:9\ (\div 3)$	(h) $8:12\ (\div 4)$
(i) $10:8\ (\div 2)$	(j) $18:9\ (\div 9)$	(k) $8:5\ (\times 7)$	(l) $12:11\ (\times 5)$
(m) $48:18\ (\div 6)$	(n) $20:6\ (\times 5)$	(o) $60:54\ (\div 3)$	(p) $15:13\ (\times 5)$

4 Write the following ratios in simplest form, by dividing by the highest common factor (HCF).

(a) $2:10$	(b) $3:12$	(c) $4:18$	(d) $6:21$
(e) $21:14$	(f) $24:15$	(g) $32:20$	(h) $48:36$
(i) $54:64$	(j) $56:68$	(k) $72:81$	(l) $38:95$
(m) $75:50$	(n) $120:80$	(o) $105:60$	(p) $240:66$

5 There are 14 girls and 9 boys in a Year 7 maths class.
   (a) The ratio of boys to girls is:
   A $14:9$ B $9:14$ C $14:23$ D $9:23$
   (b) The fraction of the whole class that are girls is:
   A $\frac{9}{23}$ B $\frac{9}{14}$ C $\frac{14}{23}$ D $\frac{14}{9}$

6 A packet of M&M's® contains three colours: 12 red, 7 blue and 16 brown.

(a) Write the ratio of red to blue to brown M&M's.

(b) How many M&M's are in the packet?

(c) What fraction of the M&M's in the packet are red?

(d) What percentage of the M&M's are blue?

7 During a cricket season, Erika's team won 15 matches, drew 2 and lost 7.

(a) Write the won : lost : drew ratio for Erika's team for that season.

(b) What fraction of their games did Erika's team lose?

(c) Write your answer to **(b)** as a percentage, rounded to the nearest whole number.

8 The students at Hillview High School are voting for a school captain. The results show that for every 4 students who voted for Guy, 5 students voted for Dominic and 3 for Melissa.

(a) Write the ratio of the number who voted for Dominic to the number who voted for Guy, to the number who voted for Melissa.

(b) What fraction of the total vote did Dominic receive?

(c) Write your answer to **(b)** as a percentage, rounded to the nearest whole number.

9 There must be 1 instructor for every 6 students on a kayaking course.

(a) Write the comparison of students to instructors as a ratio.

(b) If 24 students want to do the course, use an equivalent ratio to determine how many instructors will be required.

(c) If 5 instructors are available, what is the maximum number of students that can do the course?

10 Tom's recipe for making concrete is to add gravel, sand and cement in the ratio 3 : 2 : 1 before adding water. Tom has 4 buckets of cement. Use an equivalent ratio to show how many buckets of sand and gravel he should add to get the correct concrete mixture.

11 'White gold' is commonly made by mixing gold and nickel in the ratio 9 : 1.

(a) Write the amount of nickel as a percentage of the total mixture.

(b) A jeweller wants to make 20 g of white gold. How much nickel will he require?

12 A survey found that 75% of Year 7 students were studying music.

(a) What percentage were not studying music?

(b) Use the two percentages to write a ratio that compares the group studying music to the group not studying music, in simplest form.

Reasoning, communicating and open-ended

13

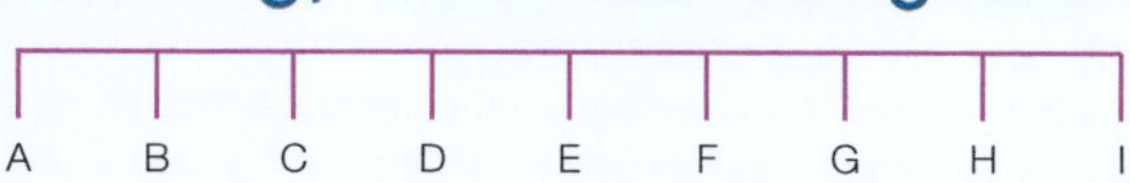

Which point divides the above line so that the ratio shorter part : longer part is:

(a) 1 : 7 **(b)** 5 : 3 **(c)** 1 : 3 **(d)** 1 : 1

14 There must be at least 1 teacher for every 20 students on a school excursion.

(a) If 68 students are going on an excursion, how many teachers are required? Give reasons for your answer.

(b) Use your answer to **(a)** to write the actual ratio of teachers : students, in simplest form.

15 Chloe made a cordial drink by mixing 30 mL of concentrate with 150 mL of water. Her friend Hannah made a cordial drink by mixing 80 mL of concentrate with 200 mL of water. Hannah said her cordial drink was 'stronger' than Chloe's. Chloe could not tell by looking.

(a) Write the ratio of concentrate : water for each girl's drink, in simplest form.

(b) Use the ratios to state whether Hannah is correct, giving a reason for your answer.

16 Mia has three cats and five dogs as pets.

(a) What fraction of her pets are dogs?

(b) What fraction of her pets are cats?

(c) If Mia adopts another cat, what fraction of her pets are cats now?

(d) After adopting the new cat, Mia moves into a share house with Shona who also has some cats and dogs. If the final ratio of cats to dogs is now 1 : 1, what is the smallest number of cats and dogs that Shona brings to the house?

17 (a) What is the ratio of girls to boys in your class?

(b) Would the ratio be the same in all the Year 7 classes in your school? Explain your answer by giving examples.

(c) Would the ratio be the same for all school classes in Australia? Give some reasons for your answer.

18 Write three ratios that are equivalent to 4 : 7.

19

What mistake has Tenille made? What is the correct answer?

Rates

4.10

Need to Know

A **rate** is a way of comparing two amounts of different types of quantities. A rate is written as the amount of the first quantity for every 1 unit of the second quantity.

For example: Speed is the distance travelled over a unit of time.

Kilometres per hour (km/h) or metres per second (m/s) are the most common metric units used for speed, while miles per hour or feet per second are used in the imperial system.

The unitary method

Solving problems by first calculating a number (such as cost) per unit is called the **unitary method.**

The unitary method can be used to compare prices.

For example, \$7.21 for an 820 g box of cereal compared to \$5.99 for a 500 g box. Supermarkets are required to show the price per 100 g (or 100 mL or 1 kg or 1 L) for many products. This is called the **unit price**.

The unit price for the 820 g box = 7.21 ÷ 820 × 100
= \$0.88 per 100 g

The unit price for the 500 g box = 5.99 ÷ 500 × 100
= \$1.20 per 100 g

Worked Example 25

WE 25

(a) If 8 kg of oranges costs \$24, find the cost of 6 kg of oranges.

(b) A machine in a soft drink factory produces 210 bottles every hour. How many bottles would be produced in 40 minutes?

Working

(a) \$24 for 8 kg
= \$3 for 1 kg (÷ 8)
= \$18 for 6 kg (× 6)

(b) 210 ÷ 60 = 3.5 bottles/min
3.5 bottles/min × 40 min
= 140 bottles in 40 min

Worked Example 26

WE26

Calculate the unit price (price per 100 g) of the following pair of products, and so determine which one is better value for money: A 500 g packet of pasta for \$2.42 or a 750 g packet for \$3.44.

Working

Number of units $= \frac{500}{100}$

$= 5$

Unit price $= \frac{\$2.42}{5}$

$= \$0.48/100\text{g}$

Number of units $= \frac{750}{100}$

$= 7.5$

Unit price $= \frac{\$3.44}{7.5}$

$= \$0.46/100\text{g}$

The 750 g packet is slightly better value for money (2c less per 100 g).

Answers page 467

4.10 Rates

Navigator

Q1, Q2, Q3, Q7, Q8, Q9, Q12, Q13, Q14, Q15, Q16, Q19	Q1, Q2, Q3, Q5, Q6, Q7, Q8, Q9, Q10, Q11, Q12, Q13, Q14, Q15, Q16, Q17, Q20	Q1 (d)–(f), Q2 (d)–(f), Q4, Q5, Q6, Q8, Q10, Q11, Q12, Q14, Q15, Q17, Q18, Q19, Q20

Fluency and understanding

WE25

1 (a) If 15 kg of potatoes costs \$12, find the cost of 7 kg of potatoes.

(b) A machine in a food canning factory produces 250 cans every hour. How many cans would be produced in 25 minutes?

(c) If 14 pens cost \$5.60, find the cost of 3 pens.

(d) Farmer Harry buys 8 cows for \$6800. How much would 10 cows have cost him?

(e) If 21 bags of rice weigh 35 kg, how much do 15 bags of rice weigh?

(f) A printing machine prints 70 pages in 8 minutes. How many pages could it print in one hour?

WE26

2 Calculate the unit price (the price per 100 g) of the following pairs of products, and so determine which one is better value for money.

(a) a 600 g jar of jam for \$4.09 or a 250 g jar for \$2.68

(b) a 500 g packet of pasta for \$2.81 or a 750 g packet for \$3.45

(c) a 375 g jar of tomato paste for \$1.28 or a 500 g jar for \$3.25

(d) a 200 g jar of peanut butter for \$3.21 or a 375 g jar for \$4.93

(e) a 500 mL bottle of soy sauce for \$2.71 or a 920 mL bottle for \$4.29

(f) a 450 g bag of cat food for \$3.78 or a 1 kg bag for \$8.79

3 Nazeh has saved \$2000 in 4 months. If he continues to save at this rate, how much will he have by the end of the year?

A \$6000 B \$8000 C \$12 000 D \$24 000

4 An entertainer receives $450 000 for a season of five performances, each of which lasts $1\frac{1}{2}$ hours. The rate of pay per minute is closest to:

A $60 000 **B** $5000 **C** $1500 **D** $1000

5 A caterer knows that 4 loaves of bread can make enough sandwiches for 25 people. How many loaves will he need to make enough sandwiches for 120 people?

6 **(a)** Which is better value for money, a 280 g jar of jam for $5.15, or a 400 g jar for $6.42?

(b) The 280 g jar goes on special for $4.49. Does this change your answer to **(a)**?

7 A mobile phone carrier charges 75c for 30 seconds. How much will a 4-minute phone call cost?

8 Rajid's cricket team required 85 runs from 12 overs to win the game. At what rate, in runs per over, must they score to win? (Answer to one decimal place.)

9 **(a)** Calculate the price per item in each pack of the following products, and so determine which pack is better value for money.

(i) a packet of 6 muesli bars for $4.83, or a packet of 10 for $7.51

(ii) a 12-pack of cans of soft drink for $14.20, an 18-pack for $20.21, or a 30-pack for $27.51

(iii) a box of 50 tea bags for $3.20, or a box of 200 for $7.99

(b) Describe another way you could compare the prices of the tea bags in part **(a) (iii)**.

Hint
200 is a multiple of 50.

10 Lucy has bought a 1-year subscription to *Australian Explorer* magazine. She paid $102 for 12 monthly magazines. The regular price at the newsagent for one copy of the magazine is $9.95. How much will Lucy save:

(a) per magazine **(b)** over the 1-year period?

11 Andrea has a leaky tap in her bathroom. She places a container underneath it and collects 1.2 litres in 8 hours. If Andrea does not fix the tap, use a mental or written method to calculate how much water would be lost in:

(a) 1 day **(b)** 1 week?

12 A 200 g serve of yoghurt contains 6.5 g of fat. Use a mental or written method to calculate how much fat there is in:

(a) a 100 g serve **(b)** a 400 g serve **(c)** a 500 g serve.

13 A 5-minute phone call costs $3.00. If the same call rate is charged, use a mental or written method to calculate the cost of:

(a) a 10-minute call **(b)** a 1-minute call **(c)** a 13-minute call.

Reasoning, communicating and open-ended

14 Charlie's bakery sells hot cross buns in packets of 6. One packet costs $7.20. A customer wants to buy exactly 10 buns. Charlie is happy to take 4 buns from a second packet, but how much should he charge the customer?

15 Imagine that you are standing in a supermarket, comparing the following pairs of products. Do some mental calculations to determine which product is the better value for money. Rounding prices slightly may help to simplify some calculations. Show how you arrived at your answer.

(a) A 1 kg box of washing powder for $5.49, or a 4 kg box for $15.97.

(b) Four 250 mL cartons of orange juice for $1.25 each, or a 1 L carton for $3.99.

(c) Buying 500 g of ham from the deli counter where the price is $13 per kg, or buying it in a packet for $5.83.

4.10

(d) A 400 g jar of honey for $7.99, or a 300 g jar for $6.54.

(e) Give some reasons why people might prefer to buy a product even though it is not the best value for money.

16 A mobile phone carrier charges a rate of 85 cents per minute for pre-paid calls. Calls made on a phone plan are charged at 75 cents per minute, plus a 35 cent connection fee. Which of the two types of charge is cheaper for a 5-minute phone call?

17 During the month of January, the water level in a Northern Territory dam increased by 78 000 kilolitres (1 kilolitre = 1000 litres).

(a) How many kilolitres per day was this? (Round your answer to two decimal places.)

(b) If the dam continued to fill at this rate, use your answer to **(a)** to calculate how much water will have flowed into the dam by the end of the year. (Assume there are 365 days in a year.)

(c) Would it be sensible to use your answer to **(b)** as a prediction for the amount of water that will flow into the dam over the year? Give some reasons for your answer.

18 'Trim and Tasty' cheese comes in three different-sized blocks:

250 g for $4.50
500 g for $7.42
750 g for $9.76

(a) How many times bigger than the 250 g block is the 500 g block?

(b) If the prices of the 500 g block and the 750 g block were in the same proportion to the price of the 250 g block as their sizes, what would you expect their prices to be?

(c) Often (but not always) larger sizes of a product have cheaper unit prices. Suggest a reason why this is the case.

19 List four supermarket products whose unit price would be usefully calculated per 100 mL, or per L.

20 Kumar needs 4.5 kg of rice for the students who are attending his cooking classes. It is available in the following sizes and prices.

500 g for $1.24 1 kg for $2.35 1.5 kg for $3.59 2 kg for $4.52

(a) List at least three different combinations of sizes that would give Kumar 4.5 kg of rice.

(b) Which combination is the cheapest?

Challenge 4

1 What percentage of whole numbers from 6 to 25 inclusive (i.e. including 6 and 25) are exact multiples of 6?

A 5 B 20 C 25 D 30

2 The numbers 5, 8, 7 and 3 are written on cards. Where should they be placed so that the following product has the greatest possible value?

□ × 0.□□□

3 Evaluate $0.1^2 - 0.1^3$.

4 Zinadene has $\frac{3}{4}$ of a dollar and Katrine has $\frac{3}{10}$ of a dollar. Together they have:

A \$0.95 B \$1.00 C \$1.05 D \$1.10

5 $200 \div 2.5 = 80$. Therefore, $20 \div 0.25 =$

A 0.80 B 8.0 C 80 D 800

6 When doing a series of additions using his calculator, Trent noticed that he added 49 095 instead of 49.95. In order to correct his error with a single entry, he should now:

A add 49.95 B subtract 49 045.05

C add 49 045.05 D subtract 49 095

7 A decimal with 2 decimal places is multiplied by a decimal with 3 decimal places. Explain with number examples how the solution could have:

(a) 5 decimal places (b) 4 decimal places (c) 3 decimal places.

8 In Felicity's first six netball games she averaged 7.5 goals per game. If the least number of goals she scored was 4, what is the lowest possible value for the highest number of goals she scored?

9 Four holes are to be drilled along the centre line of a strip of metal so that their centres are 2.25 cm apart. The centres of the two end holes are to be 3.35 cm from their corresponding ends. What is the length of the strip of metal?

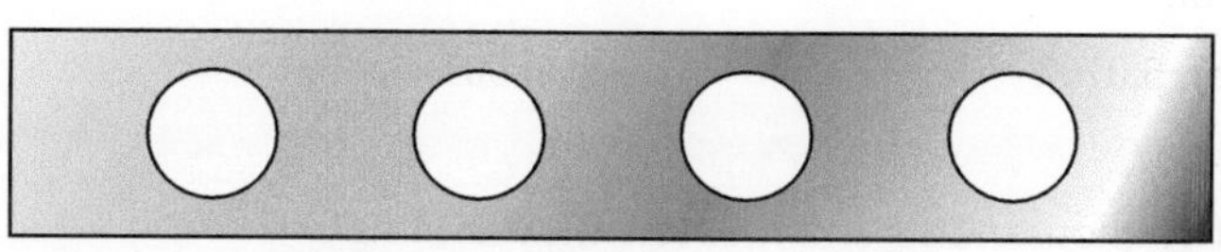

10 What is the quotient when 0.01 is divided by 0.002?

11 Mikaela had her salary reduced by 10%. She was later promoted and her salary was increased by 10%. If her original salary was \$30 000, her present salary is:

A \$24 300 B \$27 000 C \$29 700 D \$30 000

12 $\frac{2009}{20.09}$ equals

A 100 B 10 C 0.1 D 0.01

13 A town has 5000 residents, of whom 60% voted in a local council election. The result was that, of those who voted, 37% voted for *P*, 32% for *Q* and 31% for *R*. Under the voting system, *P* was elected. The number of residents who voted for *P* was:

A 1850 B 1110 C 960 D 930

Chapter review 4

D.I.Y. Summary

Key Words

decimal places	per cent	round
decimal point	place value	terminating decimal
digit	rate	unit price
equivalent ratios	ratio	unitary method
part : part ratio	rational numbers	
part : whole ratio	recurring decimal	

Copy and complete the following using the words and phrases from this list, where appropriate, to write a summary for this chapter. A word or phrase may be used more than once.

1 4.87 has three ___________s and two ___________ .

2 When adding or subtracting decimals, it is important to line up the ___________s.

3 ___________ are created by multiplying each part in the ratio by the same number.

4 ___________s compare two or more parts of a whole.

5 A ___________ has a finite number of decimal places, a ___________ has an infinite number of decimals, in a repeating pattern.

6 ___________ literally means 'for every hundred'.

7 Calculating a ___________ helps to compare the prices of different sizes and brands of products.

8 Explain the difference between 'round down' and 'round up'.

Fluency and understanding

1 Write in expanded fractional form. **Ex 4.1**

(a) 0.968 (b) 5.0702 (c) 6.005

2 Write as decimals: **Ex 4.1**

(a) six tenths, five hundredths, two thousandths and four ten-thousandths

(b) eight hundredths, nine hundred-thousandths and three millionths

3 Write as decimals: **Ex 4.1**

(a) $4 + \frac{3}{10} + \frac{8}{100} + \frac{6}{10000}$ (b) $50 + 7 + \frac{3}{100} + \frac{2}{1000} + \frac{9}{100\,000}$

4 Write each of the following in expanded word form. **Ex 4.1**

(a) 1.8531 (b) 0.070 06 (c) 61.0009

5 Write the value of the 9 in each decimal as (i) a fraction (ii) in words. **Ex 4.1**

(a) 5.091 (b) 0.0659 (c) 25.291

6 Write < or > in each of the following pairs of numbers to make a true statement. **Ex 4.1**

(a) 3.0427 ______ 3.0274 (b) 0.009 95 ______ 0.01

7 Write in order from smallest to largest. **Ex 4.1**

(a) 0.6055, 0.5506, 0.607 (b) 0.071, 0.701, 0.71

8 Copy this number line and show the positions of the following numbers: 0.25, 1.45, 0.09, 2.05. **Ex 4.1**

9 Round each of the following to the number of decimal places shown in the brackets. **Ex 4.2**

(a) 4.398 (1) **(b)** 34.6503 (2) **(c)** 23.098 (1) **(d)** 102.4728 (2)

10 Round the following to the nearest 5 cents. **Ex 4.2**

(a) \$41.21 **(b)** \$10.03 **(c)** \$79.97 **(d)** \$98.98

11 Write each of the following as a single fraction or mixed number in simplest form. **Ex 4.3**

(a) 3.9 **(b)** 0.62 **(c)** 2.45 **(d)** 0.0018

12 Write the following fractions as decimals, using the correct notation for recurring decimals if necessary. **Ex 4.3**

(a) $\frac{4}{5}$ **(b)** $\frac{19}{20}$ **(c)** $\frac{2}{9}$ **(d)** $\frac{10}{12}$

13 Calculate: **Ex 4.4**

(a) $23.6 + 4.79$ **(b)** $3.768 + 10.9205 + 0.77$ **(c)** $6 + 0.408 + 35.025$

(d) $10.367 - 2.65$ **(e)** $2.082 - 0.394$ **(f)** $7 - 2.2198$

14 Find the following products. **Ex 4.5**

(a) 3.376×4 **(b)** 2.99×35 **(c)** 6.09×33

(d) 23.6×2000 **(e)** 0.548×300 **(f)** 0.069×50

(g) 0.6×0.9 **(h)** 0.009×0.04 **(i)** 5.2×3.8

15 Calculate the following. Round answers to three decimal places where necessary. **Ex 4.6**

(a) $10.32 \div 4$ **(b)** $7.028 \div 7$ **(c)** $3.75 \div 8$

(d) $480.6 \div 2000$ **(e)** $8.23 \div 500$ **(f)** $54.63 \div 300$

(g) $0.48 \div 0.04$ **(h)** $1.2 \div 0.5$ **(i)** $11.9 \div 0.002$

16 Write the following fractions as percentages. **Ex 4.7**

(a) $\frac{7}{10}$ **(b)** $\frac{5}{4}$ **(c)** $\frac{1}{8}$ **(d)** $\frac{2}{3}$

17 Convert the following decimals to percentages. **Ex 4.7**

(a) 0.38 **(b)** 0.96 **(c)** 3.55 **(d)** 0.09 **(e)** 3.234 **(f)** 0.5432

18 Calculate the following percentages. **Ex 4.8**

(a) 40% of \$60 **(b)** 15% of 85 L **(c)** 83% of 200 m **(d)** 12.5% of \$600

19 Write the following results as percentages, rounding answers to one decimal place if necessary. **Ex 4.8**

(a) 13 out of 20 **(b)** 35 out of 70

(c) 14 out of 40 **(d)** 28 out of 30

20 There are 56 girls and 64 boys in Year 7. **Ex 4.9**

(a) Write the ratio boys : girls in simplest form.

(b) Write the number of girls as a fraction of the total number of students, in simplest form.

(c) Write the number of boys as a percentage of the total number of students, to the nearest whole number.

21 **(a)** A 10 kg box of tomatoes costs $23.50. Use the unitary method to calculate the cost of a 3 kg box.

Ex 4.10

(b) If a 250 g block of cheese costs $7.85, find the unit price (the price per 100 g).

22 Write the populations of the following countries as whole numbers.

(a) China: 1.326 billion
(b) India: 1.14 billion
(c) Indonesia: 228.2 million
(d) New Zealand: 4.27 million

23 An employer wants to buy each of her staff a uniform. If each uniform costs $75.59, how much will the uniforms cost:

(a) for 40 staff
(b) for 100 staff?

24 Kayla bought the following items while out shopping: A $30 T-shirt discounted by 10%, and a $60 pair of jeans discounted by 40%.

(a) Calculate the dollar value of each of the discounts.

(b) Subtract the discount amount to find how much Kayla paid for each item.

25 **(a)** Which is better value for money: a 250 g jar of honey for $2.98, or a 400 g jar for $4.23?

(b) The 250 g jar goes on special for $1.98. Does this change your answer to **(a)**?

(c) Give one or two reasons why people might prefer to buy a certain size or product, even though it may not have the cheapest unit price.

26 A salad dressing is made by mixing oil and vinegar in the ratio 3 : 1.

(a) How much oil should be mixed with 25 mL of vinegar?

(b) Helen mixes 60 mL of oil with 40 mL of vinegar. Is this an equivalent ratio to the one given above? Demonstrate your answer by writing it in simplest form.

27 In a springboard diving competition, Claudia needed a score of 52 or more to take first place. For her final dive she chose a dive with a degree of difficulty of 2.7. The scores she received from the judges were 7.0, 8.0, 7.5, 8.5, 7.5, 8.0 and 8.0.

(a) Remove the highest and lowest scores and calculate the average of the remaining five scores (add them up and then divide by 5).

(b) Calculate the 'three-judge total' by multiplying this average by 3.

(c) Calculate her final score by multiplying the 'three-judge total' by the degree of difficulty. Did she do well enough to win?

Reasoning and communicating

28 Tim buys 6 individual cans of drink for $0.87 per can.

(a) If he pays in cash, find the total cost of the cans.

(b) A pack of 6 cans costs $4.80. How much could Tim have saved by buying the 6 pack?

(c) What is the individual cost of the cans in the 6 pack?

(d) Tim needs to buy 40 cans for a party. What is the cheapest way to buy them?

29 There must be at least one supervisor for every 12 children at a childcare facility. If there are 54 children booked in for a day, how many supervisors will be needed? Give a reason for your answer.

30 Sally is shopping for shoes. She has found the pair she wants in two different stores. Store A has them priced at $65, with a 'take 25% off' tag on them. Store B has them priced at $75, but the store is having a '30% off everything' sale. Where should Sally buy her shoes? Explain your choice by showing the relevant calculations.

NAPLAN practice 4

Numeracy: Non-calculator

1 Which number is greater than 0.06?

A 0.0069 **B** 0.007 **C** 0.05 **D** 0.1

2 Geraldine buys 400 snack bars for the school camp. If each bar costs \$0.64, how much would Geraldine pay for the 400 bars?

A \$25.60 **B** \$256 **C** \$464 **D** \$25 600

3 What is the answer to $5.4 \div 0.9$?

A 0.06 **B** 0.6 **C** 6.6 **D** 6

4 20% of a class have the flu. The class has 25 students. How many have the flu?

A 4 **B** 5 **C** 10 **D** 20

Numeracy: Calculator allowed

5 A pack of 10 lollipops costs \$4.60.
A pack of 6 costs \$3.20.
You need to buy 22 lollipops.

What is the least amount you can pay?

6 The price of grapes at the supermarket is \$14 per kg. If Jack paid \$3.50 for his grapes, how many grams did he buy?

A 0.25 g **B** 25 g **C** 100 g **D** 250 g

7 Lucy is adding a row of square tiles across the top of a mirror. Each tile is 4.5 cm wide. The mirror is 56 cm wide. Lucy wants to tile right across the top, with no gaps between the tiles. What calculation should she do to work out how many tiles she needs?

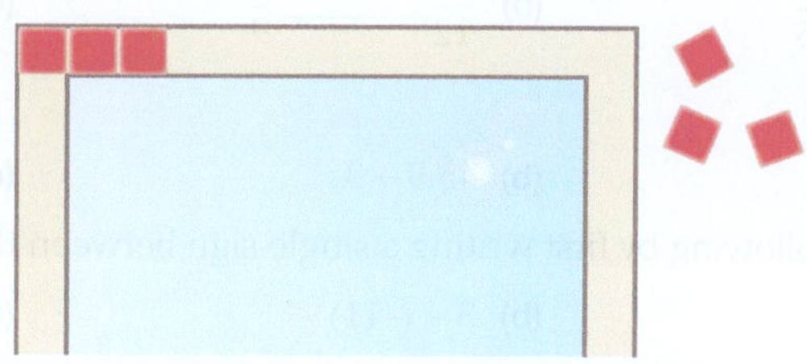

A $4.5 \div 56$ **B** $5.6 \div 4.5$ **C** $56 \div 4.5$ **D** 4.5×56

8 The material for the curtains in the classroom costs \$21.30 per metre. 8.4 metres is needed. How much will the material cost?

Mixed review B

Fluency and understanding

1 Write the following fractions in simplest form. Ex 3.2

(a) $\frac{12}{40}$ (b) $\frac{15}{50}$ (c) $\frac{48}{64}$ (d) $\frac{20}{65}$

2 Write a positive or a negative integer to describe the following. Ex 2.4

(a) a withdrawal from a bank account of $74 (b) a weight gain of 3 kg

(c) 215 m below sea level (d) an increase in length of 8 m

3 For each of the following numbers: Ex 2.3

(i) use a factor tree to find the prime factors

(ii) write the number as a product of its prime factors, using index notation.

(a) 76 (b) 108 (c) 250

4 Calculate approximate answers to the following by rounding to a convenient multiple of 5 or 10. Ex 1.4

(a) 238×21 (b) 107×31 (c) $478 \div 39$ (d) $604 \div 153$

5 For each pair of fractions, determine which one is larger, then write a > or < sign between them. Ex 3.3

(a) $\frac{2}{3}$ ______ $\frac{2}{5}$ (b) $\frac{5}{12}$ ______ $\frac{5}{9}$ (c) $\frac{11}{15}$ ______ $\frac{3}{4}$

6 Find: Ex 4.5, 4.6

(a) 0.45×0.6 (b) $45.9 \div 3$ (c) $43.5 \div 0.1$

7 Calculate the following by first writing a single sign between the integers. Ex 2.7

(a) $-1 + (-5)$ (b) $3 - (-11)$ (c) $-6 - (+4)$

(d) $9 + (-17)$ (e) $-8 - (-14)$ (f) $21 - (+30)$

8 Write each of the following as a fraction in simplest form. Ex 4.3

(a) 0.32 (b) 1.24 (c) 3.237

9 Write the following in index form. Ex 1.2

(a) $7 \times 7 \times 7 \times 7$ (b) 4 squared

(c) $2 \times 2 \times 3 \times 3 \times 3$ (d) 11 cubed

10 Find the lowest common multiples of the following numbers. Ex 2.1

(a) 5 and 6 (b) 3 and 6 (c) 2, 3 and 5

11 Write < or > between the following pairs of numbers. Ex 2.4

(a) -5 ______ 3 (b) -4 ______ -11 (c) -7 ______ 0

12 Find the HCF of the following numbers. Ex 2.1

(a) 12 and 30 (b) 16 and 40 (c) 24 and 60

13 Calculate the following. Write your answers in simplest form. Ex 3.4

(a) $\frac{1}{5} + \frac{3}{4}$ (b) $1\frac{2}{3} + 2\frac{3}{8}$ (c) $\frac{5}{6} - \frac{3}{8}$ (d) $3\frac{1}{4} - 2\frac{1}{2}$

14 Use an appropriate mental strategy to calculate the following.

(a) 29×7 **(b)** 44×11 **(c)** 13×15

(d) $260 \div 4$ **(e)** $2 \times 4 \times 7 \times 5$ **(f)** $440 \div 8$

15 Erin calculates that she needs 9 lots of 0.35 m of ribbon for her textiles project. She wants to round up to the nearest metre so that she makes sure she has enough.

(a) Calculate how many metres she should buy.

(b) If the ribbon costs $0.75 per m, how much will she pay?

16 Write **(i)** a fraction in simplest form and **(ii)** a percentage to represent the following situations.

(a) 12 black jelly beans in a bag of 40

(b) 120 Year 7 students in a school of 520 students

17 A 250 g serve of tinned soup contains 7 g of fat. Calculate the amount of fat contained in:

(a) a 500 g serve **(b)** a 50 g serve **(c)** a 150 g serve.

18 7 pizzas were each sliced into 6 equal pieces.

(a) If $4\frac{5}{6}$ pizzas were eaten, how many slices is that?

(b) How many slices are left over?

19 Camilla is on an athlete's diet where $\frac{2}{3}$ of her daily energy must come from carbohydrates, $\frac{1}{4}$ from protein, and the rest from fats. Her daily energy allowance is 4800 kJ.

(a) Calculate how many kJ of carbohydrates she must have every day.

(b) Calculate how many kJ of protein she must have every day.

(c) What fraction of her daily kJ intake is fat?

20 Explain why $3^2 = 9$, but $3 \times 2 = 6$.

Reasoning and communicating

21 A lighting display in the city has a set of blue lights that flash every 6 minutes, and a set of red lights that flash every 8 minutes. How many times an hour will the lights flash together?

22 There is a large block of chocolate sitting in the Williams family pantry. Chloe breaks off and eats $\frac{1}{6}$, and returns the rest. Later, her brother Eamon breaks off and eats $\frac{2}{3}$ of the rest.

(a) What fraction of the original block is left?

(b) If the original block contained 54 squares of chocolate, who ate more, Chloe or Eamon?

23 Shania surveyed 130 Year 7 students to find out which charity they wanted to fundraise for. Shania reported back to the Student Council that 65% of the Year 7s wanted to support the Red Cross.

(a) What is 65% of 130?

(b) Considering that we are working with numbers of students, what is 'strange' about your answer to **(a)**?

(c) What might have been the actual number of students who said they wanted to support the Red Cross? Explain how Shania obtained her figure of 65%.

24 How many whole numbers lie on the number line between $\frac{4}{3}$ and $\frac{38}{4}$?

25 Jeffery has 46 footy cards in his collection. His little brother Joseph has half as many in his collection, whereas his friend Shahin's collection is twice the size of Jeffery's. How many more cards has Shahin than Joseph?

2 Fractions, decimals and percentages

P Recall 2

Prepare for this chapter by attempting the following questions. If you have difficulty with a question, download the Recall Worksheet.

1 Evaluate the following, writing your answers in simplest form.

(a) $\frac{2}{3}+\frac{3}{5}$ (b) $\frac{5}{6}-\frac{1}{4}$ (c) $\frac{4}{9}\times\frac{3}{8}$ (d) $\frac{4}{15}\div\frac{4}{5}$

2 Evaluate the following.

(a) $2.6+3.04+5.3+2.101$ (b) $5.29-4.831$ (c) 12.7×5

(d) 4.2×3.1 (e) $3.72\div8$ (f) $12.64\div0.04$

3 Round the following decimals to the number of decimal places shown in brackets.

(a) 2.0335 (2) (b) 3.1884 (2)

(c) 2.9096 (3) (d) 4.086 19 (3)

4 Simplify the following fractions.

(a) $\frac{27}{45}$ (b) $\frac{16}{92}$ (c) $\frac{56}{21}$

5 Write the following mixed numbers as improper fractions.

(a) $2\frac{1}{2}$ (b) $4\frac{2}{9}$ (c) $3\frac{7}{20}$

6 Write the following improper fractions as mixed numbers in simplest form.

(a) $\frac{17}{7}$ (b) $\frac{40}{9}$ (c) $\frac{95}{10}$

7 Evaluate:

(a) 7.08×10 (b) 0.49×1000 (c) 0.8×100

(d) $5.21\div100$ (e) $34.65\div10$ (f) $4.037\div1000$

8 Calculate the following.

(a) $\frac{5}{6}$ of \$90 (b) $\frac{7}{12}$ of 180 kg (c) $\frac{9}{8}$ of 200 m

9 Write the following fractions as percentages.

(a) $\frac{1}{2}$ (b) $\frac{3}{4}$ (c) $\frac{2}{5}$ (d) $\frac{7}{10}$

10 Write the following decimals as percentages.

(a) 0.35 (b) 0.9 (c) 0.04 (d) 1.27

Syllabus references

Number and Algebra: Fractions, Decimals and Percentages

✓ Compare fractions using equivalence; locate and represent positive and negative fractions and mixed numerals on a number line **(ACMNA152)**
✓ Solve problems involving addition and subtraction of fractions, including those with unrelated denominators **(ACMNA153)**
✓ Multiply and divide fractions and decimals using efficient written strategies and digital technologies **(ACMNA154)**
✓ Express one quantity as a fraction of another, with and without the use of digital technologies **(ACMNA155)**
✓ Round decimals to a specified number of decimal places **(ACMNA156)**
✓ Investigate terminating and recurring decimals **(ACMNA184)**
✓ Connect fractions, decimals and percentages and carry out simple conversions **(ACMNA157)**
✓ Investigate the concept of irrational numbers, including π **(ACMNA186)**
✓ Find percentages of quantities and express one quantity as a percentage of another, with and without the use of digital technologies **(ACMNA158)**
✓ Solve problems involving the use of percentages, including percentage increases and decreases, with and without the use of digital technologies **(ACMNA187)**

Number and Algebra: Financial Mathematics

✓ Investigate and calculate the Goods and Services Tax (GST), with and without the use of digital technologies
✓ Investigate and calculate 'best buys', with and without the use of digital technologies **(ACMNA174)**
✓ Solve problems involving profit and loss, with and without the use of digital technologies **(ACMNA189)**

Working with fractions and decimals 2.1

Need to Know

Converting fractions to decimals

Fractions and decimals are used to represent parts of wholes. They are two different ways of writing the same number. A fraction has a line or bar that separates the top number (the numerator) from the bottom number (the denominator). This line can be thought of as a division sign, ÷.

To convert a fraction to a decimal, perform the division:

For example: $\frac{3}{4} = 3 \div 4$
$= 0.75$

Common fractions and their decimal equivalents

$\frac{1}{2}$	$\frac{1}{3}$	$\frac{2}{3}$	$\frac{1}{4}$	$\frac{3}{4}$	$\frac{1}{5}$	$\frac{1}{8}$	$\frac{1}{10}$	$\frac{1}{100}$	$\frac{1}{1000}$
0.5	$0.\dot{3}$	$0.\dot{6}$	0.25	0.75	0.2	0.125	0.1	0.01	0.001

Notice that the decimal equivalents for $\frac{1}{3}$ and $\frac{2}{3}$ have a dot above the decimal digit. This is to show that the digit is repeated forever, i.e. $\frac{1}{3} = 0.333\ldots$

Fractions and decimals can most easily be compared by first converting the fractions to decimals.

Converting terminating decimals to fractions

To convert a decimal to a fraction:

- write all the digits of the decimal part in the numerator
- use the place-value column of the last digit to get the denominator of the fraction
- simplify the fraction if possible.

Rounding decimals

Some decimal numbers have a lot of decimal places that are not required. In these cases, we 'round' the number to a certain number of decimal places. It is important to state how many decimal places a number has been rounded to by writing 'correct to' a certain number of decimal places (abbreviated to 'd.p.'). It is important to note that these 'rounded' answers are no longer exact. Rounding should never be done before the final calculation.

For example:

$$\frac{2}{3} \times \frac{11}{16} = \frac{11}{24}$$
$$= 0.458 \text{ (correct to 3 d.p.)}$$

However, rounding early would give $0.667 \times 0.688 = 0.459$ (correct to 3 d.p.)

If this measurement was in kilometres, rounding early would create an error of 1 metre.

To round a number to a given number of decimal places, look at the digit to the right of the one being rounded. If it is:

- 0, 1, 2, 3 or 4, round down. Keep the digit in the place to be rounded the same, and drop the digits following it:
  e.g. 1.697 214 = 1.697 (correct to 3 d.p.).
- 5, 6, 7, 8 or 9, round up. Increase the digit in the place to be rounded by 1, and drop the digits following it. If it is a 9, write the zero of the 10, and carry the one across to the next column:
  For example: 1.789 24 = 1.79 (correct to 2 d.p.)
  24.5999 = 24.600 (correct to 3 d.p.).

Hint

Remember, if the digit to the right of the one being rounded is 5 or more, round up. If it is less than 5, round down.

Technology Exploration: Calculator

Fractions and decimals on the calculator

Scientific calculators have keys that enable you to enter fractions and convert between fraction and decimal form. They will usually look something like this $\frac{\square}{\square}$ or this $a\frac{b}{c}$.

To enter a fraction, either:

- press $\frac{\square}{\square}$, then key in the numerator and the denominator, using the cursor to move between them; or key in the numerator, press $a\frac{b}{c}$, then key in the denominator and press =.

For example: To enter $\frac{13}{25}$ you would press $\frac{\square}{\square}$ 1 3 ⊙ 2 5 = or

press 1 3 $a\frac{b}{c}$ 2 5 =.

To enter a mixed number, such as $1\frac{3}{7}$ either:

- press SHIFT $\frac{\square}{\square}$ 1 ⊙ 3 ⊙ 7 = or
- press 1 $a\frac{b}{c}$ 3 $a\frac{b}{c}$ 7 =.

Pressing S⇔D or $a\frac{b}{c}$ will switch a number entered in fraction form to decimal form, and vice versa.

Pressing SHIFT S⇔D or SHIFT $a\frac{b}{c}$ will convert a mixed number entered in fraction form to decimal form, and vice versa.

2.1

Worked Example 1

WE1

Arrange the following list into ascending order (smallest to largest) by converting the fraction values to decimals.

$\frac{5}{8}, \frac{3}{5}, \frac{3}{4}$, 0.69, 0.686

Working

$$\frac{5}{8} = 5 \div 8$$
$$= 0.625$$

$$\frac{3}{5} = 3 \div 5$$
$$= 0.6$$

$$\frac{3}{4} = 3 \div 4$$
$$= 0.75$$

0.6, 0.625, 0.686, 0.69, 0.75

$\frac{3}{5}, \frac{5}{8}$, 0.686, 0.69, $\frac{3}{4}$

Worked Example 2

WE2

Write the following decimal as a fraction in simplest form: 0.384

Working

$$0.384 = \frac{384}{1000}$$
$$= \frac{96}{250}$$
$$= \frac{48}{125}$$

2.1

Worked Example 3

WE3

Use a calculator, if necessary, to answer the following question. A length of wood is $2\frac{1}{2}$ m long. How many pieces, each 600 mm long, can be cut from this length, and how much will be left over? Ignore any wastage that might occur in cutting.

Working

$2\frac{1}{2}$ m ÷ 600 mm

$= 25\not{0}\not{0} \div 6\not{0}\not{0}$
$= 25 \div 6$
$= 4.166\,666\,667$
= 4 whole pieces

$4 \times 600 = 2400$
$2500 - 2400 = 100$

We will get 4 pieces 600 mm long and will have 100 mm of wood left over.

Answers page 487

2.1 Working with fractions and decimals

Navigator

Q1 Column 1, Q2 Columns 1 & 2, Q3 (a) & (b), Q4 (a)–(d), Q5, Q6, Q7, Q9 Columns 1 & 2, Q11, Q12, Q15, Q18, Q21	Q1 Column 1, Q2 Columns 3 & 4, Q3, Q4, Q5, Q8, Q9 Columns 2 & 3, Q11, Q12, Q14, Q15, Q16, Q18, Q20, Q21	Q1 Column 2, Q2 Columns 3 & 4, Q3, Q4 (e)–(h), Q5 (e)–(h), Q6, Q8, Q9 Columns 2 & 3, Q10, Q12, Q13, Q15, Q16, Q17, Q19, Q20, Q21

Equipment required: A calculator may be used for Questions **1**, **3**, **4**, **9**, **13–15** and **17**

Fluency and understanding

WE1

1 Arrange the following lists into ascending order (smallest to largest) by converting the fraction values to decimals.

(a) $\frac{2}{5}$, 0.399, $\frac{4}{5}$, 0.382, $\frac{3}{4}$

(b) $\frac{9}{10}$, $\frac{9}{8}$, 0.88, 0.89, 0.899

(c) $\frac{1}{8}$, 0.112, 0.099, $\frac{1}{4}$, 0.07

(d) $\frac{2}{5}$, $\frac{1}{3}$, 0.3, $\frac{3}{8}$, 0.2

(e) $\frac{1}{2}$, 0.555, 0.58, $\frac{3}{5}$, 0.55

(f) 0.291, 0.302, $\frac{3}{10}$, $\frac{2}{9}$, $\frac{2}{3}$

(g) $2\frac{4}{5}$, $2\frac{3}{4}$, 2.278, 2.932, $2\frac{9}{10}$

(h) $1\frac{1}{5}$, $1\frac{2}{3}$, $1\frac{3}{8}$, 1.029, 1.243

(i) $4\frac{1}{5}$, $4\frac{1}{4}$, 4.295, 4.199, 4.201

(j) $3\frac{1}{5}$, 3.45, 3.439, $3\frac{2}{3}$, 3.482

2 Write the following decimals as fractions in simplest form. WE2

(a) 0.8 **(b)** 0.05 **(c)** 0.002 **(d)** 0.0009

(e) 0.14 **(f)** 0.62 **(g)** 0.31 **(h)** 0.85

(i) 0.711 **(j)** 0.684 **(k)** 0.625 **(l)** 0.128

(m) 0.203 **(n)** 0.094 **(o)** 0.1560 **(p)** 0.7009

3 Use a calculator, if necessary, to answer the following questions. WE3

(a) A length of wood is 2.6 m long. How many pieces, each 40 cm long, can be cut from this length, and how much wood will be left over?

(b) A bag holds 750 g of flour. How many cups, each containing 120 g of flour, can be filled from the bag, and how much will be left over?

(c) A container holds $3\frac{1}{4}$ litres of juice. How many 200 mL cups can be filled from the container, and how much juice will be left over? (1 litre = 1000 mL)

(d) A bus can carry 52 passengers. If every bus except the last one is filled to capacity, how many buses will be needed to transport 650 people, and how many people will be on the last bus?

4 Write the following mixed numbers as decimals using a calculator where necessary.

(a) $51\frac{7}{10}$ **(b)** $5\frac{3}{20}$ **(c)** $7\frac{1}{8}$ **(d)** $67\frac{7}{25}$

(e) $39\frac{1}{16}$ **(f)** $28\frac{24}{200}$ **(g)** $14\frac{13}{80}$ **(h)** $24\frac{31}{400}$

5 Write the following decimals as mixed numbers in simplest form.

(a) 7.5 **(b)** 3.4 **(c)** 1.25 **(d)** 5.64

(e) 13.02 **(f)** 27.96 **(g)** 9.045 **(h)** 124.706

6 **(a)** $1\frac{5}{8}$ written as a decimal is:

A 0.16 **B** 1.58 **C** 1.625 **D** 1.63

(b) $2\frac{1}{3}$ written as a decimal, correct to two decimal places, is:

A 0.34 **B** 2.13 **C** 2.33 **D** 2.34

7 For each set of numbers, draw the number line below and indicate the position of each number with a labelled arrow.

(a) $0.5, 1.3, 1\frac{1}{2}, \frac{1}{5}$ **(b)** $\frac{2}{2}, 0.8, 1.6, 1\frac{4}{5}$

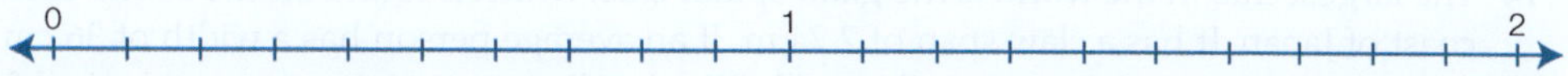

Hint

First, determine the value of the smallest interval on the number line.

8 For each set of numbers, draw the number line below and indicate the position of each number with a labelled arrow. (You may need to estimate the position of some numbers.)

(a) $\frac{3}{4}, \frac{7}{10}, 0.34, 0.47, 1\frac{1}{3}$ **(b)** $1.05, \frac{7}{8}, 0.58, \frac{8}{5}, \frac{6}{4}$

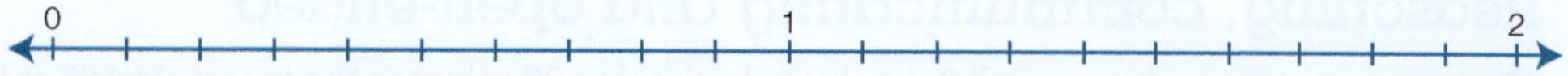

2.1

9 Use a calculator to convert the fractions to decimal form, then place a > (greater than), < (less than) or = (equal to) symbol between the pairs of numbers to make the following statements correct.

(a) $0.84 \_\_\_ \frac{18}{21}$ **(b)** $2.29 \_\_\_ 2\frac{12}{39}$ **(c)** $0.912 \_\_\_ \frac{114}{125}$

(d) $0.64 \_\_\_ \frac{16}{25}$ **(e)** $1.83 \_\_\_ 1\frac{18}{23}$ **(f)** $0.97 \_\_\_ \frac{98}{99}$

(g) $\frac{1}{3} \_\_\_ 0.3$ **(h)** $\frac{2}{9} \_\_\_ 0.23$ **(i)** $\frac{2}{3} \_\_\_ 0.67$

Hint

For Question **10**, you can turn decimal numbers into whole numbers by mutiplying by 10.

10 **(a)** Express each first quantity as a fraction of the second quantity. Write the fraction in simplest terms.

(i) first quantity 1.5 hours; second quantity 2.4 hours

(ii) first quantity 200 m; second quantity 322.5 m

(iii) first quantity 80.6 mL; second quantity 200 mL

(iv) first quantity 20.6 kg; second quantity 12.5 kg

(v) first quantity 18.6 m^2; second quantity 12 m^2

(vi) first quantity 14.4 g; second quantity 1.5 g

(b) Express each fraction from part **(a)** as a decimal, rounding your answers to three decimal places where necessary.

11 Mr Scully is out buying materials to build the set for the school production. Calculate how much he will pay for the following materials. Round your answers to the nearest 5 cents.

(a) 6 lengths of timber, at $8.99 per length

(b) 9.4 metres of canvas, at $11.20 per metre

(c) 5.6 metres of ribbon, at $0.95 per metre

12 The exchange rate between the Australian and American dollars varies daily. Suppose that according to the current rate A$1 is worth US$0.72. How much are the following amounts worth in US dollars?

(a) A$10 **(b)** A$50 **(c)** A$100 **(d)** A$267 **(e)** A$1845

13 The most overdue library book ever was a copy of *Febrile Diseases* by Dr J. Currie. The book was borrowed from the University of Cincinnati Medical Library in 1823 by Mr M. Dodd, and was returned by his great-grandson in 1968. If the fines for late returns were $18.30 a year, how big was the fine his great-grandson had to pay?

Hint

Convert measurements to centimetres first.

14 The largest crab in the world is the giant spider crab, which is found off the south-eastern coast of Japan. It has a claw span of 2.74 m. If an average person has a width of 36 cm across the waist, how many people could fit in the claw span of the giant spider crab?

15 The highest wave ever recorded in Australia was 24.9 m, off Macquarie Island. If the average surfer is 180 cm tall, how many surfers standing on top of one another would it take to reach the top of the wave?

Reasoning, communicating and open-ended

16 An insect was climbing a wall 2.7 m high. In the first 20 minutes the insect started at the bottom and climbed $\frac{1}{3}$ of the height of the wall. In the second 20 minutes it climbed $\frac{1}{4}$ of the remaining height and in the third 20 minutes it climbed $\frac{1}{5}$ of the remaining height.

(a) Calculate the distances climbed in each 20-minute period.

(b) Calculate how far the insect still had to climb to reach the top of the wall.

(c) Express the distance remaining as a fraction of the height of the wall.

17 Angela has to fill 50 gift bags with an equal amount of lollies in each. She has 4 packets of lollies that weigh 375 g each. Angela knows that the average mass of one lolly is 2.8 g.

(a) Explain how Angela can use this information to determine how many lollies she can put in each gift bag.

(b) Use your method from **(a)** to find the answer.

(c) About how many lollies will Angela have left over? Explain why this answer may not be exact.

18 The force of gravity varies from planet to planet and so the weight of an object can vary, depending on which planet it is on. Objects on Jupiter would weigh 2.6 times their weight on Earth. Objects on Mars would weigh 0.38 times their 'Earth weight'. Calculate the weight of a 5 kg bag of potatoes on **(a)** Jupiter **(b)** Mars.

19 Write three fractions with three different denominators that have decimal values between 0.2 and 0.4.

20 Write three decimals, each with a different number of decimal places, that have fraction values between $\frac{9}{10}$ and $\frac{9}{8}$.

21

(a) Explain to Leo why his reasoning is incorrect.

(b) Describe to Leo how to compare two decimal numbers.

2.2 Types of decimals

Need to Know

Terminating decimals

Decimals with a clearly defined number of digits are called **terminating decimals.**

Examples of terminating decimals:

$\frac{18}{5} = 18 \div 5$ $\quad$ $\frac{9}{8} = 9 \div 8$ $\quad$ $\frac{88}{200} = 88 \div 200$

$= 3.6$ $\quad$ $= 1.125$ $\quad$ $= 0.44$

Non-terminating decimals

Decimals with an infinite number of digits are called non-terminating decimals. If some or all of the decimal digits follow a repeating pattern, they are called **recurring decimals.**

Examples of recurring decimals:

$\frac{1}{3} = 1 \div 3$ $\quad$ $\frac{4}{15} = 4 \div 15$ $\quad$ $\frac{11}{7} = 11 \div 7$

$= 0.333\,333\,3$ $\quad$ $= 0.266666$ $\quad$ $= 1.571\,428\,571\,428$

Writing recurring decimals

- Write any non-recurring digits, then the digits that form the repeating pattern.
- Place a dot on top of the first and last digit of the repeating pattern or draw a line over the whole of the repeating pattern.

Using a dot or a dash notation is called writing the number in **exact decimal form.**

For example:

$\frac{1}{3} = 0.3 \text{ or } 0.\bar{3}$ $\quad$ $\frac{4}{15} = 0.26 \text{ or } 0.2\bar{6}$ $\quad$ $\frac{11}{7} = 1.571428 \text{ or } 1.\overline{571428}$

Writing recurring decimals as fractions

To write a recurring decimal as a fraction, create two expressions with the same decimal part and subtract one from the other.

For example:

If 0.22222… is the recurring decimal, $10 \times 0.22222\ldots = 2.22222\ldots$

$2.22222\ldots - 0.2222\ldots = 2$

and $2.22222\ldots - 0.22222\ldots = 9 \times 0.22222\ldots$

so $9 \times 0.22222\ldots = 2$

$0.22222\ldots = \frac{2}{9}$

Rational numbers

Terminating and recurring decimals are **rational numbers** because they can be written in the form of $\frac{a}{b}$ where a and b are integers and $b \neq 0$.

Irrational numbers

Some non-terminating decimals have no repeating pattern. These numbers cannot be written as fractions. **Irrational numbers** include special numbers, such as π, and the square roots of numbers such as 2, 3 and 5. Square roots that are irrational numbers are called **surds** and are written in **surd form** as $\sqrt{2}$, $\sqrt{3}$ or $\sqrt{5}$.

Because they are **non-terminating decimals**, the only way to write the exact value of a surd is to write it in surd form. An irrational number can only be written as an approximate value in decimal form. correct to a certain number of decimal places.'≈' means 'is approximately equal to'.

Example:

$$\begin{aligned}\sqrt{2} &\approx 1.4 \text{ (1 d.p.)}\\ &\approx 1.41 \text{ (2 d.p.)}\\ &\approx 1.414 \text{ (3 d.p.)}\\ &\approx 1.414\,213\,562 \text{ (9 d.p.)}\end{aligned}$$

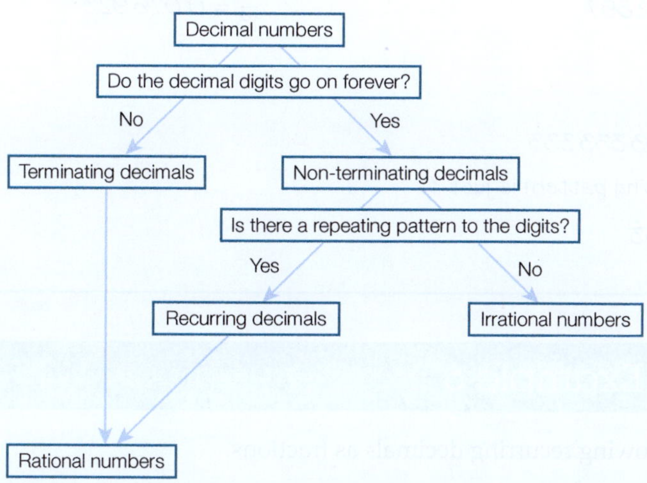

Worked Example 4

WE4

Use a calculator to express the following as decimals and classify them as either terminating, recurring, or irrational.

(a) $12 \div 7$ **(b)** $\frac{5}{16}$ **(c)** $\sqrt{50}$ **(d)** $\sqrt{156.25}$

Working

(a) $12 \div 7 = 1.714\,285\,714\ldots$

(The repeating pattern is 714 285)
Recurring

(b) $\frac{5}{16} = 0.3125$

Terminating

(c) $\sqrt{50} = 7.071\,067\,812\ldots$

(No repeating pattern)
Irrational

(d) $\sqrt{156.25} = 12.5$

Terminating

Worked Example 5

WE5

Use a calculator to write each of the following fractions in exact decimal form.

(a) $\frac{1}{7}$ (b) $\frac{1}{13}$ (c) $\frac{1}{12}$

Working

(a) $\frac{1}{7} = 1 \div 7$

$= 0.142\,857\,142$

The repeating pattern is 142 857.

$\frac{1}{7} = 0.\dot{1}42\,85\dot{7}$

(b) $\frac{1}{13} = 1 \div 13$

$= 0.076\,923\,076$

The repeating pattern is 076 923.

$\frac{1}{13} = 0.\dot{0}76\,92\dot{3}$

(c) $\frac{1}{12} = 1 \div 12$

$= 0.083\,333\,333$

The repeating pattern is just 3.

$\frac{1}{12} = 0.08\dot{3}$

Worked Example 6

WE6

Write the following recurring decimals as fractions.

(a) $0.\dot{1}\dot{5}$ (b) $1.2\dot{4}$

(a) $x = 0.151\,515\ldots$ (1)

$100x = 15.151\,515\ldots$ (2)

(2) − (1):

$$\begin{aligned} 100x &= 15.151\,515\ldots \\ -\quad x &= 0.151\,515\ldots \\ \hline 99x &= 15 \\ x &= \frac{15}{99} \\ &= \frac{5}{33} \end{aligned}$$

(b) $x = 1.24444\ldots$ (1)

$10x = 12.4444\ldots$ (2)

(2) − (1):

$$\begin{aligned} 10x &= 12.4444\ldots \\ -\quad x &= 1.2444\ldots \\ \hline 9x &= 11.2 \\ x &= \frac{11.2}{9} \\ &= \frac{112}{90} \\ &= \frac{56}{45} \text{ or } 1\frac{11}{45} \end{aligned}$$

2.2 Types of decimals

Answers page 488

Navigator

Q1, Q2, Q4, Q5, Q6, Q7, Q9 (a) & (b), Q10, Q12, Q14	Q1, Q2 Columns 2–4, Q3 Columns 1 & 2, Q4, Q5, Q6, Q7, Q9, Q10, Q11, Q12, Q14, Q15	Q1, Q2 Columns 3 & 4, Q3 Columns 2–4, Q5, Q6, Q7, Q8, Q9, Q11, Q12, Q13, Q14, Q15

Equipment required: Calculator for Questions **1**, **2**, **4(b)**, **5**, **6**, **7**, **9–13** and **15**

Fluency and understanding

1 Use a calculator to express the following as decimals and classify them as either terminating, recurring, or irrational. WE4

(a) $9 \div 8$ **(b)** $\frac{26}{4}$ **(c)** $\sqrt{449}$ **(d)** $\sqrt{84}$

(e) $\frac{5}{9}$ **(f)** $\sqrt{58.4}$ **(g)** $\frac{2}{11}$ **(h)** $22 \div 3$

(i) $\sqrt{216}$ **(j)** $15 \div 6$ **(k)** $\sqrt{533.61}$ **(l)** $\frac{11}{12}$

(m) $\sqrt{342.25}$ **(n)** $\sqrt{360}$ **(o)** $35 \div 9$ **(p)** $\sqrt{96.4}$

2 Use a calculator to write each of the following fractions in exact decimal form. WE5

(a) $\frac{1}{6}$ **(b)** $\frac{5}{6}$ **(c)** $\frac{1}{9}$ **(d)** $\frac{5}{9}$

(e) $\frac{1}{11}$ **(f)** $\frac{2}{11}$ **(g)** $\frac{3}{11}$ **(h)** $\frac{2}{13}$

(i) $\frac{1}{15}$ **(j)** $\frac{2}{15}$ **(k)** $\frac{4}{15}$ **(l)** $\frac{7}{15}$

(m) $\frac{1}{18}$ **(n)** $\frac{5}{18}$ **(o)** $\frac{7}{18}$ **(p)** $\frac{11}{18}$

> **Hint**
> The repeating pattern does not always begin at the start of the decimal.

3 Write the following recurring decimals as fractions. WE6

(a) $0.\dot{4}$ **(b)** $0.\dot{1}$ **(c)** $0.\dot{2}$ **(d)** $0.\dot{7}$

(e) $0.\dot{4}\dot{2}$ **(f)** $0.\dot{3}\dot{2}$ **(g)** $0.\dot{1}\dot{7}$ **(h)** $0.\dot{2}\dot{4}$

(i) $0.\dot{0}1\dot{2}$ **(j)** $0.\dot{2}1\dot{3}$ **(k)** $0.\dot{3}3\dot{2}$ **(l)** $0.\dot{2}2\dot{4}$

(m) $0.12\dot{5}$ **(n)** $0.1\dot{2}\dot{6}$ **(o)** $0.3\dot{2}1\dot{4}$ **(p)** $4.1\dot{0}2\dot{6}$

4 **(a)** The recurring decimal 4.563 213 213 21… written in exact decimal form is:

A $4.\overline{56321}$ **B** $4.5632\dot{1}$ **C** $4.563\overline{21}$ **D** $4.56\overline{321}$

(b) The fraction $\frac{27}{55}$ written in exact decimal form is:

A $0.\overline{490}$ **B** $0.\dot{4}\dot{9}\dot{0}$ **C** $0.4\overline{90}$ **D** $0.\dot{4}9\dot{0}$

5 If we evaluate $\sqrt{2}$ on a calculator, we get a non-terminating decimal value of 1.414 213…. Although we cannot write an exact decimal value, we can state that the value of $\sqrt{2}$ lies in between the values of 1.4 and 1.5, and is closer to 1.4. State the two decimal values that the following lie between, correct to one decimal place, also stating which of the two values is closer to the exact value.

(a) $\sqrt{3}$ **(b)** $\sqrt{5}$ **(c)** $\sqrt{12}$ **(d)** $\sqrt{20}$

(e) $\sqrt{8}$ **(f)** $\sqrt{7}$ **(g)** $\sqrt{47}$ **(h)** $\sqrt{68}$

2.2

6 The diagonal of a square is equal to its side length multiplied by $\sqrt{2}$. Jason is strengthening a large house frame by placing two diagonals across it, as shown. Using a calculator, determine the length of wood Jason needs for the diagonals, correct to three decimal places.

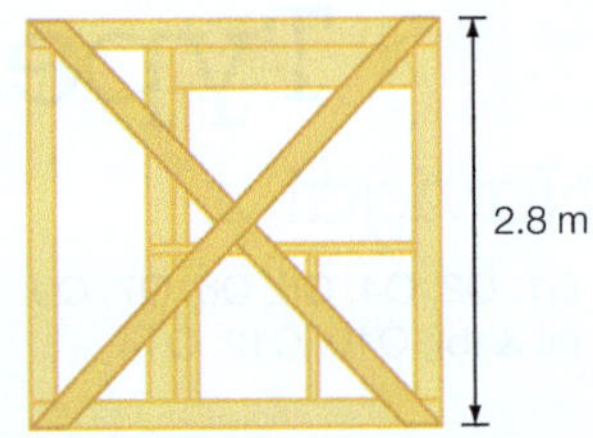

7 A shop is selling a packet of 11 lollipops for $7.

(a) How much does each lollipop cost? Write your answer in exact decimal form, then as a price (rounded to two decimal places).

(b) If you can buy one lollipop individually, how much will you pay?

8 Find the fraction equivalent of $0.\dot{9}$.

9 Arrange the following in ascending order.

(a) $\frac{1}{3}$, $\sqrt{9}$, $\sqrt{3}$, 1.3, 0.9

(b) 0.5, $\sqrt{5}$, $\frac{5}{9}$, $\sqrt{59}$, $\frac{9}{5}$

(c) $\frac{4}{11}$, 0.4, $\sqrt{4}$, 0.36, $\sqrt{11}$

(d) 1.6, $\frac{1}{6}$, $\sqrt{6}$, $\frac{1}{16}$, $\sqrt{16}$

Reasoning, communicating and open-ended

10 For each set of numbers, draw the number line below and indicate the position of each number with a labelled arrow. (You may need to estimate the position of some numbers.)

(a) $\frac{4}{3}$, $0.\dot{6}$, $\frac{11}{6}$, 0.3, $\frac{3}{5}$

(b) 1.04, 0.4, $\frac{4}{7}$, $0.\dot{4}$, $\frac{7}{4}$

11 For each set of numbers, draw the number line below, and indicate the approximate position of each number with a labelled arrow.

(a) $\sqrt{3}$, $\frac{13}{7}$, $\frac{7}{6}$, $\sqrt{6}$, $\sqrt{7}$

(b) $\sqrt{5}$, $\frac{15}{14}$, $\sqrt{11}$, $\frac{40}{11}$, $\sqrt{15}$

12 (a) The value of $\frac{11}{19}$ is closest to:

A 0.5 **B** 0.579 **C** 0.58 **D** 0.6

(b) The value of $\sqrt{75}$ is closest to:

A 8 **B** 8.66 **C** 8.7 **D** 9

13 (a) Use a calculator to write $\frac{1}{9}$ and $\frac{1}{11}$ as decimals with eight digits after the decimal point.

(b) Add together the two decimal values.

(c) Add the two fraction values together.

(d) Do you get the same answer?

(e) Repeat parts **(a)–(c)** with $\frac{2}{3}$ and $\frac{5}{9}$.

(f) Explain the difference between the fraction and decimal sums.

14 Which is bigger, 0.3 or $0.\dot{3}$? Explain your answer.

15 Write five proper fractions, with varying numbers of digits in both numerator and denominator, that have only 9s in the denominator. Can you predict the recurring digits in the decimal equivalent? Check your prediction with your calculator.

Negative fractions and decimals

2.3

Need to Know

Number line symmetry

The number line is symmetrical about zero, which means that the negative equivalent of a number will be the same distance from zero as the positive number.

For example:

-1.4 will be the same distance to the left of zero as 1.4 is to the right of zero

$-\frac{3}{4}$ will be the same distance to the left of zero as $\frac{3}{4}$ is to the right of zero.

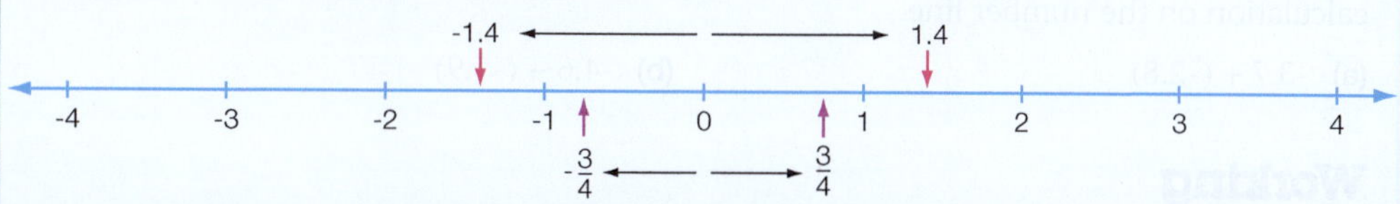

Adding and subtracting negative decimals

The symmetry of the number line can also help with addition and subtraction of negative decimal numbers. Any calculation on one side of the number line has an equivalent 'mirror image' calculation on the opposite side.

Adding and subtracting negative fractions

To add and subtract negative fractions apply the methods previously learnt for adding and subtracting fractions, along with the rules for adding and subtracting negative numbers.

Multiplying and dividing negative fractions and decimals

To multiply or divide negative fractions and decimals:

1. Ignore any positive or negative signs, and apply the usual methods for multiplication and division.
2. Determine the sign of the answer by applying the rules for multiplying and dividing integers.

Two 'like' signs give a positive answer: $+ \times +$ or $- \times - = +$
Two 'unlike' signs give a negative answer: $+ \times -$ or $- \times + = -$

Negative numbers on the calculator

To enter a negative fraction or decimal on your calculator, press the (-) or +/- key, then enter the fraction or decimal in the usual way.

2.3

Worked Example 7

WE7

Place the following sets of numbers on separate number lines.

(a) $2\frac{1}{2}, -\frac{3}{4}, \frac{1}{4}, -2\frac{1}{3}$

(b) -2.7, 0.4, 1.2, -0.9

Working

(a)

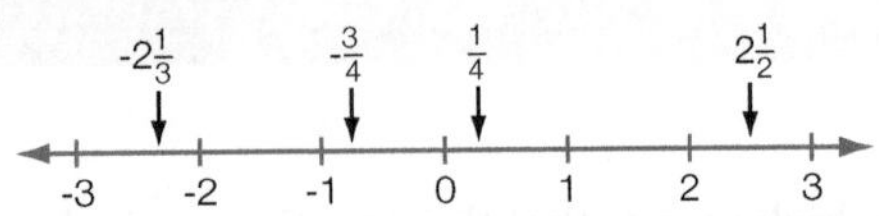

(b)

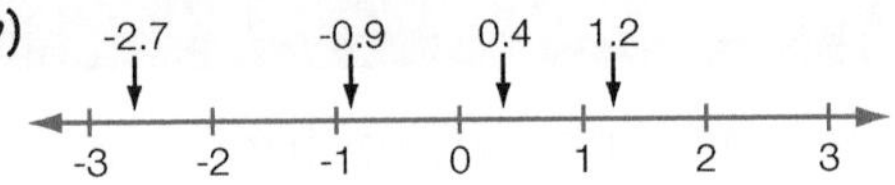

Worked Example 8

WE8

Calculate the following. It may be useful to visualise the equivalent 'mirror image' calculation on the number line.

(a) -3.7 + (-2.8)

(b) -4.6 – (-1.9)

Working

(a) $-3.7 + (-2.8)$
$= -3.7 - 2.8$

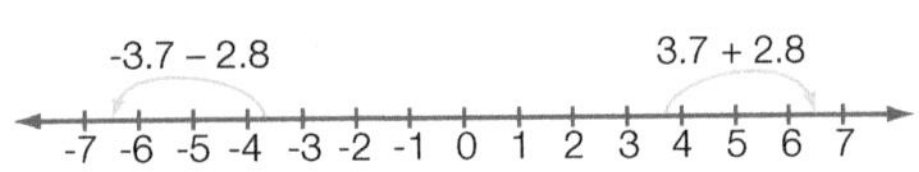

$3.7 + 2.8 = 6.5$
$-3.7 - 2.8 = -6.5$

(b) $-4.6 - (-1.9)$
$= -4.6 + 1.9$

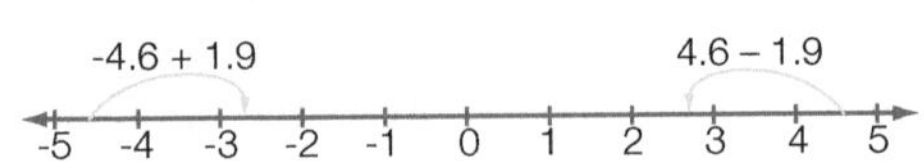

$4.6 - 1.9 = 2.7$
$-4.6 + 1.9 = -2.7$

Worked Example 9

WE9

Calculate the following, writing your answers in simplest form.

(a) $-\frac{4}{5} + \frac{1}{2}$

(b) $-1\frac{3}{4} + \left(-\frac{5}{6}\right)$

Working

(a) $-\frac{4}{5} + \frac{1}{2}$

$= -\frac{8}{10} + \frac{5}{10}$

$= -\frac{3}{10}$

(b) $-1\frac{3}{4} + \left(-\frac{5}{6}\right)$

$= -1\frac{3}{4} - \frac{5}{6}$

$= -1\frac{3}{4} - \frac{5}{6}$

$= -\frac{21}{12} - \frac{10}{12}$

$= -\frac{31}{12}$ or $-2\frac{7}{12}$

2.3 Negative fractions and decimals

Answers page 488

Navigator

Q1 (a)–(d), Q2 Columns 1–3, Q3 Columns 1 & 2, Q4 Columns 1 & 2, Q5 Columns 1 & 2, Q6, Q7, Q8, Q9, Q11, Q13 (a)–(f), Q16	Q1 (c)–(f), Q2 Columns 2–4, Q3 Columns 2 & 3, Q4 Columns 2 & 3, Q5 Columns 2 & 3, Q6, Q8, Q9, Q10, Q11, Q13 Column 1, Q14, Q15, Q16, Q18	Q1 (c)–(f), Q2 Columns 3 & 4, Q3 Columns 3 & 4, Q4 Columns 3 & 4, Q5 Columns 3 & 4, Q6, Q8, Q9, Q10, Q12, Q13 Column 2, Q14, Q15, Q17, Q18

Equipment required: Calculator for Question **13**

Fluency and understanding

1 Place the following sets of numbers on separate number lines. WE7

(a) $-3\frac{1}{2}, -2\frac{3}{4}, -\frac{1}{2}, 1\frac{1}{4}$ **(b)** -1.4, -1.9, 0.4, 2.2 **(c)** $-\frac{4}{5}, \frac{1}{5}, -2\frac{1}{4}, 1\frac{2}{5}$

(d) -3.6, -2.6, -6.1, 1.6 **(e)** $-\frac{6}{5}, -\frac{5}{6}, -2\frac{1}{6}, \frac{1}{5}$ **(f)** -0.5, 0.05, -5.2, 0.75

2 Calculate the following. It may be useful to visualise the equivalent 'mirror image' calculation on the number line. WE8

(a) $-0.5 + 0.4$ **(b)** $-0.3 - 0.7$ **(c)** $-1.5 + (-1.6)$ **(d)** $-2.4 - (-1.3)$

(e) $-7.2 + 6.8$ **(f)** $-5.9 - 8.3$ **(g)** $-4.6 + (-3.9)$ **(h)** $5.2 - (-4.3)$

(i) $7.61 + (-5.92)$ **(j)** $3.47 - (-10.98)$ **(k)** $0.8 - 2.34$ **(l)** $-9.87 + 11.3$

(m) $19.3 - 21.27$ **(n)** $-31.4 + 39.2$ **(o)** $-26.04 - (-15.97)$ **(p)** $13.92 - (-17.64)$

3 Calculate the following, writing your answers in simplest form. WE9

(a) $-\frac{3}{4} - \frac{3}{4}$ **(b)** $-\frac{1}{5} + \left(-\frac{3}{5}\right)$ **(c)** $\frac{2}{7} - \frac{5}{7}$ **(d)** $-\frac{8}{9} - \left(-\frac{1}{9}\right)$

(e) $-\frac{3}{5} + \left(-\frac{1}{4}\right)$ **(f)** $\frac{7}{8} - \frac{9}{10}$ **(g)** $-\frac{1}{6} - \left(-\frac{1}{5}\right)$ **(h)** $-\frac{2}{3} + \frac{5}{6}$

(i) $-\frac{2}{5} + 1$ **(j)** $-\frac{1}{2} + 3$ **(k)** $-1\frac{1}{3} - (-2)$ **(l)** $1 + \left(-1\frac{1}{4}\right)$

(m) $-1\frac{1}{2} + 2\frac{1}{4}$ **(n)** $-3\frac{1}{3} - 2\frac{1}{2}$ **(o)** $4\frac{1}{5} - \left(-\frac{7}{10}\right)$ **(p)** $1\frac{3}{4} + \left(-\frac{5}{6}\right)$

4 Calculate the following, writing your answers in simplest form.

(a) $-\frac{3}{7} \times \frac{1}{2}$ **(b)** $-\frac{1}{2} \div \frac{1}{4}$ **(c)** $\frac{5}{2} \times -\frac{1}{3}$ **(d)** $\frac{3}{5} \div -\frac{1}{3}$

(e) $-4 \div \frac{1}{8}$ **(f)** $-\frac{1}{6} \times -\frac{3}{4}$ **(g)** $6 \div -\frac{2}{3}$ **(h)** $-\frac{5}{8} \times -\frac{4}{5}$

(i) $-2\frac{1}{2} \times -1\frac{3}{5}$ **(j)** $3\frac{2}{3} \div -\frac{1}{2}$ **(k)** $-4\frac{1}{6} \times 1\frac{3}{5}$ **(l)** $-6\frac{3}{4} \div \frac{9}{10}$

5 Calculate the following, rounding answers to two decimal places where necessary.

(a) -0.4×0.8 **(b)** $-0.6 \div 2$ **(c)** -3×1.5 **(d)** $2.7 \div -3$

(e) $-0.72 \div -0.3$ **(f)** -12.5×-7.2 **(g)** $-1.2 \div 0.5$ **(h)** 0.05×-7.4

(i) 14.25×6.3 **(j)** $24 \div -3.4$ **(k)** 29.5×-11.23 **(l)** $-56.2 \div 10.1$

6 Copy the following number line. Use arrows to show the locations of the following fractions. (You may need to estimate the position of some numbers.)

$-\frac{2}{3}, \frac{4}{3}, -1\frac{1}{2}, \frac{11}{6}, -\frac{7}{3}$

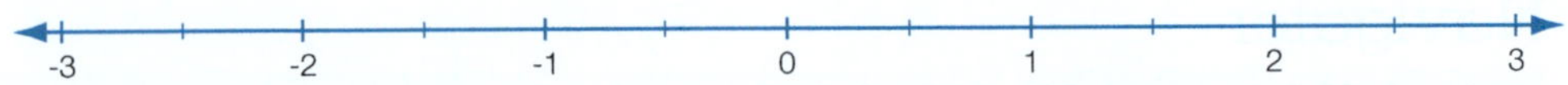

7 Copy the following number line. Use arrows to show the locations of the following decimals. (You may need to estimate the position of some numbers.)

-0.7, 1.3, -1.1, 0.65, -0.05

8 The number closest to the position indicated by the arrow on the number line is:

A -2.6 **B** -1.4 **C** -0.3 **D** 1.4

9 The following table shows the maximum and minimum temperatures recorded at Thredbo during 5 days in winter.

Temp (°C)	Mon	Tue	Wed	Thu	Fri
Maximum	5.2	-0.8	6.1	3.5	-1.2
Minimum	-1.1	-4.3	-1.7	-2.9	-5.6

(a) On which day was the coolest maximum recorded?

(b) On which day was the warmest minimum recorded?

(c) On which day was the difference between the maximum and minimum temperatures the greatest?

(d) Calculate **(i)** the mean maximum temperature and **(ii)** the mean minimum temperature, to one decimal place.

Hint

To find a mean, add up all the temperatures and divide by how many there are.

10 The executives of a bank meet every month to decide whether to raise or lower interest rates. Their decisions over an 8-month period were reported as follows.

Increase by 1.2%, no change, increase by $\frac{3}{4}$%, decrease by 0.8%, decrease by $1\frac{1}{4}$%, no change, decrease by $\frac{1}{2}$%, decrease by 0.6%.

(a) Write each of the percentage changes, using a positive or a negative sign to show increases or decreases.

(b) Work out the overall increase or decrease in interest rates at the end of the 8-month period. Write your answer in **(i)** fraction form and **(ii)** decimal form, using a negative sign if necessary.

11 Write the following in order from smallest to largest: -3, 0, $\frac{3}{10}$, 1.7, -3.4, $-\frac{3}{4}$, -0.9, 1. Use a number line, if necessary, to help you.

12 Write the following in order from smallest to largest. Use a number line, if necessary, to help you. $-2\frac{1}{5}$, 0, -1.8, $-\frac{5}{4}$, 1.1, $\frac{4}{5}$, -2, $\frac{3}{4}$

2.3

13 Use your calculator to evaluate the following. Write answers correct to three decimal places.

(a) $(78.9 \times 3.4) - (54.8 \times 7.77)$ **(b)** $(54.9 \times 5.6) - (6.2 \times 103.7)$

(c) $(-14.7 + 5.8) \times (4.3 - 8.25)$ **(d)** $(56 - 72.8) \times (-12.3 + 17.1)$

(e) $\sqrt{7} - \sqrt{3} - \sqrt{6}$ **(f)** $\sqrt{3} - \sqrt{2} - (\sqrt{5})^2$

(g) $(2.3)^4 - (5.9)^5 + (1.5)^2$ **(h)** $(\sqrt{6})^2 - (10.5)^4 - (1.005)^2$

(i) $(3.8 - 5.7)^2 - (\sqrt{8})$ **(j)** $(2\sqrt{10})^2 - (53.66)^2$

Reasoning, communicating and open-ended

14 Chelsea receives a statement from her bank every month. The statement is a list of all the transactions (deposits and withdrawals) made on the account, and the account balance (how much money is in the account after each transaction). Here is part of Chelsea's statement:

Date	Transaction	Balance
1 May		$257.89
7 May	ATM Withdrawal $180.00	$77.89
12 May	Cheque deposit $65.00	$142.89
16 May	Shopping paid for by EFTPOS $97.63	$45.26
23 May	Insurance, paid by direct debit $58.17	-$12.91
30 May	Salary deposit ________	$460.94

(a) Locate the negative number on Chelsea's statement. What is the '-' symbol used to indicate here?

(b) On 27 May, Chelsea went to an ATM and tried to withdraw $50, but the machine would not give her any money! Why not?

(c) What was the value of Chelsea's salary that was deposited on 30 May?

15 Use a number line to explain why $\frac{1}{2} > \frac{1}{4}$ but $-\frac{1}{2} < -\frac{1}{4}$.

16 Write down three decimal numbers that are greater than -2.1 but less than -1.2.

17 Write down three numbers in fraction form that are greater than $-\frac{7}{8}$ but less than 0.

18 Douglas is having trouble with negative decimals. He can add and subtract positive numbers successfully, but often gets questions with negatives wrong. Here are two questions that Douglas has done incorrectly:

$-8.3 + (-4.5) = -3.8$

$-6.7 - (-2.9) = -9.6$

(a) Explain the errors that Douglas has made.

(b) Give him some tips as to how he can do these types of questions successfully.

2.4 Estimating percentages

Need to Know

The term **per cent** means 'for every hundred' or 'out of a hundred'. A **percentage** is a value out of 100. The symbol % is used for per cent.

It's sometimes useful to estimate a percentage. A strategy for estimating a percentage of an amount is to mentally split the object up into 10 lots of 10%, as 100% represents the whole object.

Example:

The shaded area of the rectangle is about 53%.

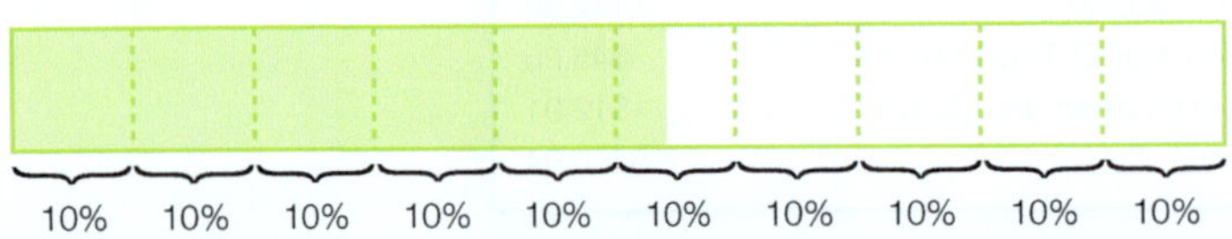

Worked Example 10

WE10

Which of the following water tanks is approximately:

(a) 50% full **(b)** 15% full **(c)** 80% full?

(i) **(ii)** **(iii)** **(iv)** **(v)**

Working

(a) 50% = half
Tank (ii) is 50% full.

(b) 15% < 25% (a quarter)
Tank (iii) is 15% full

(c) 80% > 75% (three-quarters)
Tank (v) is 80% full.

2.4 Estimating percentages

Answers page 489

Navigator

Q1, Q2, Q3, Q4, Q5, Q7, Q9, Q11	Q2, Q3, Q4, Q5, Q6, Q7, Q8, Q9, Q11	Q2, Q3, Q4, Q6, Q7, Q8, Q9, Q10, Q11

Fluency and understanding

WE10

1 Which of the following bottles of soft drink is approximately:

(a) 5% full (b) 90% full (c) 80% full (d) 20% full?

2 Which of the following fuel gauges shows a petrol tank that is approximately:

(a) 40% full (b) 15% full (c) 60% full (d) 75% full?

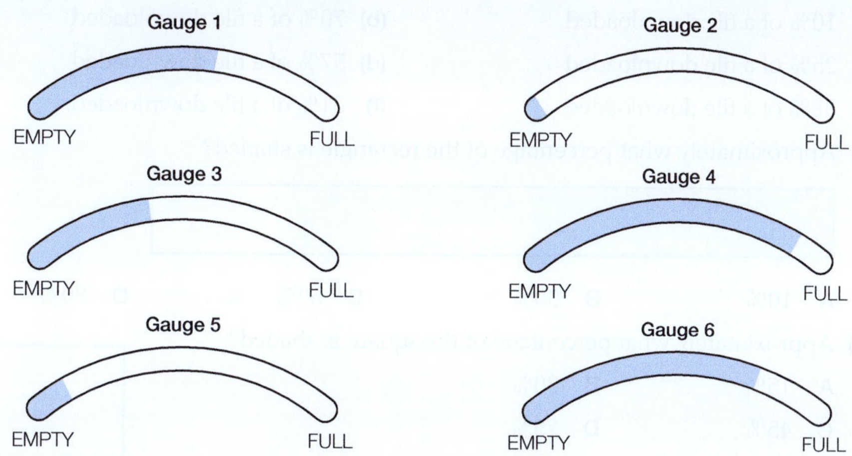

3 A computer displays the amount of a file it has downloaded by gradually filling in a bar until the file has been fully saved. Which display shows approximately:

(a) 33% saved **(b)** 70% saved **(c)** 82% saved **(d)** 22% saved?

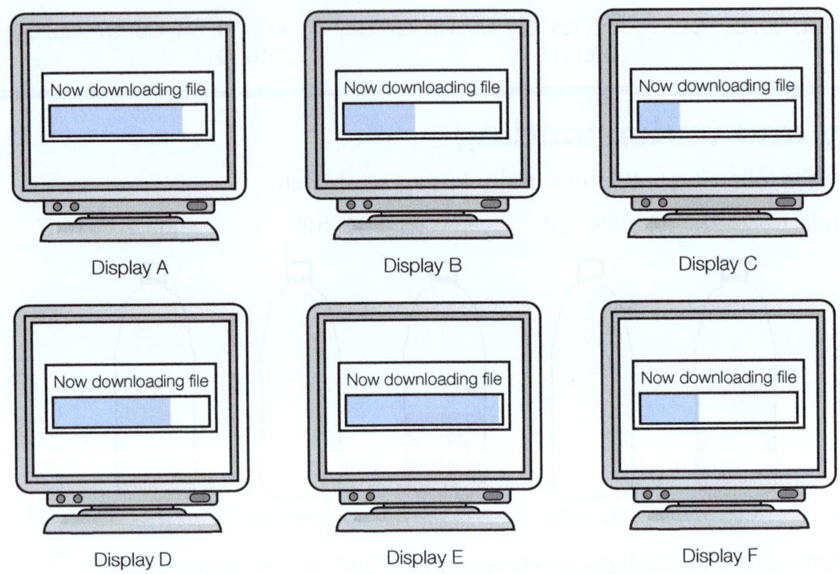

4 Draw six displays similar to those in Question **3** that show approximately:

(a) 10% of a file downloaded **(b)** 70% of a file downloaded

(c) 25% of a file downloaded **(d)** 57% of a file downloaded

(e) 88% of a file downloaded **(f)** 31% of a file downloaded.

5 **(a)** Approximately what percentage of the rectangle is shaded?

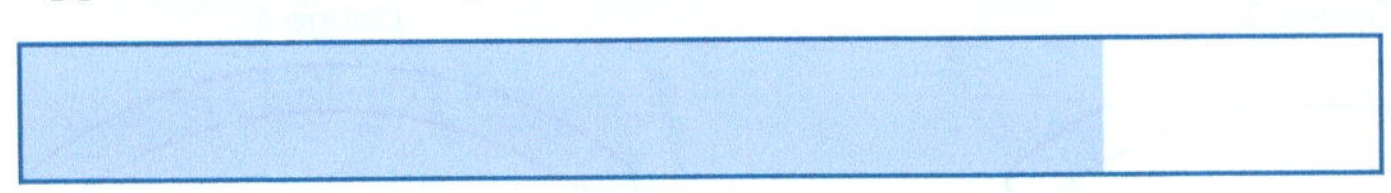

A 10% **B** 30% **C** 60% **D** 80%

(b) Approximately what percentage of the square is shaded?

A 15% **B** 30%

C 45% **D** 75%

(c) Approximately what percentage of the circle is shaded?

A 16% **B** 27%

C 55% **D** 78%

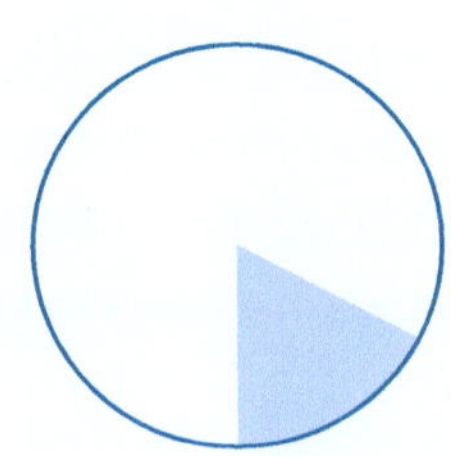

6 Estimate the percentage of lawn John has left to mow.

7 The water dispenser in Larry's office appears as shown. Estimate how full the dispenser is, using a percentage value.

Reasoning, communicating and open-ended

8 The graph shows the risk of premature death for 15-year-old smokers.

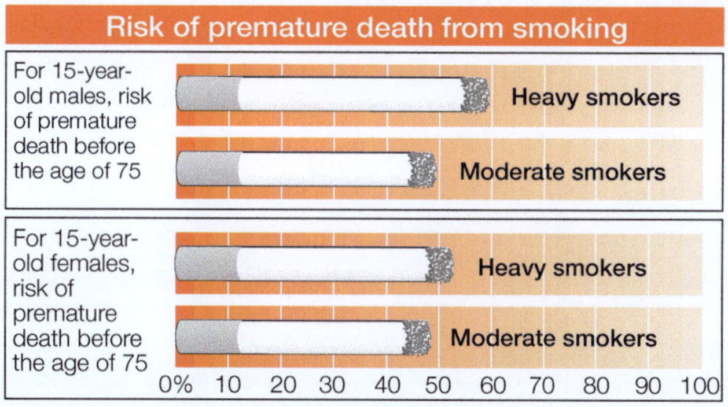

Estimate the percentage risk of premature death for:

(a) 15-year-old male heavy smokers

(b) 15-year-old male moderate smokers

(c) 15-year-old female heavy smokers

(d) 15-year-old female moderate smokers.

(e) Calculate the percentage difference between the risks for heavy and moderate smokers in males and females.

2.4

9 28% of this circle is shaded.

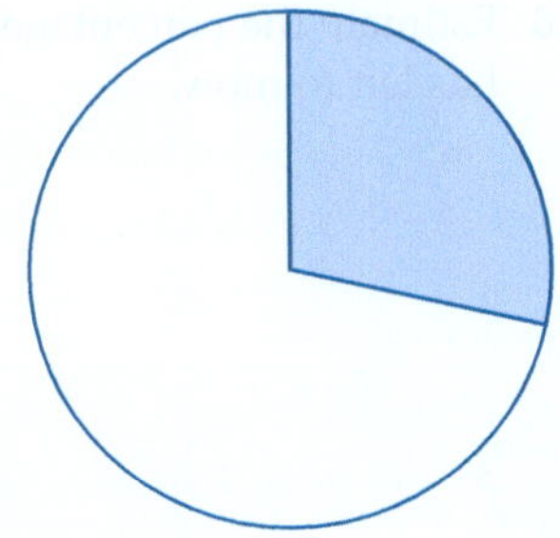

(a) Draw three similar sized circles, then shade in your estimate of:

(i) 40% of the circle

(ii) $\frac{5}{6}$ of the circle

(iii) 0.7 of the circle.

(b) In part **(a)**, did you find it easier to estimate the percentage, the decimal or the fraction amount? Why do you think this was the case?

10 **(a)** Draw at least three different shapes and shade 25% of each of them.

(b) Redraw the shapes from **(a)** and shade a different 25% of them.

(c) Redraw the shapes from **(a)** and shade 65% of them.

11 Helga has worked out the percentage of the areas shaded as follows.

(a)

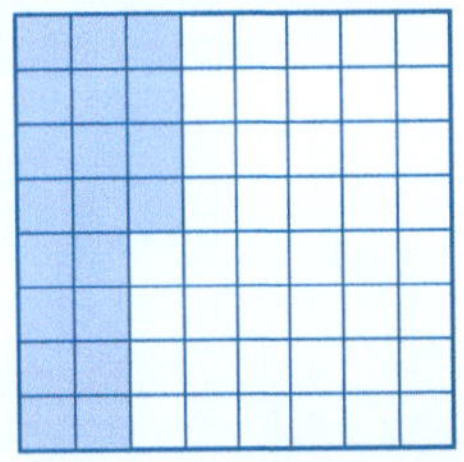

24% shaded

(b)

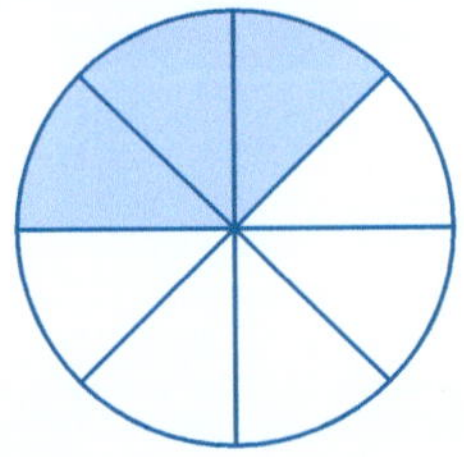

30% shaded

Explain Helga's mistake in her calculations for each shape.

Writing fractions and decimals as percentages

2.5

Need to Know

As percentages are easy to compare and are often easier than fractions for people to visualise, it is useful to be able to convert fractions and decimals into percentages. Some percentages are used so frequently that it is useful to be able to automatically recall their fraction or decimal equivalent.

Fraction	$\frac{1}{2}$	$\frac{1}{4}$	$\frac{3}{4}$	$\frac{1}{3}$	$\frac{2}{3}$	$\frac{1}{5}$	$\frac{1}{10}$	$\frac{1}{20}$	$\frac{1}{50}$	$\frac{1}{100}$
Decimal	0.5	0.25	0.75	$0.\dot{3}$	$0.\dot{6}$	0.2	0.1	0.05	0.02	0.01
Percentage	50%	25%	75%	$33.\dot{3}\%$	$66.\dot{6}\%$	20%	10%	5%	2%	1%

Converting fractions and decimals to percentages

To write a fraction or decimal as a percentage, we multiply the fraction or decimal by 100%.

With the exception of $\frac{1}{6}$ and $\frac{1}{3}$, writing the fractions in the above table as percentages is straightforward. Because the denominators are factors of 100 simply convert them to fractions out of 100, then multiply by 100%.

For example:

$$\frac{3}{4} \overset{\times 25}{=} \frac{75}{100} = 75\% \qquad \frac{1}{5} \overset{\times 20}{=} \frac{20}{100} = 20\%$$

(numerator and denominator each × 25; numerator and denominator each × 20)

However for fractions that have denominators other than 2, 5, 10, 25 or 50 (or fractions that can be simplified to these denominators), we need to use another method.

There are two methods we can use to multiply a fraction by 100% ($\frac{100}{1}$ %).

Method 1: Multiply two fractions together (a convenient 'pen and paper' method).

Method 2: Convert the fraction to a decimal then multiply by 100% (useful if you have a calculator).

Multiplying by 100% is equivalent to multiplying by 1. If we multiply a fraction or a decimal by 100%, the value of the fraction or decimal is not being changed, it is just being represented in a different way.

2.5

Worked Example 11

WE11

Write the following fractions as percentages. Write any answers that are not whole numbers in both fraction and exact decimal form.

(a) $\frac{7}{8}$

(b) $\frac{11}{15}$

Working

Method 1: Fraction multiplication

(a) $\frac{7}{8} \times 100\%$

$= \frac{7}{{}_{2}\not{8}} \times \frac{{}^{25}\not{100}}{1}\%$

$= \frac{175}{2}\%$

$= 87\frac{1}{2}\%$ or 87.5%

(b) $\frac{11}{15} \times 100\%$

$= \frac{11}{{}_{3}\not{15}} \times \frac{{}^{20}\not{100}}{1}\%$

$= \frac{220}{3}$

$= 73\frac{1}{3}\%$ or $73.\dot{3}\%$

Working

Method 2: Convert to a decimal, then multiply

(a) $\frac{7}{8} \times 100\%$

$= 7 \div 8 \times 100\%$

$= 0.875 \times 100\%$

$= 87\frac{1}{2}\%$ or 87.5%

(b) $\frac{11}{15} \times 100\%$

$= 11 \div 15 \times 100\%$

$= 0.7333... \times 100\%$

$= 73\frac{1}{3}\%$ or $73.\dot{3}\%$

Worked Example 12

WE12

Write the following as a percentage: $3\frac{1}{2}$

Working

Method 1: Fraction multiplication

$3\frac{1}{2} \times 100\%$

$= \frac{7}{{}_{1}\not{2}} \times \frac{{}^{50}\not{100}}{1}\%$

$= \frac{350}{1}\%$

$= 350\%$

Working

Method 2: Convert to a decimal, then multiply

$3\frac{1}{2} \times 100\%$

$= 3.5\% \times 100\%$

$= 350\%$

2.5

Worked Example 13

WE13

Find the percentage equivalent of each of the following.

(a) 0.8 (b) 1.65 (c) 0.032

Working

(a) $0.8 = 0.8 \times 100\% = 80\%$

(b) $1.65 = 1.65 \times 100\% = 165\%$

(c) $0.032 = 0.032 \times 100\% = 3.2\%$

2.5 Writing fractions and decimals as percentages

Answers page 490

Navigator

Q1 Columns 1–3, Q2 Columns 1 & 2, Q3 Columns 1–3, Q4, Q5, Q6, Q7, Q8, Q9, Q11, Q12, Q13, Q17, Q18	Q1 Columns 2–4, Q2 Columns 2 & 3, Q3 Columns 3 & 4, Q5, Q6, Q8, Q9, Q10, Q11, Q12, Q13, Q15, Q17, Q18	Q1 Columns 3 & 4, Q2 Columns 3 & 4, Q3 Columns 3 & 4, Q5, Q6, Q8, Q9, Q11, Q12, Q14, Q15, Q16, Q18, Q19

Equipment required: A calculator may be used for Questions **9**, **11**, **15**, **16**, **17** and **19**

Fluency and understanding

1 Write the following fractions as percentages. Write any answers that are not whole numbers in both fraction and exact decimal form. WE11

(a) $\frac{9}{100}$ (b) $\frac{1}{10}$ (c) $\frac{7}{50}$ (d) $\frac{19}{20}$

(e) $\frac{3}{5}$ (f) $\frac{1}{4}$ (g) $\frac{3}{2}$ (h) $\frac{16}{8}$

(i) $\frac{1}{3}$ (j) $\frac{2}{3}$ (k) $\frac{1}{6}$ (l) $\frac{2}{9}$

(m) $\frac{32}{40}$ (n) $\frac{9}{16}$ (o) $\frac{56}{80}$ (p) $\frac{14}{32}$

(q) $\frac{73}{80}$ (r) $\frac{13}{60}$ (s) $\frac{55}{66}$ (t) $\frac{23}{90}$

2 Write the following as percentages. Write any answers that are not whole numbers in exact decimal form. WE12

(a) $1\frac{1}{4}$ (b) $1\frac{2}{5}$ (c) $3\frac{1}{5}$ (d) $5\frac{1}{2}$

(e) $2\frac{3}{8}$ (f) $2\frac{3}{4}$ (g) $5\frac{7}{10}$ (h) $4\frac{5}{8}$

(i) $4\frac{2}{3}$ (j) $5\frac{1}{3}$ (k) $1\frac{7}{9}$ (l) $7\frac{5}{6}$

2.5

WE13

3 Find the percentage equivalent of each of the following.

(a) 0.9 **(b)** 0.4 **(c)** 0.8 **(d)** 0.6
(e) 0.17 **(f)** 0.47 **(g)** 0.82 **(h)** 0.53
(i) 0.051 **(j)** 0.438 **(k)** 0.007 **(l)** 0.342
(m) 9.2 **(n)** 5.1 **(o)** 2.02 **(p)** 9.01

4 **(a)** The percentage equivalent of 0.7 is:

A 0.07% **B** 0.70% **C** 7% **D** 70%

(b) The percentage equivalent of 3.45 is:

A 0.0345% **B** 0.345% **C** 34.5% **D** 345%

5 **(a)** Which of the following is equivalent to $\frac{9}{50}$?

A 9% **B** 18% **C** 45% **D** 90%

(b) Which of the following is equivalent to $\frac{3}{20}$?

A 3% **B** 12% **C** 15% **D** 30%

6 Copy and complete the table of commonly used fractions, decimals and percentages.

Percentage	5%			20%	25%	$33\frac{1}{3}$%	40%		60%	$66\frac{2}{3}$%		80%	100%
Fraction			$\frac{1}{8}$								$\frac{3}{4}$		
Decimal		0.1						0.5					1

7 Lou estimated that $\frac{7}{10}$ of the crowd at a rugby match were Broncos supporters. What percentage is this?

8 Due to a cyclone wiping out most of the crop, the price of bananas rose to $3\frac{3}{4}$ times the original price. What percentage of the old price was this?

9 In one term, Patrick achieved the following test results in maths.

Measurement $\frac{33}{40}$ Algebra $\frac{26}{30}$ Geometry $\frac{12}{15}$

(a) Convert Patrick's test results to percentages. Round your answers to the nearest whole percentage.

(b) In which topic did Patrick achieve his best result?

10 A teacher calculates her students' scores for a test by dividing the marks obtained by the total marks to get a decimal score. Part of her mark book appears as follows.

Find the percentage mark obtained by:

(a) Agnes **(b)** Basil **(c)** Carlotta
(d) Daphne **(e)** Elmer **(f)** Felix

Student	**Result**
Agnes	0.675
Basil	0.95
Carlotta	0.9
Daphne	0.55
Elmer	0.425
Felix	0.825

2.5

11 Rebecca needs to score at least 80% on the theory test for her Learner's Permit. She knows that the test contains 40 questions. What is the minimum number of questions Rebecca must get correct in order to receive her Learner's Permit?

12 A chemist testing a rock sample found that it contained 0.246 parts of calcium, 0.034 parts of magnesium and 0.0012 parts of potassium. Determine the percentage equivalent of each of these three elements.

13 For each of these sets of numbers, construct the percentage number line shown and indicate the position of each number with an arrow. (You may need to estimate the position of some numbers.)

(a) $25\%, \frac{2}{5}, 0.04, \frac{13}{20}$ **(b)** $3\%, 0.3, \frac{3}{4}, 34\%$

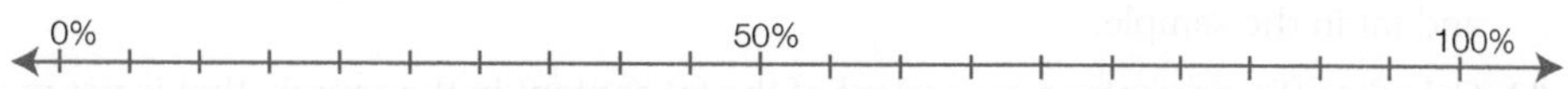

14 For each of these sets of numbers, construct the number line shown and indicate the position of each number with an arrow. (You may need to estimate the position of some numbers.)

(a) $120\%, 0.8, 52.5\%, 1.95, \frac{11}{10}$ **(b)** $115\%, 1.01, \frac{5}{3}, 270\%, \frac{14}{25}$

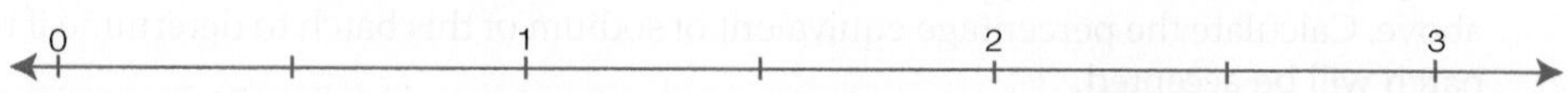

15 An advertisement for a particular brand of bread claims that it now has '20% more fibre!' The nutritional information table on the packaging shows that a 40 g serve of bread contains 3.4 g of fibre. Before the improvement was made, the same 40 g serving contained 2.8 g of fibre.

(a) Write the increase in the amount of fibre as a fraction of the original amount, in simplest form.

(b) Convert this fraction to a percentage, rounding your answer to two decimal places.

(c) Is the bread manufacturer's claim of '20% more fibre' justified? Explain your answer.

> **Hint**
> In Question **15**, you can turn decimal numbers into whole numbers by multiplying by 10.

Reasoning, communicating and open-ended

16 Imogen was comparing two toy stores to determine in which store the greater percentage of people entering the store actually bought goods. She wanted to know if the suburban store had a greater percentage of paying customers than the store in the city. She wrote the number of customers who bought goods as a fraction of the total number of customers who entered the store. The results for the suburban store over a 3-week period were $\frac{122}{200}, \frac{143}{220}$ and $\frac{275}{330}$. The results for the city store were $\frac{192}{400}, \frac{168}{320}$ and $\frac{336}{480}$.

(a) Calculate the percentage of paying customers for each week for the suburban store.

(b) Calculate the percentage of paying customers for each week for the city store.

(c) Which store had the greater percentage of paying customers, and what was the percentage difference for each week?

(d) For each store, calculate the fraction of the total number of people entering the store who bought goods over the 3-week period.

(e) Which store had the greater percentage over the 3-week period, and by what percentage?

17 A nutritionist analysing a sample of savoury biscuits obtained the following results.

Nutrient	Proportion
Protein	0.133 parts
Dietary fibre	0.03 parts
Sodium	0.0073 parts
Carbohydrate (of which 0.005 is sugar)	0.594 parts
Fat (of which 0.084 is saturated)	0.172 parts

(a) Calculate the percentage equivalent of protein, dietary fibre, sodium, carbohydrate and fat in the sample.

(b) Calculate the percentage equivalent of the fat content in the sample that is not made up of saturated fats.

(c) Calculate the percentage equivalent of the carbohydrate content in the sample that is not sugar.

(d) A batch is rejected if it is found to contain greater than 1% sodium. The next batch analysed was found to have an extra 0.0052 parts of sodium compared with the batch above. Calculate the percentage equivalent of sodium of this batch to determine if this batch will be accepted.

18 Fill in each box below with one of the digits 0–9 to make the equation correct. Try to find three different combinations.

□.3□ = □□□%

19 Write at least three different fractions that have equivalent percentage values between 68% and 75%. Only one of them can have a denominator of 100.

Half-time 2

1 Estimate the percentage of the jug that contains cordial. **Ex 2.4**

2 Express the following in decimal form, then state whether they are terminating decimals, recurring decimals, or irrational numbers. **Ex 2.2**

(a) $15 \div 6$ (b) $\frac{7}{11}$

(c) $\sqrt{54}$ (d) $\sqrt{174.24}$

(e) $83 \div 12$ (f) $\sqrt{75}$

3 Hasharan is cutting ribbon to wrap around gift boxes. She has a 4-m roll of ribbon, and needs 35 cm of ribbon per box. **Ex 2.1**

(a) How many gift boxes will Hasharan be able to wrap?

(b) What length of ribbon will she have left over?

4 Write each of the following fractions and decimals as percentages. **Ex 2.5**

(a) 4.56 (b) $\frac{3}{4}$ (c) $\frac{71}{80}$ (d) 0.432 (e) $1\frac{2}{5}$

5 Write the following recurring decimals in fraction form. **Ex 2.2**

(a) $1.\dot{3}$ (b) $3.2\dot{2}\dot{6}$ (c) $5.1\overline{902}$

6 Write the following in order from smallest to largest. **Ex 2.1–2.3**

(a) $-\frac{5}{9}, \sqrt{5}, \frac{3}{8}, -0.7, \frac{1}{2}, 1.099$ (b) $-1.23, 0.874, -\frac{2}{3}, \frac{17}{8}, \sqrt{13}$

7 Write each of the following in exact decimal form. **Ex 2.2**

(a) $\frac{2}{3}$ (b) $\frac{5}{7}$ (c) $\frac{1}{9}$ (d) $\frac{5}{11}$

8 Write the following decimals as fractions or as mixed numbers in simplest form. **Ex 2.1**

(a) 0.438 (b) 1.072 (c) 2.0506 (d) 13.7005

9 Calculate the following. **Ex 2.3**

(a) $-\frac{1}{5}+\frac{1}{2}$ (b) $-\frac{1}{4}\times\left(-\frac{6}{7}\right)$ (c) $-\frac{3}{10}-\left(-\frac{7}{12}\right)$ (d) $-\frac{5}{4}\div\frac{2}{3}$

(e) $-7.8 + (-5.49)$ (f) $0.3 - (-11.46)$ (g) -2.6×3.05 (h) $18 \div -1.2$

10 The fraction of the human body that is water varies between $\frac{11}{20}$ and $\frac{13}{20}$. Write this fraction range as a percentage range. **Ex 2.5**

2.6 Writing percentages as fractions and decimals

Need to Know

Writing a percentage as a fraction or a decimal

- Divide the value of the percentage by 100.
- For a fraction, write the division by 100 in fraction form, then simplify if possible.
- For a decimal, divide by 100. Show this by moving the decimal point two places to the left.

It is important to remember the difference between $\frac{1}{2}$ % and $\frac{1}{2}$. $\frac{1}{2}$% means 'one half of 1%'. This is one half of 0.01, or 0.005. $\frac{1}{2}$ means 'one half of the whole', which is 50%.

Worked Example 14

WE 14

Write the following percentages as fractions in simplest form.

(a) 16% **(b)** 120%

Working

(a) 16%

$= 16 \div 100$

$= \frac{\overset{4}{\cancel{16}}}{\underset{25}{\cancel{100}}}$

$= \frac{4}{25}$

(b) 120%

$= 120 \div 100$

$= \frac{\overset{6}{\cancel{120}}}{\underset{5}{\cancel{100}}}$

$= \frac{6}{5}$ or $1\frac{1}{5}$

2.6

Worked Example 15

WE15

Write the following 'fractional percentages' as fractions in simplest form.

(a) $\frac{1}{4}\%$ (b) $6\frac{2}{3}\%$

Working

(a) $\frac{1}{4}\% = \frac{1}{4} \div 100$

$= \frac{1}{4} \div \frac{100}{1}$

$= \frac{1}{4} \times \frac{1}{100}$

$= \frac{1}{400}$

(b) $6\frac{2}{3}\% = 6\frac{2}{3} \div 100$

$= \frac{20}{3} \div \frac{100}{1}$

$= \frac{20}{3} \times \frac{1}{100}$

$= \frac{20}{300}$

$= \frac{1}{15}$

Worked Example 16

WE16

Write the following percentages as fractions in simplest form.

(a) 48.2% (b) 3.75%

Working

(a) $48.2\% = 48.2 \div 100$

$= \frac{48.2}{100}$

$= \frac{482}{1000}$

$= \frac{241}{500}$

(b) $3.75\% = 3.75 \div 100$

$= \frac{3.75}{100}$

$= \frac{375}{10000}$

$= \frac{75}{2000}$

$= \frac{3}{80}$

Worked Example 17

WE17

Write the following percentages as decimals.

(a) 23% (b) 171% (c) 3.49%

Working

(a) $23\% = 23 \div 100$

$= 0.23$

(b) $171\% = 171 \div 100$

$= 1.71$

(c) $3.49\% = 3.49 \div 100$

$= 0.0349$

2.6

Worked Example 18

WE18

Write the following 'fractional percentages' as decimals.

(a) $\frac{1}{2}\%$ (b) $3\frac{1}{5}\%$

Working

(a) $\frac{1}{2}\% = 0.5\%$
$= 0.5 \div 100$
$= 0.005$

(b) $3\frac{1}{5}\% = 3.2\%$
$= 3.2 \div 100$
$= 0.032$

Answers page 491

2.6 Writing percentages as fractions and decimals

Navigator

Q1, Q2 Columns 1 & 2, Q3 Columns 1 & 2, Q4 Columns 1–3, Q5 Columns 1 & 2, Q8, Q9, Q10, Q11, Q12, Q14 (a) & (b), Q15, Q16, Q19	Q1 Columns 2–4, Q2 Columns 1–3, Q3 Columns 1–3, Q4 Columns 3 & 4, Q5 Columns 1–3, Q6, Q7, Q10, Q12, Q13, Q14, Q15, Q16, Q17, Q18, Q19	Q1 Columns 3 & 4, Q2 Columns 2–4, Q3 Columns 2–4, Q4 Columns 3 & 4, Q5 Columns 2–4, Q7, Q10, Q12, Q13, Q14 (c) & (d) Q15, Q16, Q17, Q18, Q19

Equipment required: A calculator may be used for Questions **5** and **18(a)**

Fluency and understanding

WE14

1 Write the following percentages as fractions in simplest form.

(a) 17% **(b)** 48% **(c)** 9% **(d)** 65%
(e) 58% **(f)** 76% **(g)** 117% **(h)** 129%
(i) 240% **(j)** 315% **(k)** 138% **(l)** 360%

WE15

2 Write the following 'fractional percentages' as fractions in simplest form.

(a) $\frac{1}{2}\%$ **(b)** $\frac{1}{5}\%$ **(c)** $\frac{3}{7}\%$ **(d)** $\frac{3}{8}\%$
(e) $\frac{1}{10}\%$ **(f)** $\frac{2}{3}\%$ **(g)** $\frac{5}{6}\%$ **(h)** $\frac{2}{9}\%$
(i) $2\frac{1}{4}\%$ **(j)** $6\frac{1}{2}\%$ **(k)** $5\frac{3}{5}\%$ **(l)** $8\frac{3}{4}\%$

WE16

3 Write the following percentages as fractions in simplest form.

(a) 8.3% **(b)** 12.29% **(c)** 50.5% **(d)** 62.5%
(e) 2.1% **(f)** 3.4% **(g)** 0.55% **(h)** 0.4%
(i) 10.25% **(j)** 87.5% **(k)** 150.2% **(l)** 420.5%

2.6

4 Write the following percentages as decimals. WE17

(a) 80% (b) 27% (c) 40% (d) 15%
(e) 66% (f) 92% (g) 3% (h) 5%
(i) 110% (j) 265% (k) 700% (l) 304%
(m) 0.5% (n) 7.2% (o) 40.5% (p) 6.25%

5 Write the following 'fractional percentages' as decimals. WE18

(a) $\frac{1}{4}\%$ (b) $\frac{1}{8}\%$ (c) $\frac{7}{10}\%$ (d) $\frac{2}{5}\%$
(e) $2\frac{1}{2}\%$ (f) $10\frac{3}{4}\%$ (g) $37\frac{1}{5}\%$ (h) $18\frac{4}{5}\%$
(i) $25\frac{1}{3}\%$ (j) $5\frac{2}{3}\%$ (k) $2\frac{5}{9}\%$ (l) $23\frac{5}{6}\%$

6 2.8% in its simplest fraction form is:

A $\frac{14}{500}$ B $\frac{28}{1000}$ C $\frac{7}{250}$ D $2\frac{4}{5}$

7 $\frac{3}{4}\%$ expressed as a fraction is:

A $\frac{3}{400}$ B $\frac{34}{100}$ C $\frac{3}{4}$ D $3\frac{4}{100}$

8 The decimal equivalent of 500% is:

A 50 000 B 5 C 0.5 D 0.05

9 A TV advertisement stated that 86% of dentists recommend using 'Squeaky Clean' toothpaste. What fraction of dentists is this?

10 Gina has a $9\frac{1}{2}\%$ pay rise. What fraction of her salary does this represent?

11 Monica travelled 45% of the journey across Europe from Paris to Istanbul by train. What is the decimal equivalent of the percentage travelled by train?

12 Jacob received 92.5% on his Australian History test. What is the decimal equivalent of Jacob's result?

13 In rugby league, bruisings and strains are 30% of injuries, whereas more serious injuries such as fractures are 16% and concussions are 9%.

(a) Write each of these percentages as fractions.

(b) What fraction of the total number of injuries do bruisings, strains, fractures and concussions make up altogether?

(c) What fraction of the total number of injuries are due to other causes?

14 Place the following in ascending order.

(a) 20%, $\frac{1}{4}$, 0.02, $\frac{2}{5}$ (b) 0.45, $\frac{4}{5}$, 4.5%, $\frac{5}{4}$

(c) $\frac{3}{10}$, 0.03, $\frac{1}{3}$, 33% (d) 0.72, $7\frac{1}{2}\%$, 7.2%, $\frac{7}{2}$

15 Employment figures show that 28% of those employed are tradespeople. What fraction of the employed workforce is tradespeople?

2.6

Reasoning, communicating and open-ended

16 One-quarter of the 200 Year 8 students surveyed at St Gabriel's College said that they were driven to school and 30% said that they arrived by bus. What fraction of students are not driven to school and do not take the bus?

17 The Burandoc Water Authority is trying to estimate the volume of water available in the Burandoc Shire Reservoir. At present, it is at 85% capacity and it is estimated that if there is no significant rainfall over the next 6 months the volume would be reduced to 48% of capacity. However, if the Shire receives a normal amount of rainfall during this period the volume would only decrease by 17% of capacity.

(a) What fraction of the reservoir currently contains water?

(b) What fraction of the reservoir will contain water if there is no significant rainfall during the next 6 months?

(c) By what fraction would the water volume be reduced if there is no significant rainfall during the next 6 months?

(d) What fraction of the reservoir would remain unfilled if there is a normal amount of rainfall during this 6 months?

18 (a) Write at least three fractions with denominators less than 100 that have a percentage equivalent between 50% and 60%.

(b) Write at least three decimals that have a percentage equivalent between 25% and 26%. Each of the decimals given should have a different number of decimal places.

19 Valentino writes: $3\frac{1}{4} = 3.25\%$

(a) Explain to Valentino the difference between $3\frac{1}{4}$ and 3.25%.

(b) Correctly convert $3\frac{1}{4}$ into its percentage form.

Writing one amount as a percentage of another

2.7

Need to Know

Writing one amount as a percentage of another

- Make sure both amounts are the same type, or measured in the same units; convert units if necessary.
- Write a fraction with the 'part amount' as the numerator and the 'whole amount' as the denominator.
- Convert this fraction into a percentage by multiplying by 100%.

Not all calculations of percentage give whole number answers. Remainders can be written in fraction form, exact decimal form or as rounded decimals, depending on the situation. The S⇔D or $a\frac{b}{c}$ button on your calculator is useful for converting between answers in decimal form and fraction form.

Worked Example 19

WE19

Express the first amount as a percentage of the second. Give your answers in **(i)** fractional and **(ii)** exact decimal form.

(a) 23, 40 **(b)** 35, 30

Working

Method 1: Multiply the fraction by 100%

(a) $\frac{23}{\cancel{40}_{2}} \times \frac{\cancel{100}^{5}}{1}\%$

$= \frac{115}{2}\%$

$=$ **(i)** $57\frac{1}{2}\%$ or **(ii)** 57.5%

(b) $\frac{\cancel{35}^{7}}{\cancel{30}_{6}} \times \frac{100}{1}\%$

$= \frac{7}{\cancel{6}_{3}} \times \frac{\cancel{100}^{50}}{1}\%$

$= \frac{350}{3}\%$

$=$ **(i)** $116\frac{2}{3}\%$ or **(ii)** $116.\dot{6}\%$

Working

Method 2: Convert to a decimal, then multiply by 100%

(a) $\frac{23}{40}$

$= 0.575$

$= 0.575 \times 100\%$

$= 57.5\%$ or $57\frac{1}{2}\%$

$=$ **(i)** $57\frac{1}{2}\%$ or **(ii)** 57.5%

(b) $\frac{35}{30}$

$= 1.1666...$

$= 1.1666... \times 100\%$

$= 116.\dot{6}\%$ or $116\frac{2}{3}\%$

$=$ **(i)** $116\frac{2}{3}\%$ or **(ii)** $116.\dot{6}\%$

2.7

Worked Example 20

WE20

Express the first amount as a percentage of the second: 40 cents, $5.

Working

Method 1: Fraction multiplication

40 cents, 500 cents

$\frac{40}{500} \times \frac{100}{1}\%$

$= \frac{40}{5} \times \frac{1}{1}\%$

$= \frac{8}{1} \times \frac{1}{1}\%$

$= 8\%$

40 cents is 8% of $5.

Working

Method 2: Convert to a decimal, then multiply

40 cents, 500 cents

$\frac{40}{500}$

$= 0.08$

$= 0.08 \times 100\%$

$= 8\%$

40 cents is 8% of $5.

Answers page 492

2.7 Writing one amount as a percentage of another

Navigator

Q1 Columns 1–3, Q2 Column 1, Q3, Q4, Q8, Q9, Q11, Q12, Q13, Q16, Q18, Q19	Q1 Columns 2–4, Q2 Column 2, Q3, Q4, Q6, Q7, Q9, Q10, Q12, Q15, Q16, Q17, Q18, Q19	Q1 Columns 3 & 4, Q2 Column 2, Q3, Q5, Q7, Q8, Q9, Q14, Q15, Q16, Q17, Q18, Q19, Q20

Equipment required: Calculator

Fluency and understanding

WE19

1 Express the first amount as a percentage of the second. Give your answers in **(i)** fractional and **(ii)** exact decimal form.

(a) 13, 50 **(b)** 16, 20 **(c)** 27, 45 **(d)** 42, 48
(e) 35, 80 **(f)** 15, 18 **(g)** 54, 96 **(h)** 15, 27
(i) 86, 54 **(j)** 90, 70 **(k)** 72, 66 **(l)** 26, 22

WE20

2 Express the first amount as a percentage of the second. Round your answer to two decimal places where necessary.

(a) 25 cents, $4 **(b)** 40 m, 2 km
(c) 7 mm, 2 cm **(d)** 30 seconds, 4 minutes
(e) 750 g, 2.5 kg **(f)** 5 days, 4 weeks
(g) 80 kg, 1 tonne **(h)** 85 mL, 2 L
(i) 600 mm, 7 m **(j)** 50 minutes, 40 hours

Hint

Remember that units have to be the same, so convert both values to the smaller unit first.

3 (a) Which of the following shows how to calculate 17 as a percentage of 30?

A $\frac{17}{30} \times \frac{100}{1}\%$ B $\frac{30}{17} \times \frac{100}{1}\%$ C $\frac{17}{100} \times \frac{30}{1}\%$ D $\frac{30}{100} \times \frac{17}{1}\%$

(b) Which of the following is 40 cm as a percentage of 2 m?

A 2% B 5% C 20% D 500%

4 Craig received 38 marks out of 40 for a maths test. Find his result as a percentage.

5 In an election for house captain, Jessica received 324 votes. If 540 students voted, what percentage of them voted for Jessica?

6 Karen planted 30 small trees in her new garden, but only 19 survived after a year. What percentage of the new trees survived? (Answer in exact decimal form.)

7 Stuart improved his best time for a 200 m sprint by 2 seconds during a training session. His previous best time was 25 seconds. Write Stuart's improvement as a percentage of his previous best time.

8 Hamish improved his shot-put throw from 20 m to 23 m. By what percentage of his 20 m throw did he increase his distance?

9 Lara has a part-time job that pays $160 per week. She receives a pay rise of $24 per week. Express Lara's pay rise amount as a percentage of her original pay.

10 During one particular season, Cuong records 18 wins from 27 sets of tennis played. What percentage of sets played did Cuong win that season? (Answer in fractional percentage form.)

11 Harry scored 5 of his team's 16 goals during a football match. What percentage of his team's goals did he score? (Answer in fractional percentage form.)

12 Adila spends $90 each week on food. If her total income is $400 per week, what percentage of it is spent on food? (Answer in fractional percentage form.)

13 According to *Guinness World Records*, Antonio Gomesdos Santos stood motionless (apart from involuntary blinking) for approximately 15 hours on 30 July 1988. What percentage of a day was this? (Answer in decimal form.)

14 According to *Guinness World Records*, the tallest man in reliably recorded history was Robert Wadlow of Illinois, USA, whose height was measured to be 272 cm. If the average male height today is about 1.8 m, express Robert Wadlow's measured height as a percentage of the average male height. (Answer in exact decimal form.)

15 A 30 gram serving of a breakfast cereal contains 2.5 grams of dietary fibre. What percentage of the serving of cereal is this? (Answer in fractional percentage form.)

16 A sofa is advertised in a furniture sale catalogue at the discounted price of $699, a saving of $350. Find the saving as a percentage of the original price of the sofa. Round your answer to the nearest whole number.

Reasoning, communicating and open-ended

17 In their basketball game on Friday night, Nathan scored 12 goals from 16 shots, and his team mate Dayo scored 14 goals from 18 shots.

- **(a)** Estimate which boy is the more accurate scorer, then use percentages to check your estimate (the more accurate scorer will have turned a greater percentage of their shots into goals). Round your answers to one decimal place if necessary.
- **(b)** In their game the following week, Nathan scored 13 goals from 18 shots, and Dayo scored 11 goals from 14 shots. For each boy, add the number of goals and the number of shots to those of the previous week. Use your total to calculate a new, two-game percentage for Nathan and for Dayo. (Again, round your answers to one decimal place if necessary.)
- **(c)** Based on your answer to **(b)**, which boy improved his scoring percentage from one game to the next?

18 Round your answers to the following to one decimal place.

- **(a)** Jaime estimates that she spends 2 hours a day online. What percentage of a day is this?
- **(b)** Jaime also estimates that she spends $\frac{1}{3}$ of a day sleeping. What percentage of her waking hours does she spend online?
- **(c)** Jaime does not include the 7 hours that she spends at school as part of her 'online time'. Recalculate her 'online' time as a percentage of her day, excluding school time and sleeping time.
- **(d)** Describe what happens to the value of the percentage as you move from **(a)** to **(b)** to **(c)**. Explain why this happens.

19 Write three different amounts of money that when expressed as a percentage of $4 are greater than 32% and less than 39%.

20 Bella has calculated that she is 89% of her dad's height (to the nearest whole percentage). Give three possible pairs of heights for Bella and her dad, if they are both between 1 m and 2 m tall.

Problem solving

League tables

In many sports, percentage is used to separate teams that have the same number of wins and losses. It is calculated by writing the points the team has scored as a fraction of the points scored against them, then multiplying by 100.

$$\% = \frac{\text{points for}}{\text{points against}} \times \frac{100}{1}$$

Halfway through round 4 of the competition, the Cabramatta Comets are third on the ladder, whereas the Merrylands Mozzies are fourth.

Team	Matches played	Points for	Points against	%
Comets	4	192	147	130.6
Mozzies	3	104	81	128.4

The Comets have already played their match for the round, whereas the Mozzies are yet to play. What would the final scores of the Mozzies match need to be if they are to overtake the Comets' percentage?

Strategy options

- Guess and check.
- Solve a simpler problem.

Finding a percentage of an amount

2.8

Need to Know

Finding a percentage of an amount

- Replace the 'of' in the expression with '×'.
- Convert the percentage to a fraction or a decimal.
- Perform the multiplication and simplify your answer.

If the percentage you are finding is less than 100%, then your answer will always be smaller than the original amount. It is important to check the reasonableness of your answer.

Worked Example 21

WE21

Find the following, rounding answers to two decimal places where necessary.

(a) 12% of $350 **(b)** 5.2% of 30 kg

Working

Method 1: Fraction multiplication

(a) 12% of $350

$= \frac{12}{{}_2\cancel{100}} \times \frac{{}^7\cancel{350}}{1}$

$= \frac{{}^6\cancel{12}}{{}_1\cancel{2}} \times \frac{7}{1}$

$= 6 \times 7$

$= \$42$

(b) 5.2% of 30 kg

$= 5.2\% \times 30$

$= \frac{{}^{13}\cancel{52}}{{}_{25}\cancel{1000}} \times \frac{\cancel{30}}{1}$

$= \frac{13}{25} \times 3$

$= \frac{39}{25}$

$= 1.56$ kg

Working

Method 2: Decimal multiplication

(a) 12% of $350

$= 12\% \times 350$

$= 0.12 \times 350$

$= \$42$

(b) 5.2% of 30 kg

$= 5.2\% \times 30$

$= 0.052 \times 30$

$= 1.56$ kg

2.8

Answers page 492

2.8 Finding a percentage of an amount

Navigator

Q1 Columns 1 & 2, Q3, Q5, Q6, Q7, Q9, Q11(a), Q12, Q15, Q19	Q1 Columns 2 & 3, Q2, Q3, Q4, Q5, Q6, Q7, Q8, Q9, Q10, Q11, Q12, Q14, Q16, Q18, Q19	Q1 Column 3, Q2, Q3, Q4, Q6, Q7, Q8, Q9, Q10, Q11, Q13, Q14, Q16, Q17, Q18, Q19

Equipment required: A calculator may be used for Questions **1** and **8–18**

Fluency and understanding

WE21

1 Find the following, rounding answers to two decimal places where necessary.

(a) 6% of \$200 **(b)** 20% of \$150 **(c)** 64% of \$50
(d) 80% of \$25.50 **(e)** 75% of \$279.95 **(f)** 120% of \$35.74
(g) 150% of 30 kg **(h)** 160% of 25 kg **(i)** 90% of 4 kg
(j) 23.2% of 40 min **(k)** 42.5% of 70 min **(l)** 15.7% of 60 min
(m) 3.4% of 50 L **(n)** 9.3% of 500 L **(o)** 0.6% of 3000 L

2 **(a)** Which of the following shows how to calculate 33% of 70?

A $\frac{33}{1} \times \frac{100}{1} \times \frac{70}{1}$ **B** $\frac{33}{100} \times \frac{70}{1}$ **C** $\frac{33}{70} \times \frac{100}{1}$ **D** $\frac{70}{30} \times \frac{100}{1}$

(b) Which of the following shows how to find $7\frac{1}{2}$% of 600?

A $\frac{7}{200} \times \frac{600}{1}$ **B** $\frac{3}{40} \times \frac{600}{1}$ **C** $\frac{15}{2} \times \frac{600}{1}$ **D** 7.5×600

3 **(a)** Which of the following shows how to calculate 3.5% of 15?

A 3.5×15 **B** 35×15 **C** 0.035×15 **D** 0.035×0.15

(b) Which of the following shows the calculation for finding 19% of 6000?

A 19×6000 **B** 19×60 **C** 0.19×6000 **D** 0.19×6

4 Nick got 85% of the questions right in a test containing 80 questions. How many questions did Nick answer correctly?

5 A 750 gram packet of breakfast cereal is only 20% full. How many grams of cereal are in the packet?

6 About 65% of the mass of an adult human is water. Ollie weighs 64 kg. How much of Ollie's mass is water, to the nearest kilogram?

Hint
4 in every 5 means $\frac{4}{5}$.

7 A survey of 5500 households reveals that 4 in every 5 regularly sort their rubbish for recycling.

(a) Write this result as a percentage.

(b) Calculate the number of households that sort their rubbish.

8 The Cancer Council estimates that 2 in 7 cancer deaths are related to tobacco (smoking).

(a) Write this figure as a percentage, rounded to one decimal place.

(b) Of 412 000 cancer deaths, how many will be related to tobacco?

9 A geologist finds a rock of mass 600 g containing a substantial amount of gold. She has the rock analysed and finds that the rock is 19.4% gold.

(a) How many grams of gold are in the rock?

(b) If the current price of gold is $45 per gram, how much is her find worth?

10 Forty-five percent of the residents in a particular suburb read their local newspaper. If the suburb contains 23 000 people:

(a) how many people read the local paper

(b) how many people don't read the local paper?

11 (a) A furniture store advertises a 40% reduction on the price of its beds.

(i) Calculate the dollar value of the percentage reduction on a bed priced at $1800.

(ii) Calculate the new, reduced price of the bed.

(b) The same store advertises a $33\frac{1}{3}$ % reduction on the price of sofas.

(i) Calculate the dollar value of the percentage reduction on a sofa priced at $450.

(ii) Calculate the new, reduced price of the sofa.

12 The population of a small town increases by 6.5% during the summer holidays. If the town's population is normally 1600, by how many does it increase during the summer holidays?

13 The Republic of Banano increases its banana crop by 4.4% one year. If the previous year's crop was 4200 tonnes, how many tonnes were produced in the new crop? Answer to the nearest tonne.

14 A meeting of 700 residents votes on a proposed road closure. If 2.1% of the residents voted for the closure:

(a) how many residents voted for the closure

(b) what percentage did not vote for the closure

(c) how many residents did not vote for the closure?

15 A 250 mL bottle of fruit juice drink contains 25% real fruit juice. The rest of the juice is water and sugar.

(a) What is the volume of real fruit juice in the bottle?

(b) What is the volume made up by water and sugar?

Reasoning, communicating and open-ended

16 Aaron is studying dance. Sixty per cent of his overall marks for the subject are allocated to the practical exam, and the remaining 40% is allocated to the theory exam. Aaron scored 84% in his practical exam and 77% in the theory exam. Calculate Aaron's overall percentage for the subject, correct to one decimal place.

17 A chemical company is planning to increase its production of photographic solution by 20%. At present, it produces 11 200 litres of solution per week. The Environmental Protection Agency is concerned that this increase in production will increase the company's waste water output to greater than the $1\frac{1}{2}$% of the volume of the total production allowable by law. At present, waste water makes up only $\frac{3}{5}$% of the volume of the total production. The company claims that with new production procedures the total waste water produced will amount to only 168 litres per week.

(a) (i) Calculate the amount of photographic solution represented by the 20% increase in production.

(ii) Calculate the total amount of solution that will be produced after the increase.

(b) Calculate the amount of waste water that is produced at present.

(c) Calculate the maximum amount of waste water that would be allowed after production is increased.

(d) What percentage of the total production does the company claim the waste water will amount to?

(e) Is the chemical company's planned increase in its waste water production illegal?

18 There are 60 books on Naomi's bookshelf.

(a) Naomi says that 21% of her books are recipe books. Explain why this is not an accurate statement.

(b) Find at least three percentage values that are possible for the percentage of recipe books.

19 Two Year 8 classes each have 25 students. In 8A, 20% of the students play basketball. In 8B, 20 of the students play basketball.

(a) Which of the two classes has more basketball players?

(b) Explain how '20% of the students' is different to '20 students'.

Problem solving

Cordial contents

Three jugs of cordial have volumes of 500 mL, 1250 mL and 1800 mL and they are each filled to capacity. The contents of the three jugs are combined together into another large jug. After this is completed the percentage of cordial in the large jug is 9.507%. Tilly knows that one of the original jugs contained 5% cordial, the second contained 7% cordial and the third contained 12.5% cordial. Work out which jug contained which percentage of cordial prior to mixing.

Strategy options

- Solve a simpler problem.
- Break problem into manageable parts.

Increasing or decreasing by a given percentage

2.9

Need to Know

Calculating a percentage increase or decrease

A percentage can be used to represent a change in a quantity. The result of a percentage increase or decrease can be calculated by using the fact that the original amount represents 100%, which has a decimal equivalent of 1.0.

To increase or decrease an amount by a given percentage:

- add the percentage to, or subtract it from, 100%, then write it as a decimal scale factor
- multiply this decimal by the amount to be increased or decreased.

Expressing an increase or a decrease as a percentage

Often an increase or a decrease is described as the percentage change of an original amount.

To calculate this, express the amount of increase or decrease as a fraction of the original amount, and convert it to a percentage.

Worked Example 22

WE 22

Increase or decrease the following amounts by the given percentages. Give answers in decimal form, rounding to two decimal places where necessary.

(a) Increase \$2700 by 22%.

(b) Decrease 163 kg by 45%.

Working

(a) $100\% + 22\%$
$= 122\%$
$= 1.22$

1.22×2700
$= \$3294$

(b) $100\% - 45\%$
$= 55\%$
$= 0.55$

0.55×163
$= 89.65$ kg

Worked Example 23

WE23

Tori earns $950 per week as an electrician. She receives a pay increase of $32 per week. Write Tori's pay increase as a percentage of her original pay, correct to one decimal place.

Working

$\frac{32}{950} \times 100\%$

$= 0.03368\ldots \times 100\%$

$= 3.4\%$ (1 d.p.)

Answers page 493

2.9 Increasing or decreasing by a given percentage

Navigator

Q1 (a) & (b), Q2, Q3, Q4, Q5, Q6, Q8, Q10, Q11, Q13, Q15, Q16	Q1 (b) & (c), Q2, Q3, Q5, Q6, Q7, Q8, Q9, Q10, Q11, Q13, Q15, Q16	Q1 (c) & (d), Q2, Q3, Q6, Q7, Q8, Q10, Q11, Q12, Q13, Q14, Q15, Q16

Equipment required: Calculator

Fluency and understanding

WE22

1 Increase or decrease the following amounts by the given percentages. Give answers in decimal form, rounding to two decimal places where necessary.

(a) Increase 1500 by: **(i)** 15% **(ii)** 25% **(iii)** 45% **(iv)** 120%

(b) Decrease 2200 by: **(i)** 10% **(ii)** 40% **(iii)** 66% **(iv)** 95%

(c) Increase $83.75 by: **(i)** 23% **(ii)** 340% **(iii)** 3.5% **(iv)** $3\frac{2}{3}\%$

(d) Decrease 5.2 km by: **(i)** 6% **(ii)** 74% **(iii)** 2.8% **(iv)** $1\frac{3}{4}\%$

WE23

2 **(a)** Georgia earns $1253 per week working as a designer. She receives a pay increase of $49 per week. Write Georgia's pay increase as a percentage of her original pay, correct to one decimal place.

(b) Jayden earns $67 982 per year working as a lawyer. He receives a pay increase of $2850 per year. Write Jayden's pay increase as a percentage of his original pay, correct to one decimal place.

(c) This month, the Ling family have reduced their electricity bill by $15.76 compared to last month. If last month's bill was $109.43, calculate the reduction in this month's bill as a percentage of the previous bill, correct to one decimal place.

(d) This month, the Edwards family have used 13 kL (kilolitres) less water than last month. If they used 190 kL last month, calculate the reduction in this month's water usage as a percentage of last month's usage, correct to one decimal place.

3 Which of the following calculations decreases \$129.50 by 40%?

A 129.5×0.4 **B** 1.295×40 **C** 0.6×129.5 **D** 1.4×129.5

4 Summer's monthly rent is set to increase by 15%. If she currently pays \$750 a month, what will she pay after the increase?

5 Last year, 87 453 people attended the Royal Show. This year, only 79 014 people attended. Write the drop in attendance as a percentage decrease.

6 Yang earns \$834.72 a week working on a building site. His pay increases to \$886.70 a week. Write the amount of Yang's pay rise as a percentage increase of his original pay, correct to one decimal place.

7 Sam buys a new sports car for \$45 000. As soon as he drives it out of the showroom its value decreases by 12%.

(a) Find the value of the car after it leaves the showroom.

After just 3 years, the value of the car will have decreased by 35% of its original value.

(b) What will the car be worth in 3 years' time?

(c) Write the 3-year decrease in value as an average yearly loss, in dollars per year.

8 Lizzie is a swimmer whose personal best time for her favourite race is 1 min 8.23 s.

(a) Write Lizzie's time in seconds.

(b) Lizzie is aiming to decrease her time by 5%. Use your answer to **(a)** to calculate this decrease as a number of seconds, correct to the nearest hundredth of a second.

(c) At the state titles, Lizzie lowered her time by 2.65 seconds. Write this improvement as a percentage of her previous best time, to the nearest whole number.

9 **(a)** Margaret purchased a three-bedroom house for \$385 000. A year later it was reported that house prices had increased by 9%. What is Margaret's house worth now?

(b) The following year, it was reported that house prices had risen by 7.5% compared to the previous year. What is Margaret's house worth now?

(c) Margaret later sold her house for \$469 000. What percentage increase on the original price is this, to the nearest whole number?

Reasoning, communicating and open-ended

10 At the end of the first week of trading on the share market, shares in Deepak's company had increased by 12%. The shares were worth \$2.50 each at the beginning of the week.

(a) What was the value of a share at the end of the first week?

(b) By the end of the second week, the share price had decreased by 7% compared to its price at the start of that week. Calculate the value of one share at the end of the second week.

(c) Calculate the overall increase or decrease at the end of the 2 weeks as a percentage of the original price of \$2.50 per share.

11 The population of the small town of Greenhills decreased in one year by 4%. The population at the beginning of the year was 17 000.

(a) Calculate the population at the end of the year.

(b) If the population continues to drop by 4% for the next 2 years, calculate the population:

(i) at the end of the second year **(ii)** at the end of the third year.

(c) Do the three 4% decreases in population give the same result as one 12% decrease? Explain why or why not.

2.9

12 Pheap improves his maths test result by 20%.

- **(a)** If his first result was 27 out of 60 marks, what was his improved result, if the second test was also out of 60?
- **(b)** If Pheap continues to improve by this percentage with each test and all tests are out of 60 marks, how many more tests must he take to achieve over 90%?

13 Melissa is 165 cm tall, and her younger sister Laura is 150 cm tall.

- **(a)** Calculate Melissa's height as a percentage increase of Laura's height, to the nearest per cent.
- **(b)** Calculate Laura's height as a percentage decrease of Melissa's height, to the nearest per cent.
- **(c)** Explain how the same height difference of 15 cm can produce two different answers for **(a)** and **(b)**.

14 The amount of metal ore produced on a remote island mine is 6900 tonnes per year.

- **(a)** Calculate the reduced amount of ore mined if production drops by 15%.
- **(b)** If production continues to drop by 15% each year, how many more years will it be before the amount mined falls below 4000 tonnes?
- **(c)** At this rate, will the mine ever stop producing ore? Explain.

15
- **(a)** Calculate the number of girls in your class as a percentage of the whole class.
- **(b)** Calculate the increase in this percentage if one extra girl joins the class.
- **(c)** Repeat **(a)** for the boys in the class.
- **(d)** Repeat **(b)** for the boys in the class.
- **(e)** Were the percentage increases you calculated in **(b)** and **(d)** the same? Explain why or why not.

16 How many centimetres taller would you be if your height increased by 8%? (Answer to the nearest cm.)

Financial applications of percentages 2.10

Need to Know

Businesses selling goods are called retailers. People buying these goods are called consumers. Retailers either manufacture their products or purchase them from suppliers. The cost of buying or manufacturing the goods (including things such as the cost of materials and transporting the goods) is called the **cost price (CP).**

In order to make a **profit,** the retailer needs to sell their product at a higher price than the cost price. This higher price is called the **selling price (SP).** The amount added onto the cost price is called the **mark up,** and is usually calculated as a percentage of the cost price. If the selling price is less than the cost of the item plus any associated costs (such as transport), then a **loss** is made.

Cost price (CP): the total cost of making or buying the goods

Selling price (SP): cost price + mark up

Mark up: amount added to cost price in order to make a profit

Calculating the selling price after mark up

Method 1: Calculate the mark up as a percentage of the cost price. Add this mark up amount to the cost price.

Method 2: Calculate the percentage increase after mark up by adding the mark up percentage to 100%. Find this percentage of the cost price.

For mark ups:

$$\text{Cost price (CP)} \xrightarrow{\times (100\% + \text{mark up }\%)} \text{Selling price (SP)}$$

$$\text{Cost price (CP)} \xleftarrow[\div (100\% + \text{mark up }\%)]{} \text{Selling price (SP)}$$

Convert percentages to their decimal form before multiplying or dividing.

Profit and loss

The amount of profit or loss made on a sale is the difference between the cost price and the selling price. Profit or loss is usually expressed as a percentage of the cost price or the selling price.

Profit: difference between SP and CP, where SP > CP

Loss: difference between SP and CP, where SP < CP

Finding percentage profit

Profit % can be calculated using the cost price or the selling price.

$$\%\ \text{Profit} = \frac{\text{profit in }\$}{\text{CP}} \times 100\% \text{ or } \frac{\text{profit in }\$}{\text{SP}} \times 100\%$$

where CP = cost price, SP = selling price.

Replace 'profit' with 'loss' in these formulas to calculate the percentage loss.

Hint

You make a profit when you make money on the price you paid. You make a loss when you lose money on the price you paid.

Warning

Make sure you don't confuse 'selling price' (SP) with 'sale price'. Selling price is the price of the goods after a mark up has been added. Sale price is the price of goods on sale after they have been discounted.

2.10

Discount

A business will sometimes offer goods at a lower price, or **sale price,** to make them more attractive to consumers. This reduction in price is called a **discount.**

Discount: decrease in the selling price to attract more buyers

Sale price = selling price – discount amount

Calculating the sale price after discount

Method 1: Calculate the discount as a percentage of the selling price. Subtract this discount amount from the selling price.

Method 2: Calculate the percentage remaining after discount by subtracting the discount percentage from 100%. Find this percentage of the selling price.

For discounts:

Marked price $\xrightarrow{\times (100\% - \text{discount }\%)}$ Sale price

Sale price $\xrightarrow{\div (100\% - \text{discount }\%)}$ Marked price

Goods and Services Tax

As well as adding their mark up to the cost of a product, a business must also add a government tax called the **Goods and Services Tax (GST)**, which is an extra 10% of the final selling price (the selling price after any discount has been applied).

GST or Goods and Services Tax: 10% is added to pre-GST SP or $\frac{1}{11}$ of the inclusive GST SP

Inclusive SP = pre-GST SP + GST
= pre-GST SP + 10% of pre-GST SP
= 110% of pre GST SP

Pre-GST SP = $\frac{10}{11}$ of inclusive SP

GST exemptions

Some items sold are GST exempt. (No GST is charged to the selling price.) No GST is added to fresh food and beverages, health and medical care, educational supplies and childcare. Private sales less than $75 000 p.a. and the sale by registered charities of donated second-hand goods are also GST exempt.

Worked Example 24

WE 24

Calculate the sale price of the following item, rounding your answer to the nearest cent.
A $350 stereo discounted by 25%.

Working

Method 1:

Discount = 25% of $350
= 0.25 × 350
= $87.50

Sale price = $350 – $87.50
= $262.50

Working

Method 2:

Percentage remaining = 100% – 25%
= 75%

Sale price = 75% of $350
= 0.75 × 350
= $262.50

Worked Example 25

WE 25

An electrical goods store determines the selling price of appliances by applying a 68% mark up.

(a) Calculate the pre-GST selling price of a toaster bought at a cost price of \$23.20.

(b) Find the marked price of the toaster after GST has been applied.

Working

Method 1:

(a) Mark up = 68% of \$23.20
= 0.68 × 23.20
= \$15.78

Pre-GST SP = CP + mark up
= \$23.20 + \$15.78
= \$38.98

(b) Marked price (GST included) $= \frac{11}{10} \times \$38.98$
= \$42.66

Working

Method 2:

Percentage increase = 100% + 68%
= 168%

Pre-GST SP = 168% of CP
= 1.68 × 23.20
= \$38.98

Worked Example 26

WE 26

(a) 'Great Groceries' supermarket has a policy of marking up all of its bakery products by 75%. If the selling price of a loaf of bread is \$3.49, calculate the cost price.

(b) The 'Discount Computers' store is having a 20% sale on all items. Calculate the original marked price of a hard drive that has been discounted to \$135.

Working

(a) SP = \$3.49
= 175% of CP
= 1.75 × CP

CP = SP ÷ 1.75
= 3.49 ÷ 1.75
= \$1.99

(b) Sale price = \$135
= 80% of marked price
= 0.8 × marked price

Marked price = sale price ÷ 0.8
= 135 ÷ 0.8
= \$168.75

2.10

Worked Example 27

WE 27

(a) Laura's wedding dress cost \$1450. After her wedding, she sold it for \$900.

(i) Calculate the loss Laura made on the dress.

(ii) Express the loss as a percentage of the cost price, rounded to the nearest per cent.

(b) Ajay bought a tent for \$125 and then resold it on eBay for \$139.

(i) How much profit did Ajay make on the tent?

(ii) Express Ajay's profit as a percentage of the selling price, rounded to the nearest per cent.

Working

(a) Loss = CP − SP

$= \$1450 - \$900$

$= \$550$

$= \dfrac{550}{1450} \times 100\%$

$= 38\%$

(b) Profit = SP − CP

$= 139 - 125$

$= \$14$

$= \dfrac{14}{139} \times 100\%$

$= 10\%$

Technology Exploration: Spreadsheet

Calculating profit

A spreadsheet such as Microsoft Excel® can be a very useful tool to help businesses keep track of their sales and profits. To demonstrate this, a spreadsheet for a shop selling skateboards, protective gear, skate clothing and skating DVDs will be used.

1 Open a new spreadsheet file and enter the headings and the amounts in columns B, C and I. The 'Volume sold' figures in column I are based on a typical week of sales. (Do not enter \$ or % signs into these cells.)

D2 fx =B2*(1+C2/100)

	A	B	C	D	E	F	G	H	I	J
1	Item	Cost price(\$)	Markup (%)	Selling price pre-discount (\$)	Discount (%)	Selling price pre-GST (\$)	Profit (\$)	Marked price (\$) (GST added)	Volume sold	Volume profit (\$)
2	Skateboard	103.00	65	169.95	0				9	
3	Jeans	35.00	110		0				27	
4	Hoodie	20.00	95		0				42	
5	Skate shoes	42.00	125		0				38	
6	DVD	8.50	50		0				86	
7	Helmet	17.00	76.4		0				15	

2 To calculate the selling price before discount, enter the formula =B2*(1+C2/100) into cell D2. (The * sign must be used to show multiplication.) To copy this formula down the column, you can 'Fill Down' by placing the cursor over the bottom right corner of the cell until the black cross appears, then clicking and dragging down.

3 Because the calculations involve money, you should format the cells in column D and F to show two decimal places. Highlight the column by clicking on the column letter. Right-click on the highlighted column and select Format Cells from the pop-up menu. Select the Number tab and in the Number category set the Decimal places value to 2, then click OK.

4 Click on cell F2 and enter a formula that will calculate the selling price after discount but pre-GST for that item. Make sure the formula begins with the = sign. Copy this formula down the column, then format the cells as in step **3**.

5 Click on cell G2 and enter a formula that will calculate the profit on that item. Make sure the formula begins with the = sign. Copy this formula down the column, then format the cells as in step **3**.

6 Click on cell H2 and enter a formula that will calculate the marked price with GST added. Make sure the formula begins with the = sign. Copy this formula down the column, then format the cells as in step **3**.

7 To calculate the profit made on the volume of each item sold, enter a formula into cell J2 that will multiply the profit made on one item by the volume of that item sold. Format the cells as in step **3**.

8 Find the total profit on all the items sold for the week by calculating the sum of the 'Volume profit' cells. To do this, highlight (by clicking and dragging) the cells from J2 to J9. Click on the Σ symbol (on the far right of the toolbar under the Home tab) and select Sum. The total profit will appear at the bottom of the column.

9 If all the formulas have been entered correctly, your table should look like the one below.

D2 =B2*(1+C2/100)

	A	B	C	D	E	F	G	H	I	J
1	Item	Cost price($)	Markup (%)	Selling price pre-discount ($)	Discount (%)	Selling price pre-GST ($)	Profit ($)	Marked price ($) (GST added)	Volume sold	Volume profit ($)
2	Skateboard	103.00	65	169.95	0	169.95	66.95	186.95	9	602.55
3	Jeans	35.00	110	73.50	0	73.50	38.50	80.85	27	1039.50
4	Hoodie	20.00	95	39.00	0	39.00	19.00	42.90	42	798.00
5	Skate shoes	42.00	125	94.50	0	94.50	52.50	103.95	38	1995.00
6	DVD	8.50	50	12.75	0	12.75	4.25	14.03	86	365.50
7	Helmet	17.00	76.4	29.99	0	29.99	12.99	32.99	15	194.82
8										
9	Total profit:									4995.37

10 During a sale week, a discount of 10% on skateboards, 20% on jeans and hoodies and 5% on skate shoes was applied, and the volume of sales increased as shown below. By changing the numbers in the 'Discount (%)' and 'Volume sold' columns of the spreadsheet, calculate the new total profit.

11 To compare the two sets of results, copy and paste the spreadsheet cells to create a second set of formulas and data below the original set. Enter the original data in the first set of cells and the discounted values in the second set. Then, insert a formula into an empty cell to calculate the difference between the profits of the two sets.

12 If all the formulas have been entered correctly, your table should look like the one below.

F12 =D12*(1-E12/100)

	A	B	C	D	E	F	G	H	I	J
1	Item	Cost price($)	Markup (%)	Selling price pre-discount ($)	Discount (%)	Selling price pre-GST ($)	Profit ($)	Marked price ($) (GST added)	Volume sold	Volume profit ($)
2	Skateboard	103.00	65	169.95	0	169.95	66.95	186.95	9	602.55
3	Jeans	35.00	110	73.50	0	73.50	38.50	80.85	27	1039.50
4	Hoodie	20.00	95	39.00	0	39.00	19.00	42.90	42	798.00
5	Skate shoes	42.00	125	94.50	0	94.50	52.50	103.95	38	1995.00
6	DVD	8.50	50	12.75	0	12.75	4.25	14.03	86	365.50
7	Helmet	17.00	76.4	29.99	0	29.99	12.99	32.99	15	194.82
8										
9	Total profit:									4995.37
10										
11	Item	Cost price($)	Markup (%)	Selling price pre-discount ($)	Discount (%)	Selling price pre-GST ($)	Profit ($)	Marked price ($) (GST added)	Volume sold	Volume profit ($)
12	Skateboard	103.00	65	169.95	10	152.96	49.96	168.25	15	749.33
13	Jeans	35.00	110	73.50	20	58.80	23.80	64.68	47	1118.60
14	Hoodie	20.00	95	39.00	20	31.20	11.20	34.32	63	705.60
15	Skate shoes	42.00	125	94.50	5	89.78	47.78	98.75	41	1958.78
16	DVD	8.50	50	12.75	0	12.75	4.25	14.03	86	365.50
17	Helmet	17.00	76.4	29.99	0	29.99	12.99	32.99	15	194.82
18										
19	Total profit:									5092.62
20										
21	Increased profit:									97.25

13 In another sale week when '20% off everything' discounts were applied, the volume of sales again increased. By changing the discounts and volume columns again, calculate the new total profit and the increased profit from the original.

14 If all the formulas have been entered correctly, your table should look something like the one below. (If your new 'Volume sold' data is different, then the profit will be different too.)

	Item	Cost price($)	Markup (%)	Selling price pre-discount ($)	Discount (%)	Selling price pre-GST ($)	Profit ($)	Marked price ($) (GST added)	Volume sold	Volume profit ($)
11										
12	Skateboard	103.00	65	169.95	20	135.96	32.96	149.56	26	856.96
13	Jeans	35.00	110	73.50	20	58.80	23.80	64.68	47	1118.60
14	Hoodie	20.00	95	39.00	20	31.20	11.20	34.32	63	705.60
15	Skate shoes	42.00	125	94.50	20	75.60	33.60	83.16	56	1881.60
16	DVD	8.50	50	12.75	20	10.20	1.70	11.22	105	178.50
17	Helmet	17.00	76.4	29.99	20	23.99	6.99	26.39	26	181.75
18										
19	**Total profit:**									4923.01
20										
21	**Increased profit:**									-72.36

15 Use the spreadsheet to find whether there would still be a profit if sales had not increased from the original volume with either of the sales discount plans. Find the profit or loss that would have been incurred under each discount plan.

16 If a loss was made, change the volume figures to see how much would need to be sold to ensure the same profit as would have been made in a typical week.

17 Why is the profit based on the selling price pre-GST and not on the marked price?

Answers page 493

2.10 Financial applications of percentages

Navigator

Q1, Q2, Q3, Q4, Q5, Q6, Q7, Q8, Q9, Q11, Q12, Q13, Q14, Q17, Q19, Q20, Q21, Q24, Q26, Q27, Q28, Q29	Q1, Q2, Q3, Q4, Q5, Q6, Q7, Q8, Q9, Q11, Q13, Q14, Q15, Q16, Q17, Q18, Q19, Q20, Q21, Q23, Q24, Q26, Q27, Q28, Q29	Q1 (c) & (d), Q2 (c) & (d), Q3 (c) & (d), Q4, Q6, Q7, Q8, Q9, Q10, Q11 (c) & (d), Q13, Q14, Q15, Q16, Q17, Q18, Q19, Q20, Q21, Q22, Q23, Q24, Q25, Q26, Q28, Q29

Equipment required: Calculator

Fluency and understanding

WE24

1 Calculate the sale price of the following items, rounding your answers to the nearest cent.

(a) A $520 surfboard discounted by 20%.

(b) A DVD originally priced at $18 discounted by 10%.

(c) A $34.50 book discounted by 12.5%.

(d) A digital camera with a marked price of $399.95 discounted by 35%.

Warning

When you're working with amounts of money, remember to include $ signs as well as 2 decimal places (cents) in your answer.

2.10

2 (a) A clothing store determines the selling price of T-shirts by applying a 53% mark up. WE 25

(i) Calculate the pre-GST selling price of a T-shirt bought at a cost price of $15.

(ii) Find the marked price of a T-shirt after GST has been applied.

(b) A mark up of 74% applies to all goods at Shelley's Shoe Shop.

(i) Calculate the pre-GST selling price of a pair of shoes with a cost price of $25.

(ii) Find the marked price of the shoes after GST has been applied.

(c) Glen adds a mark up of 60% to the cost price of the wooden toys he makes before selling them at his market stall.

(i) Calculate the pre-GST selling price of a toy that cost $12.50 to make.

(ii) Find the marked price of the toy after GST has been applied.

(d) A computer games store adds a mark up of 78.5% to the cost price of all of its console games.

(i) Calculate the pre-GST selling price of a game that cost the store $30.25.

(ii) Find the marked price of the game after GST has been applied.

3 (a) Deb's Designer Dresses has a policy of marking up the cost price of all of its dresses by 82%. If the pre-GST selling price of a dress is $200, calculate the cost price. WE 26

(b) Amanda's DVD store offers a storewide discount of 15% on all DVDs. Calculate the original marked price of a boxed set that has been discounted to $49.30.

(c) Calculate the cost price of a pair of sunglasses that has been marked up by 76% to a pre-GST selling price of $69.95.

(d) Calculate the original marked price of a cricket bat discounted by 30% and on sale for $169.

4 (a) Rosa buys a ticket to a concert for $65, but is unable to go and sells it to a friend for $39. WE 27

(i) Calculate the loss Rosa made on the ticket.

(ii) Express the loss as a percentage of the cost price, rounded to the nearest per cent.

(b) Travis bought an old car for $1500 and then resold it for $1950 without spending any money on it.

(i) How much profit did Travis make on the car?

(ii) Express Travis' profit as a percentage of the selling price, rounded to the nearest per cent.

5 For each of the following, express the profit or loss as a percentage of the cost price, rounded to the nearest whole number.

(a) A painting purchased for $750 and sold for $2000 + GST.

(b) A book purchased for $45 and sold for $20 + GST.

(c) Sporting goods purchased for $320 and sold for $520 + GST.

(d) Jewellery purchased for $980 and sold for $1750 + GST.

6 An electrical goods retailer buys a washing machine for $580 and applies a standard mark up of 70% to all of its goods. The pre-GST selling price for one of its washing machines is:

A $174 **B** $406 **C** $754 **D** $986

2.10

7 A magazine is advertised for 20% off the marked price and you are charged \$7.60. The original marked price of the magazine is closest to:

A \$6.10 **B** \$7.80 **C** \$9.10 **D** \$9.50

8 A bookstore applies a mark up of 92% to the cost price of all of its books. The cost price of a book with a pre-GST selling price of \$35 is closest to:

A \$2.80 **B** \$32.20 **C** \$18.23 **D** \$67.20

9 Calculate **(i)** the dollar value of the discount and **(ii)** the discount sale price of the following items. (Round your answers to the nearest cent.)

(a) A Wii console originally priced at \$379 on sale at a 10% discount.

(b) A suitcase with an original price of \$199 discounted by 40%.

(c) A \$20 T-shirt discounted by 30%.

(d) A discount of 25% on a \$99.95 pair of sunglasses.

10 Sanjay invests \$30 000 in the sharemarket only to see his shares plummet in value to \$22 000.

(a) Find the amount of Sanjay's loss.

(b) Find his percentage loss by expressing the loss as a percentage of the original amount invested.

11 In Australia, a Goods and Services Tax (GST) of 10% is added to the price of all products and services purchased by consumers (with the exception of basic grocery items).

Calculate **(i)** the amount of GST and **(ii)** the new, GST-inclusive price of the following items.

(a) a \$560 bike **(b)** a \$64 pair of jeans

(c) a \$19 calculator **(d)** a \$2640 TV

12 The amount calculated by a builder for a roof restoration is \$2450 before adding GST. Calculate:

(a) the 10% GST to be added **(b)** the final amount (including GST).

13 Mitchell buys an antique desk for \$150. He spends \$45 to restore it to its original condition and sells it for \$320.

(a) How much profit did Mitchell make, taking the cost of the restoration into account?

(b) Calculate the value of the profit as a percentage of the total cost. (Write your answer rounded to one decimal place.)

14 It costs a furniture company \$180 to manufacture a sofa. The cost of transport to the company's stores is \$7.50 per sofa. The company then sells the sofas for \$350 + GST.

(a) How much profit does the company make on the sofas, taking transport and manufacturing costs into account?

(b) Express the profit as a percentage of the selling price to two decimal places.

> **Hint**
> Don't forget to add GST.

15 An exclusive boutique applies a 200% mark up to all of the clothing it sells. What would the selling price be of a gown that had a cost price of \$150?

> **Hint**
> Calculate pre-GST selling price first.

16 A store purchases a bulk order of 72 photo frames for \$324.

(a) Calculate the profit made if the photo frames are sold individually for \$7.70 each.

(b) Calculate the profit as a percentage of the cost price to two decimal places.

17 Most of the time, the marked prices we see on goods in shops have the GST already added. Find the original, pre-GST price of:

(a) a book with a price tag of $32.95 **(b)** a microwave with a price tag of $279

(c) a boat priced at $29 999 **(d)** a T-shirt priced at $39.95.

18 A store takes delivery of a large container with 1000 packets of soap which cost the store $2750. The store policy is to add a 70% mark up and then add the GST. Find the selling price of each packet of soap.

Reasoning, communicating and open-ended

19 When out shopping, you may have noticed that some percentages are used more commonly than others, such as 10%, 20%, 25% and 50%. Explain why these percentage values are used, rather than, say, 17%, 36% or 42%.

> **Hint**
> What are the fraction equivalents of the percentages in Question **19**?

20 Julia makes and sells jars of jam to a local cafe, which then sells the jam to customers. Julia adds a 70% mark up to the cost of making the jam before selling it to the cafe. The cafe adds its own mark up of 85% to obtain the price at which the jam is sold to cafe customers. Express your answers to the following to the nearest 5 cents.

(a) If a jar of jam costs Julia $1.35 to make, at what price does she sell each jar to the cafe? Do not add GST.

(b) At what price does the cafe sell the jam to its customers? Add GST.

(c) Express the cafe's selling price of the jam as a percentage of Julia's initial cost price, correct to the nearest whole number.

21 **(a)** For the following situations, calculate the profit made as a percentage of the cost price, rounding your answers to two decimal places where necessary. (Here SP is the pre-GST selling price.)

(i) CP = $45, SP = $54 **(ii)** CP = $360, SP = $504

(iii) CP = $4200, SP = $4350 **(iv)** CP = $480, SP = $600

(b) Repeat **(a)**, this time calculating the profit as a percentage of the selling price.

(c) For each of the situations **(i)–(iv)**, compare your answers to **(a)** and **(b)**. Which gives a higher percentage—calculating profit using the cost price or using the selling price?

(d) Explain your answer to **(c)**.

22 Billy Bob is having a garage sale of toys he has made. He has been selling them with a mark up of 50% on the price it cost him to make them. As he has only one left, he decides to discount the selling price by 15%.

Find the:

(a) discounted sale price of the toy that had cost him $45 to make

(b) percentage profit made by Bob on the sale of the discounted toy

(c) decrease in percentage profit made by discounting the sale price of the toy.

23 Tariq sees an item in a bargain store initially marked down by 50% with a sign that says 'take an extra 20% off'. If, after these two discounts, the price Tariq paid was $11.20, what was the original price of the item?

24 Martha's favourite clothing shop is offering a 'buy two, get the third half price' offer on T-shirts. The shop next door is offering 20% off the same brand of T-shirts. If the regular retail price of the T-shirts in both stores is $19.95, which store is the cheaper place for Martha to buy three T-shirts? How much would she save, compared to the other store?

25 A number of people own an earth-moving business. Each person owns a certain percentage of the business. Profits are divided according to the percentage of the money they invested. For example, someone investing 7% of the operating capital (the amount of money needed to run the business) would receive 7% of the profits. The operating capital is $675 000.

(a) Calculate the amount contributed by a person who has invested 7% of the operating capital.

(b) If the company makes $150 000 yearly profit, how much would the person who invested 7% receive?

(c) If the company continued to make this profit each year, how many years would it be before the person has received back their investment?

26 It is very useful to be able to calculate some common percentages mentally. Knowing their fraction equivalents is very helpful.

(a) Write down the fraction equivalents of:

(i) 25% **(ii)** 10% **(iii)** 20% **(iv)** $33\frac{1}{3}\%$

(b) Chantal knows that $50\% = \frac{1}{2}$. When she sees a '50% off' sale sign, she halves the original price to find the new price. Use your answers to part **(a)** to help Chantal extend her method so she can calculate 25%, 10%, 20% and $33\frac{1}{3}\%$ off.

27 A sports store is having a sale on all sporting equipment. If a tennis racquet is normally priced at $299, write three different, common discount percentages and find the sale price for each.

28 Many stores, such as department stores, advertise sale items using percentage discounts, as seen in this catalogue:

At other times, stores advertise sale items using dollar disounts as seen in this catalogue:

Why do you think this is? Try to give a couple of reasons for your answer.

29

The boys think the camera should only cost $75. The shop assistant says the camera now costs $84. What mistake have the boys made? Explain how the shop assistant calculated the new price of $84.

Problem solving

The plummeting price

Frida puts a price tag of $45 on a Superman costume for sale in her fancy-dress store. Unfortunately, there has been little interest in the costume since the death of Superman in his comic, so she reduces the price to $27. As the costume still does not sell at the reduced price, Frida reduces the price again, by the same percentage as the first reduction.

1 What is the new reduced price of the costume?

2 How many reductions at the same rate would reduce the price to less than one-quarter of the original price?

3 If the initial price was different would it take the same number of reductions?

Strategy options

- Break problem into manageable parts.
- Test all possible combinations.

Challenge 2

1 The average (mean) of $\frac{2}{3}$, 0.7 and 55% is:

A $\frac{23}{36}$ B 175% C $1\frac{11}{12}$ D 5.75

2 Jake bought a large tub of yoghurt. On the first day he ate $\frac{1}{4}$ of the yoghurt. The next day he ate $\frac{1}{3}$ of what was left, and on the third day he ate $\frac{2}{3}$ of what was left. What fraction of the yoghurt remained after the third day?

3 What fraction is halfway between $\frac{1}{4}$ and $\frac{1}{2}$?

4 A piece of fabric cost Michael \$25. He purchases another piece of similar fabric that is twice as long and twice as wide as the first piece. How much should this second piece cost him?

5 Which of the following is the best value deal for a box of chocolates that normally sells for \$5.04?

A save \$1.70 B cost price of \$3.20, +5%

C discount of $33\frac{1}{3}$ % D buy 2, get 1 free

6 Write an expression using four 8s and any of the operation symbols (+, −, ×, ÷) that equals 65.

7 Find the sum of $\frac{1}{2} + \frac{0.1}{2} + \frac{1}{0.2}$, expressing your answer in decimal form.

8 Michael picks three *different* digits from the set {1, 2, 3, 4, 5} and forms a mixed number by placing the digits in the spaces $\square\frac{\square}{\square}$. The fractional part of the mixed number must be less than 1. What is the difference between the largest and the smallest mixed numbers that can be formed?

A $4\frac{4}{15}$ B $4\frac{3}{10}$ C $4\frac{7}{20}$ D $4\frac{3}{5}$

9 Suppose that x^* means $\frac{1}{x}$ which is called the inverse of x. For example, $5^* = \frac{1}{5}$. How many of the following statements are true?

(i) $2^* + 4^* = 6^*$ (ii) $3^* \times 5^* = 15^*$ (iii) $7^* - 3^* = 4^*$ (iv) $12^* \div 3^* = 4^*$

A 0 B 1 C 2 D 3

10 The sum of seven consecutive integers is always:

A odd B a multiple of 7 C a multiple of 3 D even.

11 Find the following sum, without using a calculator.

$$\frac{1}{2}+\frac{1}{4}+\frac{1}{8}+\frac{1}{16}+\frac{1}{32}+\frac{1}{64}+\frac{1}{128}+\frac{1}{256}+\frac{1}{512}+\frac{1}{1024}+\frac{1}{2048}+\frac{1}{4096}+\frac{1}{8192}$$

12 One liquid contains 22.5% water, another liquid contains 27% water. A glass is filled with 5 parts of the first liquid and 7 parts of the second liquid. What percentage of the liquid in the glass is water?

Chapter review 2

D.I.Y. Summary

Key Words

cost price (CP)	loss	profit	surd
discount	mark up	rational numbers	surd form
exact decimal form	non-terminating decimal	recurring decimal	terminating decimal
Goods and Services Tax (GST)	per cent	sale price	
irrational number	percentage	selling price (SP)	

Copy and complete the following using the words and phrases from this list, where appropriate, to write a summary for this chapter. A word or phrase may be used more than once.

1. The word __________ literally means per hundred or 'out of a hundred'.
2. A decimal such as $2.6\dot{3}$ is written in __________.
3. A decimal that has an infinite number of digits after the decimal point is called a __________.
4. To convert a decimal to a __________, you multiply by 100 and add the percentage symbol.
5. The amount added to the cost price of goods before selling them is known as the __________.
6. A number with a defined number of digits after the decimal point is called a __________.
7. An __________ is a non-terminating decimal that cannot be written as a fraction.
8. A __________ is applied to goods by decreasing the marked price by a given percentage.
9. An irrational number written in the form of a square root is called a __________.
10. To make a profit, the __________ must be higher than the __________.

Fluency and understanding

Equipment required: Calculator for Questions **2**, **5**, **6**, **15–32** and **34–37**

1. Write the following decimals as fractions or mixed numbers in simplest form. **Ex. 2.1**
   (a) 0.58 (b) 0.0124 (c) 2.605
2. Put the following into ascending order. **Ex. 2.1–2.3**
   (a) $\frac{2}{9}, -\frac{3}{10}, -0.283, 0.26, \frac{3}{4}, \sqrt{7}$ (b) $-\frac{1}{3}, \frac{7}{10}, \sqrt{2}, 0.22, -0.26, 0.253$
3. Write the following mixed numbers as decimals. **Ex. 2.1**
   (a) $28\frac{5}{8}$ (b) $15\frac{3}{40}$ (c) $53\frac{261}{320}$
4. How many 60 cm lengths of wood can be cut from a 2.5 m long plank? What length of wood is left over? **Ex. 2.1**
5. Use a calculator to express the following numbers in decimal form, then classify them as terminating decimals, recurring decimals, or irrational numbers. **Ex. 2.2**
   (a) $53 \div 4$ (b) $\sqrt{85.2}$ (c) $\frac{11}{9}$ (d) $\sqrt{125}$ (e) $\sqrt{30.25}$

6 The fraction $\frac{19}{11}$, written in exact decimal form, is: **Ex 2.2**

A $\overline{1.72}$ **B** $1.\overline{72}$ **C** $1.\dot{7}2$ **D** $1.7\dot{2}$

7 Write the following recurring decimals in fraction form. **Ex 2.2**

(a) $0.\dot{2}$ **(b)** $0.3\dot{2}$ **(c)** $0.\dot{3}\dot{2}$

8 Calculate the following, giving your answers as mixed numbers in simplest form. **Ex 2.3**

(a) $-\frac{1}{2} + \frac{3}{4}$ **(b)** $\frac{8}{9} - 1\frac{1}{6}$ **(c)** $-\frac{3}{10} \times -\frac{5}{6}$ **(d)** $-5 \div \frac{1}{3}$

9 Calculate the following. **Ex 2.3**

(a) $-3.4 + 2.8$ **(b)** $8.6 - 10.2$ **(c)** 4.2×-6.1 **(d)** $20.4 \div -5$

10 Approximately what percentage of the shape is shaded? **Ex 2.4**

A 20% **B** 34% **C** 67% **D** 75%

11 Convert the following to percentages. Express any remainders as fractions. **Ex 2.5**

(a) $\frac{23}{50}$ **(b)** $\frac{6}{5}$ **(c)** $2\frac{4}{5}$ **(d)** $\frac{25}{40}$

12 Convert the following to percentages. **Ex 2.5**

(a) 0.6 **(b)** 0.35 **(c)** 1.12 **(d)** 0.03

13 Convert the following to fractions in simplest form. **Ex 2.6**

(a) 65% **(b)** 154% **(c)** $5\frac{5}{6}\%$ **(d)** 4.7%

14 Convert the following to decimals. **Ex 2.6**

(a) 80% **(b)** 7% **(c)** 2.09% **(d)** $6\frac{1}{4}\%$

15 Express the first amount as a percentage of the second, giving answers rounded to two decimal places if necessary. **Ex 2.7**

(a) 27, 30 **(b)** 35, 16 **(c)** 4 cm, 2 m **(d)** 30c, $4.50

16 Find the following, giving answers rounded to two decimal places if necessary. **Ex 2.8**

(a) 35% of 84 m **(b)** $4\frac{2}{5}\%$ of 2000 kg **(c)** $7\frac{1}{3}\%$ of 4500 L

(d) 14% of $48 **(e)** 0.59% of $4000 **(f)** 9.5% of $90

17 **(a)** Increase $450 by **(i)** 35% and **(ii)** 67.2%. **Ex 2.9**

(b) Decrease $19.95 by **(i)** 20% and **(ii)** 7.5%.

Round answers to two decimal places where necessary.

18 Calculate the: **Ex 2.10**

(a) sale price of a $90 tennis racquet discounted by 15%

(b) pre-GST selling price of a $950 ring that will be marked up by 80%

(c) original marked price of a pair of sunglasses discounted by 30% to $62.30

(d) cost price of a dress that has been marked up by 85% to $143 (GST inclusive).

19 Kayla bought a concert ticket for $145, but was unable to go, so she sold it online for $120. Calculate Kayla's profit or loss as a percentage of what she paid for the ticket, correct to two decimal places.

Ex 2.10

20 The number of pre-booked tickets for a one-day cricket match is 62% of the 75 242 tickets available. How many tickets were pre-booked?

A 4665 **B** 28 592 **C** 46 650 **D** 121 358

21 Michelle made deposits of $210, $25, $45.50 and $66.75 into her bank account during one month and withdrawals of $35.75, $56.90, $214 and $102.50 during the same period. Use a negative number to show how much her balance decreased during this period.

22 Franco has a 5 kg bag of nuts that he needs to use to fill smaller 375 g bags.

(a) How many bags can he fill?

(b) What mass of nuts will he have left over?

23 The value of a property increases by 17.5% one year. By what fraction did the property's value increase?

24 Approximately 85% of an iceberg lies under water. The average mass of an iceberg is 150 000 tonnes. What mass of the iceberg sits above water?

25 A geography book states that 0.7092 of the Earth's surface is covered by ocean. Write this value as a percentage.

26 The Australian desert is 42.556% as large as the Sahara Desert in Africa, which has an area of 9 000 000 square kilometres. Find the area of the Australian desert to the nearest square kilometre.

27 Tom bought an old car for $3500. He spent $2700 to repair and restore it, then sold it for $9400.

(a) Calculate the profit Tom made, taking all costs into account.

(b) Express the profit as a percentage of the selling price, rounding your answer to the nearest per cent.

28 Tye buys a neighbouring 600 m^2 property to add to his 3000 m^2 farm. By what percentage has the area of his farm increased?

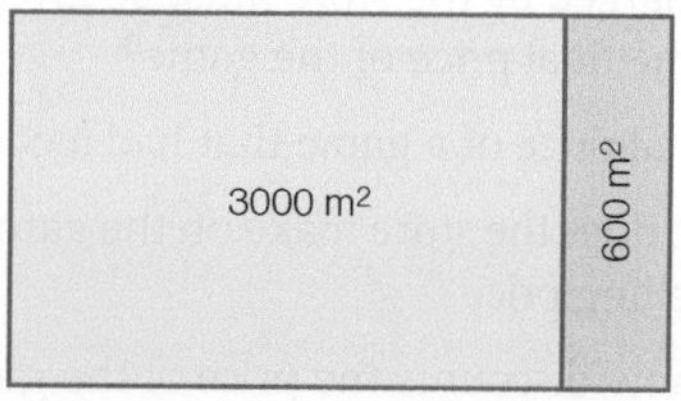

29 Gao currently earns $600 a week. If he receives a pay rise of $2\frac{1}{3}$%, what will his new weekly pay be?

30 Arala runs a newsagency. She buys boxes of pens, then sells the pens individually. A box of 60 pens costs her $17.

(a) To work out what she will sell them for individually, Arala needs to know what she paid per pen. Calculate the cost of each pen, to the nearest cent.

(b) Arala wants to sell each pen for 25% more than what she paid for them. Use your answer from **(a)** to calculate the price Arala should sell the pens for.

31 Nazim had 62 friends on his social networking web page. One month later, he had 97 friends. Write this increase as a percentage of his initial number of friends. Round your answer to one decimal place.

32 The minimum overnight temperatures for one week at Thredbo were -2°C, -3°C, -1°C, 2°C, -3°C, -2°C, 0°C. Find the mean minimum overnight temperature for the week, rounding your answer to one decimal place. (The mean is found by adding all values, then dividing the result by the number of values.)

33 Copy the number line below and place the following percentages on it: 14%, 140%, 4%, 104%, $\frac{1}{4}\%$

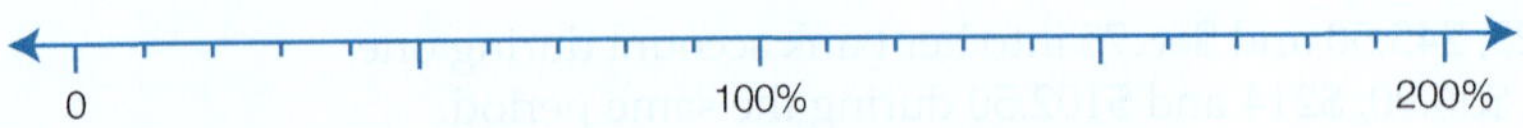

Reasoning and communicating

34 A certain cheese contains $7\frac{1}{2}\%$ fat. How much fat is in a 600 gram piece?

35 **(a)** The decimal value of $\frac{4}{7}$ is closest to:

A 0.57 **B** 0.571 **C** 0.58 **D** 0.6

(b) The decimal approximation of $\sqrt{23}$ is closest to:

A 4.7 **B** 4.796 **C** 4.8 **D** 5

36 The combined capacity of all of the dams that supply water to Melbourne is 1 810 500 ML (1 ML = 1 000 000 L). In May 2010, the dams were 33.3% full of water, higher than the 27.2% recorded on the same day the previous year.

(a) How many ML of water was in the dams in May 2010?

(b) How many more ML was present in May 2010 than in May 2009?

(c) How many more ML would need to flow into the dams to achieve a capacity of 35% full?

37 A games store applies a mark up of 80% to the cost price of all of its products, then adds 10% GST.

(a) Find the pre-GST price of a game that had a cost price of $33.25, to the nearest 5 cents.

(b) The GST charged by the store must be given to the government. What is the amount of GST in the final price of the game?

(c) Find the final price of a game that had a cost price of $33.25, to the nearest 5 cents.

(d) What profit does the store make on the game, remembering that profit is based on the pre-GST selling price.

(e) The store now discounts the price of the game by 25%. What is the new price of the game?

(f) What is the amount of GST in this new price?

(g) What profit does the store make on the game now, remembering that profit is based on the pre-GST selling price.

38 A packet of 12 chocolate frogs is on sale for $4.75.

(a) How much does each frog in the packet cost? Write your answer in exact decimal form.

(b) Write the cost of each frog rounded to the nearest cent.

(c) How much extra will you pay if you buy 12 frogs individually at this price?

(d) Where does this 'extra' amount come from?

NAPLAN practice 2

Numeracy: Non-calculator

1 Jye cuts a pizza into 12 equal slices. He eats 75% of the slices.

How many slices of pizza are left? ___________

2 A ticket costs \$75. A fee of 10% is added to the price. Which calculation will give the new price?

A $75 + 10$ B $75 + 0.1$ C 75×0.1 D 75×1.1

3 Vicky bought a house for \$400 000. At the end of 1 year, its value had increased by 7%. At the end of the next year, the value had increased by another 5%. What was Vicky's house worth after 2 years?

4 A skateboard is on sale for 20% off the marked price. If the marked price is \$140, how much is the sale price?

A \$102 B \$112 C \$120 D \$160

Numeracy: Calculator allowed

5 At midday during winter in Canberra the temperature was 11.7°C. At 6 p.m., it had dropped by 5.4°C. By midnight, the temperature was 15.2°C cooler than at midday. What was the temperature at midnight?

A 6.3°C B -3.5°C C -35°C D -9.8°C

6 There were only 13 students in Joseph's class on Wednesday. The other 7 were absent.

What percentage of Joseph's class was absent?

A 7% B 35% C 54% D 65%

7 Justin bought a multi-bag of chocolate bars. The bag contained 15 chocolate bars and it cost \$3.70. What is the price of each chocolate bar, correct to the nearest cent?

8 Maryam scored 28 goals from 40 shots during netball training. What percentage was this?

Mixed review (A)

Fluency and understanding

1 Calculate the following. **Ex 1.1, 1.2, 1.3**

(a) $7 - (-9)$ (b) $-3 + (-10)$ (c) $-14 - (+8)$ (d) $-21 + 9$

(e) -4×-9 (f) 7×-3 (g) $-60 \div 15$ (h) $-\frac{48}{8}$

2 Write the following decimals as fractions in simplest form. **Ex 2.1**

(a) 0.36 (b) 0.05 (c) 0.75 (d) 0.002

3 A TV has been bought for a cost price of $560 by a retailer, who will add a mark up of 78% before selling it in the store. Calculate: **Ex 2.10**

(a) the value of the mark up

(b) the pre-GST selling price after the mark up has been applied

(c) the selling price after GST has been added.

4 Evaluate the following by simplifying first where possible. **Ex 1.5**

(a) $2^3 \times 5^2$ (b) $3^4 \times 10^3$ (c) $-4^2 \times 5$

(d) $-3 \times (-2)^4$ (e) $(5^2 \times 3^3) \div (5 \times 3)$ (f) $(7^4 \times 8^4) \div (7^2 \times 8^2)$

(g) $\frac{4^2 \times 6}{2^3}$ (h) $\frac{8^2 \times 2^3}{4^2}$ (i) $\frac{3^3 \times 4^4}{2^6}$

5 Convert the following fractions to percentages. (Round answers to two decimal places if necessary.) **Ex 2.5**

(a) $\frac{17}{20}$ (b) $\frac{3}{5}$ (c) $\frac{5}{8}$ (d) $\frac{20}{35}$

6 Write the following fractions as recurring decimals using the correct notation. **Ex 2.2**

(a) $\frac{1}{3}$ (b) $\frac{2}{3}$ (c) $\frac{1}{6}$ (d) $\frac{5}{9}$

7 Write the following percentages **(i)** as decimals and **(ii)** as fractions in simplest form. **Ex 2.6**

(a) 70% (b) 45% (c) 120% (d) 8%

8 Find the following percentages. (Round answers to two decimal places if necessary.) **Ex 2.8**

(a) 20% of $240 (b) 65% of 700 kg (c) 42% of 8.5 m

(d) 3% of 16 seconds (e) 78% of $19.40 (f) 110% of 400 m

9 For each of the following, draw a number line from -3 to 3 and indicate the positions of the numbers (you may need to estimate the positions of some). **Ex 2.3**

(a) $-\frac{1}{2}, -\frac{4}{3}, 1\frac{1}{2}, \frac{2}{3}$ (b) 0.7, -1.2, -0.5, 2.1, 0.02

10 Which of the following is equivalent to 10^4? **Ex 1.5**

A 40 B 100 000 C 10 000 D 400

11 Evaluate: **Ex 1.4**

(a) $-9 + 2 \times -5 + 7$ (b) $-4 + (-3)^2 + 10 \div 5$

12 A pair of sunglasses that has a marked price of \$74 will be discounted by 40%. Calculate:

Ex 2.10

(a) the value of the discount

(b) the sale price after the discount has been applied.

13 Simplify the following, then evaluate.

Ex 1.5, 1.6

(a) $6^2 \times 5^2 \times 5^4 \times 6^1 \times 5^3 \times 6^2$

(b) $(7^2 \times 10^1 \times 10^3 \times 7^6) \div (7^2 \times 5^2 \times 5 \times 2^2)$

(c) $(10^2 \times 10^1 \times 10^0) \div (2^0 \times 5^0 \times 5 \times 2)$

(d) $\dfrac{3^0 \times 4 \times 5^2}{2^0} \times \dfrac{(2^3 \times 5^2)^0}{(12 \times 10 \times 5)^0}$

14 Dhanya has a group of 30 insects she is studying in her science lab. After 1 month, the number of insects had doubled, giving 30×2 (60) insects. After 2 months, the number had doubled again, giving $30 \times 2 \times 2$, or 30×2^2 insects. After 3 months, there were $30 \times 2 \times 2 \times 2$, or 30×2^3 insects. The group continues to double in number every month.

(a) Show, using a number in index form, how many insects there will be after **(i)** 6 and **(ii)** 12 months.

(b) Evaluate each of your answers to **(a)** to give an actual number of insects after 6 and 12 months.

15 An ice cube tray of water at room temperature (20°C) is placed in a freezer, where it takes 3 hours to freeze solid. What is the average hourly drop in temperature? Give your answer as a negative fraction in simplest form.

16 Show by evaluating that $(2 \times 3)^2 = 2^2 \times 3^2$.

17 Ella works in a clothing shop. Her manager has asked her to place the following items on sale at a 20% discount. Find the sale price for the following.

(a) Jeans, \$65 (b) Jacket, \$89 (c) T-shirts, \$24 (d) Socks, \$8.50

Reasoning and communicating

18 Casey is keeping track of how many goals she scores in her netball matches. She wants to steadily improve her accuracy.

Match 1: 10 goals out of 14 shots
Match 2: 15 goals out of 20 shots
Match 3: 13 goals out of 18 shots
Match 4: 12 goals out of 16 shots

Convert each of Casey's results to a percentage, then use them to rank the matches from her best performance to her worst. Round the answers to one decimal place where necessary. Would you say that Casey is steadily improving her accuracy?

19 (a) As a decimal, the value of $\frac{8}{17}$ is closest to:

A 0.4 B 0.47 C 0.48 D 0.5

(b) As a decimal, the value of $\sqrt{78}$ is closest to:

A 8 B 8.8 C 8.9 D 9

20 Ali has performed the following calculation in a maths assignment.

$2^4 + 3^3 + 5^2$
$= 8 + 9 + 10$
$= 27$

His teacher has marked this as incorrect. Explain to Ali what he has done wrong, and what he needs to remember when working with numbers in index form.

SECTION 2

Algebra, Linear and Non-Linear Modelling

3 Algebra

P Recall 3

Prepare for this chapter by attempting the following questions. If you have difficulty with a question, download the Recall Worksheet.

1 Write algebraic expressions to represent each of the following.

(a) ten more than x **(b)** m divided by three

(c) two lots of l **(d)** y less than twenty

2 Calculate:

(a) $6 + 3 \times 4$ **(b)** $2 \times -3 - 6 \times 5$

(c) $5 \times (14 - 10) \div 10$ **(d)** $5 - 11 + (16 - 15) \times 8 + 2$

3 Substitute each of the values of m into the formula $l = 2m + 1$ to find l.

(a) $m = 4$ **(b)** $m = -2$

4 Answer TRUE or FALSE for each of the following statements.

(a) $y = 3x - 4$ is an expression **(b)** $3m$ is a term

(c) $2x + 3 - 4a$ is an equation **(d)** $7y - 10$ is an expression

5 Express each number as a product of its prime factors.

(a) 18 **(b)** 28 **(c)** 42 **(d)** 660

6 Write each of the following in index form.

(a) $3 \times 3 \times 3 \times 3$ **(b)** six cubed

(c) fourteen squared **(d)** five to the power seven

7 Write these numbers in expanded form, then evaluate.

(a) 9^2 **(b)** $(-3)^4$ **(c)** $10^4 - 2^5$

8 Write down the highest common factor (HCF) of each of the following.

(a) 6 and 10 **(b)** 15 and 55 **(c)** 10, 30 and 42

Syllabus references

Number and Algebra: Algebraic Techniques 1

✓ Introduce the concept of variables as a way of representing numbers using letters **(ACMNA175)**
✓ Extend and apply the laws and properties of arithmetic to algebraic terms and expressions **(ACMNA177)**
✓ Simplify algebraic expressions involving the four operations **(ACMNA192)**

Number and Algebra: Algebraic Techniques 2

✓ Create algebraic expressions and evaluate them by substituting a given value for each variable **(ACMNA176)**
✓ Extend and apply the distributive law to the expansion of algebraic expressions **(ACMNA190)**
✓ Factorise algebraic expressions by identifying numerical factors **(ACMNA191)**
✓ Factorise algebraic expressions by identifying algebraic factors

Variables and expressions

3.1

Need to Know

The language of algebra

A **variable** describes an unknown amount in an algebraic expression or equation. Variables are represented by a letter or a symbol.

For example: The average amount of water used by a Year 8 student per day is a variable that can be represented by symbols such as a, x, θ or β.

These symbols can also be referred to as pronumerals.

A **term** can be:

- a number or a variable by itself: 7, x
- a number multiplied by a variable: $12c$, $5k$
- different variables multiplied together: mn, abc
- a variable multiplied by itself a number of times: a^3, b^2
- the product of a number and several variables: $3xy$, $4a^2b$.

An **expression** is formed when terms are added or subtracted.

For example: $3x + 4y$.

A **coefficient** of x is a number by which the variable is multiplied. The coefficient is written in front of any variable and includes any negative sign in front of the number.

For example: 8 is the coefficient of $8x$ and -6 is the coefficient of $-6y$.

Note that the coefficient of x is 1.

A **constant** is a term that does not contain any variable factors. It is a number written by itself.

For example: 6 is the constant in $3x + 6$.

An **equation** joins two expressions with an equals sign.

For example: $3x + 6 = 2x - 4$.

Algebraic conventions

- A multiplication sign is not required between variables and numbers or other variables: $5 \times b = 5b$, $a \times d \times f = adf$
- Division can be shown as a fraction: $3a \div b = \dfrac{3a}{b}$
- Indices show variables being multiplied by themselves: $c \times c = c^2$
- Numbers are written in front of a variable: $a \times 4 = 4a$
- Variables are written in alphabetical order: $y \times z \times x = xyz$
- Algebraic terms when multiplied are treated as if in brackets: $a \div bc = a \div (bc)$

3.1

Worked Example 1

WE1

For $6a + 3b - 4 = 2ab - c + 7$:

(a) identify whether this is an equation or an expression

(b) write down the coefficient of b

(c) write down all the terms in the equation or expression

(d) list the variables

(e) write down any constants.

Working

(a) It is an equation. (It has an equals sign.)

(b) 3 is the coefficient of b.

(c) $6a$, $3b$, -4, $2ab$, $-c$ and 7

(d) a, b and c are variables.

(e) -4 and 7 are constants.

Worked Example 2

WE2

Rewrite the following using algebraic conventions.

$t \times m \times 6 \times m \div 5$

Working

$$t \times m \times 6 \times m \div 5$$

$$= \frac{t \times m \times 6 \times m}{5}$$

$$= \frac{6 \times t \times m \times m}{5}$$

$$= \frac{6tmm}{5}$$

$$= \frac{6mmt}{5}$$

$$= \frac{6m^2t}{5}$$

3.1

Worked Example 3

WE3

Pete has b blue pens and r red pens. Sue has three times as many blue pens as Pete. She has 2 fewer red ones than Pete.

(a) Write an expression for the number of pens Pete has altogether.

(b) If Pete has 16 pens to start with, write an equation to show this information.

(c) Write an expression for the number of blue pens Sue has.

(d) Write an expression for the number of red pens Sue has.

(e) If Sue has 28 pens altogether, write an equation to show this information.

(f) If Pete loses 4 of his pens, write an expression in terms of b and r to show how many pens Pete has now.

Working

(a) $b + r$ (b) $b + r = 16$ (c) $3b$

(d) $r - 2$ (e) $3b + r - 2 = 28$ (f) $b + r - 4$

3.1 Variables and expressions

Answers page 496

Navigator

Q1, Q2, Q3, Q4, Q5, Q6, Q7, Q8, Q9, Q10, Q11, Q13, Q17	Q1, Q2, Q3, Q4, Q5, Q6, Q7, Q8, Q9, Q10, Q11, Q13, Q15, Q16	Q1, Q2, Q3, Q4, Q6, Q7, Q8, Q9, Q10, Q11, Q12, Q13, Q14, Q15, Q16

Fluency and understanding

WE1

1 For each of the following:

(i) $3d + 4f - 3bd - 3 + 7b$ **(ii)** $5g - 3r + 5rf = 7 - 3b$ **(iii)** $b + f - 3bf + 8$

(a) identify whether this is an equation or an expression

(b) write down the coefficient of b

(c) write down all the terms in the equation or expression

(d) list the variables

(e) write down any constants.

WE2

2 Rewrite the following using algebraic conventions.

(a) $d \times 4 \times c$ **(b)** $2 \times t \div 7$ **(c)** $k \times g \times 4 \times k$ **(d)** $6 \div (f \times g)$

WE3

3 Nerida has w watches and r rings. Donna has two times as many watches and 2 fewer rings than Nerida.

(a) Write an expression for the number of watches and rings Nerida has altogether.

(b) If Nerida has 8 watches and rings to start with, write an equation to show this information.

(c) Write an expression for the number of watches Donna has.

(d) Write an expression for the number of rings Donna has.

(e) If Donna has 14 watches and rings altogether, write an equation to show this information.

(f) If Nerida loses 2 of her rings, write an expression in terms of w and r to show how many watches and rings Nerida has now.

3.1

4 (a) The coefficient of x in $6y + 7xy + 5x$ is:

A 5 B 7 C $5x$ D $6y$

(b) The coefficient of xy in $5 + 8xy - 2y + 4x$ is:

A 2 B 4 C 8 D $8xy$

(c) The coefficient of a in $5b + 2ab + 6$ is:

A 0 B 2 C 5 D 6

(d) The coefficient of x in $4xy - 6y + x + 8$ is:

A 0 B 1 C 4 D 6

5 (a) The constant in the expression $\frac{1}{2}x^2y - 3 + 5xy$ is:

A -3 B $\frac{1}{2}$ C 3 D 5

(b) The constant in $3ef + 7efg + 12 + 11ef + 4e$ is:

A 2 B 7 C 11 D 12

6 Write the following without division and multiplication signs or brackets.

(a) $x \div 6$
(b) $h \div 9$
(c) $6 \times a \div 11$
(d) $15 \div (3 \times r)$
(e) $21 \div (12 \times v)$
(f) $4 \times s \div 19$
(g) $8 \div x - u \div 6$
(h) $h \div 5 + 4 \div i$
(i) $c \times u \div 5 + 9 \times y$
(j) $q \div (7 \times c) - g \times h \div 4$
(k) $v \times z \div 6 - 8 \div (f \times s)$
(l) $3 \div (t \times r) + 6 \times w \div (y \times z)$
(m) $4 \times h \times b \div (2 \times r)$
(n) $6 \times c \times a \div (5 \times e \times u)$

7 There are a apples and p pears in a fruit dish. There are 5 apples and 4 pears in a second dish. The total number of pieces of fruit is:

A $a + p$ B $5a + 4p$ C $a + 5a + p + 4p$ D $a + p + 9$

8 Andrew has y number of pencils in his pencil case.

(a) Dina has $y + 7$ pencils in her pencil case. What does this mean?

(b) Simon has $2y$ pencils in his pencil case. What does this mean?

(c) Suppose Cindy has $2y - 2$ pencils in her pencil case. Does she have more or fewer pencils than Simon?

(d) If Andrew took out 3 pencils from his pencil case, how many would he have left?

(e) If Andrew gave the 3 pencils to Dina and she put them in her pencil case, how many would she have?

(f) If Simon took 3 pencils out of his pencil case, how many would he have left?

(g) If Simon gave the 3 pencils to Cindy and she put them in her pencil case, how many would Cindy now have?

9 Suppose the number of cups you have is m and the number of glasses you have is n.

(a) How many cups and glasses do you have altogether?

(b) If you broke 2 glasses and threw them away, how many glasses would you now have?

(c) How many cups and glasses would you now have altogether?

(d) If you then bought another set of 8 cups, how many cups would you have?

(e) How many cups and glasses would you now have altogether?

10 Suppose x is the number of people in Sweats Sports Store at 3 p.m. on a certain Friday afternoon.

(a) If there are twice as many people at 3:30 p.m., write this in terms of x?

(b) If the number of people at 4:00 p.m. is $3x$, what does this mean?

(c) If five people leave the store between 4:00 p.m. and 4:15 p.m., and no one comes in, write an expression for the number of people in the store at 4:15 p.m.

(d) If nine people come into the store and two leave between 4:15 p.m. and 4:30 p.m., write an expression for the number of people in the store at 4:30 p.m.

11 In a glass jar there are some different coloured jelly beans. There are g green jelly beans.

(a) There are twice as many red jelly beans as green ones. Write an expression for the number of red jelly beans in terms of g.

(b) The number of blue jelly beans is five less than the number of green ones. Write an expression for the number of blue jelly beans in terms of g.

(c) The number of black jelly beans is four more than the number of green ones. Write an expression for the number of black jelly beans in terms of g.

12 On a supermarket shelf there are different varieties of chocolate. There are m kinds of milk chocolate.

(a) The number of soft-centred varieties is one more than twice the number of milk chocolate varieties. Write an expression for the number of soft-centred varieties in terms of m.

(b) There are three more varieties of chocolate-nut than soft-centred varieties. Write an expression for the number of chocolate-nut varieties in terms of m.

(c) The number of varieties of white chocolate is two more than one-third of the amount of varieties of milk chocolate. Write an expression for the number of white chocolate varieties in terms of m.

13 Lauren takes h hours and m minutes to finish her mathematics project. Tim takes twice as long as Lauren to complete his project. The time Tim takes in minutes is:

A $2h + 2m$ **B** $2h + 120m$ **C** $60h + 2m$ **D** $120h + 2m$

Reasoning, communicating and open-ended

14 Raymond is playing Crazy Maze on his games console. He starts with p points. Raymond plays the game for 45 minutes and every 5 minutes he notes how many points he has. The number of points he has is shown in the table below.

(a) At what time did Raymond have the most points?

(b) How many more points than he started with would Raymond have had if he had stopped playing at this time?

(c) At what time did Raymond have the fewest points?

(d) How many points would Raymond have lost if he had stopped playing at this time?

(e) At what times should Raymond have stopped playing if he was happy with breaking even?

Time	Number of points
10:00 p.m.	p
10:05 p.m.	$p + 2$
10:10 p.m.	$p + 7$
10:15 p.m.	$p + 1$
10:20 p.m.	$p + 4$
10:25 p.m.	p
10:30 p.m.	$p - 4$
10:35 p.m.	p
10:40 p.m.	$p - 10$
10:45 p.m.	$p - 2$

3.1

15 Suppose e stands for the number of people expected to be at a party. The table on the right tells us how many people were at the party during the evening.

Write a paragraph describing what happened during the evening.

Time	Number of people
6:00 p.m.	$e - 20$
6:10 p.m.	$e - 18$
6:15 p.m.	$e - 15$
6:20 p.m.	$e - 11$
6:25 p.m.	$e - 12$
6:30 p.m.	$e - 6$
7:00 p.m.	$e - 3$
8:00 p.m.	$e - 13$
9:30 p.m.	$e - 18$
11:00 p.m.	$e - 20$

16 Ruth has been answering some questions involving algebra, with k representing the variable in the expression.

(a) What could k represent?

(b) In this case, what would $2k$ represent?

(c) What would $k + 4$ represent?

(d) What would $\frac{k}{2} - 1$ represent?

17 (a) Write a question of your own that is similar to Question **3**.

(b) Answer the question you have written.

Problem solving

Tricky algebra

Instructions

1. Pick a number (keep it secret).
2. Multiply your number by 3.
3. Add 6.
4. Take away the number you started with.
5. Divide by 2.
6. Again, take away the number you started with.

You have finished with 3.
Use algebra to show how this works.

Create your own algebra trick and give this to a friend for them to work out.

Strategy options
- Work backwards.
- Make a model.

Hint
Let your number be 'x'.

Substitution for variables

3.2

Need to Know

If the value for an expression is to be found when the variables have certain values, replace the variables in the expression with those values. This process is called **substitution**.

For example:

In the expression $3x + 2y$, x and y are variables. When $x = 5$ and $y = 1$, the expression can be evaluated by substituting 5 for x and 1 for y so:

$$3x + 2y = 3 \times (5) + 2 \times (1)$$
$$= 17$$

Because the variables can vary, different values for the variable will give different values for the expression. When $x = -3$ and $y = 7$, the expression can be evaluated by substituting -3 for x and 7 for y so:

$$3x + 2y = 3 \times (-3) + 2 \times (7)$$
$$= 5$$

Worked Example 4

WE4

Evaluate the expression $2b - a$ when $a = 3$ and $b = 10$.

Working

When $a = 3$ and $b = 10$

$$2b - a = 2 \times 10 - 3$$
$$= 20 - 3$$
$$= 17$$

Worked Example 5

WE5

Evaluate the expression $2x^2 - 3y$ when $x = -3$ and $y = -1$.

Working

When $x = -3$ and $y = -1$

$$2x^2 - 3y = 2 \times (-3)^2 - 3 \times (-1)$$
$$= 2 \times 9 + 3$$
$$= 18 + 3$$
$$= 21$$

3.2

Worked Example 6

WE6

Use the following rules to complete the following tables.

(a) $y = 3x$

x	-1	0	2
y			

(b) $y = x^2$

x	-2	0	2
y			

Working

(a) $y = 3x$
$= 3 \times (-1), x = -1$
$= -3$

$y = 3x$
$= 3 \times (0), x = 0$
$= 0$

$y = 3x$
$= 3 \times (2), x = 2$
$= 6$

x	-1	0	2
y	-3	0	6

(b) $y = x^2$
$= (-2)^2, x = (-2)$
$= 4$

$y = x^2$
$= 0^2, x = 0$
$= 0$

$y = x^2$
$= 2^2, x = 2$
$= 4$

x	-2	0	2
y	4	0	4

Answers page 496

3.2 Substitution for variables

Navigator

Q1 Column 1, Q2 Column 1, Q3 Column 1, Q4 Column 1, Q5 Column 1, Q7, Q9, Q10, Q11, Q13, Q14	Q1 Column 2, Q2 Column 2, Q3 Column 2, Q4 Column 2, Q5 Column 2, Q6, Q7, Q9, Q10, Q12, Q13, Q14, Q15	Q1 Column 3, Q2 Column 3, Q3 Column 3, Q4 Column 3, Q5 Column 3, Q6, Q8, Q9, Q10, Q11, Q12, Q13, Q14, Q15, Q16

Fluency and understanding

WE4

1 Evaluate the following expressions when $a = 3$ and $b = 6$.

(a) $a + 2b$ (b) $3a + 2b$ (c) $2a + 5b$
(d) ab (e) $3ab$ (f) $10ab - a$
(g) $8b - 16a$ (h) $10ab - b$ (i) $6b - ab$
(j) $12a^2$ (k) $2b^3$ (l) $4a^3$
(m) $a^2 - b^3$ (n) $3a^3 - 2b^2$ (o) $2a^5 - 2b^3$
(p) $\frac{2a^2}{b}$ (q) $\frac{a^3}{b^2}$ (r) $\frac{144a^2}{3b^3}$
(s) $a^2 - b - \frac{a}{3}$ (t) $a^3 + 2b^2 - \frac{b}{2}$ (u) $3b^2 + \frac{a^5}{2} - 5$

3.2

2 Evaluate the following expressions when $x = -2$ and $y = -5$. WE5

(a) $x + y$ **(b)** $2x + 12y$ **(c)** $10x + 5y$

(d) $y - 4x$ **(e)** $2y - 3x$ **(f)** $2x - 3y$

(g) $-5y - 3 + x$ **(h)** $8 - 6y + 2x$ **(i)** $7y - 4 - 4x$

(j) $y^2 - x^2$ **(k)** $x^3 + y^3$ **(l)** $-5y^2$

3 Use the rules given to complete the following tables. WE6

(a) $y = x - 2$

x	-2	0	-4	-8
y				

(b) $y = x - 4$

x	-5	-3	-2	-7
y				

(c) $y = x - 9$

x	-7	-1	-5	-9
y				

(d) $y = 4x - 1$

x	1	2	3	4
y				

(e) $y = 2x - 4$

x	-2	0	1	2
y				

(f) $y = 3x + 14$

x	-4	-7	-20	11
y				

(g) $y = -6x$

x	-3	-2	-1	-5
y				

(h) $y = -5x - 3$

x	-1	0	3	5
y				

(i) $y = 3 - 4x$

x	-3	-1	2	4
y				

(j) $y = \frac{x}{2}$

x	8	6	12	20
y				

(k) $y = \frac{x}{3} - 2$

x	15	9	12	30
y				

(l) $y = \frac{x}{4} - 1$

x	12	-4	8	-24
y				

(m) $y = x^2$

x	-2	-1	0	1
y				

(n) $y = 2x^3 - 1$

x	-2	0	2	4
y				

(o) $y = x^2 + 3x + 1$

x	-4	-2	0	2
y				

4 Evaluate these expressions when $d = -2$ and $e = 5$.

(a) $-3de - 6$ **(b)** $-5de + 100$ **(c)** $-8de + 15$

(d) $\frac{10}{e}$ **(e)** $\frac{40}{d}$ **(f)** $\frac{e}{d}$

(g) $\frac{2}{d} - \frac{5}{e}$ **(h)** $\frac{12}{d} - \frac{30}{e}$ **(i)** $\frac{15}{e} - \frac{18}{d}$

(j) $e^3 - 3d^2$ **(k)** $e^2 - 3d^2$ **(l)** $2e^3 - 3d^2$

5 Evaluate the following expressions for $m = 3$ and $n = -2$.

(a) $2(3m - n)$ **(b)** $4n(m - 3)$ **(c)** $m(6 + 3n)$

(d) $6m(8 - n)$ **(e)** $3m(n - 2)$ **(f)** $6n(3m - 4)$

(g) $\frac{m + n}{3}$ **(h)** $\frac{2m + n}{5}$ **(i)** $\frac{6m - n}{4}$

(j) $\frac{2m + n}{n}$ **(k)** $\frac{5(m + 2n)}{5m}$ **(l)** $\frac{3(m - 4n)}{4n}$

3.2

6 When $u = 6$ and $v = 8$ are substituted into the expression $\frac{v(3u + v - 2)}{u}$, the result is:

A 16 **B** 20 **C** 25 **D** 32

7 When the values $a = -1$, $b = 2$ and $c = -3$ are substituted into the expression $ab + bc$, the answer is:

A -8 **B** -4 **C** -2 **D** 4

8 If $d = 3$ and $e = -4$, then $de(d + e)$ is:

A -12 **B** 84 **C** 12 **D** -84

9 Boon buys d DVDs at \$$y$ each.

(a) Write an expression to show how much Boon spent.

(b) Write an expression to show how much Boon received if he sold k DVDs for \$$m$ each.

(c) Write an expression to calculate the profit Boon made.

(d) If $d = 7$, $y = 12$, $k = 5$ and $m = 21$, find the profit Boon made.

(e) One DVD is returned to Boon as it is faulty. Does Boon still make a profit? Explain why or why not.

10 Asha has a dollars, Mayoarini has $2b$ dollars and Danni has $3c$ dollars.

(a) What is the difference between Danni's amount of money and Asha's amount of money if Danni has more money than Asha?

(b) Mayoarini gives her money to Asha. What is the difference now between Danni's amount of money and Asha's amount of money if Asha now has more money than Danni?

(c) How much money is there altogether if $a = 3$, $b = 5$ and $c = 4$?

11 Michael has scored r runs in a cricket game. His friend, Ravi, has scored $2r - 4$ runs.

(a) What is the smallest value that r could be?

(b) Show that Ravi and Michael score the same number of runs if $r = 4$. Explain.

(c) If $r > 4$, did Ravi or Michael score more runs? Explain.

Reasoning, communicating and open-ended

12 It takes $\frac{3N}{P+1}$ minutes to package biscuits, where N is the number of biscuits and P is the number of packing machines.

(a) How long does it take to package 1500 biscuits if 99 machines are used?

(b) Forty-nine machines are to be used to package 2200 biscuits in 2 hours. Show why this is not possible.

(c) Five faulty machines have been repaired so that now there are 54 machines available. Show whether it is now possible to package the 2200 biscuits in 2 hours.

13 **(a)** By substituting $a = 60$, $b = 5$ and $c = 2$, evaluate each of these expressions:

(i) $\frac{a}{bc}$ **(ii)** $a \div (b \times c)$

(iii) $a \div b \times c$ **(iv)** $\frac{a}{b} \times c$

(b) Which expressions are equivalent?

(c) Which expression(s) are equivalent to $a \div bc$?

3.2

14 Find two different sets of values for the pronumerals M and N so that $\frac{3M + N}{2M - N}$ gives a whole number when M and N are substituted and the expression evaluated.

15 Lisa, Anna and Saji each choose an integer between -10 and +10.

Let a represent Lisa's number, b represent Anna's number and c represent Saji's number.

(a) Find one set of possible values for a, b and c if the sum of the three numbers is 0.

(b) Find three different sets of possible values for each girl's number if the product of the three numbers is -100 and a, b and c take different values.

16 Find three sets of whole number values for a and b so that $a^3 - 3b^2$ has a result that is less than 20.

3.3 Using formulas

Need to Know

Formulas

Formulas (or formulae) are algebraic equations that are used in practical situations.

Formulas are used to calculate one quantity from others. For instance, formulas are used to find the perimeter and area of shapes.

For example:

The perimeter P of a rectangle with length l and breadth b is given by the formula:

$P = 2(l + b)$ or $P = 2l + 2b$.

The area A of the rectangle is given by the formula:

$A = lb$, where l is the length and b is the breadth.

Formulas are also used to convert from one unit to another, such as converting between imperial and metric measurements or converting currency.

For example:

- Converting mass in pounds to mass in kilograms
- Converting distance in miles to distance in kilometres
- Converting temperature in degrees Fahrenheit to temperature in degrees Celsius.

Worked Example 7

WE7

The formula $F = \frac{9C}{5} + 32$ is used to convert temperatures in degrees Celsius, C, to degrees in Fahrenheit, F. Use this formula to convert 29°C into degrees Fahrenheit.

Working

$$F = \frac{9C}{5} + 32$$
$$= \frac{9(29)}{5} + 32$$
$$= 52.2 + 32$$
$$= 84.2$$

29°C = 84.2°F

3.3

3.3 Using formulas

Answers page 497

Navigator

Q1, Q2, Q3, Q4, Q5, Q6, Q7, Q8, Q9 (a) & (b), Q10 (a) & (b), Q12	Q1, Q2, Q3, Q4, Q5, Q6, Q7, Q8, Q9, Q10, Q11, Q12	Q1, Q2, Q3, Q4, Q5, Q6, Q7, Q8, Q9, Q10, Q11, Q12, Q13

Fluency and understanding

WE7

1 The formula $F = \frac{9C}{5} + 32$, is used to convert temperatures in degrees Celsius, C, to degrees in Fahrenheit, F. Use this formula to convert the following degrees Celsius into degrees Fahrenheit.

(a) 10°C **(b)** 5°C **(c)** 35°C

2 The formula $C = \frac{5F - 160}{9}$ converts degrees Fahrenheit, F, to degrees Celsius, C. Find the value of C when F is:

(a) 32°F **(b)** 41°F **(c)** 50°F

3 The distance travelled by a moving object travelling at a constant speed can be calculated using the formula $d = st$ where

d = distance travelled in metres
s = speed of object in metres/second
t = time in seconds that the object is in motion.

Calculate the value of d when:

(a) $s = 8, t = 4$ **(b)** $s = 4\frac{1}{2}, t = 1\frac{1}{2}$

4 The time, t seconds, needed for an elevator of power w kilowatts to lift m kilograms h metres is given by the formula $t = \frac{mh}{25w}$. If $w = 20$, the time in seconds it takes to lift a group of people of total mass 400 kilograms to a height of 50 metres is:

A 20 seconds **B** 25 seconds **C** 35 seconds **D** 40 seconds

5 The amount, C dollars, to be paid for a mobile phone bill is found from the formula: $C = R + 0.5L + 0.2T$

where R is the rental of the equipment
L is the number of calls made
T is the number of text messages sent.

When the rental is $32 and there are 110 calls and 25 text messages, the amount of the bill is:

A $55.40 **B** $92.00 **C** $66.50 **D** $42.50

6 The salespeople who work at Crazy Joe's Motors are paid according to the formula $P = 250 + 100c$ where P is the total pay for one week in dollars and c is the number of cars the salesperson sells in the week.

(a) Warren didn't sell any cars one week. How much money would he have earnt that week?

(b) Caitlin sold five cars in the same week. How much would she have earnt?

(c) Claudio wants to earn over $1000 a week. What is the minimum number of cars he would have to sell to earn this much?

3.3

7 The formula $d = 4.9t^2$ tells you how far a dropped object has travelled after a certain time, where d is the distance the object has fallen in metres and t is the time it has been falling in seconds.

(a) Suppose a football has been dropped from Sydney Tower. What distance would it fall in 2 seconds?

(b) Divers on the high cliffs in Acapulco, Mexico, usually take 4 seconds to hit the water. Approximately how high are the cliffs?

(c) Wayne the skydiver always counts out 10 seconds just after he jumps out of the plane before he pulls the rip cord to open his parachute. How far does he fall in this time?

8 Temperatures can be converted from degrees Celsius, C, to degrees Fahrenheit, F, using a less accurate but simpler formula than the formula used in Question **1**. The formula is $F = 2C + 30$.

(a) Use this formula to convert 10°C to degrees Fahrenheit.

(b) Compare your answer to part **(a)** to your answer to Question **1 (a)**.

(c) Use this formula to convert 5°C to degrees Fahrenheit.

(d) Compare your answer to part **(c)** to your answer to Question **1 (b)**.

(e) Use this formula to convert 35°C to degrees Fahrenheit.

(f) Compare your answer to part **(e)** to your answer to Question **1 (c)**.

(g) In most parts of Australia daily minimum temperatures rarely fall below 0°C and daily maximum temperatures rarely exceed 40°C. For the temperature range of 0°–40°, how close to the accurate temperatures in Fahrenheit, calculated using $F = \frac{9}{5}C + 32$, are the approximate temperatures calculated using $F = 2C + 30$?

9 The formula for finding how much water goes into a rectangular fish tank is $V = lbd$

where V = the volume of water (in cubic centimetres)
l = the length (in cm) of the tank
d = the depth (in cm) of the water in the tank
b = the breadth (in cm) of the tank.

(a) How much water does a tank hold if its length is 80 cm, its width is 30 cm and its depth is 20 cm?

(b) How much water would you put in the tank if you want the water level to be 5 cm from the top?

(c) Lucia is pumping the water from her smaller fish tank into her larger one. The smaller tank is 50 cm long, 20 cm wide and 20 cm deep. The larger tank is 100 cm long, 50 cm wide and 40 cm deep. After Lucia has finished, how much *more* water will she need to add to the large tank to fill it completely?

3.3

Reasoning, communicating and open-ended

10 Banks use the formula $I = PRT$ to calculate simple interest, which is the amount of money in interest that the money in your account earns.

I stands for the *simple interest*.

P stands for the *principal*, which is the amount of money you have in an account.

R stands for the *rate* of interest the bank is offering. This is given as a percentage per annum 'p.a.' (for a year). (Percentages need to be converted to a decimal value.)

T stands for the *time* the money has been in the bank account. This is usually measured in years and is the term of the investment.

(a) How much interest would you earn on \$100 if you had it in an account for 1 year at an interest rate of 7% p.a.?

(b) How much interest would you earn on \$1000 if you had it in an account for 5 years at an interest rate of 12% p.a.?

(c) Suppose you have \$1000 that you want to invest for 5 years, and you have a choice between two banks: Aussie Bank and Matilda Bank which both offer a 5-year term investment deal. Aussie Bank offers a simple interest rate of 11% p.a., but charges a once-only set-up fee of \$100. Matilda Bank's simple interest rate is 12% p.a., but it charges a once-only set-up fee of \$160.

(i) Which bank should you invest your money in?

(ii) How much better off would you be with this bank?

(iii) Suppose you have \$2000 that you want to invest for 5 years. Which bank should you invest your money in? How much better off would you be with this bank?

11 The volume V cubic metres of a pyramid of height h metres that has a rectangular base is given by the formula $V = \frac{lbh}{3}$, where l is the length of the base and b is the breadth.

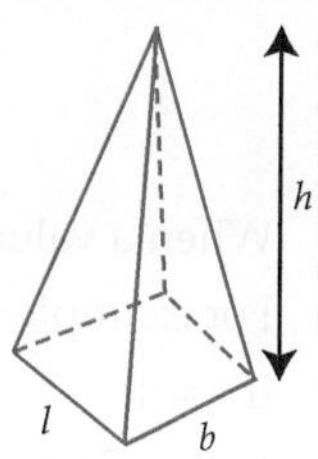

(a) The Miza pyramid in Egypt is 30 metres long, 25 metres wide, and 60 metres high. What is its volume?

(b) The pyramid is made from stone blocks with a volume of five cubic metres each. If it is known that Miza is completely solid, how many blocks were used to build the pyramid?

12 The formula $A = l \times b$ is used to find the area of a rectangle. Given that the area is 20 cm^2 and the length, l, is longer than the breadth, b, find the possible dimensions of two different rectangles.

13 The volume of a triangular prism is given by $V = \frac{bhH}{2}$ where b is the base length, h is the perpendicular height of the triangle and H is the height of the prism.

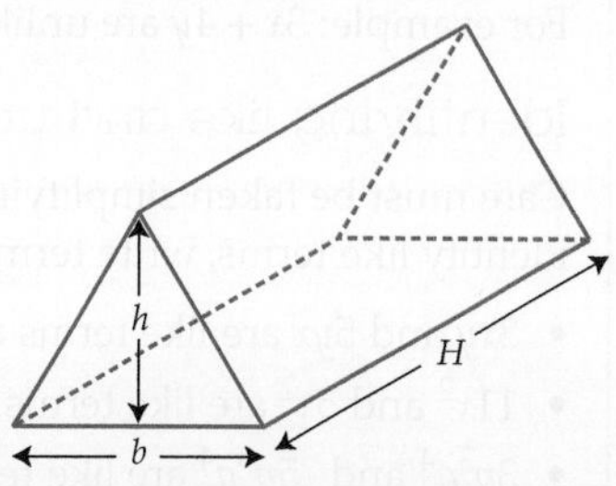

Find two different sets of whole-number values for b, h and H if V is a whole number between 12 000 and 25 000, and each of b, h and H must be at least 20.

3.4 Simplifying expressions

Need to Know

Like and unlike terms

Algebraic terms that have exactly the same variables are called **like terms**.

Like terms can be simplified by adding their coefficients.

For example:

$3x$ and $5x$ are like terms and can be simplified by adding the coefficients of x.

$3x + 5x = 8x$

This can be demonstrated by adding two rectangles of the same length, x m, together if one has a width of 3 m and the other a width of 5 m.

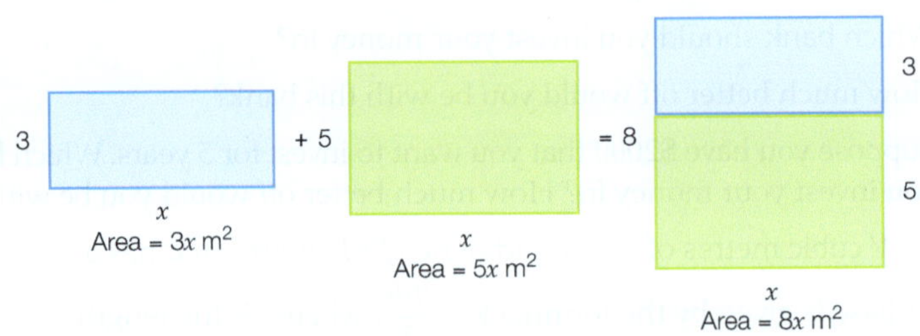

When a value is substituted for x in both expressions, the result will be the same.

For example:

If $x = 4$,

$$\begin{aligned} \text{LHS} &= 3x + 5x \\ &= 3 \times 4 + 5 \times 4 \\ &= 12 + 20 \\ &= 32 \end{aligned} \qquad \begin{aligned} \text{RHS} &= 8 \times x \\ &= 8 \times 4 \\ &= 32 \end{aligned}$$

An expression can be simplified by collecting like terms. **Unlike terms** cannot be simplified.

For example: $3x + 4y$ are unlike terms and the expression cannot be simplified.

Identifying like and unlike terms

Care must be taken simplifying expressions when several variables are involved. To identify like terms, write terms with more than one variable in alphabetical order.

- $3xy$ and $5yx$ are like terms as $5yx = 5xy$.
- $11x^2$ and $5x^2$ are like terms.
- $3p^3q^4$ and $-5p^3q^4$ are like terms.

However:

- $5ab$ and $2ac$ are unlike terms, as $ab \neq ac$. ($\neq$ means is not equal to.)
- $3xy$ and $5xyz$ are unlike terms, as $xy \neq xyz$.
- $6x$ and $4x^2$ are unlike terms, as $x \neq x^2$.

3.4

Worked Example 8

WE 8

In each of the following, write down whether the pair of terms is like or unlike.

(a) $5t$ and $3t$

(b) $2xy$ and $2x$

(c) $6m^2n$ and $4mn^2$

(d) $3xy^2$ and $7y^2x$

Working

(a) $5t$ and $3t$ are like terms.
(Both terms have t as their only variable.)

(b) $2xy$ and $2x$ are unlike terms.
(The first term has $x \times y$, but the second term only has x.)

(c) $6m^2n$ and $4mn^2$ are unlike terms.
(The first term has $m \times m \times n$ whereas the second term has $m \times n \times n$.)

(d) $3xy^2$ and $7y^2x$ are like terms.
(The first term has $x \times y \times y$ and the second term has $y \times y \times x$. So, it has exactly the same variables, just in a different order.)

Worked Example 9

WE 9

Simplify the following expressions, if possible, then check your answers by substituting the value for the variables given in brackets.

(a) $5x + 3y - x + 7y$ ($x = 1$, $y = 2$)

(b) $7x^2 + 4xy + 8yx - 2x$ ($x = 3$, $y = 2$)

Working

(a)

$$5x + 3y - x + 7y$$
$$= 5x - x + 3y + 7y$$
$$= 4x + 10y$$

$$5x + 3y - x + 7y$$
$$= 5 \times 1 + 3 \times 2 - 1 + 7 \times 2$$
$$= 5 + 6 - 1 + 14$$
$$= 24$$

$$4x + 10y$$
$$= 4 \times 1 + 10 \times 2$$
$$= 24$$

As both expressions equal 24, the simplification is correct.

(b)

$$7x^2 + 4xy + 8yx - 2x$$
$$= 7x^2 - 2x + 4xy + 8yx$$
$$= 7x^2 - 2x + 4xy + 8xy$$
$$= 7x^2 - 2x + 12xy$$

$$7x^2 + 4xy + 8yx - 2x$$
$$= 7 \times 3^2 + 4 \times 3 \times 2 + 8 \times 2 \times 3 - 2 \times 3$$
$$= 63 + 24 + 48 - 6$$
$$= 129$$

$$7x^2 - 2x + 12xy$$
$$= 7 \times 3^2 - 2 \times 3 + 12 \times 3 \times 2$$
$$= 63 - 6 + 72$$
$$= 129$$

As both expressions equal 129, the simplification is correct.

3.4

Answers page 497

3.4 Simplifying expressions

Navigator

Q1 Column 1, Q2 Column 1, Q3 (a–d), Q4, Q5, Q6, Q7, Q10, Q11, Q12, Q16	Q1 Column 2, Q2 Column 2, Q3 (c–f), Q4, Q5, Q6, Q7, Q8, Q10, Q11, Q12, Q14	Q1 Column 2, Q2 Column 2, Q3 (e–h), Q4, Q5, Q6, Q9, Q10, Q11, Q12, Q13, Q14, Q15

Fluency and understanding

WE8

1 In each of the following, write down whether the pair of terms is like or unlike.

(a) $4w$ and $6w$ (b) $7u$ and $7w$

(c) $7t$ and $6i$ (d) $3e$ and $5e$

(e) $6d$ and $3de$ (f) yx and $33xy$

(g) $9xyz$ and $5xyz$ (h) $6x$ and $52xyz$

(i) $6xy$ and $8yx$ (j) $2xyz$ and $4zyx$

(k) $4x^2y$ and $7x^2y$ (l) $8xy^2$ and $12yx^2$

WE9

2 Simplify the following expressions, if possible, then check your answers by substituting the value for the variables given in brackets.

(a) $6a + 11a$ $(a = 1)$ (b) $15f + 6f$ $(f = 6)$

(c) $7v - 2v$ $(v = 4)$ (d) $5v - 8v$ $(v = 2)$

(e) $-7w + 17w$ $(w = -2)$ (f) $-12d + 15d$ $(d = -3)$

(g) $8t + 7t + 2d + 5d$ $(d = 1, t = 3)$ (h) $40w + 50w + 100v + 34v$ $(v = 2, w = 1)$

(i) $6a + 5b + 4a - 3b$ $(a = 2, b = 1)$ (j) $18f - 5f + 6g - 3g$ $(f = 1, g = 4)$

(k) $-6b - 8b + 10b$ $(b = 5)$ (l) $-9b - 4b + 18b$ $(b = 4)$

(m) $7d + 8d + 9 - 3d$ $(d = 3)$ (n) $9f + 5 - 2f + 5f$ $(f = 3)$

(o) $-3r + 6u + 10r - 9u$ $(r = 2, u = 5)$ (p) $6m - 9n - 13m - 2n$ $(m = -1, n = 4)$

(q) $7f + 8fg - 4f - 6fg$ $(f = 4, g = 2)$ (r) $8ij + 14j - 2ij + 6j$ $(i = 2, j = 7)$

3 Simplify these expressions.

(a) $3hdw + 6hd + 7d - 4hd + 8 - 2d$ (b) $4x + 15 + 6x + 7xyz - 7 - 3xyz$

(c) $5x^2 - 6y - 13x^2 - 4y + 2$ (d) $7x^3 - y - 12x^3 - 5y + xy$

(e) $28ab + 40 + 30a - 10ab - 6a + 15b$ (f) $9s + 40st + 5 - 100st - 5s + 15$

(g) $-9b - 4b^3 + ab - b^3 + 8b - 6ab$ (h) $6a^3 - 8a^2 - 4a - 5a - 4a^3 - 61$

4 Which of these is a like term for $6y$?

A $6x$ B $14y$ C $6 + y$ D 6

5 Which of these is a like term for mn?

A $9n$ B mnp C amn D $8nm$

6 Which of these is a like term for $5xyz$?

A $22xy$ B $2xzj$ C $xy + z$ D $6zxy$

3.4

7 Answer TRUE or FALSE for each of these statements.

(a) These two rules are really the same:

Rule 1: 'Take a number and multiply it by three. Take the same number and multiply it by seven. Then add the two answers together.'

Rule 2: 'Take a number and multiply it by ten.'

(b) These two rules are really the same:

Rule 1: 'Take a number and multiply it by ten. Take another number and multiply it by five. Then add the two answers together.'

Rule 2: 'Take a number and multiply it by fifteen.'

(c) These two rules are really the same:

Rule 1: 'Take a number and multiply it by fourteen. Take the same number and multiply it by ten. Then subtract the second answer from the first.'

Rule 2: 'Take a number and multiply it by four.'

8 A store is selling all its CDs for c dollars and all its DVDs for d dollars each. Tom bought eight CDs and five DVDs. His friend Ayman bought seven CDs and six DVDs.

(a) Write the expression for the amount Tom paid for his collection.

(b) Write the expression for the amount Ayman paid for his collection.

(c) Write the expression for the amount they paid altogether. Simplify this expression.

9 For each of her birthdays, Alysia is given as a present three times as much money as she was given for her previous birthday. On her first birthday she received d dollars.

(a) Write the expression for the amount of money Alysia received on her fourth birthday.

(b) Write an expression for the total amount of money Alysia received for her first four birthdays.

(c) How much money did Alysia receive altogether for her first four birthdays if $d = 5$?

10 (a) By drawing three rectangles each with a width of 2 cm, show that $2 \times 3 + 2 \times 4 = 2 \times 7$.

(b) Substitute 'y' for '2' and rewrite the statement above in algebra.

Reasoning, communicating and open-ended

11 (a) Which of the following expressions cannot be simplified by collecting like terms?

(i) $5x^2 + 4x$ (ii) $3x^2 + 2x^2$ (iii) $2xy + 6yx$

(iv) $4xy^2 + 5x^2y$ (v) $5xy + xy$ (vi) $4xy - xy$

(b) Simplify the remaining expressions. Explain why these expressions can be simplified.

12 Jess has x ten-cent coins and $10x$ fifty-cent coins.

(a) Write an expression for the total number of coins Jess has.

(b) Ryan has $44x$ coins. Explain what this means, in terms of the coins that Jess has.

(c) Write an expression for the value of Jess' coins in dollars.

(d) Can you find the value of Ryan's coins? Why or why not?

3.4

13 A number of coins are placed on a square of a chessboard. Twice as many coins are then placed on the next square. On the third square, there are twice as many coins stacked as there are on the second square. This process is continued for six squares.

(a) If x is the number of coins on the first square, how many are stacked on the sixth square?

(b) How many coins in total are there on the board in these six squares?

(c) If each coin's value is $\$v$, what is the total value of the coins on the board?

(d) Find a number for x and a number for v that will make the total value of the coins between \$300 and \$400.

14 Consider the algebraic expressions $a + b$, $2a + b$, $3a + b$.

Find two sets of values for a and b so that the sum of the three expressions is a multiple of ten.

15 The sequence of terms: $x + y$, $3x + 2y$, $5x + 3y$, $7x + 4y$, $9x + 5y$, ... follows a pattern.

Find two sets of values for x and y so that the sum of the first five terms in this sequence is a multiple of ten.

16 Lindy was simplifying $7xy - xy$ and wrote: $7xy - xy = 7$.

(a) What mistake did Lindy make?

(b) Give Lindy a suggestion to help her avoid this mistake in the future.

Half-time 3

1 If Michael has n \$2 coins and m \$1 coins, write an expression for how much money he has altogether. **Ex 3.1**

2 Simplify: **Ex 3.4**

(a) $7x + 10x$ (b) $14y - 9y$ (c) $3x + 2y - x + 4y$

(d) $4x^2 + 3xy - 2x^2 + 4x^2$ (e) $10a + 13b - 3b - 26a$ (f) $4x + 2a - 3x + 2a^2$

3 If $x = 3$, $y = -4$ and $z = 5$, find: **Ex 3.2**

(a) $3x + 2y$ (b) $2z - 3y$ (c) $7x + 2y - z$

(d) $6x^2 - y$ (e) $10xy^2 + z$ (f) $2xyz + 3z^3$

4 Use the rules to complete the following tables. **Ex 3.2**

(a) $y = 3x - 4$

x	-2	0	1	2
y				

(b) $y = 3 + 2x$

x	-3	-1	2	4
y				

(c) $y = 2x^2 + 3$

x	-1	0	3	5
y				

5 The following formulas are often used in mathematics and science. **Ex 3.3**

(a) For $V = IR$, find V when $I = 1.8$ and $R = 20$.

(b) For $v = u + at$, find v when $u = 20$, $a = 1.2$ and $t = 15$.

(c) For $t = a + (n - 1)d$, find t when $a = 2.6$, $n = 10$ and $d = -0.4$.

(d) For $f = \frac{vu}{v + u}$, find f when $v = 25$ and $u = 20$.

(e) For $s = ut + \frac{1}{2}at^2$, find s when $u = 3$, $t = 10$ and $a = 9.8$.

6 Susan gets a ticket standing in a queue at the deli section of her local supermarket. The number on her ticket is t. Assume the tickets are given out in consecutive order (i.e. 1, 2, 3, ...). **Ex 3.1**

(a) John gets the ticket after Susan. Write an expression for the number on John's ticket.

(b) Jill is the person who is served just before Susan. Write an expression for the number on Jill's ticket.

7 Simplify the following expressions. **Ex 3.4**

(a) $4xy + 3x - 2y + 5xy$ (b) $2x^2 - 4x - 3xy + 5x^2$

(c) $8m^2 - 12m + m^2 + 8m + 5$ (d) $9pq - qp + 4p^2q - 3p^2q^2 + 4pq^2$

8 For each expression state the: **Ex 3.1**

(i) number of terms it contains

(ii) coefficient of x

(iii) constant.

(a) $2x + 3$ (b) $a - 5x - 4$ (c) $x + 7 + 4x^2$

(d) $-2x - 14xy$ (e) $6ax - 5 - \frac{1}{2}x$ (f) $3x + 4y + 6$

(g) $4x^2y - 2 + x$ (h) $-10 - 12x$ (i) $7x$

3.5 Multiplying and dividing algebraic terms

Need to Know

Multiplying algebraic terms

The **associative law** can be used to simplify multiplication of algebraic terms.

The associative law tells us that the order in which numbers are multiplied together is not important.

- Rearrange the multiplication to group numbers and variables separately, writing the variables in alphabetical order.
- Multiply the coefficients (numerical factors) then write the variables (algebraic factors) omitting multiplication signs.

For example:

$$5a \times 2 = 5 \times a \times 2 = 5 \times 2 \times a = 10a$$

$$2d \times 3c = 2 \times d \times 3 \times c = 2 \times 3 \times c \times d = 6cd$$

Dividing algebraic terms

- Write as a fraction first.
- Simplify the fraction by dividing both the numerator and the denominator by the **Highest Common Factor (HCF)** of the numerator and the denominator.
- Common factors can be numbers and/or variables.

Worked Example 10

WE10

Simplify:

(a) $6a \times 3$ (b) $5a \times -2c \times b \times 3$

Working

(a) $6a \times 3 = 18a$ (b) $5a \times -2c \times b \times 3 = -30abc$

Worked Example 11

WE11

Simplify: $\frac{-15ef}{10f}$

Working

$$\frac{^{3}\cancel{-15}}{_{2}\cancel{10}} \times \frac{e\cancel{f}^{1}}{_{1}\cancel{f}}$$

$$= -\frac{3e}{2}$$

Worked Example 12

WE12

Simplify: $14ab \div 7bc$

Working

$$14ab \div 7bc = \frac{14ab}{7bc}$$

$$= \frac{^{2}\cancel{14}a\cancel{b}^{1}}{_{1}\cancel{7}_{1}\cancel{b}c}$$

$$= \frac{2a}{c}$$

3.5 Multiplying and dividing algebraic terms

Answers page 498

Navigator

Q1 Column 1, Q2 Column 1, Q3 Column 1, Q4, Q5, Q6, Q7, Q9, Q10, Q11	Q1 Column 2, Q2 Column 2, Q3 Column 2, Q4, Q5, Q6, Q7, Q8, Q9, Q10, Q12	Q1 Column 3, Q2 Column 3, Q3 Column 3, Q4, Q5, Q6, Q7, Q8, Q9, Q11, Q12

Fluency and understanding

WE10

1 Simplify:

(a) $5 \times 3a$ (b) $6 \times 2a$ (c) $9g \times 2$

(d) $7 \times 8z$ (e) $3z \times 12$ (f) $9p \times 6$

(g) $11e \times 6f$ (h) $9m \times 8n$ (i) $12j \times 11k$

(j) $6gh \times k \times 3$ (k) $8gh \times 3 \times 2k$ (l) $6b \times 5eh \times 3$

(m) $4 \times -5y$ (n) $-3 \times -7y$ (o) $-2a \times -8$

(p) $11a \times -4b$ (q) $-3r \times -6uq$ (r) $-7u \times 8rq$

(s) $-4j \times -2 \times 10k$ (t) $10j \times -2k \times 3$ (u) $8 \times -u \times 3v \times 2$

3.5

WE11

2 Simplify:

(a) $\frac{14a}{7}$ (b) $\frac{12b}{3}$ (c) $\frac{63e}{9}$

(d) $\frac{15f}{5f}$ (e) $\frac{10x}{12x}$ (f) $\frac{x}{15x}$

(g) $\frac{15ab}{5}$ (h) $\frac{24ef}{8e}$ (i) $\frac{26ab}{13a}$

(j) $\frac{ef}{8e}$ (k) $\frac{7mn}{28m}$ (l) $\frac{11ef}{77f}$

(m) $\frac{-22ab}{11ab}$ (n) $\frac{-6ab}{3ba}$ (o) $\frac{-16ef}{-4ef}$

(p) $\frac{-9h}{-45gh}$ (q) $\frac{-7g}{-77gh}$ (r) $\frac{-9i}{-36ij}$

WE12

3 Simplify:

(a) $20c \div 5$ (b) $40d \div 4$ (c) $12g \div 6$

(d) $60h \div 10h$ (e) $18x \div 27x$ (f) $20x \div 22x$

(g) $18cd \div 9$ (h) $56cd \div 8$ (i) $30gh \div 6$

(j) $gh \div 3g$ (k) $45gh \div 18g$ (l) $50gh \div 16g$

(m) $8cd \div -4c$ (n) $-50ab \div 30a$ (o) $60cd \div -12d$

(p) $12ab \div -8ba$ (q) $-ij \div -6ij$ (r) $-11ij \div -ji$

4 $-2 \times 5g \times -a$ simplifies to:

A $-7ga$ B $-10ag$ C $7ga$ D $10ag$

5 $\frac{-12mad}{-9md}$ simplifies to:

A $\frac{4a}{3}$ B $-\frac{4m}{3}$ C $\frac{4m}{3}$ D $-\frac{4a}{3}$

6 $-a \times -2b \times -3c \div -6bad$ simplifies to:

A $-12a$ B $\frac{c}{d}$ C $6cd$ D $-\frac{6d}{c}$

7 Write and simplify the product of the three terms $3abc$, $-5d$ and -4.

8 On the right are two toolboxes. The volume of each is given by the product of the length, width and height.

(a) What is the volume of Toolbox 1?

(b) What is the volume of Toolbox 2?

(c) Find the volume of each toolbox if $w = 4$ and $j = 12$.

(d) How many small toolboxes will fit inside the large toolbox?

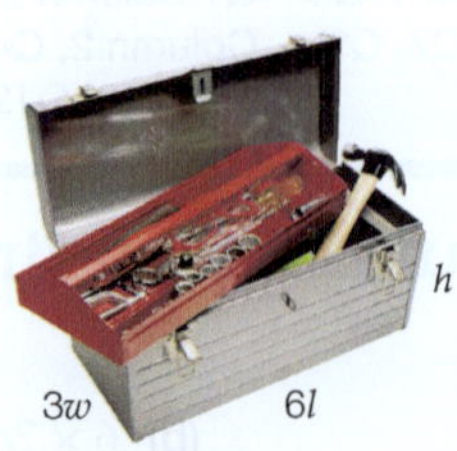

Toolbox 1

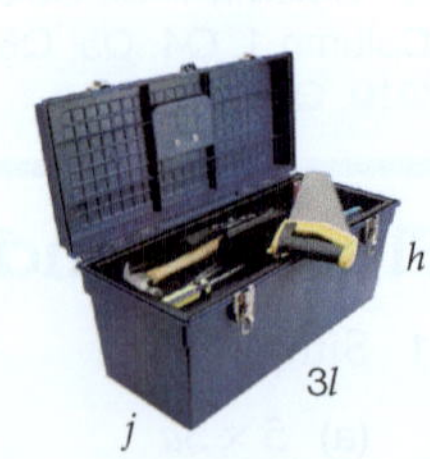

Toolbox 2

Reasoning, communicating and open-ended

9 A necklace is to be made from 3 strands of gold wire. The shortest strand is x mm long. Each strand is 1.2 times longer than the previous one.

(a) Write expressions to show the lengths in mm of each strand.

(b) Write an expression in terms of x to show the total length in mm of gold wire required.

(c) If gold wire costs $\$d$ per mm, write an expression to show the total cost of gold used.

(d) There are $6y$ small diamonds to be added to the necklace and each strand will contain the same number of diamonds. Write an expression to show how many diamonds will be inserted in each strand.

(e) If each diamond costs $\$c$, write an expression to show the total cost of all the diamonds.

(f) Write an expression to show the total cost of materials required to make the necklace if a fastener costs $\$f$.

(g) If $x = 30$, $d = 5$, $y = 4$, $c = 40$ and $f = 34$, find the total cost of materials required to make the necklace.

10 Joshua simplified an expression. His work is shown below. Explain the mistake he has made and write the correct answer, so that he does not make a similar mistake in the future.

$3k \times -2k = k$

11 Choose two terms, each containing the variables p and r, so that the quotient of the two terms gives a result of $-\frac{3p}{7r}$. (Remember, the quotient is the result after one term has been divided by the other.)

12 Choose three terms, each containing the variables m and n, so that the product of two terms divided by the third gives a result of $6mn$.

Problem solving

Make a profit

A T-shirt manufacturer in Australia pays fixed expenses of $400 per week to run his factory. In addition to this, it costs $12 per T-shirt manufactured. The information in the table shows how many T-shirts he will sell at certain prices.

Profit = income from selling T-shirts – cost of manufacturing the T-shirts

Sale price per T-shirt	Number of T-shirts sold
$10	45
$15	40
$20	35
$25	30
$30	25
$35	20
$40	15
$45	10

1 What is the number of T-shirts that need to be manufactured in order to make a profit?

2 What is the number of T-shirts that need to be manufactured to make the largest profit?

Strategy options

- Guess and check.
- Make a table.
- Look for a pattern.

3.6 Expanding brackets

Need to Know

Removing brackets

What is meant by $3(x + 2)$?

Let x be the number of counters in a cup, so that $x + 2$ is the number of counters in the cup plus two more.

$x + 2$

$3(x + 2)$ represents 3 lots of $(x + 2)$ or $3 \times (x + 2)$ and can be shown by 3 cups with x counters in each and 2 more counters with each cup.

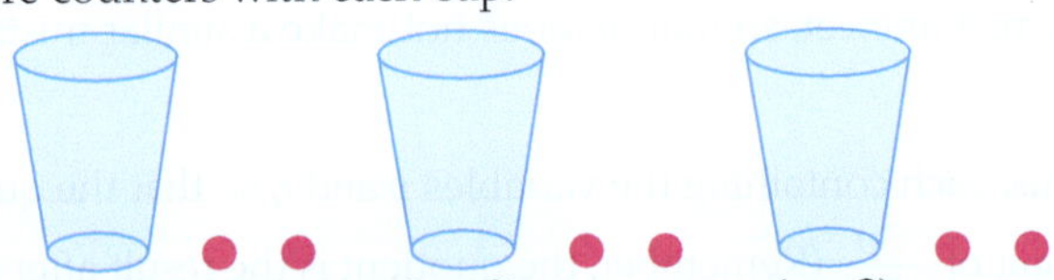

$(x + 2) \quad + \quad (x + 2) \quad + \quad (x + 2)$

This can be rearranged to give 3 cups, each with x counters, and 6 more counters, to give a total of $3x + 6$ counters.

$(x \quad + \quad x \quad + \quad x) \quad + \quad (2 \quad + \quad 2 \quad + \quad 2)$

Therefore, $3(x + 2) = 3x + 6$.

Expanding brackets

Expanding is the process of removing brackets using the **distributive law**.

For example:

$$\begin{aligned} 3 \times 102 &= 3 \times (100 + 2) \\ &= 3 \times 100 + 3 \times 2 \\ &= 300 + 6 \\ &= 306 \end{aligned}$$

Remember that variables in algebra are simply representing unknown numbers, so the distributive law must work in algebra if it works with numbers.

So, we can use the distributive law to expand algebraic expressions.

$$\begin{aligned} 3(x + 2) &= 3 \times x + 3 \times 2 \\ &= 3x + 6 \end{aligned}$$

Expanding using the distributive law:

$a(b + c) = ab + ac$ and $a(b - c) = ab - ac$

Expanding and collecting like terms

After an expression is expanded using the distributive law, it may contain like terms. These can be collected and the expression simplified.

3.6

Worked Example 13

WE13

Use the distributive law to find 24×99.

Working

24×99
$= 24 \times (100 - 1)$
$= 24 \times 100 - 24 \times 1$
$= 2400 - 24$
$= 2376$

Worked Example 14

WE14

Expand each of the following expressions.

(a) $5(a - 2)$ **(b)** $2a(a + 3b)$ **(c)** $-3(a - b)$

Working

(a) $5(a - 2)$
$= 5 \times a - 5 \times 2$
$= 5a - 10$

(b) $2a(a + 3b)$
$= 2a \times a + 2a \times 3b$
$= 2a^2 + 6ab$

(c) $-3(a - b)$
$= -3 \times a - 3 \times -b$
$= -3a + 3b$

Worked Example 15

WE15

Simplify the following expressions by expanding the brackets and collecting like terms.

(a) $3(r - 4) - 8r$ **(b)** $5(a + 6b) + a(9 - 10a)$

Working

(a) $3(r - 4) - 8r$
$= 3r - 12 - 8r$
$= -5r - 12$

(b) $5(a + 6b) + a(9 - 10a)$
$= 5a + 30b + 9a - 10a^2$
$= 14a + 30b - 10a^2$

Worked Example 16

WE16

Simplify the following expressions by expanding the brackets and collecting like terms.

(a) $8d - (2d + 3)$ **(b)** $7(a - 3b) - 5(a + 3b)$ **(c)** $3m(3n - 4t) - 4n(2m - t)$

Working

(a) $8d - (2d + 3)$
$= 8d - 1 \times (2d + 3)$
$= 8d - 2d - 3$
$= 6d - 3$

(b) $7(a - 3b) - 5(a + 3b)$
$= 7a - 21b - 5a - 15b$
$= 2a - 36b$

(c) $3m(3n - 4t) - 4n(2m - t)$
$= 9mn - 12mt - 8mn + 4nt$
$= mn - 12mt + 4nt$

3.6

Answers page 498

3.6 Expanding brackets

Navigator

Q1 Column 1, Q2 Columns 1 & 2, Q3 Column 1, Q4 Column 1, Q5, Q6, Q7, Q8, Q9, Q12	Q1 Column 2, Q2 Columns 2 & 3, Q3 Column 2, Q4 Column 2, Q5, Q6, Q7, Q8, Q9, Q10, Q12	Q1 Column 3, Q2 Columns 2 & 3, Q3 Column 2, Q4 Column 2, Q5, Q6, Q8, Q10, Q11, Q12, Q13

Fluency and understanding

WE13

1 Use the distributive law to find:

(a) 21×101 **(b)** 17×102 **(c)** 16×101

(d) 23×98 **(e)** 41×99 **(f)** 37×97

(g) 54×11 **(h)** 38×1010 **(i)** 64×1100

WE14

2 Expand each of the following expressions.

(a) $5(x + 12)$ **(b)** $2(s - 5)$ **(c)** $12(a + 5)$

(d) $s(4 + t)$ **(e)** $l(6 - r)$ **(f)** $a(2 + h)$

(g) $3(4x + 5)$ **(h)** $9(6x - 1)$ **(i)** $2(3h + 4)$

(j) $2m(n + r)$ **(k)** $3p(r - z)$ **(l)** $6p(r + t)$

(m) $5t(t + 2)$ **(n)** $2p(5p - 7)$ **(o)** $4m(4 + 2m)$

(p) $3m(5m + 7n)$ **(q)** $4k(p - 5k)$ **(r)** $10x(2x - 5y)$

WE15

3 Simplify the following expressions by expanding the brackets and collecting like terms.

(a) $5(7 - d) + 11d$ **(b)** $8(3 - e) - 12e$

(c) $6n(m - 4) - 12nm$ **(d)** $4x(y - 2) - 7xy$

(e) $5t(7 - 4k) - 40tk$ **(f)** $9v(5 - 10w) - 50v$

(g) $6m(7n - 4) + 4(8mn - 4m)$ **(h)** $2a(5b - 8) + 2(6a - 12ab)$

(i) $9x(5 - y) + 3(10x - 4xy)$ **(j)** $3t(4 - 8z) + 6(7tz - 20t)$

WE16

4 Simplify the following expressions by expanding the brackets and collecting like terms.

(a) $3x - 4(x + 2)$ **(b)** $5x - (2x + 2)$

(c) $8 - 3(x + 2)$ **(d)** $14 - (3x - 7)$

(e) $5x - 3(5x - 2)$ **(f)** $7x - 3(x - 2)$

(g) $5(x + 2) - 3(2x + 1)$ **(h)** $4(3x - 2) - 12(x + 5)$

(i) $3(3x + 4) - 5(7x - 3)$ **(j)** $5(x - 3) - (5x - 1)$

(k) $5(2x - 3y) - 3(3x + 4y)$ **(l)** $4(3x - 5y) - 2(5x + 2y)$

(m) $c(3d + 4e) - 2c(d + 5e)$ **(n)** $d(5e + 2f) - 4d(e + 4b)$

(o) $m(2n - 3t) - t(2m - n)$ **(p)** $a(2p - q) - q(2p - 3a)$

Hint

If there's a minus in front of a bracket such as $(x + 5) - (3x + 1)$, put a 1 and a $\times$ sign in between the minus and bracket: $(x + 5) - 1 \times (3x + 1)$.

5 The expression that can be modelled by the cups and counters in the diagram is:

(a)

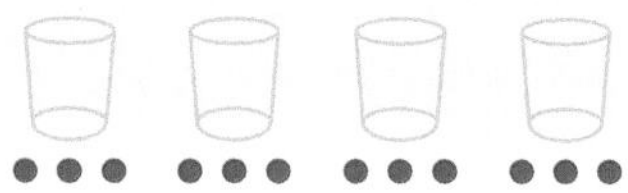

A $4x + 3$ **B** $4(x + 12)$ **C** $4(x + 3)$ **D** $2x + 12$

(b)

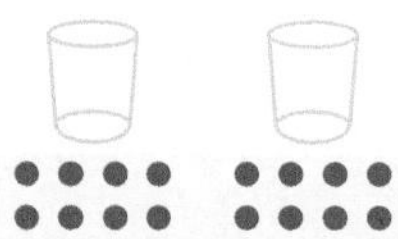

A $x + 16$ **B** $2x + 8$ **C** $2x + 16$ **D** $2(x + 16)$

6 Expanding and simplifying $3y(2y + a) - a(3y - a)$ gives:

A $6y^2 - ay$ **B** $6y^2 - 3ay + a^2$ **C** $3ay - 6y^2$ **D** $6y^2 + a^2$

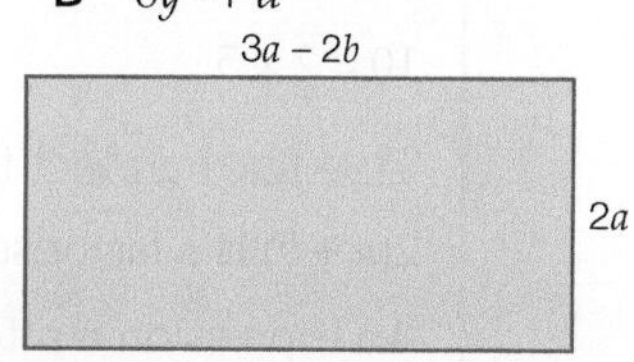

7 A rectangle has width $2a$ metres and length $3a - 2b$ metres. Write the expression for the area and expand the brackets.

8 **(a)** **(i)** Expand $33(x + 5)$. **(ii)** Substitute $x = 100$ and evaluate.

(b) **(i)** Expand $105(y + 3)$. **(ii)** Substitute $y = 30$ and evaluate.

(c) Explain why your answers to **(a) (ii)** and **(b) (ii)** are the same.

Reasoning, communicating and open-ended

9 A rectangle has length $(5 + x)$ m and width 5 m.

(a) Draw a diagram to represent this information.

(b) Find the area of the rectangle.

(c) Split your diagram up into two parts, with one part being a square of side length 5 m.

(d) Find the area of each part of your diagram.

(e) Use your diagram and the answers to parts **(b)** and **(d)** to explain why $5 \times (5 + x) = 25 + 5x$.

10 Lucy amazed her friends with her quick computational skills. When asked to multiply 8 by 8.125, she immediately answered 65. She told her friends it was easy, and that they should investigate $20 \times 20.05 = 401$ and $4 \times 4.25 = 17$. She also said that she used the distributive law. How did she do the arithmetic so quickly?

11 Quentin was working on algebraic expressions, expanding brackets and then collecting like terms.

His answer was: $20x - xy$

Write a question that Quentin might have been answering, assuming he answered correctly.

12 Michael was simplifying by expanding and collecting like terms. His working is shown below.

$$3(2x - 1) - (5x - 2) = 6x - 3 - 5x - 2$$
$$= x - 5$$

(a) Identify the mistake that Michael made.

(b) Correct his mistake by completing the question.

(c) Suggest how Michael can avoid making this mistake in the future.

13 Write three expressions containing brackets that can be expanded to give the first term as $12ef$.

3.7 Factorising expressions

Need to Know

Factorising

Factorising means writing an expression as the product of factors.

$10 = 2 \times 5$ $\quad\quad$ $2a = 2 \times a$

The factorised form

$2(a + b)$ is a factorised expression as it is the product of 2 and $(a + b)$.

An expression written in this way is in **factorised form**.

Expanding and factorising

To find the perimeter of a rectangle:

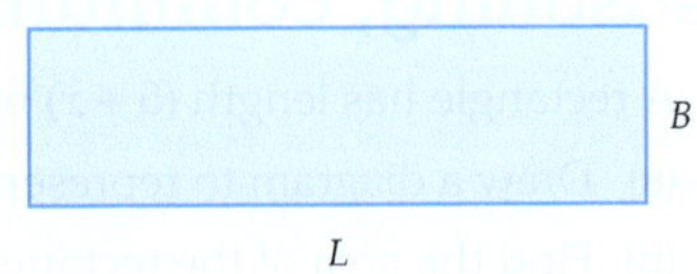

Method 1: Double the length and double the breadth, then add these two amounts.

Writing this as a formula:

$P = 2L + 2B$ where L = length and B = breadth. This is the expanded form.

Method 2: Add the length and breadth together first and then double that amount.

Writing this as a formula:

$P = 2(L + B)$. This is the factorised form.

Using brackets

Brackets have been used to change the order of operations.

This means that $2L + 2B = 2(L + B)$.

The expanded form = the factorised form.

We also know that $2(L + B) = 2L + 2B$ by expanding.

The factorised form = the expanded form.

This shows us that factorising is the reverse of expanding.

Factorising using areas

What is the sum of the areas of the two rectangles on the right?

The total area can be written in two equally correct forms.

Expanded form: $6x + 9y$

Factorised form: $3(2x + 3y)$

So $6x + 9y = 3(2x + 3y)$

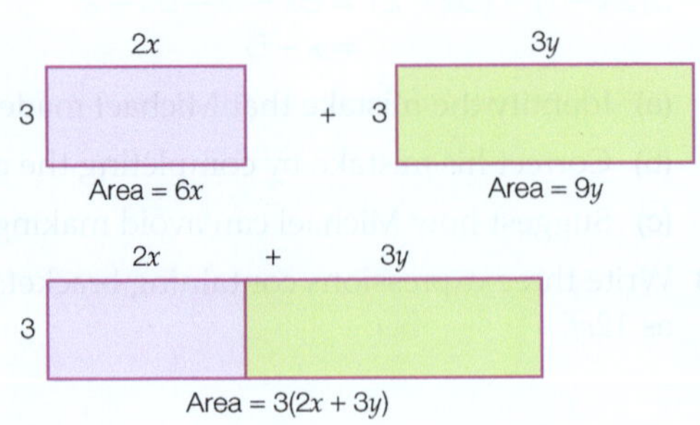

Factorising an expression

- Write each term in the expression as a product of prime factors.
- Circle all common factors to find the highest common factor (HCF) of all the terms in the expression.
- Write the HCF in front of the brackets.
- Write the sum or difference of the other factors inside the brackets.
- Write the expression in factorised form omitting multiplication signs.
- Check your answer by expanding the brackets.

For example:

$6x + 9y = 3 \times 2 \times x + 3 \times 3 \times y$ 3 is the HCF of $6x$ and $9y$

$= 3(2x + 3y)$

Check: $3(2x + 3y) = 6x + 9y$.

Worked Example 17

WE17

Find the highest common factor (HCF) of the following pairs of algebraic terms.

(a) $10x$ and 15 **(b)** $12a$ and $18b$ **(c)** $6c^2d$ and $8cd^2$

Working

(a) $10x = 2 \times 5 \times x$

$15 = 3 \times 5$

HCF = 5

(b) $12a = 2 \times 2 \times 3 \times a$

$18b = 2 \times 3 \times 3 \times b$

$\text{HCF} = 2 \times 3$

$= 6$

(c) $6c^2d = 2 \times 3 \times c \times c \times d$

$8cd^2 = 2 \times 2 \times 2 \times c \times d \times d$

$\text{HCF} = 2 \times c \times d$

$= 2cd$

Worked Example 18

WE18

Factorise the following expressions.

(a) $3y + 3$ **(b)** $10a - 12$ **(c)** $-15b + 10c$

Working

(a) $3y + 3$

$= 3 \times y + 3 \times 1$

$= 3 \times (y + 1)$

$= 3(y + 1)$

(b) $10a - 12$

$= 2 \times 5 \times a - 2 \times 2 \times 3$

$= 2 \times (5 \times a - 2 \times 3)$

$= 2(5a - 6)$

(c) $-15b + 10c$

$= -3 \times 5 \times b - -2 \times 5 \times c$

$= -5 \times (3 \times b - 2 \times c)$

$= -5(3b - 2c)$

3.7

Answers page 499

3.7 Factorising expressions

Navigator

Q1 Column 1, Q2 Column 1, Q4, Q6, Q10, Q12	Q1 Column 2, Q2 Column 2, Q4, Q5, Q6, Q9, Q10, Q12	Q1 Column 3, Q2 Column 3, Q3, Q4, Q5, Q7, Q8, Q9, Q11, Q12

Fluency and understanding

WE17

1 Find the highest common factor (HCF) of the following pairs of algebraic terms.

(a) $6m$ and 9 (b) $14p$ and 21 (c) $22d$ and 55

(d) $8x$ and $36y$ (e) $30d$ and $45c$ (f) $24a$ and $54b$

(g) $2p^2$ and $8p^4$ (h) $9t^5$ and $12t$ (i) $15d^5$ and $10d^3z^3$

(j) fm^4 and f^4m^2 (k) a^2b and a^3b^2 (l) x^4y^2 and x^2y^3

(m) $20mn^4$ and $25m^3n^2$ (n) $18s^3t^6$ and $15st$ (o) $6g^2h$ and $15gh^3i$

WE18

2 Factorise the following expressions.

(a) $3t + 15$ (b) $2h + 14$ (c) $2a - 8$

(d) $9p + 3$ (e) $20p - 5$ (f) $15z + 3$

(g) $12 - 20b$ (h) $24 + 10y$ (i) $18 - 14q$

(j) $3px + 12$ (k) $2kt - 20$ (l) $4hs + 14$

(m) $-8ab - 16$ (n) $-5mn - 15$ (o) $-12kr - 132$

Hint

You can check your answers by expanding the brackets. You should get the original expression back again.

3 Factorise the following expressions.

(a) $3lx + 15x$ (b) $10pm - 20m$ (c) $14pq + 28q$

(d) $6vx - 3x$ (e) $30ab + 20a$ (f) $18xy - 40y$

(g) $16p + 14pq$ (h) $15l - 20lm$ (i) $22j - 55jl$

(j) $30mz - 20m$ (k) $32hk + 36k$ (l) $40h + 28hi$

(m) $-12ab + 16b$ (n) $-14xy + 21x$ (o) $-72m - 56mn$

4 The highest common factor of $20x$ and $14xy$ is:

A $140xy$ B $7y$ C $2y$ D $2x$

5 The HCF of $2mn$ and $2m^6$ is:

A m^6n B $2m$ C $2m^6n$ D $2mn$

6 Fred has \$$5ab$ and Barney has \$$25c$. Write down the total amount of money they have between them, in factorised form.

7 The kinetic energy of a moving object that has a mass of m kilograms and a speed of v metres per second is given by $\frac{mv^2}{2}$. The object's gravitational potential energy h metres above the ground is given by mgh, where g is the acceleration of the object due to gravity at that point.

(a) Write down the total energy of the object (kinetic energy plus gravitational potential energy).

(b) Write down the total energy in factorised form.

8 Factorise:

(a) $3(x + 1) - y(x + 1)$ (b) $7(3x - 4) + y(3x - 4)$ (c) $5(x + 7) + 2y(x + 7)$

3.7

Reasoning, communicating and open-ended

9 The expression $ax + ay + bx + by$ contains four terms.

(a) Factorise $ax + ay$.

(b) Factorise $bx + by$.

(c) Write down the expression for the sum of your factorised answers for **(a)** and **(b)**.

(d) Factorise the expression in **(c)** by taking the expression in brackets to be the highest common factor.

(e) Use the method above to factorise $3x + 3b + xy + by$.

10 Write down three pairs of terms whose highest common factor is 8.

11 A rectangle has a breadth of x cm and an area given by the formula $A = 10x - x^2$.

(a) Factorise the expression for the area to find an expression for the length of the rectangle.

(b) If the length is greater than the breadth, find a possible value for x.

12 Stephen has factorised $9x + 3$ as shown below.

$9x + 3 = 3(3x + 3)$

His friend Mohammed tells him that he is wrong.

Mohammed has written $9x + 3 = 3(3x + 1)$.

State who is correct. Explain where the other person has made a mistake and what can be done to avoid this mistake in the future.

Challenge 3

1 When four consecutive even numbers are added the total is 140. Use algebra to find the numbers.

2 In how many different ways can you add four positive odd numbers to total 16? You can use any positive odd number as many times as you like.

3 Three large circles overlap. Six smaller circles are located on the points where the circles cross.

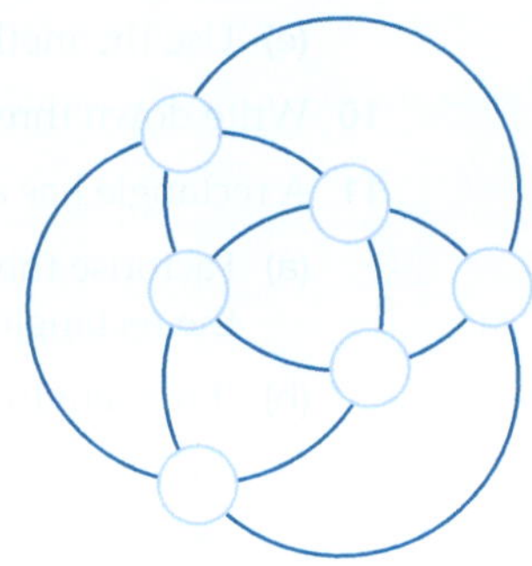

(a) Copy this diagram and then write the digits 0, 1, 2, 3, 4 and 5 in the blue circles so that the totals of the numbers in the blue circles around each large circle are the same.

(b) Copy this diagram again and then write the expressions x, $x + 1$, $x + 2$, $x + 3$, $x + 4$ and $x + 5$ in the blue circles so that the totals of the numbers in the blue circles around each large circle are the same.

(c) Find six other integers or expressions for which this will work. (Integers are positive and negative whole numbers and zero; e.g. -7, 0, 4.)

4 Given the factorisations $x^3 + y^3 = (x + y)(x^2 - xy + y^2)$ and $x^3 - y^3 = (x - y)(x^2 + xy + y^2)$, answer the following questions without using a calculator.

(a) Write 3^6 and 4^6 as perfect cubes.

(b) Show that 25 is a factor of $3^6 + 4^6$.

(c) Show that 4 is a factor of $3^6 - 5^3$.

5 Write an algebraic expression in each empty cell so that you create a magic square in which the sum of each row, column and diagonal is $(4x + 2)$.

$x + 8$	$x - 6$		$x + 5$
		$x + 2$	x
$x - 4$	$x + 6$	$x + 7$	

6 $\dfrac{a + \frac{1}{b}}{b + \frac{1}{a}}$ equals

A $\dfrac{a}{b}$ B $\dfrac{b}{a}$ C ab D 1

Chapter review 3

D.I.Y. Summary

Key Words

associative law	equation	factorising	substitution
coefficient	expanding	formulas	term
constant	expression	Highest Common Factor (HCF)	unlike terms
distributive law	factorised form	like terms	variable

Copy and complete the following using the words and phrases from this list, where appropriate, to write a summary for this chapter. A word or phrase may be used more than once.

1. $4b + 5$ is called an ___________. The letter b in $4b + 5$ is called a ___________ because it can take the place of any number. The number 5 is called a ___________ and the number 4 is the ___________.
2. When we replace x with a certain number, this is called ___________.
3. To simplify $4e + 3f - 2e + 6f$ we collect ___________. The expression then simplifies to $2e + 9f$.
4. When we obtain the expression $4c + 12$ from $4(c + 3)$ we are ___________ the bracket using the ___________.
5. When we write $6a + 18$ as $6(a + 3)$ we are ___________ the expression.
6. Algebraic equations used in practical situations are called ___________.

Fluency and understanding

1. You have x number of power points in your house. **Ex 3.1**
   (a) The house next door has $x + 3$ power points. Comment on how this compares to the number in your house.
   (b) The house on the other side of you has $x - 2$ power points. Comment on how this compares to the number in your house.
   (c) The block of flats across the road has $6x$ power points. Comment on how this compares to the number in your house.
2. Evaluate the following expressions when $a = 4$ and $b = 5$. **Ex 3.2**
   (a) $3a + 2b$ (b) $2ab - 12$ (c) $\frac{8b}{a} + 2b$ (d) $\frac{10}{b} + \frac{ab}{10}$
3. Evaluate the following expressions when $a = 2$ and $b = 3$. **Ex 3.2**
   (a) $7(2a - b)$ (b) $3a(b + 17)$ (c) $2a(3 - 5b)$
4. Evaluate the following expressions when $a = -2$ and $b = -3$. **Ex 3.2**
   (a) $-2a + b - 5$ (b) $ab - 4 - \frac{6b}{2}$
5. Evaluate the following expressions when $x = 2$ and $y = 12$. **Ex 3.2**
   (a) $\frac{x + y}{2}$ (b) $\frac{2x + 3y}{5}$ (c) $\frac{12(x - y)}{y}$

6 (a) Using the formula $F = ma$, find F when $m = 100$ and $a = 6.3$. **Ex 3.3**

(b) Using the formula $v^2 = u^2 + 2as$, find v when $u = 1$, $a = 10$ and $s = 4$.

7 Simplify the following, if possible, by collecting like terms. **Ex 3.4**

(a) $-j - 9j$ (b) $5ty - 2ty$ (c) $7ghi - 4ghi$

(d) $24ab - 5ab + 7$ (e) $6a + 12b - 7a + 11b$ (f) $2a^2 - 4a$

8 Simplify: **Ex 3.5**

(a) $15a \times 2b$ (b) $6ab \times c \times 11$ (c) $-2a \times -7 \times b$

9 Simplify: **Ex 3.5**

(a) $24a \div 20a$ (b) $\dfrac{25ab}{5a}$ (c) $-4ab \div -44a$

10 Expand: **Ex 3.6**

(a) $7(2 + a)$ (b) $a(2b + 6)$ (c) $4a(3a + 5)$

11 Simplify by expanding and collecting like terms: **Ex 3.6**

(a) $2(4 + a) + 7a$

(b) $10a(6 + 2b) + 7(2a - ab)$

(c) $2a^2(3a^4 + 11a) + 3a(5a^5 - 2a^2)$

12 Factorise: **Ex 3.7**

(a) $3a - 15$ (b) $20ab + 16$ (c) $-15p + 20q$

13 Suppose there are m jelly beans in a packet. Melissa, Roderigo and Huong have bought some packets of jelly beans. The following tells us how many jelly beans each has at certain times.

Time	Melissa	Roderigo	Huong
12.30	m	0	0
12.31	$m - 2$	$2m$	0
12.32	$m - 4$	$2m - 10$	$3m$
12.33	$m - 6$	m	$3m - 5$
12.34	$m - 8$	$m - 6$	$2m$
12.35	$m - 14$	0	$2m - 8$

Write a paragraph telling a story about what happened during the 5 minutes.

14 In a school hall there are t teachers, g female students and b male students.

(a) How many people are there in total?

(b) Eight teachers, six female students and half the boys exit the hall. How many people remain inside?

(c) Half of the number of teachers and female students who had left re-enter the hall. How many people are now in the hall?

15 In centimetres, the heights of three Year 8 students, Troy, Barry and Tenielle, are, in order, $3a + 2$, $4b - 5$, and $6c + 1$.

If $a = 52$, $b = 41$, and $c = 26$, what is each person's height?

16 Who is the tallest, and what is his/her height if $d = 6$ and the heights of three Year 8 students, William, Stuart and Yumi, are, in order, $4d^2 - 8$, $d^3 - 45$, $2d^2 + 18d$?

17 If $a = 1$, $b = 2$, $c = 3$ and so on, with the number increasing by 1 for each letter, find the value of $ab - bc + cd - de + ef - fg$.

18 Fatima draws a design consisting of straight lines. She first draws a line of length u cm straight up, then another line of length r cm to the right and a line of length d cm straight down.

(a) Write an expression for the total length of the three lines.

(b) If Fatima continues this design ten times, what is the total length of the lines?

(c) If $r = d$, $r = \frac{1}{2}u$ and $u = 7$, what is your answer to part **(b)**?

19 The speed v metres per second at which a wheel of radius r metres turns is given by the formula $v = \frac{6.28r}{t}$, where t seconds is the time the wheel takes to make one turn. Find the speed of the wheel if $r = 0.35$ metres and $t = 1.57$ seconds.

20 Apples are sold at a market. A standard bag of apples contains x apples. A bonus bag of apples has twice the number of apples of a standard bag plus an extra 4 apples. A case of apples contains 2 fewer apples than 6 standard bags of apples. Three customers each bought a bonus bag of apples and five customers each bought a case of apples.

(a) Write an expression to show the number of apples in a bonus bag.

(b) Write an expression to show the number of apples in a case of apples.

(c) Write an expression to show the number of apples sold.

(d) Expand and simplify your answer.

(e) How many apples were sold if a standard bag contained a dozen apples ($x = 12$)?

21 A special tint is used to protect glass in a rectangular window of length L in metres and breadth B in metres. The cost in dollars of applying the tint is $7 \times$ area of glass.

(a) Write an expression for the cost of tinting the window in terms of L and B.

(b) Find the cost of tinting a window 2.5 m long and 0.8 m wide.

22 The area of a triangle is half the product of its base length and its height: $A = \frac{bh}{2}$.

(a) Write the expression for the area of a triangle whose base is $4x$ cm and whose height is $(3x + 2)$ cm.

(b) Expand and simplify the expression obtained in part **(a)**.

(c) What is the area of the triangle if the height is 14 cm?

Reasoning and communicating

23 A stencil is made by cutting a triangle out of a rectangle of length l cm and breadth b cm.

(a) If the base of the triangle is half the length of the rectangle and the height of the triangle is 6 cm less than the width of the rectangle, find the area of the triangle in terms of l and b in: **(i)** factorised form and **(ii)** expanded form.

(b) Show that the area of the rectangle that is left once the triangle is cut out is given by $\frac{3lb}{4} + \frac{3l}{2}$.

(c) Factorise this answer.

(d) Find the area of the rectangle that is left once the triangle is cut out if $l = 20$ cm and $b = 16$ cm by:

(i) subtracting the area of the triangle from the area of the rectangle

(ii) substituting the values for l and b into your factorised answer in part **(c)**.

INTRODUCTION TO MATHEMATICAL THINKING
A publication from Pearson Custom Publishing

24 a, b, c represent three numbers.

(a) Write the expression that represents b being increased by 1.

(b) Write the expression for the product of a and your answer to part (a).

(c) Write the expression for the product of c and your answer to part (a).

(d) Write the expression for the sum of the answers to parts (b) and (c).

(e) Factorise the answer obtained in part (d).

(f) (i) Write the expression for the answer to part (e) divided by $a + c$.

(ii) Simplify your answer to part (i).

(iii) Subtract 1 from your answer to part (ii).

(iv) What do you notice about your answer to part (iii)?

(g) Using the values $a = 3$, $b = 7$, $c = 8$, repeat steps (a) to (f).

NAPLAN practice 3

Numeracy: Non-calculator

1 What is the value of $4a^2b$ when $a = -3$ and $b = 2$?

A -72 **B** -48 **C** 48 **D** 72

2 Which one of the following expressions is equivalent to $5(2x - 1)$?

A $10x - 1$ **B** $10x - 5$ **C** $10x + 1$ **D** $10x + 5$

3 Which one of the following expressions is equivalent to $3xy - 8x^2 + 2yx - 4x^2$?

A $3xy - 12x^2 + 2yx$ **B** $5xy - 12x^2$ **C** $-7x^2y$ **D** $5xy - 4x^2$

4 Which expression is equivalent to $4 - 8k$?

A $8k - 4$ **B** $8k + 4$ **C** $-8k + 4$ **D** $-8k - 4$

Numeracy: Calculator allowed

5 Which one of the following expressions is equivalent to $12x - 9y$?

A $9(3x - y)$ **B** $12(x - 9y)$ **C** $3(4x - 3)$ **D** $3(4x - 3y)$

6 If $a = 2$, what is the value of $\frac{3a - 4}{2a}$?

A 2 **B** $\frac{1}{2}$ **C** -2 **D** $-\frac{1}{2}$

7 A rule for y in terms of x is $y = 12 - 3x$. When $x = 2.3$ the value of y is:

A 5.1 **B** 16.6 **C** 20.7 **D** 34.5

8 Dylan is x years old. His brother, Liam, is 8 years old. John's age is three times Dylan's age, plus Liam's age. Find an expression in terms of x for John's age.

A $x + 11$ **B** $8x + 3$ **C** $3x + 8$ **D** $3x - 8$

Solving equations

7

Recall 7

Prepare for this chapter by attempting the following questions. If you have difficulty with a question, download the Recall Worksheet.

1 Which of the following are equations?

(a) $3 = 7 - 4$ **(b)** $a + 5$ **(c)** $3m + 2 = m - 4$

2 Solve each of the following equations. (Find the value of the pronumeral that makes the statement true.)

(a) $a + 7 = 12$ **(b)** $\frac{b}{2} = 10$ **(c)** $6c = -18$

(d) $d - 4 = 13$ **(e)** $e - 3 = -8$ **(f)** $3f = 24$

3 Copy and complete these flowcharts:

(a)
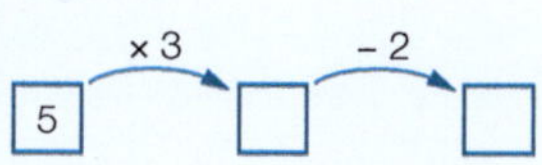

(b)
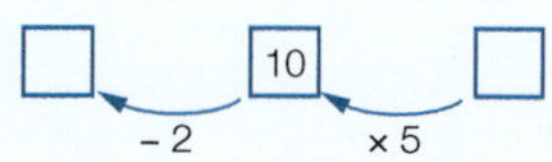

(c)
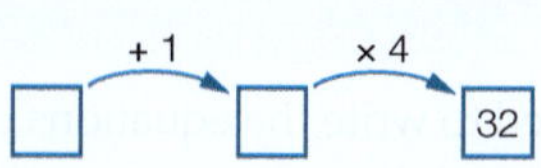

(d)
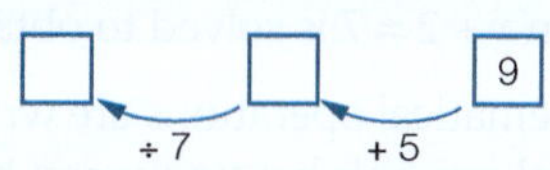

4 Find the value of each of the following expressions by substituting the value given in the brackets.

(a) $\frac{x}{2} + 9 \ (x = 10)$ **(b)** $4(x + 5) \ (x = 3)$ **(c)** $\frac{5x + 1}{7} \ (x = 4)$

5 What is the inverse (or opposite) operation to each of these?

(a) multiply by 2 **(b)** add -4 **(c)** divide by 5

6 Expand each of the following to remove the brackets.

(a) $2(x + 5)$ **(b)** $3(2x - 7)$

Syllabus references

Number and Algebra: Equations

✓ Solve simple linear equations **(ACMNA179)**

✓ Solve linear equations using algebraic techniques and verify solutions by substitution **(ACMNA194)**

✓ Solve simple quadratic equations

Number and Algebra: Linear Relationships

✓ Solve linear equations using graphical techniques **(ACMNA194)**

7.1 The language of equations

Need to Know

Equations

An equation is a mathematical sentence that contains two expressions connected by an equals sign. The equals sign shows that the two expressions have the same value.

Equations are used when some information is unknown. Unknown amounts are called variables and are represented by letters or symbols called **pronumerals**.

For example:

$x + 2 = 7$ is an equation where the pronumeral, x, represents a variable such as length.

Solving equations

Finding the value that makes the equation a true number sentence is called **solving** the equation.

For example:

The equation $x + 2 = 7$ is solved to obtain the **solution** $x = 5$.

When mathematical operations are written in words, we need to write the equations using mathematical symbols before we can begin the solution process.

Linear and quadratic equations

In **linear equations**, the largest power of the variable is 1.

For example: $x + 3 = 9$

In **quadratic equations**, the largest power of the variable is 2.

For example: $x^2 = 9$

Worked Example 1

WE1

Write an equation for each of the following. Use the letter in brackets to represent the variable.

(a) Seven is added to a number to give a result of ten. (a)

(b) A number is multiplied by five, then four is subtracted to give a result of sixteen. (b)

Working

(a) $a + 7 = 10$

(b) $b \times 5 - 4 = 16$
$5b - 4 = 16$

Hint

Remember that $b \times 5$ is written as $5b$.

7.1

Worked Example 2

WE2

Check by substitution whether the value given in the brackets is the solution for the equation $\frac{x+4}{6} = 2$ $(x = 14)$. (Does it make the equation true?) Answer yes or no.

Working

$\text{LHS} = \frac{x+4}{6}$

$= \frac{14+4}{6}$ when $x = 14$

$= \frac{18}{6}$

$= 3$

$\neq \text{RHS}$

No

The value given is not the solution.

7.1 The language of equations

Answers page 519

Navigator

Q1 (a)–(d), Q2 Column 1, Q3 (a)–(d), Q4 Column 1, Q5, Q7, Q8	Q1 (c)–(f), Q2 Column 1, Q3 (b)–(e), Q4 Column 2, Q5, Q6, Q7, Q8	Q1 (e)–(h), Q2 Column 2, Q3 (c)–(f), Q4 Column 3, Q5, Q6, Q7, Q8

Fluency and understanding

WE1

1 Write an equation for each of the following. Use the letter in brackets to represent the variable.

(a) Eight is added to a number to give a result of twelve. (a)

(b) Four is subtracted from a number to give a result of sixteen. (b)

(c) Nine times a number gives a result of sixty-three. (c)

(d) The sum of eleven and a number is zero. (f)

(e) Seven is added to three times a number to give a result of ten. (u)

(f) A number is multiplied by two, then seven is added to give a result of thirteen. (v)

(g) Nine is added to a number, then the result is divided by seven to give six. (x)

(h) The sum of six and a number is multiplied by eight to give a result of zero. (z)

(i) The square of a number is thirty-six. (n)

7.1

WE2

2 Check by substitution whether the value given in the brackets is the solution for each of the following equations. (Does it make the equation true?) Answer yes or no.

(a) $m + 2 = 9$ $(m = 7)$
(b) $l - 2 = -9$ $(l = -11)$
(c) $n - 6 = 3$ $(n = 9)$
(d) $10 - p = 4$ $(p = 14)$
(e) $5q = 55$ $(q = 11)$
(f) $\frac{r}{8} = 9$ $(r = 72)$
(g) $6s - 7 = 23$ $(s = 4)$
(h) $3a + 5 = -10$ $(a = -5)$
(i) $3(u - 8) = 6$ $(u = 6)$
(j) $4(5 - v) = 12$ $(v = 8)$
(k) $\frac{w + 3}{4} = 8$ $(w = 29)$
(l) $\frac{6 - x}{7} = 4$ $(x = 34)$
(m) $\frac{l^2 - 3}{2} = 23$ $(l = 7)$
(n) $\frac{k^2 + 4}{5} = 25$ $(k = -11)$

3 Which equation describes the sentence given in each of the following?

(a) Five is added to a number to give a result of twelve.
A $5 + 12 = n$ B $n + 12 = 5$ C $n + 5 = 12$ D $5n = 12$

(b) A number is subtracted from nineteen to give a result of eight.
A $b - 19 = 8$ B $-b - 19 = 8$ C $b - 8 = 19$ D $19 - b = 8$

(c) Six is subtracted from a number to give a result of seven.
A $b - 6 = 7$ B $6 - b = 7$ C $b - 7 = 6$ D $7 - b = 6$

(d) Five is added to three times a number to give a result of twenty.
A $3p + 20 = 5$ B $5 + 20 = 3p$ C $3(p + 5) = 20$ D $3p + 5 = 20$

(e) A number is subtracted from six; this result is multiplied by four to give eight.
A $4(w - 6) = 8$ B $4(6 - w) = 8$ C $4w - 6 = 8$ D $8 - 4w = 6$

(f) Three is subtracted from five times a number to give a result of seven.
A $(5 - 3)k = 7$ B $5(k - 3) = 7$ C $3 - 5k = 7$ D $5k - 3 = 7$

(g) Eight is subtracted from the square of a number to give a result of seventy-three.
A $8 - x^2 = 73$ B $(x - 8)^2 = 73$ C $x^2 + 8 = 73$ D $x^2 - 8 = 73$

4 Write each of these equations in words.

(a) $m + 2 = 9$
(b) $l - 2 = 9$
(c) $n - 6 = 3$
(d) $10 - p = 4$
(e) $5q = 55$
(f) $\frac{r}{8} = 9$
(g) $6s - 7 = 23$
(h) $3a - 5 = -10$
(i) $3(u - 8) = 6$
(j) $4(5 - v) = 12$
(k) $\frac{w + 3}{4} = 8$
(l) $\frac{6 - x}{7} = 4$
(m) $x^2 + 3 = 7$
(n) $d^2 - 5 = -1$
(o) $(x - 1)^2 = 25$

5 Write an equation for each of these rules, using the given variables for each of the quantities described.

(a) The area (A) of a rectangle is equal to its length (l) multiplied by its width (w).

(b) The area (A) of a triangle is equal to half its base (b) multiplied by its perpendicular height (h).

(c) The average speed (s) of a car is equal to the distance it travels (d) divided by the time taken (t).

(d) The area of a trapezium (A) is equal to half the sum of the lengths of the parallel sides (a and b) multiplied by the distance between them (h).

(e) The cost in dollars (C) of a cruise is 200 times the number of nights (n) plus 300.

(f) The tax-free price (F) of an item is $\frac{10}{11}$ of its retail price (R).

Reasoning, communicating and open-ended

6 A school wants to take a class of Year 8 students on an excursion to the zoo. The bus will cost \$150 and the zoo entrance is \$$x$ per person.

(a) Write an equation to show how much it will cost the school to take a class of 25 students and 2 staff members.

(b) If the zoo entrance fee is \$5 per person, how much will the excursion cost?

(c) Write an equation to show the total amount the students will pay for the excursion.

(d) If the students are charged \$10 each for the excursion and the staff do not pay, show that this will not cover the cost of the excursion.

(e) As the bus can carry up to 57 passengers, it was decided that another class of 25 students and another 2 staff should go on the same excursion. If the students are now to be charged \$8.50 each, use equations to show that this will cover the cost of the excursion.

(f) How much money will be left over?

7 Sarah wants to save enough money to buy a ticket to a concert. Tickets go on sale in 5 weeks and will cost \$95.00. She has saved \$20 already.

(a) Write an equation to show how much money she will have in 5 weeks if she gets \$$x$ pocket money each week for doing all her chores for the next 5 weeks.

(b) If she gets \$15 per week pocket money, use your equation to show that she will have enough money to buy the ticket.

(c) Sarah's friend, Rhiannon, wants to go to the same concert. She gets \$$y$ pocket money each week but owes her brother \$15. Write an equation to show how much money she will have in 5 weeks time if she repays her brother.

(d) If Rhiannon gets \$20 per week pocket money, use your equation to find whether she will have enough money to buy a ticket.

8 Write at least three equations that have $w = 3$ as a solution, using different operations $+, -, \times$ and $\div$ in each equation. One equation must be a quadratic.

7.2 Solving linear equations

Need to Know

Numerical solutions

To solve an equation by **inspection** we look at the equation and see the solution.

To solve an equation by **guess, check and improve** we choose a number, substitute it into the equation and repeat this process until the LHS (left-hand side) equals the RHS (right-hand side).

Algebraic solutions

Hint

Remember that we undo addition and subtraction before multiplication and division.

Flowcharts can be used to show the sequence of steps required to build an expression and then undo it by working backwards. This is called **backtracking**.

When backtracking, we can 'undo' an operation by using the **inverse** (opposite) **operation** in reverse order.

+ and – are inverse operations. × and ÷ are inverse operations.

Undo + and – before × and ÷.

For example: Solve $2x + 3 = 7$ using backtracking.

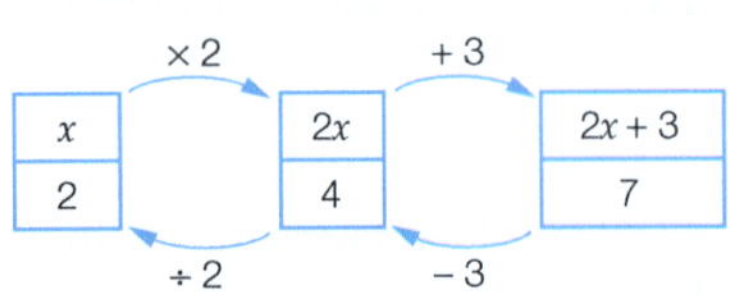

$x = 2$ is the solution.

Numerical methods and backtracking are time-consuming methods for solving equations and do not work for more complex equations. Equations can be solved using an algebraic process called equivalence. **Equivalent** equations occur when a new equation can be written that leaves the value of the variable unchanged. Equivalent equations remain **balanced** as the left-hand side of the equation is always equal to the right-hand side.

Solving equations using equivalence

- Equations can be solved by performing inverse operations on both sides of the equals sign.
- The order in which you perform the inverse operations is important and is the reverse of the order used in the construction process.
- Whatever is done to one side of the equation must be done to the other side.
- Always check the signs in front of the constants and the coefficients. If they are negative, we need to undo the negative as well as the numerical value.

For example:

As $5 - 2x = -2x + 5$, we undo by subtracting 5 and dividing by -2.

7.2

Solving equations graphically

One of the important uses of graphs is to be able to find solutions to equations.

If we draw a graph of a linear equation, all points on the graph will make the equation true.

We can use the graph to find the y-coordinate for any given x-value and the x-coordinate for any given y-value.

- A graph can be used to evaluate an expression. If a value for x is used as the x-coordinate of the graph, the y-coordinate of the point gives the value of the equation for that x-value.
- A graph can be used to solve an equation. If a value for y is used as the y-coordinate of the graph, the x-coordinate of the point gives the solution of the equation.

Linear solutions

Linear equations always have only one solution.

For example:

$2x + 3 = 7$ has only one solution $x = 2$.

Worked Example 3

WE3

Solve the following equations numerically to find x.

(a) $5x = 35$

(b) $\frac{5x-9}{2} = 8$

Working

(a) $5x = 35$

$x = 7$

(b) $\frac{5x-9}{2} = 8$

Try $x = 1$

$$\text{LHS} = \frac{5\times1-9}{2} = \frac{5-9}{2} = \frac{-4}{2} = -2 \neq \text{RHS}$$

Try $x = 2$

$$\text{LHS} = \frac{5\times2-9}{2} = \frac{10-9}{2} = \frac{1}{2} \neq \text{RHS}$$

Try $x = 5$

$$\text{LHS} = \frac{5\times5-9}{2} = \frac{25-9}{2} = \frac{16}{2} = 8 = \text{RHS}$$

$x = 5$ is the solution because LHS = RHS.

7.2

Worked Example 4

WE4

Solve each of the following equations using algebra. Check your answer by substitution.

(a) $4x - 5 = 7$ **(b)** $5 - 2x = 7$

Working

(a)

$$4x - 5 = 7$$
$$4x - 5 + 5 = 7 + 5$$
$$4x = 12$$
$$\frac{4x}{4} = \frac{12}{4}$$
$$x = 3$$

Check: LHS $= 4x - 5$
$= 4 \times 3 - 5$
$= 12 - 5$
$= 7$
$=$ RHS

(b)

$$5 - 2x = 7$$
$$5 - 5 - 2x = 7 - 5$$
$$-2x = 2$$
$$\frac{-2x}{-2} = \frac{2}{-2}$$
$$x = -1$$

LHS $= 5 - 2x$
$= 5 - 2 \times -1$
$= 5 + 2$
$= 7$
$=$ RHS

Worked Example 5

WE5

Use the following graph to find the value of:

(a) y when $x = -4$

(b) x when $y = 2$.

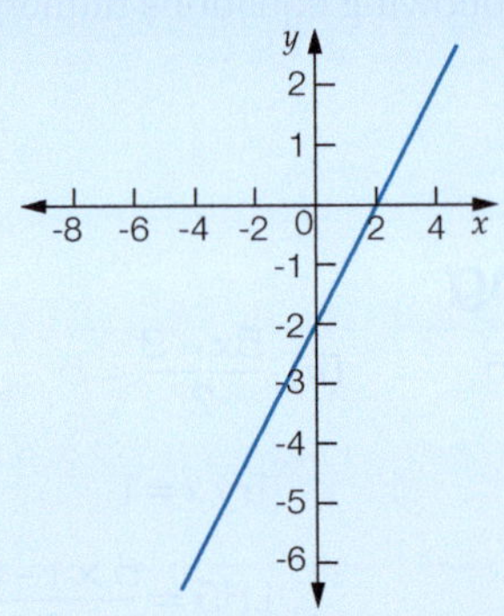

Working

(a)

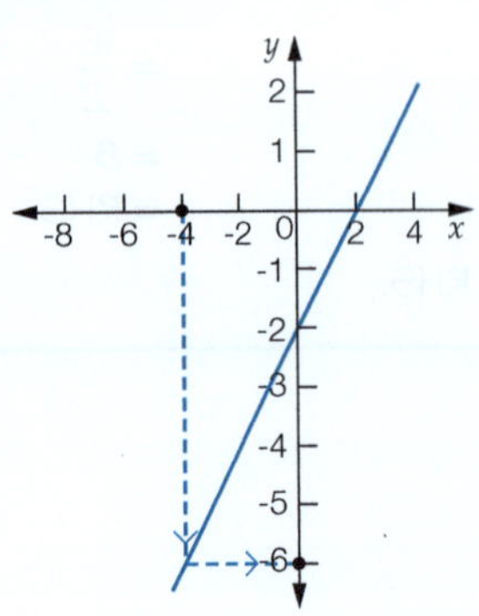

When $x = -4$, $y = -6$.

(b)

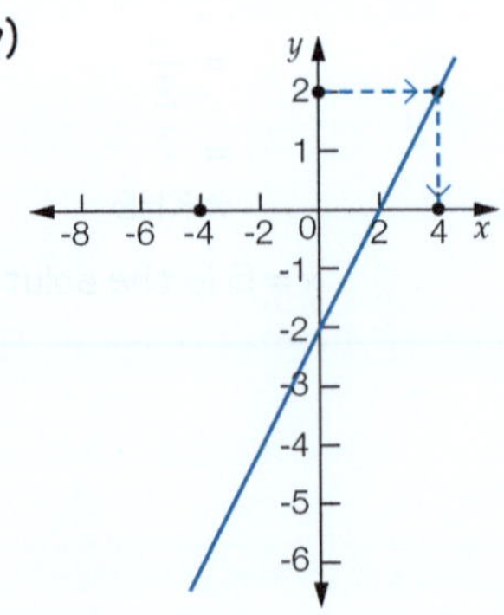

When $y = 2$, $x = 4$.

7.2

Worked Example 6

WE 6

Use the graph of $y = 2x + 1$ to solve:

(a) $2x + 1 = 11$

(b) $2x + 1 = -3$

Working

(a)

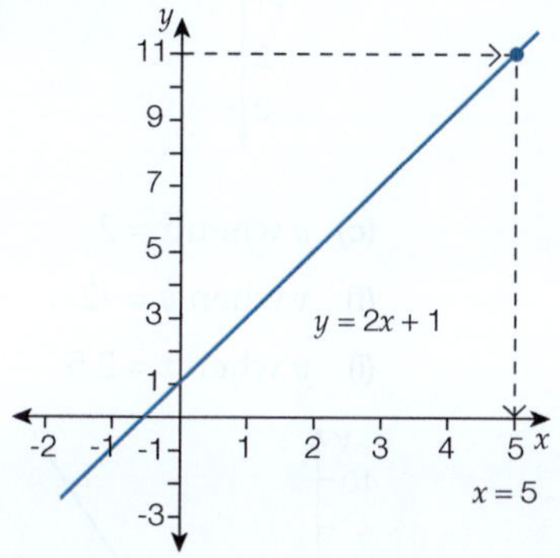

x = 5

(b)

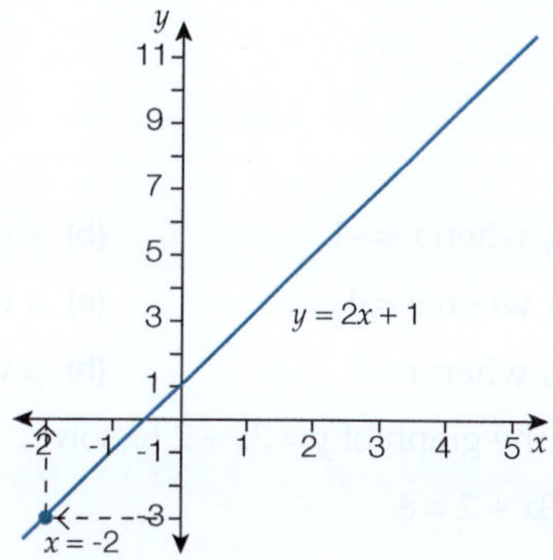

x = -2

7.2 Solving linear equations

Answers page 520

Navigator

Q1 Column 1, Q2, Q3, Q4, Q5, Q6, Q7, Q8 Column 1, Q9 Column 1, Q10, Q11, Q12(a) & (b), Q13, Q14, Q15, Q16, Q21, Q22, Q24	Q1 Column 2, Q2, Q3, Q4, Q5, Q6, Q7, Q8 Column 2, Q9 Column 2, Q10, Q11, Q12(a) & (b), Q13, Q14, Q15, Q16, Q17, Q19, Q20, Q21 Q23	Q1 Column 3, Q2, Q3, Q4, Q5, Q6, Q7, Q8 Column 3, Q9 Column 3, Q10, Q11, Q12(c) & (d), Q13, Q14, Q15, Q16, Q17, Q18, Q19, Q20, Q21, Q23

Fluency and understanding

1 Solve each of the following equations numerically to find x.

WE 3

(a) $x - 5 = 11$
(b) $x - 7 = 4$
(c) $x - 6 = -2$
(d) $2x = -16$
(e) $\frac{x}{5} = 3$
(f) $\frac{x}{7} = -11$
(g) $2x - 3 = -11$
(h) $6x + 5 = 47$
(i) $3x - 7 = -16$
(j) $\frac{3x + 7}{5} = 5$
(k) $\frac{4x - 3}{3} = 11$
(l) $\frac{7x - 4}{6} = -3$

2 Solve each of the following equations using algebra. Check your answer by substitution.

WE 4

(a) $5a + 2 = 7$
(b) $3b - 8 = 4$
(c) $3c + 10 = 1$
(d) $8 - 3x = 11$
(e) $9 - 5x = 24$
(f) $13 - 7x = 69$
(g) $7 - 3x = 4$
(h) $11 - 2x = 5$
(i) $7 - 5x = -23$

7.2

WE5

3 Use the following graph to find the value of:

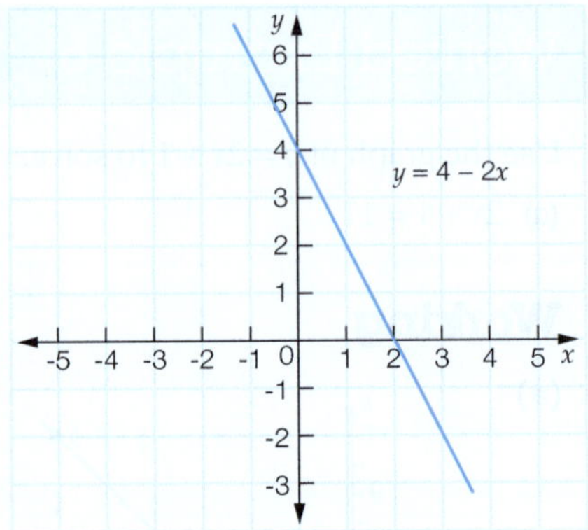

(a) y when $x = -1$
(b) y when $x = 0$
(c) y when $x = 2$
(d) x when $y = 4$
(e) x when $y = 2$
(f) x when $y = -2$
(g) y when $x = 1$
(h) y when $x = -0.5$
(i) y when $x = 2.5$

WE6

4 Use the graph of $y = 3x + 2$ to solve:

(a) $3x + 2 = 8$
(b) $3x + 2 = -1$
(c) $3x + 2 = 2$

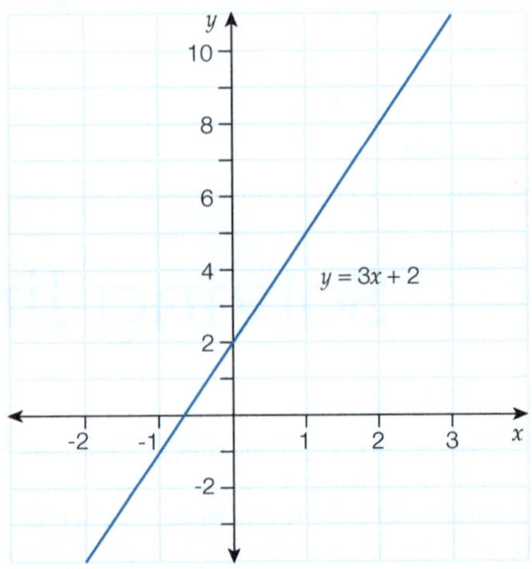

5 Use the set of scales below to write and solve equations.

(a) Write an equation that represents the scales shown.
(b) Write an equation to show the mass of the two boxes. (Let b = mass of a box.)
(c) Write an equation to show the mass of each box.
(d) What is the solution to the equation you found in part **(a)**?

6 (a) Which of the following equations is equivalent to $5z - 10 = 20$?

A $5z = 10$ **B** $5z = 30$ **C** $5z = -10$ **D** $5z = -30$

(b) Which of the following equations is equivalent to $\frac{5p-3}{2} = 4$?

A $5p - 3 = 20$ **B** $5p - 3 = 6$ **C** $5p - 3 = 7$ **D** $5p - 3 = 8$

7.2

7 **(a)** To obtain x from $4x + 1$ you would need to:

A add 1, then multiply by 4 **B** divide by 4, then subtract 1

C subtract 1, then multiply by 4 **D** subtract 1, then divide by 4

(b) To obtain k from $-3k - 5$ you would need to:

A add 5, then divide by 3 **B** subtract 5, then multiply by 3

C add 5, then divide by -3 **D** subtract 5, then divide by -3

(c) To obtain m from $4 - 2m$ you would need to:

A subtract 2, then divide by 4 **B** subtract 4, then divide by -2

C add 2, then subtract 4 **D** add 4, then divide by -2

8 Solve each of the following using algebra.

(a) $3d + 5 = -10$ **(b)** $5e + 6 = -4$ **(c)** $2f + 7 = -13$
(d) $9 + 4k = 1$ **(e)** $28 + 5h = 18$ **(f)** $22 + 3g = 7$
(g) $8 + 9n = -1$ **(h)** $12 + 7c = -16$ **(i)** $5 + 9k = -13$
(j) $2r + 7 = 10$ **(k)** $3t + 8 = 15$ **(l)** $7l + 11 = 23$
(m) $3d + 2.4 = 8.9$ **(n)** $4p + 3.8 = 7.2$ **(o)** $12a + 5.7 = 11.3$
(p) $2g + 1\frac{3}{4} = 2\frac{1}{2}$ **(q)** $3r + \frac{5}{6} = 4\frac{1}{3}$ **(r)** $8p + 3\frac{1}{3} = 4\frac{2}{9}$

9 Solve each of the following equations using algebra.

(a) $-2a + 7 = 3$ **(b)** $-5b + 11 = 6$ **(c)** $-3c - 4 = 8$
(d) $-d - 1 = 7$ **(e)** $-e + 5 = 2$ **(f)** $-f + 9 = 12$
(g) $3 - 4g = -17$ **(h)** $8 - 3h = -10$ **(i)** $-9 - 2m = 7$
(j) $7 - 3r = 4$ **(k)** $13 - 7p = -8$ **(l)** $29 - 5t = -11$
(m) $2 - 3p = 4$ **(n)** $8 - 5l = 12$ **(o)** $9 - 7k = 19$
(p) $5.3 - 2a = -7.8$ **(q)** $6.5 - 5b = -2.1$ **(r)** $7.9 - 4c = -6.7$

10 Use the graph of $y = 3x + 5$ shown to find the solution to:

(a) $3x + 5 = 2$ **(b)** $3x + 5 = 11$

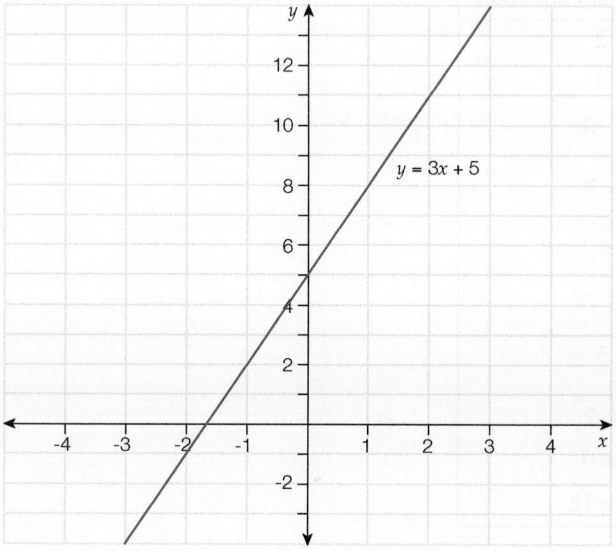

7.2

11 The graphs of $y = 3 - x$ and $3x - 2y = 4$ are drawn on the same axes.

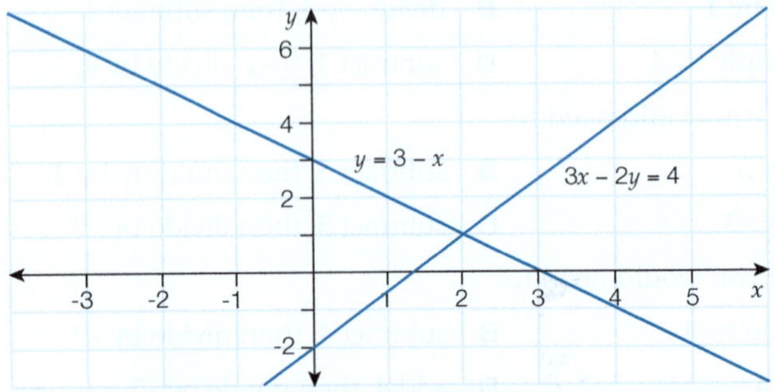

(a) What are the coordinates of the point of intersection of the two graphs (the point where the graphs cross each other)?

(b) Substitute the x-coordinate of the point of intersection into the equation $y = 3 - x$ to solve for y. What do you notice?

(c) Substitute the x-coordinate of the point of intersection into the equation $3x - 2y = 4$ and solve for y. What do you notice?

(d) What connection can you make between the point of intersection and the solution that satisfies both equations?

12 Write an equation for each of these word problems and use algebra to find the unknown number.

(a) A number is multiplied by 4, then 5 is subtracted to give a result of 23. (Let the number be n.)

(b) The sum of a number and seven is multiplied by three to give a result of 45. (Let the number be m.)

(c) The sum of five and a number is divided by nine to give a result of three. (Let the number be p.)

(d) A number is divided by four, then two is added to give a result of negative one. (Let the number be q.)

13 Match each equation in the left column with an equivalent equation in the right column.

Equation	Equivalent equation
(a) $2x + 1 = 3$	A $x = 2$
(b) $3x - 4 = 5$	B $x = 9$
(c) $4x = 8$	C $x = -3$
(d) $\frac{x}{5} = 1$	D $2x = 8$
(e) $2x - 3 = 5$	E $2x = 2$
(f) $3x = -6$	F $3x = -3$
(g) $\frac{x}{3} = 3$	G $x = 5$
(h) $3x + 7 = 4$	H $x = -2$
(i) $2x - 9 = 3$	I $3x = 9$
(j) $-4x = 12$	J $2x = 12$

14 Anita's father is twice her weight plus 16 kilograms. Her father weighs 102 kg.

(a) Write an equation using a to represent Anita's weight.

(b) Solve the equation to find Anita's weight.

15 This is an addition pyramid. Each brick is the sum of the two bricks below it.

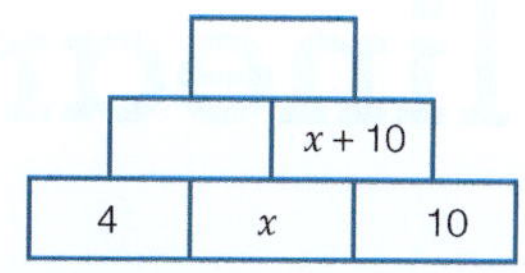

(a) Complete the rest of the pyramid.

(b) If the top brick has a value of 20, what is the value of x?

Reasoning, communicating and open-ended

16 Jeremy has a water tank in his garden. One day, his tank had only 1000 litres left in it. During the day it rained heavily and the tank filled at a rate of 400 litres per hour. Write an equation and solve it to find how many hours it took for Jeremy's tank to contain 3600 L.

17 Jamal wants to buy a puppy. He has been saving all his pocket money each week for the last 5 weeks, and his grandmother has given him \$25 for his birthday. He started with no money and now has \$62.50. Using p to represent the amount of pocket money he gets each week, write an equation and solve it to find how much pocket money Jamal receives each week.

18 A light aircraft needs to be careful of its weight limits. The pilot has been told that as long as the passengers on board are of average weight the limit will not be exceeded if the passengers consist of 2 adults and 3 children, or 1 adult and 5 children.

(a) How many children balance one adult?

(b) If the pilot was to take only children on the plane, use an equation to find how many she could take.

19 Fatima drives to and from work each day for the five-day working week, and 75 kilometres on the weekend. She manages to travel 325 kilometres every week in her car. Use an equation to find how far it is from her home to her workplace.

20 Paul has to take 3 litres of water with him on a boat trip. He has one 1.25 L container and two other containers of equal capacity. Together, the three containers will hold 3 litres. Use an equation to find the capacity of each of the other containers.

21 Write an equation using:

(a) multiplication and addition

(b) multiplication and subtraction

and then solve it.

22 The scales have 2 apples with a bunch of 20 grapes on the left and 3 apples with a bunch of 10 grapes on the right.

Claire was trying to find how many grapes had a mass equivalent to one apple. She suggests the following as possible first steps.

- Remove 20 grapes from each side.
- Take off half of everything from both sides.

(a) Explain to Claire the faults in her suggested first steps.

(b) Solve the problem in two steps.

23 Write down at least three equations that are equivalent to $2x + 3 = 11$.

24 Jill adds a number to both sides of an equation, then divides both sides by a different number.

She gets the answer: $x = 2$.
What might the original equation be?

7.3 Solving more complex linear equations

Need to Know

The equations in this section may look more complex than those solved so far. However, the process of performing inverse operations on both sides is the same.

To find a solution to these equations:

1. identify the order in which the equation was constructed around the variable by using the correct inverse order of operations
2. deconstruct (undo) the equation step by step with inverse operations in the reverse order to the way the equation was built. The number of operations that have been performed on the variable tells you how many operations need to be performed to solve the equation.

$-x$ can be thought of as $-1 \times x$.

The fraction bar means division. It also acts as though there are brackets around everything above it. For example, $\frac{2x+5}{3} = (2x+5) \div 3$.

Worked Example 7

WE7

Solve each of the following using algebra. Check your answer by substitution.

(a) $\frac{y}{6} + 9 = 7$ **(b)** $\frac{t+3}{4} = -4$

Working

(a)

$$\frac{y}{6} + 9 = 7$$

$$\frac{y}{6} + 9 - 9 = 7 - 9$$

$$\frac{y}{6} = -2$$

$$\frac{y}{6} \times 6 = -2 \times 6$$

$$y = -12$$

$$\text{LHS} = \frac{y}{6} + 9$$

$$= \frac{-12}{6} + 9$$

$$= -2 + 9$$

$$= 7$$

$$= \text{RHS}$$

(b)

$$\frac{t+3}{4} = -4$$

$$\frac{(t+3)}{4} \times 4 = -4 \times 4$$

$$t + 3 = -16$$

$$t + 3 - 3 = -16 - 3$$

$$t = -19$$

$$\text{LHS} = \frac{t+3}{4}$$

$$= \frac{-19+3}{4}$$

$$= \frac{-16}{4}$$

$$= -4$$

$$= \text{RHS}$$

7.3

Worked Example 8

WE 8

Solve each of the following equations using algebra. Check your answer by substitution.

(a) $5(2a - 9) = 13$ **(b)** $6(3 - x) = 12$

Working

(a)

$$5(2a - 9) = 13$$
$$10a - 45 = 13$$
$$10a - 45 + 45 = 13 + 45$$
$$10a = 58$$
$$\frac{10a}{10} = \frac{58}{10}$$
$$a = 5.8 \text{ or } 5\frac{4}{5}$$

$$\begin{aligned} \text{LHS} &= 5(2a - 9) \\ &= 5(2 \times 5.8 - 9) \\ &= 5(11.6 - 9) \\ &= 5 \times 2.6 \\ &= 13 \\ &= \text{RHS} \end{aligned}$$

(b)

$$6(3 - x) = 12$$
$$18 - 6x = 12$$
$$8 - 18 - 6x = 12 - 18$$
$$-6x = -6$$
$$\frac{-6x}{-6} = \frac{-6}{-6}$$
$$x = 1$$

$$\begin{aligned} \text{LHS} &= 6(3 - x) \\ &= 6(3 - 1) \\ &= 6 \times 2 \\ &= 12 \\ &= \text{RHS} \end{aligned}$$

Warning

Beware! $-6x$ means $-6 \times x$, so divide by -6 to undo.

Worked Example 9

WE 9

Solve each of the following equations using algebra. Check your answer by substitution.

(a) $\frac{2a - 3}{7} = -3$ **(b)** $\frac{6 - 4k}{5} = -2$

Working

(a)

$$\frac{2a - 3}{7} = -3$$
$$\frac{2a - 3}{7} \times 7 = -3 \times 7$$
$$2a - 3 = -21$$
$$2a - 3 + 3 = -21 + 3$$
$$2a = -18$$
$$\frac{2a}{2} = \frac{-18}{2}$$
$$a = -9$$

$$\begin{aligned} \text{LHS} &= \frac{2a - 3}{7} \\ &= \frac{2 \times -9 - 3}{7} \\ &= \frac{-18 - 3}{7} \\ &= \frac{-21}{7} \\ &= -3 \\ &= \text{RHS} \end{aligned}$$

(b)

$$\frac{6 - 4k}{5} = -2$$
$$\frac{6 - 4k}{5} \times 5 = -2 \times 5$$
$$6 - 4k = -10$$
$$6 - 6 - 4k = -10 - 6$$
$$-4k = -16$$
$$\frac{-4k}{-4} = \frac{-16}{-4}$$
$$k = 4$$

$$\begin{aligned} \text{LHS} &= \frac{6 - 4k}{5} \\ &= \frac{6 - 4 \times 4}{5} \\ &= \frac{6 - 16}{5} \\ &= \frac{-10}{5} \\ &= -2 \\ &= \text{RHS} \end{aligned}$$

Answers page 520

7.3 Solving more complex linear equations

Navigator

Q1 Column 1, Q2 Column 1, Q3 Column 1, Q4, Q5 Column 1, Q6, Q7, Q8, Q9, Q13, Q14, Q15, Q17	Q1 Column 2, Q2 Column 2, Q3 Column 2, Q4, Q5 Column 2, Q6, Q7, Q8, Q9, Q11, Q13, Q14, Q16, Q17	Q1 Column 3, Q2 Column 3, Q3 Column 3, Q4, Q5 Column 3, Q6, Q7, Q8, Q10, Q11, Q12, Q13, Q14, Q15, Q16

Fluency and understanding

WE7

1 Solve each of the following using algebra. Check your answer by substitution.

(a) $\frac{c}{6} + 8 = 10$ **(b)** $\frac{f}{7} + 5 = 8$ **(c)** $\frac{m}{9} + 3 = 12$

(d) $\frac{x}{2} - 3 = -5$ **(e)** $\frac{b}{3} - 4 = -7$ **(f)** $\frac{t}{4} - 5 = -9$

(g) $\frac{x+5}{3} = -3$ **(h)** $\frac{p+2}{5} = -7$ **(i)** $\frac{r+7}{9} = -8$

(j) $\frac{x-3}{4} = 3$ **(k)** $\frac{y-9}{5} = 7$ **(l)** $\frac{t-11}{12} = -5$

WE8

2 Solve each of the following equations using algebra. Check your answer by substitution.

(a) $4(x-2) = 20$ **(b)** $5(x+3) = -50$ **(c)** $8(x+2) = -24$

(d) $2(2x+1) = -9$ **(e)** $3(6x+4) = -16$ **(f)** $7(2x-5) = 41$

(g) $3(2-x) = 36$ **(h)** $5(9-x) = 30$ **(i)** $7(5-x) = 56$

(j) $4(2-3x) = -16$ **(k)** $5(3-2x) = -25$ **(l)** $9(5-3x) = -27$

WE9

3 Solve each of the following equations using algebra. Check your answer by substitution.

(a) $\frac{2a+5}{3} = -9$ **(b)** $\frac{3m+5}{4} = -7$ **(c)** $\frac{6r+1}{5} = -7$

(d) $\frac{3r-2}{5} = 8$ **(e)** $\frac{5n-3}{4} = 8$ **(f)** $\frac{7p-9}{12} = 8$

(g) $\frac{7-3r}{2} = 2$ **(h)** $\frac{28-5f}{4} = 2$ **(i)** $\frac{3-7m}{3} = 8$

(j) $\frac{4-5r}{3} = -2$ **(k)** $\frac{5-2p}{4} = -3$ **(l)** $\frac{7-4k}{5} = -9$

4 **(a)** To obtain x from $\frac{x}{2} + 7$ you would need to:

A multiply by 2, then subtract 7 **B** add 7, then multiply by 2

C subtract 7, then multiply by 2 **D** divide by 2, then add 7

(b) To obtain k from $\frac{k-8}{3}$ you would need to:

A add 8, then multiply by 3 **B** multiply by 3, then add 8

C subtract 8, then divide by 3 **D** multiply by 3, then subtract 8

7.3

5 Solve each of the following equations.

(a) $\frac{3a}{7} + 5 = 11$ **(b)** $\frac{5p}{3} + 8 = -7$ **(c)** $\frac{7m}{4} + 3 = -18$

(d) $5(3 - 2a) = -7$ **(e)** $4(5 - 3a) = 11$ **(f)** $3(7 - 4a) = 37$

(g) $\frac{2x + 3}{5} = 4$ **(h)** $\frac{3x + 5}{2} = 9$ **(i)** $\frac{4x - 7}{3} = 8$

(j) $\frac{1 - 2x}{3} = 3$ **(k)** $\frac{6 - 3x}{5} = 3$ **(l)** $\frac{8 - 5x}{7} = 4$

6 Match each equation (left column) with the appropriate list of inverse operations needed to solve the equation (right column).

Equation	Inverse operations needed to solve the equation
(a) $\frac{2x + 4}{3} = 5$	**A** $\times 4$, then $+ 2$, then $\div 3$
(b) $\frac{3x}{4} + 2 = 5$	**B** $- 4$, then $\times 3$, then $\div -2$
(c) $\frac{2x}{3} + 4 = 5$	**C** $\times 3$, then $- 4$, then $\div -2$
(d) $\frac{3x + 2}{4} = 5$	**D** $\times 3$, then $- 4$, then $\div 2$
(e) $\frac{2x - 4}{3} = 5$	**E** $- 2$, then $\times 4$, then $\div -3$
(f) $\frac{4 - 2x}{3} = 5$	**F** $- 2$, then $\times 4$, then $\div 3$
(g) $\frac{3x - 2}{4} = 5$	**G** $- 4$, then $\times 3$, then $\div 2$
(h) $\frac{2 - 3x}{4} = 5$	**H** $\times 3$, then $+ 4$, then $\div 2$
(i) $2 - \frac{3x}{4} = 5$	**I** $\times 4$, then $- 2$, then $\div 3$
(j) $4 - \frac{2x}{3} = 5$	**J** $\times 4$, then $- 2$, then $\div -3$

7 Write the following statements as equations and then find the value of the unknown.

(a) A number is multiplied by three, seven is added, and the result is divided by four to give an answer of four. What is the number?

(b) A number is divided by four, one is subtracted , and the result is doubled to give an answer of six. What is the number?

(c) Two is subtracted from five times a number, then the result is divided by six to give an answer of negative twelve. What is the number?

(d) Seven is added to one and a half times a number to give a result of eight. What is the number?

8 Margaret has put aside a certain amount of money to buy Christmas presents for five relatives. She spends \$10 of the money on wrapping paper and cards, and intends to divide the remaining money equally between each relative. However, she then decides to add \$20 to her budget to buy a special present for one relative. This present costs \$48.

(a) Write an equation and solve it to find how much money Margaret initially put aside.

(b) How much money did Margaret spend altogether?

9 Amir has a job as a telemarketer to sell a particular product. He will earn two-elevenths of the total value of the product he sells each day plus an extra \$50.

(a) Write an equation and solve it to find the total value of the products sold in one day if he earns \$150.

(b) If Amir makes 22 sales of this product, what is the selling price of the product?

10 Rimesh planted some daffodil bulbs in his garden. The following year he found that the number of bulbs had doubled. He also planted another six bulbs. The year after that he found the number of bulbs had tripled to give him 108 daffodil bulbs.

Write an equation and solve it to find how many bulbs he planted initially.

Reasoning, communicating and open-ended

For each of the following, write an appropriate equation and solve it to answer the question.

11 The sum of the digits of a two-digit number is 12. If the first digit is three times as large as the second digit, what is the number?

12 The sum of the digits of a three-digit number is 15. If the first digit is twice as large as the third digit and the second digit is one less than the third digit, what is the number?

13 Lyndall goes to the markets each Sunday to sell flowers. At the end of each market day she spends some of her earnings, which she keeps in a special 'market' wallet. At the beginning of July, she has a certain amount of money in her wallet. On the first weekend she doubled this amount, then spent \$20. On the second weekend, she doubled the money in the wallet, then spent \$25. On the third weekend, she trebled the money in the wallet and spent \$15. At this point, Lyndall has \$150 in her market wallet. How much did she have at the beginning of July?

14 Given that the area of the following rectangle is $106\ \text{cm}^2$, determine the value of x.

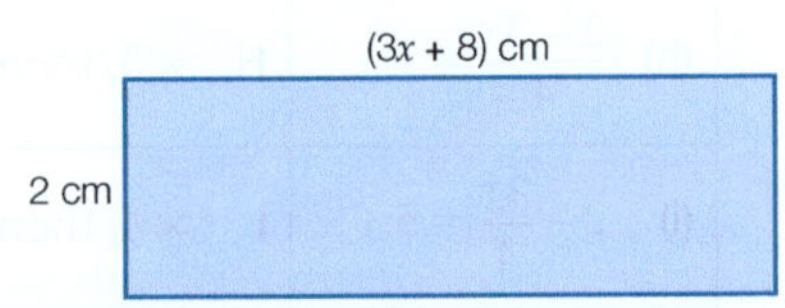

15 Write down at least three equations that can be solved by first adding 3, then dividing by $\frac{2}{3}$.

16 **(a)** Starting with the number 3, perform a sequence of three different operations on it to obtain an answer of 5. Then, write your sequence as an equation with x instead of 3.

(b) Make up two more three-step equations. Give the equations to your friends and see whether they can solve them.

17 Caitlyn tries to solve the equation $2 - 5x = 8$ as follows.

$2 - 5x = 8$	step 1
$5x = 8 - 2$	step 2
$5x = 6$	step 3
$x = \frac{6}{5}$	step 4
$x = 1\frac{1}{5}$	step 5

Her answer is incorrect. At what step did Caitlyn make a mistake? Show Caitlyn how to solve the equation.

Half-time 7

1 Solve the following equations using algebra. **Ex 7.3**

(a) $3(5 - x) = 24$ (b) $\frac{5x}{2} + 4 = 14$

(c) $\frac{3x - 6}{8} = 9$ (d) $3(x + 3) = -6$

(e) $\frac{7 - 4x}{5} = -7$ (f) $5 - \frac{x}{3} = 25$

2 Check whether the solution given in the brackets is correct for the following equations. **Ex 7.1**

(a) $x + 5 = 11$ $(x = 7)$ (b) $2x - 3 = 7$ $(x = 5)$

(c) $3(x - 9) = 15$ $(x = 14)$ (d) $\frac{x - 8}{3} = 8$ $(x = 23)$

3 Phillip, James and Sam all contribute to a present worth $130 for Chris. If James contributes half the amount that Phillip does and Sam contributes $10 less than twice the amount that Phillip contributes, find the amount that each person gives towards the present. **Ex 7.3**

4 Solve the following equations using algebra. **Ex 7.2**

(a) $x - 9 = 15$ (b) $2x + 12 = 16$

(c) $\frac{x}{4} + 7 = 12$ (d) $8 - \frac{x}{5} = 17$

5 Write the following sentences as equations and find the value of the unknown number. **Ex 7.1**

(a) Four is subtracted from a number to give a result of twelve.

(b) A number multiplied by seven is equal to thirty-five.

(c) A number multiplied by six is equal to three times twelve.

(d) A number is multiplied by seven, then three is added to give a result of twenty-four.

6 The total distance around this rectangular field is 480 metres. Find the value of x and use this result to find the length and width of the field. **Ex 7.3**

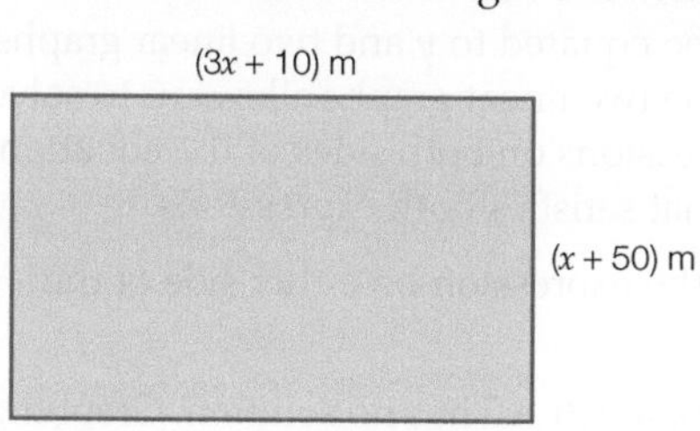

7.4 Solving equations where the unknown appears on both sides

Need to Know

Solving with algebra

To solve equations where the unknown appears on both sides of the equation:

Step 1: Perform the same inverse operation on both sides of the equation to remove the variable from the RHS (right-hand side) of the equation. This keeps the unknown on the LHS (left-hand side).

Step 2: Perform the same operation on both sides of the equation to remove any number on the LHS of the equation.

Step 3: Continue to use inverse operations on both sides of the equation until a solution is found.

Step 4: Always check by substitution that you have found the solution.

Some useful strategies

- If the coefficient of x is greater on the RHS of the equation than on the LHS, interchange the sides before beginning the solution process.
- Try to organise your equation so the variable is on the LHS and its coefficient is positive.
- If brackets are present in the equation, always expand first.
- If a fraction is present in the equation, always multiply both sides by the LCD (lowest common denominator) and expand the result.

Using graphs

The expression on either side of the equation can be equated to y and two linear graphs can be drawn. Finding the **point of intersection** of the two linear graphs allows us to solve equations where the same variable appears in expressions on both sides of the equation, as the x-coordinate of the point is the only value that satisfies both expressions.

The y-coordinate of your graph gives the value of the expression on either side of the equals sign.

Some equations have no solutions. For instance, $2x + 4 = 2x - 7$ has no solution. Graphing $y = 2x + 4$ and $y = 2x - 7$ will show that they are parallel lines.

7.4

Worked Example 10

WE10

Solve the following equations using algebra.

(a) $5x - 2 = 3x + 8$ (b) $2x + 7 = 5x - 8$

Working

(a)
$$5x - 2 = 3x + 8$$
$$5x - 3x - 2 = 3x - 3x + 8$$
$$2x - 2 = 8$$
$$2x - 2 + 2 = 8 + 2$$
$$2x = 10$$
$$\frac{2x}{2} = \frac{10}{2}$$
$$x = 5$$

LHS $= 5 \times 5 - 2$
$= 23$

RHS $= 3 \times 5 + 8$
$= 23$

$x = 5$ is the solution

(b)
$$2x + 7 = 5x - 8$$
$$5x - 8 = 2x + 7$$
$$5x - 2x - 8 = 2x - 2x + 7$$
$$3x - 8 = 7$$
$$3x - 8 + 8 = 7 + 8$$
$$3x = 15$$
$$\frac{3x}{3} = \frac{15}{3}$$
$$x = 5$$

LHS $= 2 \times 5 + 7$
$= 17$

RHS $= 5 \times 5 - 8$
$= 17$

$x = 5$ is the solution

Worked Example 11

WE11

Solve the following equations using algebra.

(a) $3(x + 4) = 2(1 - x)$ (b) $\frac{3x + 4}{2} = \frac{2x - 1}{5}$

Working

(a)
$$3(x + 4) = 2(1 - x)$$
$$3x + 12 = 2 - 2x$$
$$3x + 12 + 2x = 2 - 2x + 2x$$
$$5x + 12 = 2$$
$$5x + 12 - 12 = 2 - 12$$
$$5x = -10$$
$$\frac{5x}{5} = \frac{-10}{5}$$
$$x = -2$$

LHS $= 3(-2 + 4)$
$= 6$

RHS $= 2(1 - -2)$
$= 6$

$x = -2$ is the solution

(b)
$$\left(\frac{3x + 4}{2}\right) = \left(\frac{2x - 1}{5}\right)$$
$${}^{5}\cancel{10} \times \left(\frac{3x + 4}{{}_{1}\cancel{2}}\right) = {}^{2}\cancel{10} \times \left(\frac{2x - 1}{{}_{1}\cancel{5}}\right)$$
$$5(3x + 4) = 2(2x - 1)$$
$$15x + 20 = 4x - 2$$
$$15x - 4x + 20 = 4x - 4x - 2$$
$$11x + 20 - 20 = -2 - 20$$
$$11x = -22$$
$$\frac{11x}{11} = \frac{-22}{11}$$
$$x = -2$$

LHS $= \frac{3 \times -2 + 4}{2}$
$= -1$

RHS $= \frac{2 \times -2 - 1}{5}$
$= -1$

$x = -2$ is the solution

7.4

Worked Example 12

WE12

Solve $-x - 1 = 2x - 4$ graphically by finding the point of intersection of relevant graphs.

Working

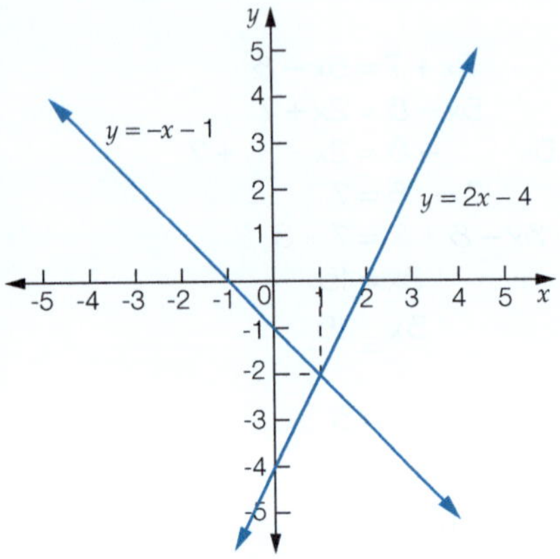

$y = -x - 1$ and $y = 2x - 4$

Point of intersection is at (1, -2).
Therefore, $x = 1$ and $y = -2$.

$x = 1$

LHS $= -1 - 1$
$= -2$

RHS $= 2 \times 1 - 4$
$= -2$

$x = 1$ is the solution

Answers page 521

7.4 Solving equations where the unknown appears on both sides

Navigator

Q1 Column 1, Q2 Column 1, Q3 Column 1, Q4, Q5, Q6, Q8, Q9, Q14	Q1 Column 2, Q2 Column 2, Q3 Column 2, Q4, Q5, Q6, Q8, Q9, Q11, Q13, Q14	Q1 Column 3, Q2 Column 3, Q3 Column 3, Q4, Q6, Q7, Q8, Q9, Q10, Q11, Q12, Q13, Q14

Equipment required: A graphing program can be used for Question **3**

Fluency and understanding

WE10

1 Solve the following equations using algebra.

(a) $7c + 2 = 6c - 5$ **(b)** $4b + 1 = 2b - 15$ **(c)** $3a - 5 = a + 7$

(d) $5d - 4 = 2d + 17$ **(e)** $5e - 3 = 2e + 9$ **(f)** $9f + 5 = 5f - 3$

(g) $3g + 12 = 2g + 5$ **(h)** $2h - 7 = h - 13$ **(i)** $4x - 3 = 7 - x$

(j) $k + 6 = 9 - k$ **(k)** $10j + 1 = 25 - 2j$ **(l)** $3y - 22 = -4 - 3y$

(m) $2m + 5 = 4m - 3$ **(n)** $5 + 4n = 7n - 1$ **(o)** $7x - 4 = 9x - 6$

(p) $2 - y = 3 - 2y$ **(q)** $p + 16 = 19 - 2p$ **(r)** $6 + 3x = 21 - 2x$

(s) $6 - 5p = 30 - p$ **(t)** $14 - 5x = 5 - 2x$ **(u)** $17 - 15a = 39 - 4a$

2 Solve the following equations using algebra. WE11

(a) $3(x-2)=2x+1$ (b) $4(x-1)=x+8$ (c) $5(x+3)=2x+9$

(d) $5(2x+3)=2(3x+1)$ (e) $7(x+4)=3(x-4)$ (f) $4(3x-2)=3(2x+9)$

(g) $2(3x+7)=3(5-x)$ (h) $3(2x+1)=4(3-x)$ (i) $4(2x+3)=3(4-5x)$

(j) $\frac{2x+3}{3}=\frac{4x-1}{5}$ (k) $\frac{4x+2}{3}=\frac{3x-4}{5}$ (l) $\frac{5x+1}{3}=\frac{5x-2}{4}$

(m) $\frac{3x+1}{5}=\frac{2x+7}{3}$ (n) $\frac{3x+2}{4}=\frac{4x+1}{5}$ (o) $\frac{2x+5}{4}=\frac{4x+1}{3}$

3 Solve the following equations graphically by finding the point of intersection of relevant graphs. WE12

(a) $x+6=-x$ (b) $5x-2=2x+7$ (c) $2x-6=-2x+10$

(d) $-2x+7=x-5$ (e) $5x-1=3x+7$ (f) $x-2=2x+7$

(g) $7x-2=2x+3$ (h) $5x-1=-x+11$ (i) $-2x+1=4x-5$

4 (a) $9x+4=2x-3$ has the solution:

A $x=1$ B $x=-1$ C $x=2$ D $x=4$

(b) $2y+7=-8-3y$ has the solution:

A $y=2$ B $y=1$ C $y=-2$ D $y=-3$

(c) $4(3k-2)=5(k+4)$ has the solution:

A $k=3$ B $k=-2$ C $k=1$ D $k=4$

5 (a) Write an expression to show 11 subtracted from four times a certain number.

(b) Write an expression to show two times the number in part **(a)** subtracted from 7.

(c) Equate the expressions in parts **(a)** and **(b)**.

(d) Solve the equation in part **(c)** to find the unknown number.

6 (a) Write an expression to show 36 added to a certain number.

(b) Write an expression to show the number in part **(a)** being added to 6 and the result doubled.

(c) Equate the expressions in parts **(a)** and **(b)**.

(d) Solve the equation in part **(c)** to find the unknown number.

7 (a) Write an expression to show five less than three-quarters of a certain number.

(b) Write an expression to show twice the result of subtracting ten from the number in part **(a)**.

(c) Equate the expressions in parts **(a)** and **(b)**.

(d) Solve the equation in part **(c)** to find the unknown number.

8 (a) Show that (-3, 2) is the point of intersection of $y=2x+8$ and $y=-x-1$.

(b) Write an equation without y that is equivalent to the two equations in part **(a)**.

(c) Without using algebra or doing any calculations, write down the solution to your equation in part **(b)**.

(d) What does the y-coordinate of the point (-3, 2) tell you about your equation in part **(b)**?

Reasoning, communicating and open-ended

Hint

Let s = the amount of money Sarah has.

9 Sarah has \$20 more than Frank, but \$40 less than Bruce. If Bruce has \$10 more than twice the amount that Frank has, write an equation to find how much money each person has.

10 Find the value of x in the following magic square. The rule for any magic square is that the sum of each row, column and diagonal are equivalent (or add to the same value).

$x+3$	$x+4$	$x-1$
$x-2$	6	$x+6$
9	x	$x+1$

11 Solve $2(x-5) = \frac{x+5}{3}$ by using graphs.

12 Three times Theo's age a year ago is the same as double his age in five years' time. Write this information as an equation and then find Theo's age using graphs.

13 Make up three equations similar to those in Question **1**, where the solution is a positive whole number.

14 Barry tries to solve the equation $6(x+6) = 2(x-4)$ as follows.

$6x + 36 = 2x - 4$	step 1
$4x + 36 = -4$	step 2
$4x = 32$	step 3
$x = 8$	step 4

His answer is incorrect. At what step(s) did Barry make a mistake? Show Barry how to solve this equation.

Problem solving

Alphametics

Alphametics are a type of mathematical puzzle in which digits are replaced by letters that form words and phrases. A famous example is:

```
    S E N D                9 5 6 7
+   M O R E   solution  +  1 0 8 5
-----------             ----------
  M O N E Y              1 0 6 5 2
```

The rules of alphametics are:

1 Each letter represents a unique digit.

2 Numbers must not start with a zero.

3 The solution is unique (unless otherwise stated).

Use the example and rules above to solve the following alphametics.

```
1       E A T        2     H E R E
    + T H A T            +   S H E
    ---------            ---------
    A P P L E            C O M E S

3     M E M O        4     C R O S S
    +   F R O M          + R O A D S
    ---------            -----------
    H O M E R            D A N G E R
```

Strategy options

- Guess and check.
- Work backwards.
- Look for a pattern.

Solving problems using linear equations

7.5

Need to Know

A simple process for solving problems using algebraic equations is:

Step 1: Identify the variable that needs to be found and represent it with a pronumeral.

Step 2: Use the information to write an equation, focusing on key phrases, such as 'is more than', 'is less than' and 'triple the amount'. Turn these phrases into corresponding operations, remembering the correct order of operations and ensuring that you convert all given amounts to the same units.

Step 3: Use inverse operations to solve the equation.

Step 4: Check that your answer is the solution by substituting it into the original equation and evaluating.

Worked Example 13

WE13

For the following situation, write an equation to show the situation, define the pronumeral clearly, and then solve the equation.

Jared has \$6.40 to spend on lollipops that cost 40 cents each. How many lollipops can he buy?

Working

Let n represent the number of lollipops.

$40n = 640$ (\$6.40 = 640c)

$$\frac{40n}{40} = \frac{640}{40}$$

$$n = 16$$

$$\begin{aligned} \text{LHS} &= 40n \\ &= 40 \times 16 \\ &= 640 \\ &= \text{RHS} \end{aligned}$$

Therefore, $n = 16$ is the solution.

Jared can buy 16 lollipops.

7.5

Worked Example 14

WE 14

A repair company charges a fixed travel fee of \$50 and an hourly rate of \$30 for labour on each job completed in a client's home. If h is the number of hours a job takes:

(a) write an expression for the total cost of the job

(b) use an equation to find the number of hours the job took if the total cost of the job was \$200.

Working

(a) Let h represent the number of hours.

$30h + 50$

(b)

$$30h + 50 = 200$$
$$30h + 50 - 50 = 200 - 50$$
$$30h = 150$$
$$\frac{30h}{30} = \frac{150}{30}$$
$$h = 5$$

The job takes 5 hours.

Substitute $h = 5$ into the original equation.

$$\begin{aligned} \text{LHS} &= 30h + 50 \\ &= 30 \times 5 + 50 \\ &= 150 + 50 \\ &= 200 \\ &= \text{RHS} \end{aligned}$$

Therefore, $h = 5$ is the solution.

Answers page 522

7.5 Solving problems using linear equations

Navigator

Q1, Q2, Q3, Q4, Q5, Q6, Q7, Q9, Q13, Q15, Q16, Q17, Q22, Q24	Q1, Q2, Q3, Q4, Q5, Q6, Q7, Q8, Q10, Q14, Q15, Q18, Q19, Q20, Q23	Q1, Q2, Q3, Q4, Q5, Q6, Q7, Q8, Q11, Q12, Q15, Q16, Q17, Q19, Q21, Q23

Fluency and understanding

WE 13

1 For each of the following situations, write an equation to show the situation, define the pronumeral clearly, and then solve the equation.

(a) Elena buys 3 apples for \$1.17. What is the cost of each apple?

(b) The mass of a shark pup is 935 kg less than that of its mother. If the pup has a mass of 9 kg, what is its mother's mass?

(c) A bus has the capacity to seat 49 passengers. If 21 passengers have boarded the bus, how many seats are still available?

(d) After a cool change, the outside temperature fell 9°C to a temperature of 27°C. What was the outside temperature before the cool change arrived?

2 Darren walked to the park while Ann rode her bicycle along the bike path to the park. If Darren walked x km while Ann rode 7 km: **WE14**

(a) write an expression for the total distance travelled by both

(b) use an equation to find the distance walked by Darren if together they travelled a total distance of 11 km.

3 Alan bought a salad roll and a container of orange juice for his lunch.

(a) If the salad roll cost \$$x$ and the orange juice cost \$2, write an expression for the total cost of Alan's lunch.

(b) Use an equation to find the cost of the salad roll if Alan spent \$4.50 on his lunch.

4 Olivia spent $\frac{3}{4}$ of her weekly pay from a part-time job on a new shirt.

(a) If Olivia's weekly pay is \$$w$, write an expression for the cost of the shirt.

(b) If the shirt cost \$24, solve an equation to find how much Olivia is paid each week.

5 The Kittens basketball team won 14 games more than it lost during the last season.

(a) If the team lost x games during the season, write an expression for the total number of games the team played.

(b) Use an equation to find the number of games won and lost by the Kittens if they played a total of 20 games in the season.

6 **(a)** Daniel is travelling at a speed of 28 km/h on his bicycle. How far does he travel in 2.5 hours if he maintains this speed?

If d represents the distance (in km) Daniel travels in 2.5 hours, the equation to solve is:

A $d + 2.5 = 28$ **B** $\frac{d}{2.5} = 28$ **C** $d = \frac{2.5}{28}$ **D** $2.5d = 28$

(b) Melita buys 5 lamingtons from the local bakery and receives \$5.50 change from \$10. Find the cost of each lamington.

If c represents the cost of a lamington (in cents), the equation to solve is:

A $5c + 10 = 5.5$ **B** $5c + 550 = 1000$ **C** $5c + 1000 = 550$ **D** $5c + 550 = 10$

(c) The perimeter of a rectangular swimming pool is 14 m. The pool is 3 m longer than it is wide, as shown in the diagram opposite.

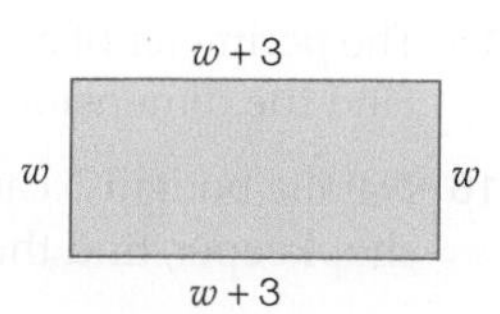

If w represents the width of the pool (in m), the equation to find the width of the pool is:

A $w + 3 = 14$ **B** $4w + 3 = 14$ **C** $4w + 14 = 6$ **D** $4w + 6 = 14$

7 Six friends share a bag of jelly beans equally. After dividing them up, there is one jelly bean left over.

(a) If each person received x jelly beans, write an expression for the total number of jelly beans in the bag.

(b) Use an equation to work out how many jelly beans each person received if there were 25 jelly beans in the bag.

7.5

8 **(a)** For the rectangle shown, where l is the length of the rectangle and w is the width, write an expression for the perimeter.

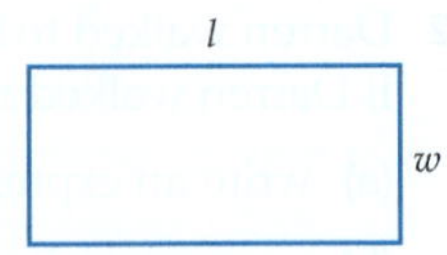

(b) Use the expression for perimeter to form an equation to solve for each of the following problems.

(i) Cameron has a 10 m length of wire fencing he wishes to use to protect a rectangular vegetable garden. If the width must be 2 m, how long can the vegetable garden be?

(ii) Cameron finds an extra 5 m of wire fencing, so, instead of having a width of 2 m, the vegetable garden is to have a width of 3 m. Find the new length of the vegetable garden.

For each of the following questions, form an equation and solve it to find the answer.

9 A birthday card and wrapping paper cost a total price of $5.65. If the card cost $2.90, how much did the wrapping paper cost? (Write 5.65 on the RHS of your equation.)

10 Samir has $70. At a sale, he buys three CDs (marked at the same price) and receives $11.50 change. How much did each CD cost? (Write 70 on the RHS of your equation.)

11 Megan buys three ice-creams and four drinks for her family at a cost of $12. If each drink costs $1.20, find the cost of each ice-cream. (Write 12 on the RHS of your equation.)

12 It took Sasha two hours and twelve minutes to get between two towns while travelling to Queensland. The computer in the car told her that her average speed was 97 km/h. Sasha forgot to reset the odometer and she wants to know how far the two towns are apart.

(a) Write an equation to show this situation, using d to represent the distance in km.

(b) Find the distance that Sasha travelled between the two towns.

For each of the following, write an equation and solve it to find the answer.

13 A farmer needs to fence off a rectangular paddock, but must use 4 layers of wire at different heights to keep her piglets in. She uses a 110 m roll of wire and has 2 m of wire left over. Find the perimeter of the piglet pen.

14 Jack spent 7 hours at the beach. If he relaxed in the shade reading for three more hours than he surfed, for how long did he surf? Hence, find the time he spent reading.

15 The perimeter of a rectangular block of land is 74 m. The width is 7 m less than the length. Find the dimensions of the block of land. (First, draw a diagram.)

16 Natalie bought 5 cartons of milk and received $2 in change. If she gave $10 to the shopkeeper, find the cost of each carton of milk.

Reasoning, communicating and open-ended

17 Yvonne buys a car that costs $23 000. She pays a deposit of $14 000 and then arranges to pay an amount each month for the next three years. Use an equation to find how much each equal monthly repayment should be if she wishes to pay off the car in three years.

18 At the end of the term each student in the class was given their average test score in mathematics. During the term the class sat three tests. For two of her tests Leonie obtained results of 93% and 85%, but she has forgotten her result in the third. Use an equation to find this unknown result for Leonie if she scored an average of 91%.

19 A rectangle has an area of 27 cm^2. Its width is 4 cm, and its length is longer than its width. Write an equation to show this situation, and find its length.

20 A boy is twice as tall as his little sister and 30 cm shorter than his father. The combined height of the three family members is 3.8 m. Find (in cm) the height of each person.

21 The sum of three consecutive even numbers is 102. Find the numbers by using an equation.

22 The angles inside a quadrilateral add up to 360°. For the following quadrilateral, write an equation to determine the value of x and then find the value of the four angles.

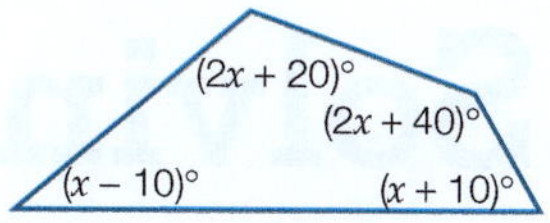

23 Two plumbers quote for a job to renovate a bathroom. Plumber Alan quotes \$1050 for material and \$50 per hour for labour. Plumber Barry quotes \$1200 for material and \$45 per hour for labour. Choose different numbers of hours and decide which quote would be cheaper for those hours. For what number of hours will the quotes be the same?

Hint

Drawing graphs may help you to answer Question **23**.

24 The formula for the area of a triangle is:

$$A = \frac{1}{2} \times \text{base} \times \text{height}$$

Identify at least three different sets of base length (b) and height (h) to give an area of 64 m^2.

Problem solving

The perfect holiday

The cost of a cruise ship holiday from Totally Oz Tours consists of a booking fee of \$150 and a charge of \$25 per day. The cost of food is an additional \$15 per day and there is a one-off beverage charge of \$100 for the duration of the cruise.

A rival company, Fun in the Sun Cruises, charges \$200 as a booking fee and \$45 per day (including food and beverage charge).

(a) After how many days does each company charge the same amount?

(b) Explain under which circumstances either company would be a cheaper option.

Strategy options

- Make a model.
- Guess and check.
- Test all possible combinations.

7.6 Solving quadratic equations

Need to Know

Solving simple quadratic equations

Quadratic equations may have 0, 1 or 2 solutions.

Simple quadratic equations can be 'undone' by taking the square root of each side. Remember that both a number and its negative can be squared to get the same result, which will always be positive.

Quadratic equations in the form of $x^2 = c$ can only be solved if c is a positive number or zero.

Therefore:

- $x^2 = c$ has no solutions if c is a negative number
- $x^2 = c$ has only one solution if $c = 0$
- $x^2 = c$ has two solutions if c is a positive number.

For example:

$x^2 = -1$ has no solution

$x^2 = 0$ has one solution, $x = 0$

$x^2 = 1$ has two solutions, $x = +1$ and $x = -1$

$x = +1$ and $x = -1$ can be written as $x = \pm 1$.

When taking the square root of a fraction make sure it is in simplest form, then take the square root of the numerator and denominator separately.

For example:

$x^2 = \frac{16}{25}$ has two solutions, $x = \frac{4}{5}$ and $x = -\frac{4}{5}$.

Mixed numbers need to be converted to improper fractions before taking the square root.

For example:

$$x^2 = 1\frac{9}{16}$$
$$= \frac{25}{16}$$
$$x = \frac{5}{4} \text{ and } x = -\frac{5}{4}$$

Exact solutions and decimal approximations

The square root may not be an integer or an exact fraction. The solution should then be left in exact form using the square root sign, unless otherwise instructed.

For example:

$$x^2 = \frac{7}{5}$$
$$x = \pm\sqrt{\frac{7}{5}}$$

When using the solution of a quadratic for practical purposes, a decimal approximation may be found using a calculator.

For example:

$$\sqrt{\frac{7}{5}} \approx 1.18321595$$
$$= 1.18 \text{ (correct to 2 decimal places)}$$

Note that a square root sign without a sign in front of it means the positive square root only.

For example: $\sqrt{2} = +\sqrt{2}$

When solving problems using quadratic equations, the context of the problem must be considered.

For example: length cannot be negative.

Worked Example 15

WE15

Solve the following quadratic equations.

(a) $x^2 = 49$ **(b)** $x^2 = \frac{32}{50}$ **(c)** $x^2 = 7\frac{1}{9}$

Working

(a) $x^2 = 49$

$x = \pm\sqrt{49}$

$x = \pm 7$

(b) $x^2 = \frac{32}{50}$

$= \frac{16}{25}$

$x = \pm\sqrt{\frac{16}{25}}$

$= \pm\frac{4}{5}$

(c) $x^2 = 7\frac{1}{9}$

$= \frac{64}{9}$

$x = \pm\sqrt{\frac{64}{9}}$

$= \pm\frac{8}{3}$

$= \pm 2\frac{2}{3}$

Worked Example 16

WE16

Solve the quadratic equation $x^2 = 43$ leaving the solution:

(a) in exact form

(b) as a decimal approximation correct to 2 decimal places.

Working

(a) $x^2 = 43$

$x = \pm\sqrt{43}$

(b) $x \approx \pm 6.557438524$

$= \pm 6.56$ (correct to 2 decimal places)

Answers page 523

7.6 Solving quadratic equations

Navigator

Q1, Q2, Q3, Q4, Q5, Q6, Q7, Q8 | Q1, Q2, Q3, Q4, Q5, Q6, Q7, Q8 | Q1, Q2, Q3, Q4, Q5, Q6, Q7, Q8

Fluency and understanding

WE15

1 Solve the following quadratic equations.

(a) $x^2 = 25$ (b) $x^2 = 81$ (c) $x^2 = 121$

(d) $x^2 = \frac{8}{72}$ (e) $x^2 = \frac{48}{75}$ (f) $x^2 = \frac{27}{147}$

(g) $x^2 = 6\frac{1}{4}$ (h) $x^2 = 3\frac{1}{16}$ (i) $x^2 = 2\frac{14}{25}$

WE16

2 Solve the following quadratic equations leaving the solution:

(i) in exact form

(ii) as a decimal approximation correct to 2 decimal places.

(a) $x^2 = 7$ (b) $x^2 = 15$ (c) $x^2 = 39$

(d) $x^2 = 11$ (e) $x^2 = 29$ (f) $x^2 = 55$

3 Solve the following quadratic equations.

(a) $x^2 = 0.36$ (b) $x^2 = 0.49$ (c) $x^2 = 1.44$

4 Solve the following quadratic equations leaving the solution:

(i) in exact form

(ii) as a decimal approximation correct to 2 decimal places.

(a) $x^2 = 1\frac{8}{11}$ (b) $x^2 = 3\frac{5}{7}$ (c) $x^2 = 5\frac{4}{5}$

5 Solving the quadratic equation $4x^2 = 9$ gives:

A $x = \frac{4}{9}$ **B** $x = \frac{3}{2}$ **C** $x = -\frac{3}{2}$ **D** $x = \pm\frac{3}{2}$

Reasoning, communicating and open-ended

6 A square garden has an area of 54 m^2. Find the length of the side of the garden in metres, correct to 1 d.p. (The area of a square = s^2 where s is the length of the side of the square.)

7 Explain why the quadratic equation $x^2 + 81 = 0$ has no solutions.

8 Anthony's solutions to the quadratic equation $x^2 = 1\frac{9}{16}$ were $x = \pm 1\frac{3}{4}$. Leonie told him that he was wrong, and that the correct solutions were $x = \pm\frac{5}{4}$. Who was correct and where did the other person go wrong?

Challenge 7

1 Mike is twice as old as his brother Jamie. When Mike turns 15, the sum of his and Jamie's ages will be 25. How old are Mike and Jamie now?

2 R, S, T and U are all whole, positive numbers. If $R \times S = 15$, $S \times T = 3$ and $T \times U = 7$, find the values of $R \times U$ and $S \times U$.

3 If $\frac{4}{h} = \frac{h}{36}$, find the value(s) of h.

4 Amber purchases 5 tyres for her car. If each tyre can be used for 16 000 km, how many kilometres can Amber travel using these five tyres?

5 The four numbers a, b, c and d add to give 30. b is double a, c is a third of b and d is 4 more than c. What are the numbers?

6 A square and an equilateral triangle have equal perimeters. If the side length of the triangle is 1 cm longer than the side of the square, what is the length of the side of the square?

7 If $d + e = 15$ and $e + f = 12$ and $d + e + f = 20$, find d, e and f.

8 The equilateral triangle has sides $2x$ cm and $(x + 15)$ cm as shown. The perimeter of the triangle is:

A 15 cm **B** 30 cm

C 60 cm **D** 90 cm

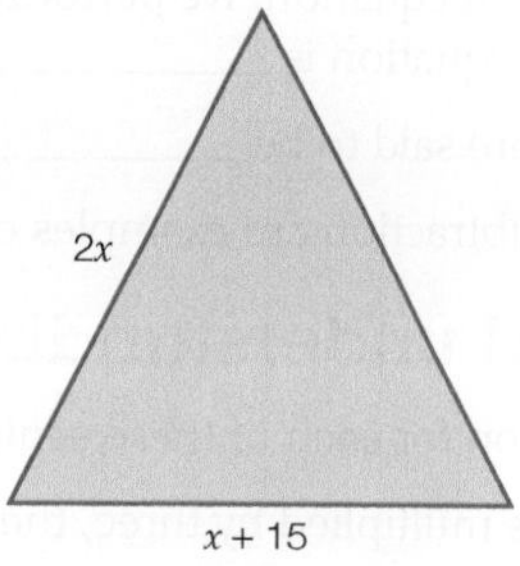

9 The height of a tree is measured annually. It was x metres high when planted, and has grown by exactly y metres between each annual measurement. At its latest annual measurement it was 23 metres high. Which of the following are possible?

A $x = 6, y = 2$ **B** $x = 7, y = 5$ **C** $x = 4, y = 7$ **D** $x = 5, y = 3$

10 In a test, Sue answered 15 out of the first 20 questions correctly. Of the remaining questions, she answered exactly one-third of them correctly. If all the questions were worth the same number of marks and Sue scored 50% on the test, how many questions were in the test?

11 ACD is a triangle in which B lies on the side AC such that $AD = DB = BC$. $\angle DAC = x°$. All angles in the diagram are a positive whole number of degrees.

(a) Find the value of x for which $\angle ADC$ has its largest value.

(b) What is the largest value of $\angle ADC$?

(c) What is the smallest value of $\angle ADC$?

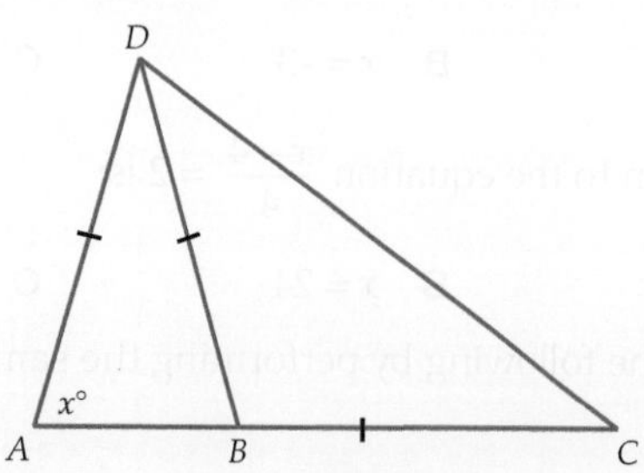

Chapter review 7

D.I.Y. Summary

Key Words

backtracking	guess, check and improve	point of intersection	solving
balanced	inspection	pronumeral	
equivalent	inverse operation	quadratic equations	
flowchart	linear equations	solution	

Copy and complete the following using the words and phrases from this list, where appropriate, to write a summary for this chapter. A word or phrase may be used more than once.

1 The __________ of an equation is the value of the variable that makes the equation true.

2 A __________ is a way of solving an equation by using a diagram and then __________ to find the solution.

3 When we solve an equation, we perform the same operations on each side of the equals sign so that the equation is __________.

4 Two equations are said to be __________ if they have the same solution.

5 Addition and subtraction are examples of __________s.

Fluency and understanding

1 Write an equation for each of these sentences. **Ex.7.1**

(a) A number is multiplied by three, then four is subtracted to give a result of two.

(b) A number is added to two, then the answer is divided by 5 to give a result of six.

2 Solve each of the following by performing the same operation on both sides of the equation. **Ex.7.2**

(a) $2x + 5 = 3$ (b) $11 - 3b = 5$ (c) $19 = -4c - 1$

3 (a) The solution to the equation $3x - 2 = 7$ is: **Ex.7.2**

A $x = 1$ B $x = 2$ C $x = 3$ D $x = 4$

(b) The solution to the equation $\frac{x}{2} + 1 = -2$ is:

A $x = -2$ B $x = -3$ C $x = -4$ D $x = -6$

(c) The solution to the equation $\frac{x-4}{4} = 2$ is:

A $x = 12$ B $x = 24$ C $x = -4$ D $x = 8$

4 Solve each of the following by performing the same operation on both sides of the equation. **Ex.7.2**

(a) $\frac{a}{3} + 7 = 10$ (b) $\frac{b-6}{5} = -1$ (c) $\frac{3c}{2} = 12$

5 Solve each of the following equations using algebra. **Ex.7.3**

(a) $\frac{4x-1}{3} = 5$ (b) $6 - \frac{x}{5} = 1$ (c) $\frac{5x}{4} + 3 = 13$ (d) $\frac{1-2x}{7} = 3$

6 Solve each of the following equations using algebra. **Ex 7.3**

(a) $\frac{3m+4}{2} = 5$ (b) $3 - \frac{2n}{5} = 1$ (c) $4(x+5) = 32$ (d) $\frac{3(m+11)}{8} = -6$

7 Solve each of the following equations using algebra. **Ex 7.4**

(a) $7x + 2 = 6x - 5$ (b) $3x + 7 = 15 - 5x$

(c) $7(x+1) = 2(x-4)$ (d) $3(5x+4) = 4(2-3x)$

8 Tony has $25 and saves $5 a week. Use an equation to find how many weeks it will take him to save $80. **Ex 7.5**

9 Use equations to solve each of the following problems. **Ex 7.5**

(a) Janis passes a $10 note to the shopkeeper to pay for 5 cans of cola. He receives $2.25 in change. Find the cost of each can of cola.

(b) Almira has set up a hot dog stand at the school fete. The ingredients for the hotdogs have cost $60. If Almira sells each hot dog for $3, how many does she need to sell to make a profit of $24?

(c) A stick 1 m long is broken into two parts. If one piece is 9 cm longer than the other, how long are the pieces?

10 Solve the following quadratic equations. **Ex 7.6**

(a) $x^2 = 25$ (b) $x^2 = \frac{49}{64}$ (c) $x^2 = 5\frac{1}{16}$

11 Sam buys a new suitcase that is 70 cm wide. He discovers that two of his folders (which are identical) will fit next to each other in the suitcase with only 5 cm to spare.

(a) Write an equation to describe this situation, using f to represent the width of the folder.

(b) Find the width of the folder.

12 Three consecutive integers add to 24. Use an equation to find the three numbers.

13 Robert has 50 coins, all 5 cent and 10 cent pieces, amounting to $3.50. By using an equation, find how many of each type of coin he has.

14 Use an equation to find x in the magic square shown. As with all magic squares the sum of each row, column and diagonal is the same.

$4x$	$6x+2$	28
$6x-2$	30	$2(2x+1)$
$4(x+2)$	$4x-2$	$6x$

15 If Peter is 7 years older than Sharon and John is twice as old as Peter, find how old Peter is if the mean of their ages is 19.

Reasoning and communicating

Use an equation to solve each of the following.

16 (a) In 5 years' time, Maja's age will be twice what it was 6 years ago. How old is Maja now?

(b) Jake rides a motorbike. He travels to and from work three times in a week. He also travels to and from his girlfriend's place, which is one-third of the distance to work, four times a week. In total, he travels 156 km each week. How far is it from his home to work?

(c) Jung spent a quarter of his money on a new shirt. He then gave a donation to charity of $3 and deposited half of the remaining money in the bank. If he is left with $15, how much did he start with? Write an equation first using s to represent the starting money.

17 The sum of two numbers is 37. If the larger number is divided by the smaller number, the result is 3, remainder 5. Find the numbers.

18 An isosceles triangle has a perimeter of 45 cm. Its equal sides each measure 6 cm longer than the non-equal base. How long is the base of this particular triangle?

19 Three students take different routes home from school. Mark needs to walk twice as far as Tom, and Fred walks four times as far as Tom. If the three cover a total distance of 3.5 km (or 3500 m), how far does each person walk?

20 The shape at right has a perimeter of 120 cm. Use the information provided in the question to generate an algebraic equation, and then use it to find the value of x cm.

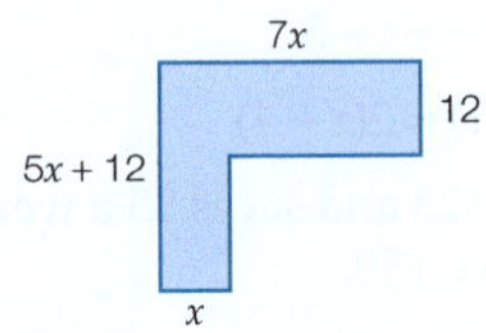

21 **(a)** Using the magic square, find the magic sum in terms of a, b and c. Hence, determine the expressions that belong in Box A and Box B. Remember, as with all magic squares, the sum of each row, column and diagonal is the same.

(b) If $a = 3$, $b = 2$ and the magic sum is 27, use an equation to find c. Hence, replace the pronumerals with numbers in each of the boxes.

$a + b$	$a + 2c$	$a + c + 2b$
$a + 2b + 2c$	$a + c + b$	Box A
$a + c$	Box B	$a + 2c + b$

22 A triangular garden bed has its three sides measured at x metres, $(x + 4)$ metres and $(x + 5)$ metres. The perimeter of the garden bed is measured to be 51 metres. Find the lengths of all three sides of the garden bed.

NAPLAN practice 7

Numeracy: Non-calculator

1 What is the value of p in this equation? $13p - 8 = 8p + 7$?

2 The equation that best represents the following 'I am thinking of a number—when I add 4 the number is 13' is:

A $x + 13 = 4$ **B** $x - 4 = 13$ **C** $x + 4 = 13$ **D** $x - 13 = 4$

3 What is the value of x in the equation $4x + 7 = 3$?

A $x = -1$ **B** $x = 1$ **C** $x = 2$ **D** $x = 11$

4 If $y = 2x - 1$ and $y = 3x + 2$, which value of x satisfies both of these equations?

A $x = -3$ **B** $x = -1$ **C** $x = 1$ **D** $x = 3$

5 The digits 4, 5 and 6 are placed in the equation $2x + y - z = 8$. If y is 4, what is the value of z?

Numeracy: Calculator allowed

6 Robert and Fergus shared a bag of Smarties. If Fergus ate 18 more Smarties than Robert did, and there were 66 Smarties in the bag, how many Smarties did Fergus eat?

7 The solution to the equation $6x - 1 = 4x + 5$ is:

A $x = -2$ **B** $x = 2$ **C** $x = 3$ **D** $x = 4$

8 Find the value of x that makes the equation $\frac{x+3}{4} = 5$ true.

9 The cost in dollars to make k cakes is $8 + 3k$.

How many cakes are made for a cost of $203?

10 Phil is 3 years older than his sister Bridget. The product of their ages is the same as their mother's age. If Phil's mother is 40 years old, how old is Bridget?

Algebra

2

Recall 2

Prepare for this chapter by attempting the following questions. If you have difficulty with a question, go to Pearson Places and download the Recall Worksheet.

1 Simplify the following expressions.

(a) $3x + 4x + 7x$ **(b)** $5y + 7y - 4y$ **(c)** $3k^2 + 5k^2 - 7k^2$

2 Simplify the following expressions.

(a) $5y + 6h - 2h - 3y$ **(b)** $6x^2 + 5y^3 - 2x + 3y^3$

3 Simplify the following expressions.

(a) $3x \times 4y$ **(b)** $5k \times 4k$ **(c)** $-5g \times 6g \times 2g$

4 Simplify, writing your answer in index form.

(a) $2^4 \times 2^9$ **(b)** $6^8 \div 6^2$ **(c)** $(5^2)^2$ **(d)** $(8.6325 \times 5.15)^0$

5 Expand:

(a) $\left(\frac{4^2}{5^3}\right)^3$ **(b)** $(2^4 \times 5^2)^4$ **(c)** $\left(\frac{7^2 \times 3^4}{4^5}\right)^2$

6 Find **(i)** the perimeter and **(ii)** the area of each of the following.

(a)

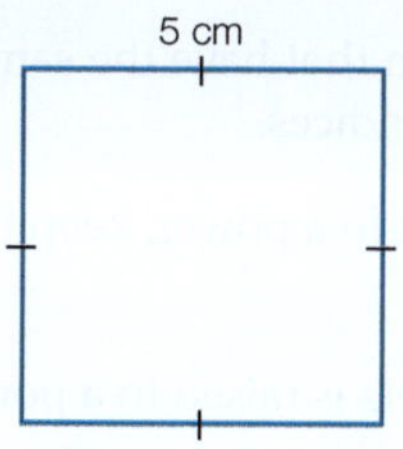

(b)

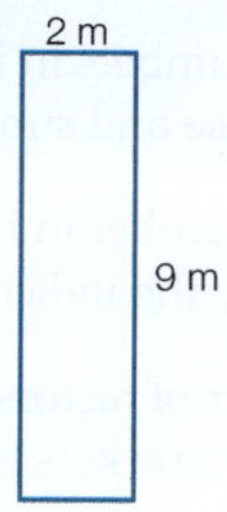

(c)

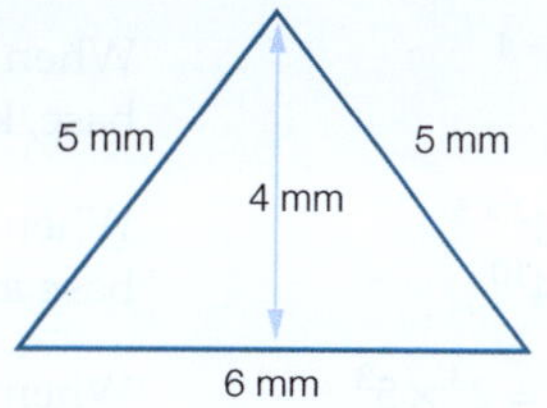

7 Expand the following expressions.

(a) $4(x + 2)$ **(b)** $6(x - 3)$ **(c)** $-2(x + 4)$

8 Factorise the following expressions.

(a) $6x - 12$ **(b)** $10x - 25$ **(c)** $-3x + 15$

Syllabus references

Number and Algebra: Indices

✓ Extend and apply the index laws to variables, using positive-integer indices and the zero index **(ACMNA212) (Stage 5.1)**

✓ Simplify algebraic products and quotients using index laws **(ACMNA231) (Stage 5.1)**

✓ Apply index laws to numerical expressions with integer indices **(ACMNA209) (Stage 5.1)**

✓ Apply index laws to algebraic expressions involving integer indices **(ACMNA209) (Stage 5.2)**

Number and Algebra: Algebraic Techniques

✓ Apply the distributive law to the expansion of algebraic expressions, including binomials, and collect like terms where appropriate **(ACMNA213) (Stage 5.2)**

✓ Factorise algebraic expressions by taking out a common algebraic factor **(ACMNA230) (Stage 5.2)**

✓ Expand binomial products and factorise monic quadratic expressions using a variety of strategies **(ACMNA233) (Stage 5.2)**

✓ Expand binomial products using a variety of strategies **(ACMNA233) (Stage 5.3§)**

✓ Factorise monic and non-monic quadratic expressions **(ACMNA269) (Stage 5.3§)**

Number and Algebra: Equations

✓ Rearrange literal equations **(Stage 5.3§)**

Measurement and Geometry: Numbers of any Magnitude

✓ Investigate very small and very large time scales and intervals **(ACMMG219) (Stage 5.1)**

✓ Express numbers in scientific notation **(ACMNA210) (Stage 5.1)**

2.1 Introducing index laws using variables

Stage 5.1

Need to Know

Index form

number → $81 = 3^4$ ← index

↑ base

81	$=$	$3 \times 3 \times 3 \times 3$	$=$	3^4
		Expanded form		**Index form**

Index laws for integers:

$2^4 \times 2^5 = 2^{4+5}$ $= 2^9$	When multiplying numbers in index form that have the same **base**, keep the base and add the indices.
$\frac{3^6}{3^4} = 3^{6-4}$ $= 3^2$	When dividing numbers in index form that have the same base, keep the base and subtract the indices.
$(4^2)^5 = 4^{2 \times 5}$ $= 4^{10}$	When raising a number in index form to a power, keep the base and multiply the indices.
$(2 \times 5)^3 = 2^3 \times 5^3$	When the product of factors in brackets is raised to a power, each factor in the brackets is raised to the same power.
$\left(\frac{7}{3}\right)^4 = \frac{7^4}{3^4}$	When the quotient of factors is raised to a power, each factor in the brackets is raised to the same power.
$2^0 = 1, 4975^0 = 1$ $6.347^0 = 1, \left(\frac{5}{7}\right)^0 = 1$	Any number, except 0, raised to the zero index is equal to 1 (0^0 is undefined).

Multiplication with indices

To calculate $a^2 \times a^3$, write each power in expanded form and count the number of factors in the multiplication string.

$a^2 = a \times a$	There are two factors of base a in the multiplication string.
$a^3 = a \times a \times a$	There are three factors of base a in the multiplication string.
$a^2 \times a^3 = a \times a \times a \times a \times a$ $= a^5$	There are five factors of base a in the multiplication string.

Notice that $2 + 3 = 5$, so $a^2 \times a^3 = a^{2+3}$
$= a^5$

$a^m = a \times a \times \ldots \times a \times a$	There are m factors of base a in the multiplication string.
$a^n = a \times a \times \ldots \times a \times a$	There are n factors of base a in the multiplication string.

To multiply a^m by a^n,

$a^m \times a^n = a^{m+n}$	There are $m + n$ factors of base a in the multiplication string.

Division with indices

As for multiplication, a pattern for the division of numbers in index form can be identified.

To calculate $a^5 \div a^3$, write the division as a fraction, write each power in expanded form and cancel the common factor of a^3.

$$\frac{a^5}{a^3} = \frac{a \times a \times a \times a \times a}{a \times a \times a}$$
$$= a^2$$

Notice that $5 - 3 = 2$ so $a^5 \div a^3 = a^{5-3}$
$$= a^2$$

To divide a^m by a^n,

$a^m \div a^n = a^{m-n}$ There are $m - n$ factors of base a in the multiplication string.

Power of a power

To calculate $(a^5)^4$, rewrite as a product and then use the index law of multiplication.

$$(a^5)^4 = a^5 \times a^5 \times a^5 \times a^5 = a^{5+5+5+5}$$
$$= a^{4 \times 5} \text{ or } a^{5 \times 4}$$
$$= a^{20}$$

To find the power of a power,

$(a^m)^n = a^{m \times n}$

Index laws for variables

$a^m \times a^n = a^{m+n}$ When multiplying **variables** in index form that have the same base, keep the base and add the indices.

$a^m \div a^n = \frac{a^m}{a^n} = a^{m-n}$ When dividing variables in index form that have the same base, keep the base and subtract the indices.

$(a^m)^n = a^{m \times n}$ When raising a variable in index form to a power, keep the base and multiply the indices. This rule is known as the 'power of a power' law.

Remember that $a = a^1$.

Worked Example 1

WE1

Simplify each of the following, leaving your answers in index form.

(a) $7^3 \times 7^8$ **(b)** $c^3 \times c^6$ **(c)** $4h^5 \times 3h^9$

Working

(a) $7^3 \times 7^8$
$= 7^{3+8}$
$= 7^{11}$

(b) $c^3 \times c^6$
$= c^{3+6}$
$= c^9$

(c) $4h^5 \times 3h^9$
$= 4 \times 3 \times h^5 \times h^9$
$= 12h^{5+9}$
$= 12h^{14}$

2.1

Worked Example 2

WE 2

Simplify each of the following, leaving your answers in index form.

(a) $\frac{11^9}{11^5}$ (b) $\frac{d^7}{d^3}$ (c) $20c^8e^7 \div 4c^3e^2$

Working

(a) $\frac{11^9}{11^5}$

$= 11^{9-5}$

$= 11^4$

(b) $\frac{d^7}{d^3}$

$= d^{7-3}$

$= d^4$

(c) $\frac{20c^8e^7}{4c^3e^2}$

$= \frac{{}^{5}\cancel{20}c^{8-3}e^{7-2}}{{}_{1}\cancel{4}}$

$= 5c^{8-3}e^{7-2}$

$= 5c^5e^5$

Worked Example 3

WE 3

Simplify each of the following, leaving your answers in index form.

(a) $(6^3)^4$ (b) $(2^3)^2 \times (2^4)^3$ (c) $(5^3)^4 \div (5^5)^2$

(d) $(r^3)^5$ (e) $(p^3)^4 \times (p^5)^2$ (f) $(m^5)^3 \div (m^2)^6$

Working

(a) $(6^3)^4$

$= 6^{3\times 4}$

$= 6^{12}$

(b) $(2^3)^2 \times (2^4)^3$

$= 2^{3\times 2} \times 2^{4\times 3}$

$= 2^6 \times 2^{12}$

$= 2^{6+12}$

$= 2^{18}$

(c) $(5^3)^4 \div (5^5)^2$

$= 5^{3\times 4} \div 5^{5\times 2}$

$= 5^{12} \div 5^{10}$

$= 5^{12-10}$

$= 5^2$

(d) $(r^3)^5$

$= r^{3\times 5}$

$= r^{15}$

(e) $(p^3)^4 \times (p^5)^2$

$= p^{3\times 4} \times p^{5\times 2}$

$= p^{12} \times p^{10}$

$= p^{12+10}$

$= p^{22}$

(f) $\frac{(m^5)^3}{(m^2)^6}$

$= \frac{m^{5\times 3}}{m^{2\times 6}}$

$= \frac{m^{15}}{m^{12}}$

$= m^{15-12}$

$= m^3$

2.1

2.1 Introducing index laws using variables

Answers page 424

Navigator

Stage 5.1

Q1 Column 1, Q2 Column 1, Q3 Column 1, Q4, Q5, Q6 Columns 1 & 2, Q7, Q8, Q9, Q10, Q11, Q14, Q18	Q1 Column 2, Q2 Column 2, Q3 Column 2, Q4, Q6 Columns 2 & 3, Q7, Q8, Q9, Q10, Q11, Q12, Q15, Q17, Q18	Q1 Column 3, Q2 Column 3, Q3 Column 3, Q4, Q6 Columns 2 & 3, Q7, Q8, Q9, Q10, Q11, Q12, Q13, Q15, Q16, Q18

Fluency and understanding

1 Simplify each of the following, leaving your answers in index form. WE1

(a) $2^4 \times 2^5$ (b) $3^7 \times 3^4$ (c) $5^{11} \times 5^7$
(d) $a^3 \times a^4$ (e) $b^6 \times b^5$ (f) $5c^3 \times c^2$
(g) $3d^3 \times 2d^2$ (h) $4b^4 \times 10b^4$ (i) $7b^2 \times 4b^2$
(j) $4a^5 \times 3a$ (k) $7a^3 \times 2a$ (l) $20a^3 \times 10a^4$
(m) $6a^4b^2 \times 4a^5$ (n) $9a^3b^3 \times 3ab^2$ (o) $5b^4c^5 \times 6b^3c^2$
(p) $-5a^4b^6 \times 7b^2$ (q) $-9e^5f^5 \times -4ef$ (r) $3a^5b^8 \times -8ab^2$

2 Simplify each of the following, leaving your answers in index form. WE2

(a) $3^8 \div 3^5$ (b) $4^7 \div 4^2$ (c) $9^5 \div 9^3$
(d) $m^5 \div m^2$ (e) $4b^5 \div b^3$ (f) $5g^7 \div g^3$
(g) $\frac{n^6}{n^4}$ (h) $\frac{14b^7}{7b^5}$ (i) $\frac{22c^6}{11c^2}$
(j) $\frac{6a^5}{3a^4}$ (k) $\frac{12a^7}{8a^3}$ (l) $\frac{14e^4f^5}{10ef^2}$
(m) $10a^7b^7 \div 5a^4b^3$ (n) $-24a^5b^3 \div 21a^2b$ (o) $144a^6b^8 \div -12a^5b^7$

3 Simplify each of the following, leaving your answers in index form. WE3

(a) $(2^3)^4$ (b) $(3^2)^5$ (c) $(5^4)^2$
(d) $(3^2)^5 \times (3^4)^2$ (e) $(4^3)^2 \times (4^5)^3$ (f) $(7^2)^5 \times (7^4)^2$
(g) $(5^2)^8 \div (5^3)^2$ (h) $(8^5)^8 \div (8^2)^3$ (i) $(11^4)^8 \div (11^3)^9$
(j) $(b^7)^3$ (k) $(k^8)^4$ (l) $(a^6)^2$
(m) $y^5 \times (y^3)^4$ (n) $(k^3)^8 \times k^3$ (o) $(m^6)^3 \times m^2$
(p) $(g^2)^5 \times (g^4)^2$ (q) $(y^2)^4 \times (y^5)^3$ (r) $(t^3)^3 \times (t^2)^6$
(s) $(p^3)^4 \div (p^5)^2$ (t) $(n^7)^2 \div (n^2)^4$ (u) $(b^8)^3 \div (b^3)^7$

4 (a) $-4r^3 \times -r^2$ can be simplified to give:

A $4r^1$ B $-4r^5$ C $4r^5$ D $4r^6$

(b) The product of $-abc$, $-5a$ and $-2a$ is:

A $-10a^3bc$ B $-7a^2bc$ C $10abc$ D $10a^2bc$

2.1

5 Copy and complete each of the following by writing the correct index number in the box.

(a) $\frac{x^7 \times x^9}{x^{12}}$

$= \frac{x^{\square}}{x^{12}}$ (Add indices)

$= x^{\square}$ (Subtract indices)

(b) $\frac{6y^4 \times y^6}{y^2 \times y^5}$

$= \frac{6y^{\square}}{y^{\square}}$ (Add indices)

$= 6y^{\square}$ (Subtract indices)

6 Simplify each of the following by completing any multiplication of variables first.

(a) $\frac{x^8 \times x^7}{x^6}$ (b) $\frac{t^7}{t^2 \times 3t^4}$ (c) $\frac{c^6 \times c^9}{c^5 \times c^3}$

(d) $\frac{j^4 \times j^8}{j^2 \times j^5}$ (e) $\frac{4a^7 \times 3a^4}{a^3}$ (f) $\frac{8q \times 3q^6}{12q^4}$

7 (a) $\frac{f^8 \times f^6}{f^{10}}$ simplifies to:

A f^4 B f^6 C f^{48} D f^{12}

(b) $\frac{15x^9 \times 3x^6}{9x^{10} \times x^4}$ simplifies to:

A $5x^9$ B $9x$ C $5x$ D $9x^9$

(c) $-36k^4n^3 \div 9n^2k^3$ can be simplified to:

A $4k^4n$ B $-4k^7n^6$ C $-4kn$ D $45k^{12}n^6$

(d) $\frac{42x^5y^7z^6}{14x^4y^2z^3}$ simplifies to:

A $3xy^5z^2$ B $28xy^5z^3$ C $3xy^5z^3$ D $3x^9y^9z^9$

8 (a) How many groups of $3a^2$ will divide into $15a^4$?

(b) Use your answer to (a) to rewrite this as a multiplication that has $15a^4$ as the product.

9 (a) $\frac{a^8 \times b^5}{a^2 \times a^6}$ simplifies to:

A $a^{16}b^5$ B $a^{10}b^{11}$ C b^5 D b^{11}

(b) $\frac{3c^7 \times 8d^4}{12c^3 \times 4c^4}$ simplifies to:

A $2d^4$ B $\frac{d^4}{2}$ C $\frac{c^4d^4}{24}$ D $\frac{d^4}{24}$

10 Simplify each of the following.

(a) $(m^5)^2 \div m^7$ (b) $(p^6)^4 \div p^{10}$ (c) $\frac{(k^4)^3}{k^9}$ (d) $\frac{(n^8)^2}{n^6}$

11 (a) $(k^2)^3 \times (k^5)^2$ simplifies to:

A k^{16} B k^{12} C k^{60} D k^{15}

(b) $(b^5)^4 \div (b^3)^5$ simplifies to:

A b^4 B b C b^5 D b^{12}

2.1

Reasoning, communicating and open-ended

12 Solve for x in the following equations.

(a) $5^{x+2} = 5^{2x-1}$ (b) $3^{x+1} = 3$ (c) $2^{2x+4} = 1$ (d) $4(7^x + 7^x) = 56$

Hint

If the bases on both sides of the equation are the same, the expression for the powers must be equal.

13 When Josh started his part-time job, he decided to save some of his wages on a weekly basis. He was paid $50.00 a week, so he decided to start by saving 5c the first week, 10c the next week, 20c the third week, continuing to double the amount he saved each week. He was shocked to find that, by the 11th week, he could not afford to keep up his plan.

(a) Copy the following table and complete the second column to show how much Josh saved each week for the first 5 weeks.

Week	Amount saved each week (c)	Amount saved each week (c) in index form
1		
2		
3		
4		
5		

(b) Complete the third column in this table by writing these amounts in index form.

(c) Write a rule in index form for finding the amount Josh saved in the nth week.

(d) Use your rule to calculate how much Josh was supposed to save in the 11th week.

(e) How much would Josh need to earn each week if he was to keep up his saving plan for 6 months (26 weeks)?

Hint

Look for a pattern. What values do you get if 1 is subtracted from a power of 2?

14 Find three different expressions for ■ and ▲ in terms of a and b to make this a true statement:

■ × ▲ = $-3a^2b^3$

15 Write three possible divisions involving indices that will give an answer of $5a^2b^3$.

16 Create three index number division expressions and simplify them using any of $24a^2b^3$, $3ab$, $48a^5b^6$, $22a^2b^2$ or 2×3^2a. Check your answers by giving the divisions to a friend to simplify.

17 Write three different sets of values for m and n so that $(a^m)^n$ simplifies to a^{24}.

18 The index laws are sometimes used incorrectly. Below is a list of equations that 'look' reasonable, but are actually incorrect.

In each case, explain the mistake that has been made and give the correct answer in its place. Explain why your answer is correct.

(a) $x^2x^4 = x^8$ (b) $\frac{x^9}{x^3} = x^3$ (c) $(x^2)^4 = x^{2^4} = x^{16}$

2.2 More index laws

Stage 5.1

Need to Know

To **expand** an **expression** is to remove the brackets in the expression.

Expanding products

Expand $(2 \times a)^3$.

$$\begin{aligned}(2 \times a)^3 &= (2 \times a) \times (2 \times a) \times (2 \times a)\\ &= 2 \times 2 \times 2 \times a \times a \times a\\ &= 2^3 \times a^3\end{aligned}$$

So $(2 \times a)^3 = 2^3 \times a^3$.

Each factor in the brackets has been raised to the power of 3.

Expand $(kp)^4$.

$$\begin{aligned}(kp)^4 &= kp \times kp \times kp \times kp\\ &= k \times k \times k \times k \times p \times p \times p \times p\\ &= k^4 \times p^4\\ &= k^4p^4\end{aligned}$$

Each factor in the brackets has been raised to the power of 4.

So $(ab)^m = a^mb^m$.

Expanding quotients

Expand $\left(\frac{m}{n}\right)^5$.

$$\begin{aligned}\left(\frac{m}{n}\right)^5 &= \frac{m}{n} \times \frac{m}{n} \times \frac{m}{n} \times \frac{m}{n} \times \frac{m}{n}\\ &= \frac{m^5}{n^5}\end{aligned}$$

Each variable in the brackets has been raised to the power of 5.

So $\left(\frac{a}{b}\right)^m = \frac{a^m}{b^m}$.

The zero index

If $2^0 = 1$, $45\,678^0 = 1$ and $0.000\,123^0 = 1$, is $a^0 = 1$ also?

To simplify $\frac{a^m}{a^m}$, subtract indices (second index law).

$$\begin{aligned}\frac{a^m}{a^m} &= a^{m-m}\\ &= a^0\end{aligned}$$

Alternatively, treating it as a fraction division,

$$\frac{a^m}{a^m} = 1$$

Therefore $a^0 = 1$.

2.2

Negative indices

Evaluating 2^{-6}.

Consider $2^6 = 64$, $2^5 = 32$, $2^4 = 16$, $2^3 = 8$, $2^2 = 4$, $2^1 = 2$ and $2^0 = 1$.

The index on the left-hand side of each equation is one less than the previous one and the number on the right-hand side of each equation is half the previous one. Continuing this pattern:

$2^{-1} = \frac{1}{2}$ $\quad 2^{-2} = \frac{1}{2^2}$ $\quad 2^{-3} = \frac{1}{2^3}$

So, $2^{-6} = \frac{1}{2^6}$

In general, $2^{-m} = \frac{1}{2^m}$.

Note that any base number can be used, e.g. $3^{-4} = \frac{1}{3^4}$, $81^{-6} = \frac{1}{81^6}$.

Index laws for powers with the same base

The following laws are true if $a, b \neq 0$ (a and b cannot be zero). Only integer values for m and n are considered here.

Multiplication of powers	$a^m \times a^n = a^{m+n}$	When multiplying numbers with the same base, keep the base and add the indices.
Division of powers	$a^m \div a^n = \frac{a^m}{a^n} = a^{m-n}$	When dividing numbers with the same base, keep the base and subtract the indices.
Power of a power	$(a^m)^n = a^{mn}$	When raising a number in index form to a power, keep the base and multiply the indices.
Expanding products	$(ab)^m = a^m b^m$	When removing brackets around a product raised to a power, raise each number to that power.
Expanding quotients	$\left(\frac{a}{b}\right)^m = \frac{a^m}{b^m}$	When removing brackets around a quotient raised to a power, raise each number to that power.
The zero index	$a^0 = 1$ $(a \neq 0)$	Any variable or product of variables raised to the power of 0 is equal to 1. Note that 0^0 is undefined.
Negative indices with a numerical base	Example $2^{-m} = \frac{1}{2^m}$	A number raised to a negative power is equal to 1 divided by the same number raised to the positive power.

Worked Example 4

WE4

Use an index law to expand each of the following, leaving your answers in index form.

(a) $(3 \times 5)^4$ **(b)** $(ab)^5$ **(c)** $(5p^4q)^3$

Working

(a) $(3 \times 5)^4$
$= 3^4 \times 5^4$

(b) $(ab)^5$
$= a^5 \times b^5$
$= a^5 b^5$

(c) $(5p^4q)^3$
$= 5^3 \times (p^4)^3 \times q^3$
$= 5^3 \times p^{4\times 3} \times q^3$
$= 5^3 \times p^{12} \times q^3$
$= 5^3 p^{12} q^3$

2.2

Worked Example 5

WE5

Use an index law to expand each of the following, leaving your answers in index form.

(a) $\left(\frac{4}{3}\right)^5$ (b) $\left(\frac{p}{q}\right)^7$ (c) $\left(\frac{2a}{c^4}\right)^3$

Working

(a) $\left(\frac{4}{3}\right)^5$
$= \frac{4^5}{3^5}$

(b) $\left(\frac{p}{q}\right)^7$
$= \frac{p^7}{q^7}$

(c) $\left(\frac{2a}{c^4}\right)^3$
$= \frac{(2a)^3}{(c^4)^3}$

Worked Example 6

WE6

Simplify each of the following, leaving your answers in index form.

(a) $2^3 \times 6^3$ (b) $5^4 \times h^4$ (c) $m^2 \times n^2$

Working

(a) $2^3 \times 6^3$
$= (2 \times 6)^3$
$= 12^3$

(b) $5^4 \times h^4$
$= (5h)^4$

(c) $m^2 \times n^2$
$= (mn)^2$

Worked Example 7

WE7

Simplify each of the following, leaving your answers in index form.

(a) $\frac{16^3}{4^3}$ (b) $\frac{t^4}{2^4}$ (c) $\frac{c^8}{d^8}$

Working

(a) $\frac{16^3}{4^3}$
$= \left(\frac{16}{4}\right)^3$
$= 4^3$

(b) $\frac{t^4}{2^4}$
$= \left(\frac{t}{2}\right)^4$

(c) $\frac{c^8}{d^8}$
$= \left(\frac{c}{d}\right)^8$

2.2

Worked Example 8

WE 8

Simplify each of the following.

(a) $4p^0$ (b) $\frac{2y^4 \times y^3}{y^5 \times y^2}$ (c) $\frac{m^3n^6}{m^3n}$

Working

(a) $4p^0$
$= 4 \times p^0$
$= 4 \times 1 = 4$

(b) $\frac{2y^4 \times y^3}{y^5 \times y^2}$
$= \frac{2y^{4+3}}{y^{5+2}}$
$= \frac{2y^7}{y^7}$
$= 2y^{7-7}$
$= 2y^0$
$= 2 \times 1$
$= 2$

(c) $\frac{m^3n^6}{m^3n}$
$= m^{3-3}n^{6-1}$
$= m^0n^5$
$= 1 \times n^5$
$= n^5$

Worked Example 9

WE 9

Simplify the following, writing your answer with positive indices.

(a) 3^{-2} (b) 5^{-3} (c) 8^{-5}

Working

(a) $3^{-2} = \frac{1}{3^2}$ (b) $5^{-3} = \frac{1}{5^3}$ (c) $8^{-5} = \frac{1}{8^5}$

2.2 More index laws

Answers page 424

Stage 5.1

Navigator

Q1 Column 1, Q2 Column 1, Q3 Column 1, Q4 Column 1, Q5 Column 1, Q6, Q7, Q8, Q9, Q10 Columns 1 & 2, Q11 Columns 1 & 2, Q12, Q13, Q14, Q15, Q16, Q19, Q20, Q21	Q1 Column 2, Q2 Column 2, Q3 Column 2, Q4 Column 2, Q5 Column 2, Q6, Q7, Q8, Q9, Q10 Columns 2 & 3, Q11 Columns 2 & 3, Q12, Q13, Q14, Q15, Q16, Q17, Q19, Q20, Q21	Q1 Column 3, Q2 Column 3, Q3 Column 3, Q4 Column 3, Q5 Column 3, Q6, Q7, Q8, Q9, Q10 Columns 3 & 4, Q11 Columns 3 & 4, Q12, Q13, Q14, Q15, Q16, Q17, Q18, Q19, Q20, Q21

Fluency and understanding

1 Use an index law to expand each of the following, leaving your answers in index form.

WE 4

(a) $(6 \times 5)^2$ (b) $(11 \times 3)^4$ (c) $(12 \times 9)^5$
(d) $(4x)^2$ (e) $(p^3q)^2$ (f) $(a^4f)^3$
(g) $(2p)^5$ (h) $(p^3q^5)^7$ (i) $(c^2d^7)^4$
(j) $(bc)^3$ (k) $(6a^4)^3$ (l) $(9p^5q^4)^2$
(m) $(7m^3n^6)^2$ (n) $(2ab^2)^3$ (o) $(3ef^4)^4$

2.2

WE5

2 Use an index law to expand each of the following, leaving your answers in index form.

(a) $\left(\frac{12}{5}\right)^2$ (b) $\left(\frac{7}{3}\right)^4$ (c) $\left(\frac{9}{2}\right)^8$

(d) $\left(\frac{y}{x}\right)^3$ (e) $\left(\frac{d}{e}\right)^5$ (f) $\left(\frac{m^3}{n}\right)^4$

(g) $\left(\frac{w^4}{z}\right)^2$ (h) $\left(\frac{r^5}{s}\right)^4$ (i) $\left(\frac{d^3}{g}\right)^2$

(j) $\left(\frac{3}{w^6}\right)^4$ (k) $\left(\frac{8}{m^2}\right)^3$ (l) $\left(\frac{2}{r^4}\right)^3$

(m) $\left(\frac{h^3}{g^2}\right)^5$ (n) $\left(\frac{a^3}{c^5}\right)^4$ (o) $\left(\frac{t^4}{v^3}\right)^3$

(p) $\left(\frac{5a}{b}\right)^3$ (q) $\left(\frac{c^4}{2x}\right)^6$ (r) $\left(\frac{f^6}{6g}\right)^3$

WE6

3 Simplify each of the following, leaving your answers in index form.

(a) $3^2 \times 4^2$ (b) $4^6 \times 5^6$ (c) $8^4 \times 7^4$

(d) $4^8 \times h^8$ (e) $3^4 \times p^4$ (f) $7^9 \times r^9$

(g) $a^5 \times m^5$ (h) $k^7 \times t^7$ (i) $p^8 \times n^8$

WE7

4 Simplify each of the following, leaving your answers in index form.

(a) $\frac{15^4}{3^4}$ (b) $\frac{24^3}{6^3}$ (c) $\frac{132^5}{11^5}$

(d) $\frac{r^5}{3^5}$ (e) $\frac{n^7}{6^7}$ (f) $\frac{g^9}{4^9}$

(g) $\frac{x^6}{y^6}$ (h) $\frac{k^{11}}{n^{11}}$ (i) $\frac{g^{10}}{h^{10}}$

WE8

5 Simplify each of the following.

(a) k^0 (b) $7g^0$ (c) $7 + g^0$

(d) a^0b^2 (e) $3m^0n^0$ (f) $c^3d^0e^0$

(g) $-3a^4b^0$ (h) $-8g^{11}h^0$ (i) $-(4pr)^0$

(j) $y^4 \div y^4$ (k) $3x^7 \div x^7$ (l) $24m^6 \div 8m^6$

(m) $\frac{3x^4z}{z}$ (n) $\frac{h^5k^8}{h^5}$ (o) $\frac{8m^3n^4}{16m^3}$

(p) $\frac{a^5b^4}{a^2b^4}$ (q) $\frac{6h^4r^3}{2h^4r^3}$ (r) $\frac{9b^4d^3}{6b^4d^3}$

(s) $\frac{12e^3f^2}{16e^3}$ (t) $\frac{15r^2s^9}{27r^2}$ (u) $\frac{25k^4p^2}{15k^4}$

Hint

Remember that any number (except 0) divided by itself equals one.

WE9

6 Simplify the following, writing your answers with positive indices.

(a) 4^{-2} (b) 2^{-3} (c) 5^{-4}

(d) 3^{-3} (e) 5^{-2} (f) 4^{-3}

2.2

7 (a) $(2k)^6$ simplifies to:

A $2k^6$ B $12k$ C $64k^6$ D $12k^6$

(b) $(3a^4)^3$ simplifies to:

A $27a^{12}$ B $9a^{12}$ C $9a^7$ D $27a^7$

(c) $\left(\frac{3}{g}\right)^4$ simplifies to:

A $\frac{12}{g^4}$ B $\frac{81}{g^4}$ C $\frac{12}{g}$ D $\frac{81}{g}$

(d) $\left(\frac{k^8}{7}\right)^2$ simplifies to:

A $\frac{k^{10}}{7}$ B $\frac{k^{16}}{49}$ C $\frac{k^{16}}{7}$ D $\frac{k^{10}}{49}$

8 Simplify each of the following by first applying the index law for expanding products, then completing the multiplication, applying other index laws if necessary.

(a) $(xy)^2 \times (xy)^4$ (b) $(gh)^7 \times (gh)^2$ (c) $(3x)^3 \times (4x)^2$

(d) $(4y)^2 \times (5xy)^2$ (e) $(cd)^3 \times (c^3d^2)^3$ (f) $(x^2y^3)^2 \times (xy)^2$

9 Simplify each of the following by first applying the index law for expanding quotients, completing the multiplication, then simplifying if possible.

(a) $\left(\frac{f}{g}\right)^3 \times \left(\frac{f}{g}\right)^2$ (b) $\left(\frac{p}{q}\right)^6 \times \left(\frac{p}{q}\right)^4$ (c) $\left(\frac{d^6}{q^4}\right)^2 \times \left(\frac{d^5}{q^2}\right)^3$

(d) $\left(\frac{a^4}{b^2}\right)^2 \times \left(\frac{b}{a}\right)^6$ (e) $\left(\frac{2a}{b}\right)^2 \times \left(\frac{a}{b}\right)^3$ (f) $\left(\frac{3x^2}{y}\right)^3 \times \left(\frac{2x}{y}\right)^2$

10 Simplify each of the following by applying the expanding products and the powers of powers index laws, then the index law for division.

(a) $\frac{(w^4z^3)^3}{(w^5z^2)^2}$ (b) $\frac{(c^7e^5)^2}{(c^4e^2)^3}$ (c) $\frac{(pq)^{12}}{(p^2q^3)^3}$ (d) $\frac{(vw)^9}{(v^3w^2)^2}$

11 Simplify each of the following.

(a) $(x^3)^0$ (b) $(d^5)^0$ (c) $(k^0)^4$ (d) $(p^0)^2$

(e) $(m^3)^4 \times (m^2)^0$ (f) $(f^5)^4 \times (f^3)^0$ (g) $(t^0)^3 \times t^4$ (h) $((q^4)^5 \times (q^5)^4)^0$

(i) $\frac{x^6 \times x^5}{x^{11}}$ (j) $\frac{3k^2}{k \times 5k}$ (k) $\frac{4a^3}{a^2 \times 7a}$ (l) $\frac{n^4 \times n^3}{n^5 \times n^2}$

(m) $\frac{2b^6 \times 5c^8}{c^7 \times 4c}$ (n) $\frac{d^6 \times e^2}{3e \times 2e}$ (o) $\frac{c^4 \times d}{2c^2 \times 5c^2}$ (p) $\frac{p^7 \times q^5}{15p^2 \times p^5}$

12 Use the negative index law to write the following with positive indices, then evaluate.

(a) 4^{-2} (b) 3^{-3} (c) 5^{-4}

(d) $\frac{1}{2^{-1}}$ (e) $\frac{1}{4^{-3}}$ (f) $\frac{1}{3^{-2}}$

(g) 4×3^{-2} (h) 3×4^{-3} (i) 2×5^{-2}

(j) $-\frac{7 \times 3^{-2}}{3}$ (k) $-\frac{6 \times 2^{-4}}{2^{-3}}$ (l) $-\frac{3 \times 3^{-4}}{3^{-6}}$

2.2

13 **(a)** $\frac{1}{2^{-4}}$ is the same as:

A 16 **B** 8 **C** $\frac{1}{16}$ **D** -16

(b) $\frac{1}{8}$ is the same as:

A 2^3 **B** 2^{-3} **C** 3^{-2} **D** $\frac{1}{2^{-3}}$

Reasoning, communicating and open-ended

14 Kellie and Eva are given the following question as part of their indices homework.

Simplify $\left(\frac{4x^2}{2x}\right)^3$.

Both girls had the correct answer, but Eva finished much faster. 'How did you do it so fast?' Kellie asked Eva.

Explain why Eva's method took fewer steps.

Here is Kellie's working:	Here is Eva's working:
$\left(\frac{4x^2}{2x}\right)^3$	$\left(\frac{4x^2}{2x}\right)^3$
$= \frac{4^3 \times x^{2\times 3}}{2^3 \times x^3}$	$= (2x)^3$
$= \frac{64x^6}{8x^3}$	$= 8x^3$
$= 8x^3$	

15 By rewriting the expression if required and then using index laws, simplify each of the following using the smallest possible base.

(a) $\frac{9^{2n}}{9^n}$ **(b)** $2^n \times 4^{n+1}$ **(c)** $\frac{125}{5^n} \times \frac{5^{2n}}{25}$

16 **(a)** Using index notation, write the numbers 125, 25, 5 and 1, using base 5.

(b) Using index notation, write the next five numbers using the pattern in part **(a)**.

(c) Write the whole sequence of nine numbers in this pattern without using index notation.

17 A model for the number of planets, N, created in a section of a galaxy is given by the formula $N = \left(\frac{n^2T^3p}{b}\right)^2$. In the formula:

n is the number of suns
T is the temperature of the surrounding space
p is the number of planets already present
b is the number of black holes.

(a) Write the formula without brackets.

(b) **(i)** How many new planets does the model predict if $T = 1$, $n = 1000$ and $b = p$?

(ii) If the temperature of the surrounding space is a negative number, will we obtain a positive or a negative answer? Explain your reasoning.

2.2

18 The intensity of sound is measured in watts per square metre. The level of sound (loudness) is measured in decibels. The threshold of human hearing is 10^{-12} watts/m^2 and is defined as a level of sound of 0 decibels (dB). All other decibel levels are calculated by comparing the sound intensity with the threshold of human hearing. To do this we follow a simple rule:

1 Divide the intensity measured in watts/m^2 by the sound intensity of the threshold of human hearing.

For example, normal conversation is measured at 10^{-6} watts/m^2.

$$\frac{10^{-6}}{10^{-12}} = 10^{-6-(-12)}$$
$$= 10^{6}$$

This means that normal conversation is 1 million times as loud as the threshold of human hearing.

2 Multiply the index by 10. This is the intensity level in decibels.

Intensity level $= 10 \times 6 = 60$ dB

To convert from decibels to find the sound intensity, we reverse the process.

For example, the sound level for a quiet library = 40 dB.

To find the sound intensity:

1 Divide the number of decibels by 10.

$$\frac{40}{10} = 4$$

2 Use this number as an index for 10 and multiply by the threshold for human hearing.

The sound intensity of a quiet library $= 10^4 \times 10^{-12}$
$= 10^{-8}$ watts/m^2

(a) Copy and complete the following table.

Source	Sound level (loudness; dB)	Sound intensity (watts/m^2)
Jet aircraft, 50 m away		100
Threshold of pain		10
Chainsaw, 1 m distance	110	
Nightclub, 1 m from speaker		0.01
Busy traffic	80	
Vacuum cleaner, 1 m distance	70	
Average home		0.000 000 1
Rustling of leaves in the distance	10	

(b) If a whisper is 20 dB, how many times as loud is normal conversation?

(c) If the intensity of a rock concert is 10^0 watts/m^2 and the level of intensity of the threshold of discomfort is 120 dB, what comment could you make comparing these two measures?

(d) Which is louder, thunder at a sound level of 110 dB or a loud TV at a sound intensity of 10^{-4} watts/m^2?

2.2

19 **(a)** Write at least two expressions involving indices and multiplication that, when simplified, equal $36m^4n^8$.

(b) Write at least two expressions involving indices and division that can be simplified to $\frac{4m^4}{5n^6}$.

20 Write three different terms that $27x^3y^7$ could be divided by so that no x appears in the answer.

21 Write a question that could be simplified to $\frac{3x^2}{y^3}$ by using:

(a) any two index laws with an expression that has only positive indices

(b) any three index laws with an expression that has only positive indices.

Show, step by step, how your question can be simplified to obtain the final answer.

Problem solving

Square cubes

Equipment required: 1 brain, calculator

'I have been studying the cube numbers. You know, some of them are also square numbers!' It was smart Alec, bragging again.

'You're kidding!' said Sue. 'A number can't be a square and a cube.'

'Yes it can,' said Alec. 'Like, 64 is both 8 squared and 4 cubed.'

If Alec is right, there may be more.

1 Can you find any other two-digit numbers that are a square and a cube?

2 With the help of your calculator, how many can you find under 1000?

Strategy options

- Test all possible combinations.
- Look for a pattern.
- Break the problem into manageable parts.

Negative indices with a variable base

2.3

Stage 5.2

Need to Know

Negative indices with a variable as the base

Consider two ways to simplify $a^2 \div a^6$.

To simplify $a^2 \div a^6$, subtract indices.

$$a^2 \div a^6 = a^{2-6}$$
$$= a^{-4}$$

Alternatively, expand the numerator and denominator and cancel the common factors.

$$a^2 \div a^6 = \frac{a \times a}{a \times a \times a \times a \times a \times a}, a \neq 0$$
$$= \frac{1}{a^4}, a \neq 0$$

The expressions $a^{-4} = \frac{1}{a^4}, a \neq 0$

a^{-4} is the **reciprocal** of a^4. Reciprocals always multiply to give a product of 1.

$$a^4 \times a^{-4} = a^0 \qquad \text{or} \qquad a^4 \times \frac{1}{a^4} = \frac{a^4}{a^4}$$
$$= 1 \qquad\qquad\qquad = 1$$

So, $a^{-m} = \frac{1}{a^m}, a \neq 0$

Negative indices	$a^{-m} = \frac{1}{a^m} \ (a \neq 0)$	A number raised to a negative power is equal to 1 divided by the same number raised to the positive power.

Worked Example 10

WE10

Simplify the following, writing your answer with positive indices.

(a) m^{-5} (b) $4p^{-3}$ (c) $6x^{-5}y^2$

Working

(a) $m^{-5} = \frac{1}{m^5}$

(b) $4p^{-3} = 4 \times p^{-3}$
$= 4 \times \frac{1}{p^3}$
$= \frac{4}{p^3}$

(c) $6x^{-5}y^2 = \frac{6y^2}{x^5}$

Answers page 425

2.3 Negative indices with a variable base

Stage 5.2

Navigator

Q1, Q2, Q3, Q4, Q5 Columns 1 & 2, Q7, Q9	Q1, Q2, Q3, Q4, Q5, Q7, Q8, Q9	Q1, Q2, Q3, Q4, Q5, Q6, Q7, Q8, Q9

Fluency and understanding

WE10

1 Simplify the following, writing your answer with positive indices.

(a) x^{-5} (b) y^{-3} (c) b^{-7}
(d) $3e^{-3}$ (e) $7x^{-5}$ (f) $3g^{-4}$
(g) $8p^{-5}q^{-2}$ (h) $2x^{-3}y^{-5}$ (i) $12a^{-3}b^{-9}$
(j) $3p^3q^{-1}$ (k) $5x^2y^{-7}$ (l) $7a^{-5}b^2$

2 Express each of the following with positive indices.

(a) $\frac{1}{d^{-2}}$ (b) $\frac{1}{k^{-5}}$ (c) $\frac{3}{a^{-1}}$
(d) $\frac{x^3}{y^{-6}}$ (e) $\frac{p^7}{q^{-2}}$ (f) $\frac{m^{-4}}{n^{-4}}$
(g) $\left(\frac{a^2}{b^3}\right)^{-4}$ (h) $\left(\frac{m^4}{n^5}\right)^{-2}$ (i) $\left(\frac{x^{-5}}{y^2}\right)^{-3}$

3 Rewrite with positive indices.

(a) $(x+4)^{-1}$ (b) $(x-y)^{-2}$ (c) $5(c-3)^{-1}$

2.3

4 (a) a^{-4} is the same as:

A $-a^4$ B $-4a$ C $\frac{1}{a^4}$ D $\frac{1}{a^{-4}}$

(b) $\frac{1}{y^2}$ is the same as:

A $-y^2$ B y^{-2} C $-y^{-2}$ D $2y$

5 Simplify, writing your answer with positive indices.

(a) $x^2 \times x^{-5}$ (b) $p^{-3} \times p$ (c) $2c^{-4} \times 6c^{-1}$

(d) $a^{-3} \div a^4$ (e) $k^{-2} \div k^{-4}$ (f) $3d^{-3} \div 4d^{-2}$

(g) $(m^{-3})^{-3}$ (h) $(xy^{-2})^{-4}$ (i) $\left(\frac{c^{-3}d}{2c^2d^{-2}}\right)^{-1}$

Reasoning, communicating and open-ended

6 Evaluate the expression $y^{-2} \div x^{-1}$ when $x = \frac{1}{2}$ and $y = \frac{2}{3}$.

7 (a) If I divide a number by its square, what is the result?

(b) If a rectangular room has one side length of $12a$ m and an area of 84 m^2, what is the length of the other side of the room? ($b = \frac{A}{w}$ where b is the breadth, A the area and w the length.)

8 Write a question that could be simplified to $\frac{3x^2}{y^3}$ by using:

(a) two index laws with an expression that has at least one negative index

(b) three index laws with an expression that has at least one negative index.

Show, step by step, how your question can be simplified to obtain the final answer.

9 Jodie made four mistakes when simplifying the following index question. Find them, explain to Jodie what she did incorrectly and redo the question to get the correct answer. Jodie's working is shown below:

$$\frac{(3x^2y^4)^2}{2x^{-2}y^5} = \frac{6x^4y^6}{2x^{-2}y^5}$$
$$= 4x^2y$$

2.4 Scientific notation and significant figures

Stage 5.1

Need to Know

Powers of 10

When multiplying by a power of 10 with a positive index, move the decimal point to the right by the same number of places as the value of the index. This makes the number bigger. Zeroes are inserted as the values for any places not filled by other digits.

When multiplying by a power of 10 with a negative index, move the decimal point to the left by the same number of places as the value of the index after the negative sign. This makes the number smaller. Zeroes are inserted as the values for any places not filled by other digits.

This is the same as dividing by the reciprocal power.

$$\begin{aligned} 6.5 \times 10^{-4} &= \frac{6.5}{10^4} \\ &= \frac{6.5}{10\,000} \\ &= 0.000\,65 \end{aligned}$$

Scientific notation

When dealing with very large or very small numbers, **scientific notation** is often used. A number is in scientific notation (also known as 'standard form') if it is written as the product of a number, a, and a power of 10, where 'a' is any number greater than or equal to 1 but less than 10:

$a \times 10^b$ where $1 \leq a < 10$ and b is an integer.

Scientific notation on a calculator

When the result of an arithmetic calculation is larger or smaller than the display on a calculator allows, it is written in scientific notation. However, it looks a little different from the way it is written.

For example:

$4\,589\,765\,125 \times 849\,235\,713 = 3.897\,792\,458 \times 10^{18}$

On a calculator, this result may be written 3.897792458E18 where E18 represents $\times 10^{18}$.

Similarly, 3.7792311E-5 would represent $3.779\,231\,1 \times 10^{-5}$.

Significant figures

Significant figures are used in measurement because all measurements are approximate as they are only as accurate as measuring devices will allow. Using significant figures is just another way of rounding. The number of significant figures indicates how accurate the measurements are.

Rounding using significant figures

1 Write the number in scientific notation: $23\,467.8 = 2.346\,78 \times 10^4$.

2 Decide how many significant figures you require.

$2.346\,78 \times 10^4 \approx 2.3 \times 10^4$ to two significant figures
$\approx 2.35 \times 10^4$ to three significant figures
$\approx 2.347 \times 10^4$ to four significant figures.

Remember to check the next digit after the one being rounded to decide whether to round up or down.

Counting significant figures

- All non-zero digits are significant. The first significant figure in a number is the first non-zero digit:

  2.346 78 has six significant figures.
- Zeroes between non-zero digits or at the end of a decimal number are significant:

  10 203 has five significant figures.

  0.03**0** has two significant figures.
- Zeroes at the end of an integer or at the beginning of a decimal number are not significant.

  1 350 000 has three significant figures.

  0.000 013 5 also has three significant figures.

Writing significant figures

- Zeroes appearing at the end of a decimal number after rounding are significant figures and must be retained to give the required number of significant figures.

  7.999 is written as 8.00 to three significant figures.
- Zeroes sometimes need to be added.

  7.2 is written as 7.20 to three significant figures.

Accuracy of measurement

All measurements are only as accurate as the measuring equipment used can measure. Very large and very small measurements are usually approximated further by rounding to a certain number of significant figures. However, all calculations should be made using the maximum number of digits possible on a calculator until a final answer is achieved, before any rounding is carried out.

Using units for very large and very small measurements

For very small measurements, 'milli', 'micro' and 'nano' are used, where 'milli' means $\times 10^{-3}$, 'micro' means $\times 10^{-6}$ and 'nano' means $\times 10^{-9}$.

For very large measurements, 'mega', 'giga' and 'tera' are used, where 'mega' means $\times 10^3$, 'giga' means $\times 10^6$ and 'tera' means $\times 10^9$.

It is always important to use the correct abbreviation for units, although some units can correctly be abbreviated in different ways. For example, kilobytes can be abbreviated to kB, KB, k or K.

2.4

Worked Example 11

WE11

Find the value of each of the following.

(a) 4.35×10^6 (b) 1.78×10^{-3}

Working

(a) $4.35 \times 10^6 = 4\,350\,000$

(Move the decimal six places to the right and insert four zeroes.)

(b) $1.78 \times 10^{-3} = 0.00178$

(Move the decimal point three places to the left and insert two zeroes.)

Worked Example 12

WE12

Express each of the following in scientific notation.

(a) 5800 (b) 0.000 603 4

Working

(a) $5800 = 5.8 \times 10^3$

(b) $0.000\,603\,4 = 6.034 \times 10^{-4}$

Worked Example 13

WE13

Round each of the following to the number of significant figures specified.

(a) 50 842 to three significant figures (b) 0.007 497 to three significant figures

Working

(a) $50\,842 = 5.0842 \times 10^4$
$\approx 5.08 \times 10^4$ or $50\,800$

(b) $0.007\,497 = 7.497 \times 10^{-3}$
$\approx 7.50 \times 10^{-3}$ or 0.00750

2.4

2.4 Scientific notation and significant figures

Answers page 425

Stage 5.1

Navigator

Q1 Columns 1 & 2, Q2 Column 1, Q3, Q4, Q5, Q6, Q7, Q8, Q9 Column 1, Q10, Q11, Q12, Q14, Q15, Q16, Q17, Q19, Q23, Q24	Q1 Columns 2 & 3, Q2 Column 2, Q3, Q4, Q5, Q6, Q7, Q8, Q9 Column 2, Q10, Q11, Q12, Q13, Q14, Q15, Q16, Q17, Q19, Q20, Q23, Q24	Q1 Columns 3 & 4, Q2 Column 3, Q3, Q4, Q5, Q6, Q7, Q8, Q9 Column 3, Q10, Q11, Q12, Q13, Q14, Q15, Q16, Q17, Q18, Q20, Q21, Q22, Q24

Equipment required: calculator for Question 9

Fluency and understanding

1 Find the value of each of the following. WE11

(a) 9.8×10^3 **(b)** 6.5×10^7 **(c)** 3.25×10^6 **(d)** 4.2071×10^5
(e) 2×10^{-3} **(f)** 5×10^{-5} **(g)** 7×10^{-8} **(h)** 5.032×10^{-2}
(i) 8.26×10^{-3} **(j)** 4.18×10^{-5} **(k)** 7.14×10^{-4} **(l)** 3.55×10^{-6}

2 Express each of the following in scientific notation. WE12

(a) 2560 **(b)** 90 020 **(c)** 614 000
(d) 190 000 **(e)** 8 050 000 **(f)** 56 000 000
(g) 0.305 **(h)** 0.03 **(i)** 0.0004
(j) 0.002 01 **(k)** 0.000 072 466 **(l)** 0.000 000 532 18

3 Round each of the following to the number of significant figures indicated in the brackets. WE13

(a) 394 024 (2) **(b)** 0.054 02 (3) **(c)** 674.5 (2)
(d) 0.003 (3) **(e)** 1 679 000 (2) **(f)** 0.899 5 (3)

4 To how many significant figures have each of the following been written?

(a) 825 **(b)** 63.73 **(c)** 145 000
(d) 0.0034 **(e)** 3.008 **(f)** 0.0040

5 **(a)** A fly's wing flap is 3 milliseconds. Express this time in seconds using scientific notation.

(b) The time for light to travel once around the Equator is 134 milliseconds. Express this time in seconds using scientific notation.

(c) The average length of a blink is 300–400 milliseconds. Express this range of times in seconds using scientific notation.

6 Express each quantity described below using scientific notation.

(a) The distance around the Earth's equator is 40 075 km.

(b) The thickness of a soap bubble is about 0.000 000 1 m.

(c) A 10-cent coin is about 0.0015 m thick.

(d) Pluto is approximately 5900 million km from the Sun.

(e) The poliomyelitis virus has a diameter of 0.000 000 3 m.

(f) The melting point of gold is 1063°C.

2.4

7 Express each quantity described below as a number.

(a) Radio waves travel at about 3×10^8 m/s.

(b) In a hydrogen atom, the average distance between the proton and the electron is 1.5×10^{-15} m.

(c) The thickness of fine paper is about 1.2×10^{-4} m.

(d) The aluminium in a soft-drink can contains around 4×10^{24} atoms.

(e) This distance from the Sun to Jupiter is approximately 7.78×10^8 km.

(f) The temperature at the centre of the Sun is about 1.5×10^7°C.

8 Find the largest number of the following:

A 1.325×10^5 B 2.853×10^2 C 9.851×10^{-2} D 6.312×10^4

9 Use a calculator to evaluate each expression. Express your answers in scientific notation.

Hint

A negative power in the denominator can be written in the numerator with a positive power.

(a) $2.4 \times 10^6 \times 5.9 \times 10^4$

(b) $7.34 \times 10^3 \times 2.9 \times 10^{-6}$

(c) $7.86 \times 10^3 \div (2.62 \times 10^{-8})$

(d) $4.38 \times 10^9 \div (2.19 \times 10^{-3})$

(e) $6.2 \times 10^{-4} - 1.34 \times 10^{-5}$

(f) $4.6 \times 10^{-10} - 3.01 \times 10^{-11}$

(g) $2(4.3 \times 10^3 - 6.9 \times 10^7)$

(h) $5(9.1 \times 10^5 - 2.7 \times 10^9)$

(i) $\dfrac{3.21 \times 10^3 \times 4.19 \times 10^4}{5 \times 10^6}$

(j) $\dfrac{5.92 \times 10^4 \times 8.01 \times 10^{12}}{3 \times 10^5}$

(k) $\dfrac{3(2.5 \times 10^6 + 4.9 \times 10^7)}{6.5 \times 10^{-4}}$

(l) $\dfrac{4(7.3 \times 10^9 + 2.1 \times 10^4)}{8.6 \times 10^{-5}}$

10 Convert the following to the units given in brackets, writing your answer in scientific notation.

(a) 63 nanoseconds (seconds)

(b) 857 megahertz (hertz)

(c) 2562 gigalitres (litres)

(d) 59 micrometres (metres)

11 Given that $2^{10} = 1024$ and 1 byte = 8 bits, 1 kilobyte = 1024 bytes, 1 megabyte = 1024 kilobytes, 1 gigabyte = 1024 megabytes and 1 terabyte = 1024 gigabytes, find the number of bits in 1 terabyte giving your answer:

(a) in index form

(b) in scientific notation.

12 Convert the following to the units given in brackets, writing your answer in scientific notation.

(a) 6 megabytes (kilobytes)

(b) 1.5 gigabytes (megabytes)

(c) 64 kilobytes (megabytes)

13 Estimate the answer to the following by first writing the numbers in scientific notation.

(a) $8\,975\,328 \times 4129$

(b) $0.023\,75 \div 178.4$

2.4

14 Express each quantity described below using scientific notation.

(a) The diameter of the nucleus of an atom is about 0.000 000 000 000 01 m.

(b) The highest Sherpa village on the slopes of Mount Everest is about 4000 m above sea level.

(c) Fine capillaries, which carry blood to the body's tissues, are about 0.000 02 m in diameter.

(d) The filament of a light bulb reaches a temperature of 2730°C.

(e) The number of athletes who attended the Sydney 2000 Olympic Games was 10 651.

(f) About 100 million million atoms could fit on the point of a household pin.

(g) The Sun is approximately 150 000 000 km from Earth. The temperature at the surface of the Sun is about 6000°C.

(h) Australia has the highest rate of skin cancer in the world. More than 140 000 cases of skin cancer are reported each year in Australia.

15 The nervous system controls movement in the body. Electrical impulses travel through nerves to the muscles causing them to contract. Here is a graph showing the time it takes for a nerve to transmit an impulse. Use the graph to answer the following questions, expressing all times in seconds using scientific notation.

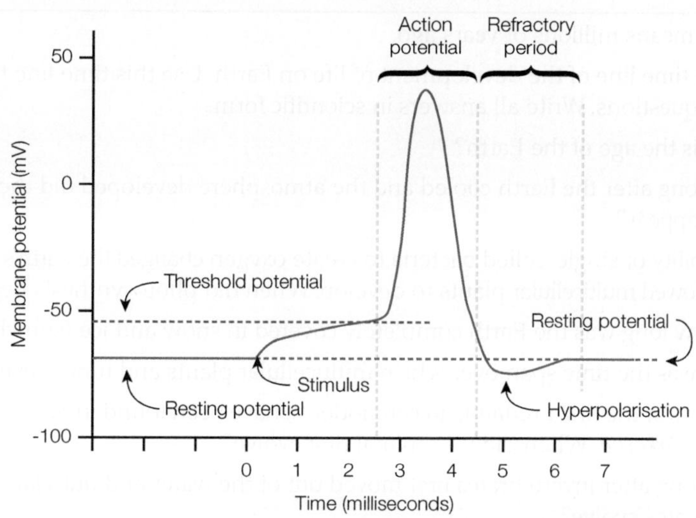

Events that characterise the transmission of a nerve impulse.

(a) How long does it takes for a nerve to return to resting potential after it has been stimulated?

(b) How long does it takes for a nerve to reach hyperpolarisation after it has reached the threshold potential?

(c) How long does the action potential last?

(d) How long does the refractory period last?

16 (a) A personal computer in 1982 required approximately 1000 nanoseconds (1 microsecond) to perform an operation. Modern computers require between 0.3 and 1 nanosecond to perform an operation. How much faster are modern computers than the computers used in 1982? Express this range in seconds using scientific notation.

(b) The RAM of a personal computer in 1982 had a response time of 200 nanoseconds. The RAM of modern computers has a response time of between 5 and 10 nanoseconds. How much faster is the response time of the RAM in modern computers compared with computers in 1982? Express this range in seconds using scientific notation.

Hint

Express numbers in scientific notation before you compare them.

2.4

17

Timescale	Event
4570 mya	Planet Earth first formed
4440 mya	Earth cooled enough to form oceans and the atmosphere
3800 mya	First life on Earth: single-celled bacteria (microbial mat)
2400 mya	First photosynthesis by plants to produce oxygen
1700–1200 mya	Multicellular plants and fungi (sea dwellers)
850–630 mya	Ice age 'snowball Earth'—extinction event
580 mya	Multicellular animals: *Charnia* (filter feeders such as sponges; sea dwellers)
550 mya	First creatures capable of movement: trilobites (first arthropod)
530 mya	First predators with mouth, teeth and stomach: echinoderms (*Arkarua*)
490–476 mya	Plants and invertebrates on land: arthropods (insects, giant dragonfly)
370 mya	Vertebrate animals on land
251 mya	Extinction event
230 mya	Dinosaurs
150 mya	First birds (type of dinosaur)
130–90 mya	Flowering plants and social insects (bees, termites, ants)
65 mya	Extinction of the dinosaurs (except birds)
6 mya	Humans (upright apes)
200 000	Modern humans

Note: mya means millions of years ago.

Above is a time line of the development of life on Earth. Use this time line to answer the following questions. Write all answers in scientific form.

(a) What is the age of the Earth?

(b) How long after the Earth cooled and the atmosphere developed did the first life on Earth appear?

(c) The ability of single-celled bacteria to create oxygen changed the Earth's atmosphere and allowed multicellular plants to develop. When did photosynthesis begin?

(d) For how long was the Earth completely covered in snow and ice (snowball Earth)?

(e) What was the time span over which multicellular plants and fungi developed?

(f) Remains of the first predator, an echinoderm, have been found in South Australia and named *Arkarua*. When did this animal first evolve?

(g) How long after invertebrates first moved out of the water and onto land did land vertebrates evolve?

(h) For how long was the Earth inhabited by dinosaurs?

(i) How many years before the first humans appeared on Earth did the first birds evolve?

(j) Extinction events are events that have extinguished almost all life on Earth. Recent extinction events are thought to have been due to volcanic eruptions or collisions with meteorites producing dust storms that blocked out the Sun, causing plants to die and disrupting the entire food chain. When were the last two extinction events?

(k) What is the time span from the formation of the Earth to the beginning of modern humans?

18 Light travels at a speed of about 3×10^8 m/s.

(a) Given that speed $= \dfrac{\text{distance}}{\text{time}}$, how far does light travel in:

(i) 2 minutes (ii) 6 weeks (iii) 1 year (365.25 days)? (This is called a *light year.*)

Express your answers in scientific form and in metres.

(b) If the mean distance of the Earth from the Sun is 1.496×10^8 km, approximately how long does light from the Sun take to reach Earth? Express your answer in minutes, correct to one decimal place.

19 Neil Armstrong sent messages back from the Moon on 20 July 1969. Radio waves travel at about 3.0×10^8 m/s. If the Moon is 384 000 km away from Earth, how long did it take, in seconds, for a message to travel from the Moon to Earth?

(Speed = $\frac{\text{distance}}{\text{time}}$)

20 The thickness of a sheet of paper is approximately 2×10^{-4} m and the thickness of a sheet of cellophane is approximately 2×10^{-3} m.

(a) Which is thinner?

(b) Find the difference in thickness between the two sheets.

Reasoning, communicating and open-ended

21 Two bacteria cultures, one containing 6.2×10^{11} cells and the other 3.5×10^9 cells, are combined in the one flask.

(a) How many cells are there in the flask?

(b) If the bacteria population doubles in size every 12 hours, how many cells will there be after 3 days?

(c) Find the value of your answer to **(b)**.

22 The planet Venus is approximately 108 million km from the Sun.

(a) Express this quantity in scientific notation.

(b) Write this distance in metres using scientific rotation.

(c) If the orbit of Venus around the Sun is assumed to be circular, find the distance, in kilometres, covered by Venus in one orbit. Answer in scientific notation.

(d) Earth is about 1.5×10^{11} m from the Sun. Find the closest distance, in metres, between Venus and Earth. Answer in scientific notation.

23 Estimate the diameter of a hair from your head and write your answer in scientific notation.

24 Josh wrote:

$3 \times 10 = 30^2$

$= 900$

Explain to Josh why he is incorrect.

25 Estimate the number of red blood cells in your body and write your answer in scientific notation. (Children typically have 3.8–5.5 million red blood cells per microlitre of blood. Children weighing 40–60 kg would have approximately 2.7–4.5 L of blood.)

Problem solving

Stairway to the Moon

A standard piece of paper is approximately 0.1 mm thick.

If you cut it in half and place the two halves on top of each other, you now have a 'stack' of two pieces 0.2 mm thick. If you now cut the stack in half and place one half on top of the other, you have four pieces 0.4 mm thick.

The Moon is 363 100 km from the Earth at its closest point. How many 'cut 'n' stacks' will you need to do to have a pile of paper that reaches all the way to the Moon? (Assuming, of course, you are able to cut the stack as many times as you like.)

Strategy options

- Act it out.
- Make a table.
- Look for a pattern.

2.5 Rearranging formulas

Stage 5.3

Need to Know

A **formula** is an **equation** showing the relationship between two or more variables or unknowns. It is written with one variable on the left-hand side of the equation and all other variables on the right-hand side. The variable on the left-hand side is called the subject of the formula.

When the variable required is not the subject of the formula, we need to rearrange the equation. The formula for finding the velocity, v, of an object after accelerating over a distance d is $v^2 = u^2 + 2ad$. We can rearrange this to make u, a or d the subject; that is, we can rearrange the formula to the form $u = \ldots$, $a = \ldots$ or $d = \ldots$.

We need to use the correct order of operations to rearrange the formula correctly.

Alternatively, the values could be substituted into the original formula and the equation solved in the usual way.

Sometimes before the required variable can become the subject of the formula, the equations need to be undone with operations such as squaring both sides or taking the square root or reciprocal of both sides.

Worked Example 14

WE14

The formula for finding the velocity, v, of an object after accelerating over a distance, d, is $v^2 = u^2 + 2ad$.

(a) Rearrange the formula to make a the subject.

(b) Find a when $v = 5$, $u = 1$ and $d = 2$.

Working

(a)

$$v^2 = u^2 + 2ad$$
$$v^2 - u^2 = u^2 + 2ad - u^2$$
$$v^2 - u^2 = 2ad$$
$$\frac{v^2 - u^2}{2d} = \frac{2ad}{2d}$$
$$\frac{v^2 - u^2}{2d} = a$$
$$a = \frac{v^2 - u^2}{2d}$$

(b)

$$a = \frac{v^2 - u^2}{2d}$$
$$a = \frac{5^2 - 1^2}{2 \times 2}$$
$$a = \frac{25 - 1}{4}$$
$$a = \frac{24}{4}$$
$$a = 6$$

2.5

Worked Example 15

WE15

Rearrange the formula in each of the following to make the variable in brackets the subject of the formula.

(a) $A = \pi r^2$ (r) (b) $T = 2\pi\sqrt{\dfrac{l}{g}}$ (l)

Working

(a)
$$A = \pi r^2$$
$$\frac{A}{\pi} = r^2$$
$$r = \sqrt{\frac{A}{\pi}}$$

(b)
$$T = 2\pi\sqrt{\frac{l}{g}}$$
$$\frac{T}{2\pi} = \sqrt{\frac{l}{g}}$$
$$\frac{T^2}{(2\pi)^2} = \frac{l}{g}$$
$$l = \frac{T^2 g}{4\pi^2}$$

2.5 Rearranging formulas

Answers page 426

Stage 5.2

Navigator

Q1, Q2 Column 1, Q3 Column 1, Q4, Q5, Q6, Q9, Q10, Q13	Q1, Q2 Column 2, Q3 Column 2, Q4, Q5, Q6, Q7, Q9, Q10, Q11, Q13	Q1, Q2 Column 3, Q3 Column 3, Q4, Q5, Q6, Q7, Q8, Q9, Q10, Q11, Q12, Q13

Equipment required: calculator for Questions **5**, **6** and **7**

Fluency and understanding

WE14

1 (a) (i) Rearrange the formula to make k the subject if $P = mk$.
(ii) Find k when $P = 32$ and $m = -4$.

(b) (i) Rearrange the formula to make b the subject if $y = mx + b$.
(ii) Find b when $y = 9$, $m = 5$ and $x = 2$.

(c) (i) Rearrange the formula to make h the subject if $C = ah - g$.
(ii) Find h when $C = 19$, $a = -24$ and $g = -7$.

(d) (i) Rearrange the formula to make x the subject if $a = mx + c$.
(ii) Find x when $a = -21$, $m = 4$ and $c = 3$.

(e) (i) Rearrange the formula to make p the subject if $k = \dfrac{mp - 2}{n}$.
(ii) Find p when $k = 42$, $m = -4$ and $n = 5$.

(f) (i) Rearrange the formula to make w the subject if $u = \dfrac{w}{v} - xy$.
(ii) Find w when $u = 6$, $x = -3$, $y = 7$ and $v = -2$.

2.5

WE15

2 Rearrange the formula in each of the following to make the variable in brackets the subject of the formula. (Assume all variables are positive.)

(a) $V = \pi r^2 h \quad (r)$ **(b)** $E = \frac{1}{2}mv^2 \quad (v)$ **(c)** $F = \frac{dh^2}{k} \quad (h)$

(d) $k = 7\sqrt{\frac{m}{n}} \quad (m)$ **(e)** $d = 2\sqrt{\frac{b}{c}} \quad (b)$ **(f)** $m = x\sqrt{\frac{y}{z}} \quad (y)$

3 Rearrange each of the following to make x the subject.

(a) $x + y = z$ **(b)** $x - 3a = 2b$ **(c)** $kx - m = n$

(d) $dx + c = f$ **(e)** $rt - x = p$ **(f)** $vy - x = -w$

(g) $\frac{2x + y}{p} = k$ **(h)** $\frac{t + rx}{w} = m$ **(i)** $\frac{mp + px}{m} = n$

(j) $cdx + e = gh$ **(k)** $axp - k = mn$ **(l)** $s - tux = vw$

(m) $b + \frac{xd}{y} = f$ **(n)** $\frac{mx}{n} + kp = t$ **(o)** $ab - \frac{dx}{c} - f = o$

4 The formula for finding the velocity, v, of a moving object is $v^2 = u^2 + 2as$. A correct rearrangement of the formula would be:

A $u^2 = v^2 + \frac{2as}{2}$ **B** $s = \frac{v^2 - u^2}{2a}$ **C** $a = \frac{2s}{v^2 - u^2}$ **D** $a = \frac{u^2 - v^2}{2s}$

5 The formula for finding the velocity, v m/s, of an object starting with a velocity of u m/s and undergoing a constant acceleration of a m/s^2 over time t s is $v = u + at$.

(a) Rearrange the formula to make t the subject.

(b) Find the value of t when:

(i) $a = 14$ m/s^2, $u = 20$ m/s and $v = 76$ m/s

(ii) $a = 2.4$ m/s^2, $u = 5.7$ m/s and $v = 12.9$ m/s.

6 The circumference (C) of a circle is related to the radius (r) by the formula $C = 2\pi r$. Use the calculator value for π and give your answer correct to two decimal places.

(a) Rearrange the formula to make r the subject.

(b) Find the value of r when:

(i) $C = 628$ **(ii)** $C = 16.4$

7 The formula for the volume (V) of a cylinder is $V = \pi r^2 h$. Rearrange the formula to find h when $r = 5$ and $V = 157$. Use the calculator value for π and give your answer correct to two decimal places.

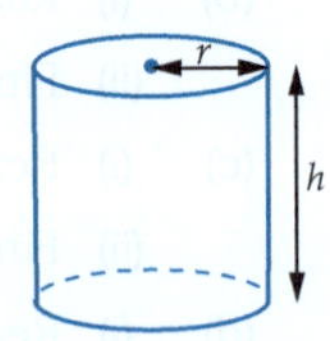

8 The perimeter of a rectangle is given by $P = 2(l + w)$, where l and w represent the length and width of the rectangle, respectively.

Rearrange the formula to make w the subject. Find the width of the rectangle if the length is 23 m and the perimeter is 84 m.

Reasoning, communicating and open-ended

9 The kinetic energy (E) of an object, in joules, is found using the formula $E = \frac{1}{2}mv^2$, where m is the mass in kg and v is the velocity of the object in m/s.

(a) Determine E when $m = 2$ kg and $v = 1.2$ m/s.

(b) Determine m when $E = 162$ joules and $v = 9$ m/s.

(c) Determine v when $E = 1.96 \times 10^6$ joules and $m = 8$ kg.

(d) Explain why there are two answers for part **(c)** and what each answer means.

10 The density (D) of a substance is related to its mass (M) and its volume (V) by the formula $D = \frac{M}{V}$.

(a) Find the difference in the mass of two objects made of the same material with a density of 7 g/cm^3 if one has a volume of 12 cm^3 and the other has a volume of 15 cm^3.

(b) A flotation device is to be made of an outer material with a density of 1.8 g/cm^3 and a mass of 2.4 kg. It is filled with an inner material of density 0.9 g/cm^3 and a mass of 7.8 kg. Will the device float in sea water of density 1.025 g/cm^3?

> **Hint**
> Find the total mass and the total volume.

11 The area of a trapezium is given by the formula $A = \frac{(a+b)h}{2}$.

(a) Rearrange the formula to make b the subject.

(b) Use this rearranged formula to find the value of b if $a = 9$, $h = 6$ and $A = 66$.

(c) We could substitute these values for a, h and A into the original formula and solve the equation $66 = \frac{(9+b) \times 6}{2}$ to find the value of b. Explain why we would rearrange the formula to find the value of b.

12 Consider this formula: $S = \frac{n}{2}[2a + (n-1)d]$

Expand all the brackets in this expression and use this form to explain why it is difficult to make n the subject.

13 Ruth has been rearranging the formula for the area of a trapezium, $A = \frac{(a+b)h}{2}$. She has produced the following working while trying to make a the subject. Where has she made a mistake? Now rearrange the formula to make b the subject.

$$A = \frac{h(a+b)}{2}$$
$$2A = h(a+b)$$
$$\frac{2A}{h} = a + b$$
$$a = b - \frac{2A}{h}$$

Half-time 2

Stage 5.1 Q1–7 **Stage 5.2 Q8–10**

Ex 2.1

1 Simplify each of the following, expressing your answer with positive indices.

(a) $3x^3y \times 4x^4y$ (b) $\dfrac{8x^3y^5}{2xy^2}$ (c) $\dfrac{3m^4n^2 \times 2m^5n^3}{8m^3n^4}$

Ex 2.4

2 Write the following numbers correct to three significant figures.

(a) 0.2999 (b) 80 (c) 62 078

Ex 2.2

3 Simplify each of the following, expressing your answer with positive indices.

(a) $(3x^4y^5)^2$ (b) $\left(\dfrac{a^4b^3}{6ab^4}\right)^2$ (c) $\dfrac{5^3x^4y^6}{3x^3y^2} \times \dfrac{x^6y}{x^7y^5}$

Ex 2.4

4 Express each quantity described below in scientific notation.

(a) The Moon is 384 000 km away.

(b) 500 mL of water is lost through the skin daily.

Ex 2.4

5 To how many significant figures have each of the following been written?

(a) 36 050 000 (b) 0.000 000 852 (c) 0.0830

Ex 2.1

6 Simplify the following, leaving your answer in index form.

(a) $\dfrac{(2^3)^4 \times (5^6)^2}{(2^5) \times (5^2)^4}$ (b) $\dfrac{(3^2 \times 5^4)^2}{25 \times 27}$ (c) $\dfrac{8^3 \times 9^2}{6^4}$

Ex 2.4

7 Express each quantity described below as a number.

(a) The number of bacteria that will fit on a pin head is approximately 1.1×10^{14}.

(b) The nucleus of a leaf cell is about 4.4×10^{-6} m in diameter.

Ex 2.3

8 Simplify each of the following, expressing your answer with positive indices:

(a) $\dfrac{(p^2q^3)^{-1} \times p^5q^{-2}}{p^3q^2}$ (b) $\dfrac{(c^{-3}d^{-4})^2 \times c^2d^6}{c^5d^3}$

(c) $\dfrac{4(a^2b)^3 \times 3a^{-5}b^{-2}}{6ab}$ (d) $\dfrac{6(mn^3)^2 \times 4m^{-4}n^{-1}}{3m^2n}$

Ex 2.3

9 Express each of the following with positive indices.

(a) x^{-5} (b) $3p^{-4}$ (c) $5m^{-3}n$ (d) $6p^{-4}q^{-1}r^2$

Ex 2.5

10 Rearrange the formula $z = 2xy - 3y$ to make x the subject and then find the value of x when $z = 7$ and $y = 5$.

Expanding the brackets

2.6

Stage 5.2

Need to Know

The distributive law

Brackets are used to change the order of operations, allowing addition or subtraction to be done before multiplication or division. The expression is written without brackets by multiplying every term inside the brackets by the number outside the brackets. This is called expanding the brackets.

$$2(3 + 4) = 2 \times 7 = 14$$

and

$$2(3 + 4) = 2 \times 3 + 2 \times 4 = 6 + 8 = 14$$

If variables replace numbers, the expression is rewritten without the brackets the same way.

$$a(b+c) = ab + ac$$

This is known as the **distributive law**. Every term inside the brackets is multiplied by the term outside the brackets.

$a(b + c) = ab + ac$

Binomial products

Consider a room 6 metres long and 4 metres wide. If it is extended by 3 metres in length and 2 metres in width, as in the diagram, find the new area.

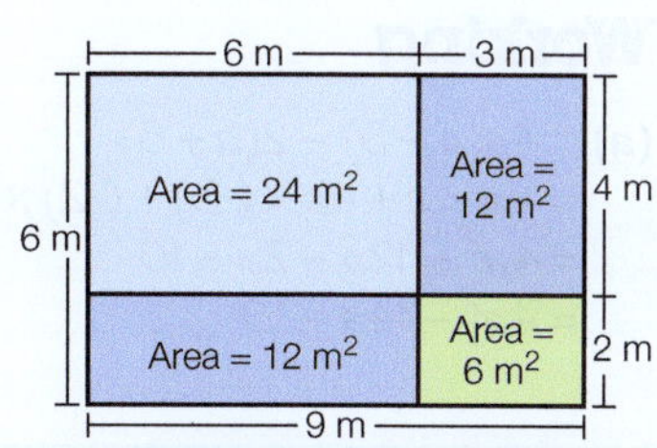

The total area can be found in two ways:

$$(6 + 3)(4 + 2) = 9 \times 6 = 54 \text{ m}^2$$

or

$$6 \times 4 + 6 \times 2 + 3 \times 4 + 3 \times 2 = 24 + 12 + 12 + 6 = 54 \text{ m}^2$$

If a room a metres long and c metres wide has been extended by b metres in length and d metres in width, we can find the new area using the diagram.

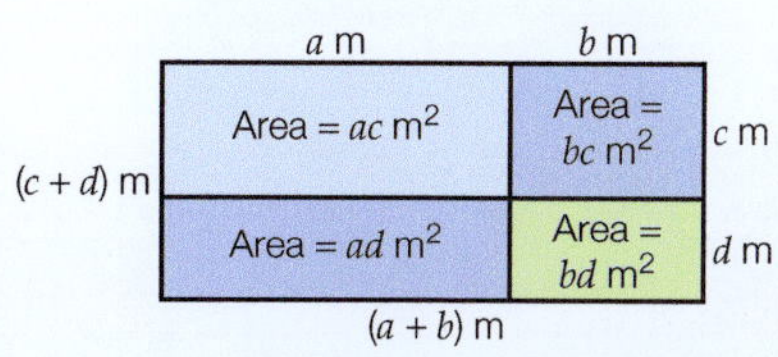

$$\text{Area} = (a + b)(c + d) = a \times c + a \times d + b \times c + b \times d = ac + ad + bc + bd \text{ m}^2$$

2.6

A **binomial expansion** is the result of multiplying two terms in one set of brackets by two terms in another set of brackets. Every term in the second bracket is multiplied by every term in the first bracket.

$$(a + b)(c + d) = ac + ad + bc + bd$$

Binomial expansion can be completed using FOIL.

First: Multiply the first terms in each bracket.

Outside: Multiply the outside terms in each bracket.

Inside: Multiply the inside terms in each bracket.

Last: Multiply the last terms in each bracket.

Often after expansion, like terms can be collected and the expression simplified, leaving three terms called a quadratic trinomial.

Worked Example 16

WE16

Expand:

(a) $5(p + 4)$ **(b)** $2m(4m - 3n)$

Working

(a) $5(p+4) = 5 \times p + 5 \times 4$
$= 5p + 20$

(b) $2m(4m - 3n) = 2m \times 4m + 2m \times -3n$
$= 8m^2 - 6mn$

Worked Example 17

WE17

Expand and simplify:

(a) $4a(a - 3) - 2(a + 5)$ **(b)** $3(2 - b) - (4 - b)$

Working

(a) $4a(a - 3) - 2(a + 5)$
$= 4a \times a + 4a \times (-3) + (-2) \times a + (-2) \times 5$
$= 4a^2 - 12a - 2a - 10$
$= 4a^2 - 14a - 10$

(b) $3(2 - b) - (4 - b)$
$= 3(2 - b) - 1(4 - b)$
$= 3 \times 2 + 3 \times (-b) + (-1) \times 4 + (-1) \times (-b)$
$= 6 - 3b - 4 + b$
$= 2 - 2b$

2.6

Worked Example 18

WE18

Expand each of the following expressions.

(a) $(c+4)(d+2)$ (b) $(2m-4)(3n-5)$ (c) $(2p-3r)(3t+4)$

Working

(a) $(c+4)(d+2)$
$= c \times (d+2) + 4 \times (d+2)$
$= c \times d + c \times 2 + 4 \times d + 4 \times 2$
$= cd + 2c + 4d + 8$

(b) $(2m-4)(3n-5)$
$= 2m \times (3n-5) + (-4) \times (3n-5)$
$= 2m \times 3n + 2m \times (-5) + (-4) \times 3n + (-4) \times (-5)$
$= 6mn - 10m - 12n + 20$

(c) $(2p-3r)(3t+4)$
$= 2p \times (3t+4) + (-3r) \times (3t+4)$
$= 2p \times 3t + 2p \times 4 + (-3r) \times 3t + (-3r)$
$= 6pt + 8p - 9rt - 12r$

Worked Example 19

WE19

Expand each of the following expressions. Simplify where possible.

(a) $(x-2)(x+5)$ (b) $(t-3)(t-7)$ (c) $(4x+3)(5x-2)$

Working

(a) $(x-2)(x+5)$
$= x \times x + x \times 5 + (-2) \times x + (-2) \times 5$
$= x^2 + 5x - 2x - 10$
$= x^2 + 3x - 10$

(b) $(t-3)(t-7)$
$= t \times t + t \times (-7) + (-3) \times t + (-3) \times (-7)$
$= t^2 - 7t - 3t + 21$
$= t^2 - 10t + 21$

(c) $(4x+3)(5x-2)$
$= 4x \times 5x + 4x \times (-2) + 3 \times 5x + 3 \times (-2)$
$= 20x^2 - 8x + 15x - 6$
$= 20x^2 + 7x - 6$

2.6

Answers page 427

Stage 5.2

2.6 Expanding the brackets

Navigator

Q1 Column 1, Q2 Column 1, Q3 Column 1, Q4 Column 1, Q5 Column 1, Q6 Column 1, Q7, Q8, Q9, Q10, Q14, Q15	Q1 Column 2, Q2 Column 1, Q3 Column 2, Q4 Column 2, Q5 Column 2, Q6 Column 2, Q7, Q8, Q9, Q10, Q14, Q16	Q1 Column 3, Q2 Column 2, Q3 Column 3, Q4 Column 3, Q5 Column 3, Q6 Column 3, Q7, Q8, Q9, Q10, Q11, Q12, Q13, Q15, Q16

Fluency and understanding

WE16

Hint

Remember $-(x+2)$ really means $-1(x+2)$.

1 Expand each of the following expressions.

(a) $2(x+5)$ (b) $4(a+9)$ (c) $8(b+3)$
(d) $8(x-2)$ (e) $9(x-2)$ (f) $3(k-4)$
(g) $-7(x+1)$ (h) $-4(d+8)$ (i) $-6(x+3)$
(j) $-9(c-6)$ (k) $-2(x-5)$ (l) $-5(x-6)$
(m) $x(x+3)$ (n) $m(m+5)$ (o) $x(x+9)$
(p) $-x(x-8)$ (q) $-x(x-4)$ (r) $-n(n-9)$
(s) $3x(x+2)$ (t) $2u(u-6)$ (u) $5x(x-1)$
(v) $-2x(x+7)$ (w) $-7q(q+1)$ (x) $-9b(b-6)$

WE17

2 Expand and simplify each expression.

(a) $3(x-2)+2(x+4)$ (b) $2(y+1)+3(y-2)$
(c) $4(2x-3)-(6x-1)$ (d) $3(4a+5)-(7a+2)$
(e) $a(2a+3)-2(3a-5)$ (f) $3x(x+2)-4(x-1)$
(g) $7y(y-3)-5(y+2)$ (h) $6k(2k+3)-3k(k-3)$

WE18

3 Expand each of the following.

(a) $(x+6)(y+5)$ (b) $(x+4)(y+7)$ (c) $(c+d)(e+3)$
(d) $(r+5)(t-8)$ (e) $(a+2)(b-1)$ (f) $(p+3)(q-2)$
(g) $(a-3)(y-4)$ (h) $(m-2)(n-6)$ (i) $(k-5)(m-4)$
(j) $(4p+q)(m+n)$ (k) $(7x+y)(a+k)$ (l) $(3p+2r)(4s+t)$
(m) $(2a+1)(b-3)$ (n) $(3x-1)(x+2)$ (o) $(3m+2)(n-1)$
(p) $(2a-3b)(5m-3n)$ (q) $(5c-2d)(3x-4y)$ (r) $(7p-2q)(2r-3t)$

WE19

4 Expand each of the following and simplify.

(a) $(a+3)(a+2)$ (b) $(c+5)(c+3)$ (c) $(y+7)(y+6)$
(d) $(x-3)(x+1)$ (e) $(y-2)(y+9)$ (f) $(a-7)(a+12)$
(g) $(b-2)(b-1)$ (h) $(x-4)(x-3)$ (i) $(p-8)(p-7)$
(j) $(3m+1)(2m+3)$ (k) $(4d+5)(2d+3)$ (l) $(3k+5)(7k+4)$
(m) $(4y-3)(2y+1)$ (n) $(5k+3)(2k-7)$ (o) $(2a+5)(4a-3)$
(p) $(6a-5)(2a-1)$ (q) $(4p-3)(3p-5)$ (r) $(3r-2)(8r-7)$

5 Expand each of the following and simplify.

(a) $(x + y)(x + y)$ **(b)** $(m + n)(m + n)$ **(c)** $(a + b)(a + b)$
(d) $(p - r)^2$ **(e)** $(c - d)^2$ **(f)** $(g - h)^2$
(g) $(2c + 3d)(2c + 3d)$ **(h)** $(3m + 2n)(3m + 2n)$ **(i)** $(5a + 3b)(5a + 3b)$
(j) $(5x - 4y)(5x - 4y)$ **(k)** $(2j - 3k)(2j - 3k)$ **(l)** $(7c - 5d)^2$

6 Expand each of the following and simplify.

(a) $(m + n)(m - n)$ **(b)** $(p + q)(p - q)$ **(c)** $(a - b)(a + b)$
(d) $(3a + 2b)(3a - 2b)$ **(e)** $(4x - 3y)(4x + 3y)$ **(f)** $(7m - 3n)(7m + 3n)$

7 When simplified, $5m(-3 + 2n) - n(10m - 1)$ is:

A $n - 15m$ **B** $10n - 15m - 1$ **C** $-15m - 20mn + n$ **D** $15m + n$

8 Consider this envelope.

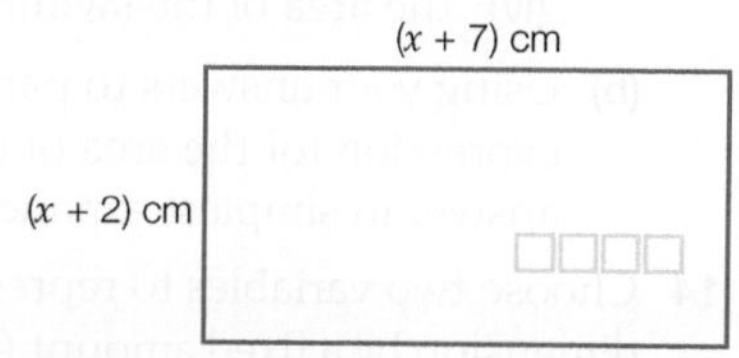

(a) Write the expression (in factorised form) for the area of this envelope.
(b) Use your answer to find the area when $x = 10$.
(c) What value would x represent if the area of the envelope is 104 cm^2? Try substituting different values for x until you find the appropriate one.
(d) Expand the expression you obtained for the area in part **(a)**.
(e) Check that the expanded expression is correct by calculating the area when $x = 10$ and comparing this with the answer obtained in part **(b)**.

9 $(3x - 2)(3x + 2)$ is the same as:

A $6x^2 + 4$ **B** $9x^2 - 12x - 4$ **C** $9x^2 + 6x - 4$ **D** $9x^2 - 4$

10 A vegetable bed, 15 m long and 3 m wide, is to be extended by x metres in both length and width.

(a) Write expressions for each of the following.
  (i) The new length of the garden bed.
  (ii) The new width of the garden bed.
  (iii) The new area of the garden bed as the product of factors.
  (iv) The new area of the garden bed as the expanded product of factors.
(b) Compare the area of the new garden bed with the area of the original garden bed. By how much has the area increased?
(c) Use your answer from part **(b)** to calculate how much extra area of garden bed is available if both the length and width are increased by 2 m.

11 The sides of a rectangle and a triangle have the lengths shown below.

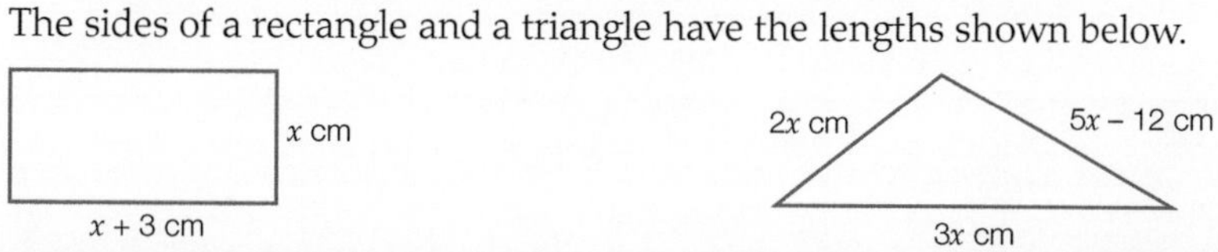

(a) Write the expression for the area of the rectangle using the given lengths.
(b) Write the expression for the perimeter of the triangle using the given lengths.
(c) Find two values for x so that the area of the rectangle is the same as the perimeter of the triangle.

2.6

Reasoning, communicating and open-ended

12 There are a number of ways to check that $x(x+2)=x^2+2x$ is a true statement. Choose two of these methods and show clearly how you are able to check that $x(x+2)=x^2+2x$ is true. Can you use these methods to show that $a(b+c)=ab+ac$? If so, show how.

13 A backyard, square in shape, is to have an in-ground rectangular swimming pool constructed in it. A path around the pool is to be laid as shown in the diagram.

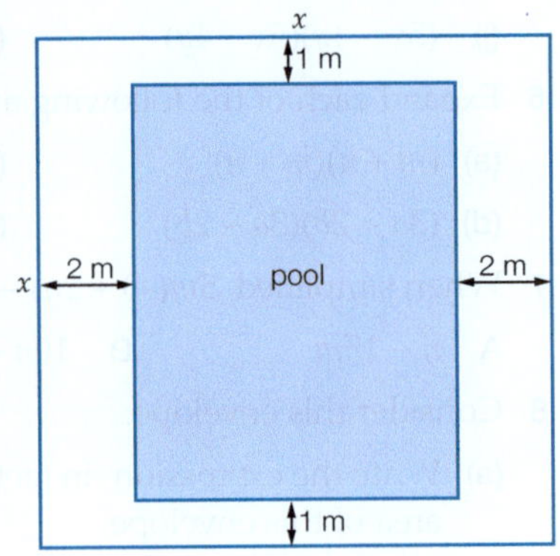

(a) Write expressions for each of the following:
   (i) the total area of the backyard
   (ii) the width of the swimming pool
   (iii) the length of the swimming pool
   (iv) the area of the swimming pool.

(b) Using your answers to parts **(i)** and **(iv)**, write an expression for the area of the path. State your answer in simplest expanded form.

14 Choose two variables to represent the length and width of a rectangle. Increase each dimension by a fixed amount (e.g. increase the length by 2 units and the width by 3 units). Find the area of the new rectangle.

15 Three students gave the following incorrect answers when expanding $-3(5x-2)$.

(a) $-15x-2$ (b) $5x+6$ (c) $-8x+5$

Explain the errors made by each student. What is the correct answer?

16 Three students obtained the following incorrect answers when expanding $(x-3)(x+4)$.

(a) $x^2+7x+12$ (b) $2x-12$ (c) x^2+x-7

Explain what errors were made by each student. What is the correct answer?

Factorising using common factors

2.7

Stage 5.2

Need to Know

To **factorise** an expression is to write it as a product of its factors.

The opposite or inverse of expanding is called factorising.

Expanding	**Factorising**
$a(b + c) = ab + ac$	$ab + ac = a(b + c)$

Factorising using the highest common factor (HCF)

- Place the HCF outside brackets.
- Divide each term by the **HCF**; this quotient goes inside the brackets.
- If the HCF is not taken out as a factor, the expression is not fully factorised (e.g. $4x + 8 = 2(2x + 4)$ is not fully factorised).
- A negative sign may be taken out as well as the HCF; be careful to change signs inside the brackets (e.g. $-5x + 10 = -5(x - 2)$).

This procedure is called 'taking out the highest common factor'. The HCF may be any positive number, a variable or the product of a number and one or more variables. Once an expression is factorised, no matter what method has been used, it will have brackets in it.

Worked Example 20

WE20

Factorise each of the following.

(a) $5x + 20$
(b) $y^2 - y$
(c) $-9y - 6z + 3$
(d) $3a^2bc - acd$
(e) $-8mn^2 - 12m^2n$
(f) $a(b + 3) + 2(b + 3)$

Working

(a) $5x + 20 = 5(x + 4)$
(b) $y^2 - y = y(y - 1)$
(c) $-9y - 6z + 3 = -3(3y + 2z - 1)$
(d) $3a^2bc - acd = ac(3ab - d)$
(e) $-8mn^2 - 12m^2n = -4mn(2n + 3m)$
(f) $a(b + 3) + 2(b + 3) = (b + 3)(a + 2)$

2.7

Answers page 428

2.7 Factorising using common factors

Stage 5.2

Navigator

Q1 Column 1, Q2 Column 1, Q3 Column 1, Q4 Column 1, Q5, Q6, Q7, Q9, Q12, Q13	Q1 Column 2, Q2 Column 2, Q3 Column 2, Q4 Column 1, Q5, Q6, Q8, Q9, Q10, Q12, Q13	Q1 Column 3, Q2 Column 3, Q3 Column 3, Q4 Column 2, Q5, Q6, Q7, Q8, Q9, Q10, Q11, Q13

Fluency and understanding

WE20

1 Factorise each of the following.

(a) $3x + 9$ (b) $12y + 24$ (c) $6 + 42a$
(d) $2k - 6$ (e) $4m - 20$ (f) $8b - 4$
(g) $7x - xy$ (h) $4s - 9st$ (i) $3a - ab$
(j) $15y + 5w + 10x$ (k) $2a + 4b - 8c$ (l) $3x - 15y + 9$
(m) $3ac + 2abc$ (n) $5xy^2 + 2x^2y$ (o) $9f^2gh - 12fg^2$
(p) $6gh^2 + 18g^2h$ (q) $16x^2y - 40xy^2$ (r) $18ab - 42a^2b$
(s) $-4m - 20$ (t) $-7d - 49$ (u) $-12r - 60$
(v) $x(a + 3) + 5(a + 3)$ (w) $m(n - 2) + 9(n - 2)$ (x) $3(x + 5) + y(x + 5)$

2 Factorise each of the following.

(a) $10x + 25$ (b) $16y + 36$ (c) $14a - 35$
(d) $12d - 18$ (e) $18 - 27m$ (f) $36 - 15k$
(g) $24abc - 10b + 7bc$ (h) $5km - 3kn + 2ghk$ (i) $c + 4bc + 2ac$
(j) $x^2 + 3x$ (k) $y^2 + 6y$ (l) $5k + k^2$
(m) $8m - 3m^2$ (n) $4p^2 - 7p$ (o) $9a^2 - 5a$
(p) $6d - 9d^2$ (q) $22g - 14g^2$ (r) $24a - 32a^2$

3 Factorise each of the following by taking out the highest negative common factor.

(a) $-2m - 12$ (b) $-4k - 24$ (c) $-5h - 25$
(d) $-x^2 - x$ (e) $-y^2 - 3y$ (f) $-p^2 - 4p$
(g) $-16a + 4$ (h) $-21b + 7$ (i) $-30r + 6$
(j) $-24p + 40$ (k) $-18w + 16$ (l) $-45k - 63$
(m) $-15x^2 - 36x$ (n) $-14y^2 - 49y$ (o) $-20r^2 - 44r$
(p) $-8d - 4f - 6g$ (q) $-15a - 10b - 5c$ (r) $-14c - 28d - 7f$
(s) $-12x^2 - 16x + 20xy + 4xz$ (t) $-2km + 4km^2 - 6k - 8k^2$ (u) $-10b^2 + 4bc - 8bc^2 + 16ab$

4 Factorise each of the following.

(a) $4(y - 1) + w(y - 1)$ (b) $5(m - 3) + n(m - 3)$ (c) $5(p - 2) + r(p - 2)$
(d) $p(q + 5) - 2(q + 5)$ (e) $y(2x + 3) - (2x + 3)$ (f) $a(3p + 5) - (3p + 5)$
(g) $d(f - 1) - 6(f - 1)$ (h) $5d(d - 2) - 4(d - 2)$ (i) $3m(m - 6) - 7(m - 6)$

5 When factorised, the expression $3ab^2c - 6abc + 3bc$ is:

A $3bc(ab - 2a)$ **B** $3abc(b - 2 + 3a)$
C $3bc(ab - 2a + 1)$ **D** $6bc(ab - a + 2bc)$

6 When fully factorised, the expression $-2x(3m - 4) + 5y(3m - 4)$ is:

A $(3m - 4)(5y - 2x)$ **B** $(5y - 4)(3m - 2x)$
C $(-3m + 4)(-2x + 5y)$ **D** $(-2x - 4)(3m + 5y)$

7 **(a)** Write the expression for 'Eight times the product of two numbers added to twelve times one of the numbers'.

(b) Factorise the expression in **(a)**.

8 A rectangular pizza is of length d cm and width $(c + 2)$ cm. Another rectangular pizza of the same width has a length of 5 cm. The two pizzas are joined to make a longer pizza.

(a) Write the area of the new shape in factorised form.

(b) What is the relationship between c and d if the pizza is square?

9 A floor-to-ceiling window of width x m is to be installed in a wall 5 m long and $(y + 3)$ m high. The remaining part of the wall is to be plastered.

(a) Find the length of the wall to be plastered after the window is installed.

(b) Find the area of the wall to be plastered after the window is installed.

Reasoning, communicating and open-ended

10 A wood template is to be made for the front of a letter box with the dimensions shown.

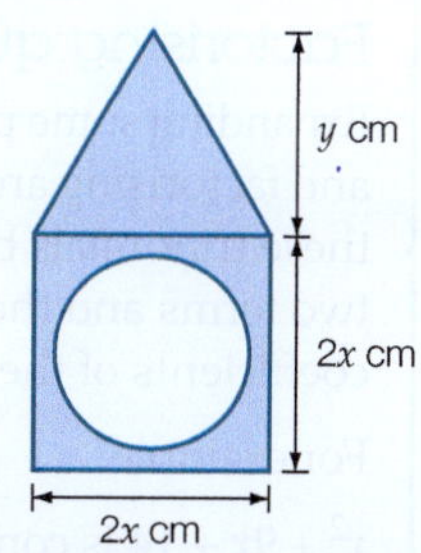

(a) Write an expression that would show the area of timber used in the construction before the circular opening is cut out.

(b) Express this answer in a factorised form.

(c) If the diameter of the circular cut out needs to be 2 cm less than the width, find the area of the timber cut out for the opening.

(d) Write this in expanded form.

(e) Show that the total area of timber that will be used is $(4 - \pi)x^2 + (y + 2\pi)x - \pi$ cm^2.

(f) What area of timber would be wasted if the template was cut from a rectangular piece of timber with the smallest possible dimensions?

11 A lasagne is baked in a tray $(3p + 2)$ cm long and $(2p - 1)$ cm wide. A slice, the full width of the tray and $(r + 2)$ long is taken from the lasagne.

(a) Find the area of the lasagne, before the slice was removed, in factorised form.

(b) Find the area of the slice taken from the lasagne in factorised form.

(c) Use your answers to parts **(a)** and **(b)** to find the area remaining after the slice has been removed. Write your answer in simplest factorised form.

(d) How could you find the area in part **(c)** more simply? Show your working.

12 Make up three different expressions whose highest common factor is $3x$.

13 In a test, students were asked 'Factorise the expression $-5x + 10$'. Three students wrote these answers; one answer was correct, the other two answers were incorrect. Identify the correct answer and explain the mistakes the other two students made.

A $-5(x + 2)$ **B** $-5(x + 10)$ **C** $-5(x - 2)$

Problem solving

Divisibility dilemma?

The number 123 456 is divisible by 2, 3 and 4. Find the smallest number that is divisible by all the digits from 2 to 9 (inclusive) if it begins with the digits 123 456.

Strategy options

- Test all possible combinations.
- Guess and check.
- Break the problem into manageable parts.

2.8 Further factorisation

Stage 5.2

Need to Know

Factorising by grouping in pairs

If there are four terms in an expression and there is no common factor for all four terms, we can sometimes group the expression in pairs of terms so that there is a common factor in each pair. If this gives us another common factor in brackets, we can factorise further. This technique is called **grouping in pairs**. Sometimes you may have to consider which two terms to pair together.

Factorising quadratic trinomials

Expanding some products gives quadratic expressions when simplified. Because expanding and factorising are opposite operations, the expanding process can be reversed to convert these trinomials back into factor form. The middle term in the trinomial can be split into two terms and then factorised by grouping in pairs as long as the product of the coefficients of the two middle terms equals the last term.

For example:

$x^2 + 9x + 14$ is converted to $x^2 + 2x + 7x + 14$ because $2 + 7 = 9$ and $2 \times 7 = 14$.

Worked Example 21

WE21

Factorise the following.

(a) $ab + 5b + 3a + 15$

(b) $3x + 12 - xy - 4y$

Working

(a) $ab + 5b + 3a + 15$
$= b(a + 5) + 3(a + 5)$
$= (a + 5)(b + 3)$

(b) $3x + 12 - xy - 4y$
$= 3(x + 4) - y(x + 4)$
$= (x + 4)(3 - y)$

Worked Example 22

WE22

Factorise $df + 6g - 6f - dg$.

Working

$df + 6g - 6f - dg = df - 6f - dg + 6g$
$= f(d - 6) - g(d - 6)$
$= (d - 6)(f - g)$

Worked Example 23

WE 23

For each of the following expressions, (i) factorise, (ii) simplify, (iii) write the relationship between your answers to parts (i) and (ii), and (iv) state the relationship between the coefficients of the two middle terms and the last term.

(a) $x^2 + 5x + 6x + 30$ (b) $p^2 - 4p - 4p + 16$ (c) $a^2 - 12a + 12a - 144$

Working

(a) (i) $x^2 + 5x + 6x + 30 = x(x + 5) + 6(x + 5)$
$= (x + 5)(x + 6)$

(ii) $x^2 + 5x + 6x + 30 = x^2 + 11x + 30$

(iii) $x^2 + 11x + 30 = (x + 5)(x + 6)$

(iv) $5 \times 6 = 30$

The product of the coefficients of the two middle terms gives the last term.

(b) (i) $p^2 - 4p - 4p + 16 = p(p - 4) - 4(p - 4)$
$= (p - 4)(p - 4)$
$= (p - 4)^2$

(ii) $p^2 - 4p - 4p + 16 = p^2 - 8p + 16$

(iii) $(p - 4)^2 = p^2 - 8p + 16$

(iv) $-4 \times -4 = 16$

The product of the coefficients of the two middle terms gives the last term.

(c) (i) $a^2 - 12a + 12a - 144 = a(a - 12) + 12(a - 12)$
$= (a - 12)(a + 12)$

(ii) $a^2 - 12a + 12a - 144 = a^2 - 144$

(iii) $a^2 - 144 = (a - 12)(a + 12)$

(iv) $-12 \times {+12} = -144$

The product of the coefficients of the two middle terms gives the last term.

Worked Example 24

WE 24

Factorise the following trinomials by splitting the middle term.

(a) $x^2 + 9x + 14$ (b) $x^2 - 7x + 12$

Working

(a) $x^2 + 9x + 14$
$= x^2 + 2x + 7x + 14$
$= x(x + 2) + 7(x + 2)$
$= (x + 2)(x + 3)$

(b) $x^2 - 7x + 12$
$= x^2 - 3x - 4x + 12$
$= x(x - 3) - 4(x - 3)$
$= (x - 3)(x - 4)$

2.8

Answers page 428

Stage 5.2

2.8 Further factorisation

Navigator

Q1 Column 1, Q2 Column 1, Q3 Column 1, Q4, Q5, Q6, Q7, Q8, Q9, Q12, Q13	Q1 Column 2, Q2 Column 2, Q3 Column 2, Q4, Q5, Q6, Q7, Q8, Q12, Q13, Q14	Q1 Column 3, Q2 Column 3, Q3 Column 3, Q4, Q5, Q6, Q7, Q8, Q9, Q10, Q11, Q12, Q13, Q14

Fluency and understanding

WE21

1 Factorise the following.

(a) $xy + 4y + 3x + 12$ (b) $pq + q + 5p + 5$ (c) $mn + 2n + 9m + 18$
(d) $ab + bf + ad + df$ (e) $pq + qt + pr + rt$ (f) $km + 3nk + 6m + 18n$
(g) $np - 7n + p - 7$ (h) $ab - 5b + a - 5$ (i) $mn - 2n + 7m - 14$
(j) $ab - 6b - 4a + 24$ (k) $x^2 + 2x - 8x - 16$ (l) $y^2 + 4y - 6y - 24$

WE22

2 Factorise the following.

(a) $2p + 3k + kp + 6$ (b) $mn + pq + np + mq$ (c) $cd + 12 + 6c + 2d$
(d) $ad - hk - dk + ah$ (e) $xy - 8 - 4y + 2x$ (f) $eg - 3 - g + 3e$
(g) $bc - 1 + c - b$ (h) $xy - 3 + 3x - y$ (i) $mk - 3n + 3m - nk$

WE23

3 For each of the following expressions, **(i)** factorise, **(ii)** simplify, **(iii)** write the relationship between your answers to parts **(i)** and **(ii)**, and **(iv)** state the relationship between the coefficients of the two middle terms and the last term.

(a) $x^2 + 7x + 2x + 14$ (b) $j^2 + 3j + 5j + 15$ (c) $z^2 + z + 8z + 8$
(d) $x^2 + 2x - 8x - 16$ (e) $y^2 + 4y - 6y - 24$ (f) $x^2 + 3x - 4x - 12$
(g) $a^2 - 4a + 5a - 20$ (h) $k^2 - 2k + 5k - 10$ (i) $m^2 - 7m + 4m - 28$
(j) $y^2 - 3y - 7y + 21$ (k) $c^2 - 9c - 5c + 45$ (l) $d^2 - 8d - 12d + 96$
(m) $m^2 + 3m + 3m + 9$ (n) $r^2 + 6r + 6r + 36$ (o) $k^2 + 11k + 11k + 121$
(p) $p^2 - 5p - 5p + 25$ (q) $d^2 - 7d - 7d + 49$ (r) $n^2 - 8n - 8n + 64$
(s) $c^2 - 3c + 3c - 9$ (t) $t^2 - 2t + 2t - 4$ (u) $b^2 - 9b + 9b - 81$

WE24

4 Factorise the following trinomials by splitting the middle term.

(a) $x^2 + 14x + 24$ (b) $x^2 + 11x + 24$ (c) $x^2 - 10x + 24$
(d) $x^2 + 8x + 12$ (e) $x^2 - 7x + 12$ (f) $x^2 - 13x + 12$
(g) $x^2 + 8x + 15$ (h) $x^2 + 11x + 18$ (i) $x^2 + 10x + 16$
(j) $x^2 - 2x - 24$ (k) $x^2 + 5x - 24$ (l) $x^2 - 10x - 24$

5 By taking out the common factor from $k(j-1) + (j-1)$ we obtain:

A $(j-1)(k+1)$ **B** $(j+1)(k-1)$ **C** $(j-1)(k)$ **D** $(j+1)(k+1)$

6 Square a number, x, subtract four times x, add 12 and subtract three times x.

(a) Write the expression for the information given.

(b) Simply the expression in **(a)**.

(c) Factorise the expression in **(a)**.

2.8

7

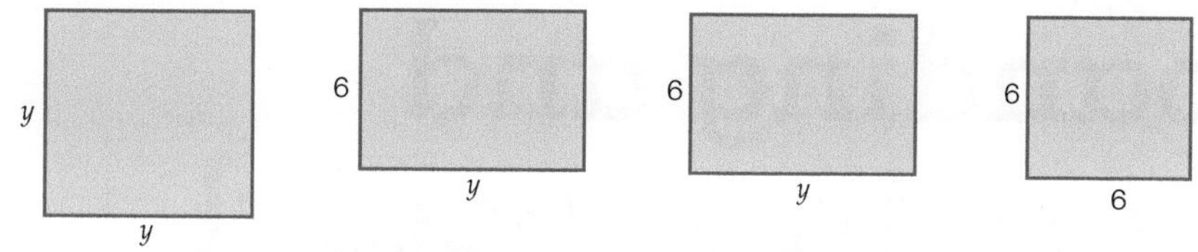

(a) Write the expression for the total area of the four shapes above.

(b) Factorise the expression in **(a)**.

(c) Draw a diagram to show how the four shapes can fit together to make a square (include dimensions on your diagram).

Reasoning, communicating and open-ended

8 (a) Factorise $ab + ac + ad + 7b + 7c + 7d$ by grouping in threes. That is, take out a common factor from three terms at a time.

(b) Show that, if $b = a$ and $c + d = 7$, the expression is a perfect square.

9 (a) Add five different positive numbers to $x^2 + 10x$ so that it factorises differently each time.

(b) Factorise each of the new expressions you have created in part (a).

(c) Explain why there are only five possible positive numbers that can be added in part (a).

(d) Explain why there are an infinite number of negative numbers that could be added to the expression so creating an infinite number of quadratic trinomials that have $x^2 + 10x$ as the first two terms.

10 (a) Factorise $35y^2 - 21y$.

(b) If this represents the area of a rectangle in cm^2 and the factors represent the width, write the dimensions of the rectangle.

(c) If an area of $15y - 9$ cm^2 is added to the area in part **(a)**, find the combined area in factorised form.

(d) Write the dimensions of the new rectangle if the factors in part **(c)** represent the length and width.

11 A trapezium has an area of $\frac{3x^2 + 2xy - 6x - 4y}{2}$ m^2 where x and y are both positive numbers and $x > 2$.

(a) Write the area in factorised form.

(b) If the smaller factor represents the height of the trapezium, what is the sum of the lengths of the two parallel non-equal sides?

12 Without expanding the factorised expression in the following, add a term to:

(a) $x^2 - 5x$ so that it factorises as $(x + 4)(x - 9)$

(b) $x^2 - 5x$ so that it factorises as $(x - 2)(x - 3)$

(c) $x^2 + 6$ so that it factorises as $(x + 1)(x + 6)$.

13 (a) Find two terms that, when added to $x^2 + ax$, can be factorised by grouping in pairs.

(b) Factorise your expression in **(a)**.

14 (a) Draw a rectangle of length l and width w. Extend the length by a certain amount and the width by a different amount. Use your diagram to write the area of the rectangle in two different ways.

(b) Draw a square of side s. Extend one pair of opposite sides by a certain amount and the pair of opposite sides by a different amount. Use your diagram to write the area of the new rectangle in two different ways.

(c) Draw a square of side s and extend the sides by the same amount to form a new square. Use your diagram to find the area of the new square in two different ways.

2.9 Expanding and factorising special products

Stage 5.3§

Need to Know

There are two special products formed from the expansion of $(a + b)(c + d)$.

Expanding perfect squares

The square of a sum or a difference of two terms is called a **perfect square**. Perfect squares are expanded in exactly the same way as other binomial products.

Square of a sum

$$\begin{aligned}(a + b)^2 &= (a + b)(a + b)\\ &= a^2 + ab + ab + b^2\\ &= a^2 + 2ab + b^2\end{aligned}$$

or

Square of a difference

$$\begin{aligned}(a - b)^2 &= (a - b)(a - b)\\ &= a^2 - ab - ab + b^2\\ &= a^2 - 2ab + b^2\end{aligned}$$

Perfect square rules:

$(a + b)^2 = a^2 + 2ab + b^2$

$(a - b)^2 = a^2 - 2ab + b^2$

Note that the only difference in these two rules is the sign of the middle term and that this sign is the same as the sign inside the brackets.

Expanding the difference of two squares

The **difference of two squares** is another useful rule found when the sum of two terms is multiplied by the difference of those terms.

$$\begin{aligned}(a + b)(a - b) &= a^2 - ab + ab - b^2\\ &= a^2 - b^2\end{aligned}$$

Difference of two squares rule:

$(a + b)(a - b) = a^2 - b^2$

As factorising is the opposite of expansion, the special product rules can be used in reverse to factorise special products.

Factorising perfect squares

$a^2 + 2ab + b^2 = (a + b)^2$ and $a^2 - 2ab + b^2 = (a - b)^2$

Factorising the difference of perfect squares

$a^2 - b^2 = (a + b)(a - b)$

Worked Example 25

WE 25

Expand each of the following expressions using a perfect square rule.

(a) $(p+4)^2$ (b) $(d-3)^2$ (c) $(2f+5)^2$ (d) $(3m-2n)^2$

Working

(a) $(p+4)^2 = p^2 + 2 \times p \times 4 + 16$
$= p^2 + 8p + 16$

(b) $(d-3)^2 = d^2 - 2 \times d \times 3 + 9$
$= d^2 - 6d + 9$

(c) $(2f+5)^2 = (2f)^2 + 2 \times 2f \times 5 + 25$
$= 4f^2 + 20f + 25$

(d) $(3m-2n)^2$
$= (3m)^2 - 2 \times 3m \times 2n + 2n \times 2n$
$= 9m^2 - 12mn + 4n^2$

Worked Example 26

WE 26

Use the difference of two squares rule to expand the following.

(a) $(x+3)(x-3)$ (b) $(5-k)(5+k)$

(c) $(3x+2)(3x-2)$ (d) $(4m-5n)(4m+5n)$

Working

(a) $(x+3)(x-3) = x^2 - 3^2$
$= x^2 - 9$

(b) $(5-k)(5+k) = 5^2 - k^2$
$= 25 - k^2$

(c) $(3x+2)(3x-2) = (3x)^2 - (2)^2$
$= 9x^2 - 4$

(d) $(4m-5n)(4m+5n) = (4m)^2 - (5n)^2$
$= 16m^2 - 25n^2$

Worked Example 27

WE 27

Factorise the following using a perfect square rule.

(a) $x^2 + 6x + 9$ (b) $x^2 - 10x + 25$ (c) $4x^2 - 12x + 9$

Working

(a) $x^2 + 6x + 9$
$= x^2 + 2 \times 3x + 3^2$
$= (x+3)^2$

(b) $x^2 - 10x + 25$
$= x^2 - 2 \times 5x + 5^2$
$= (x-5)^2$

(c) $4x^2 - 12x + 9$
$= (2x)^2 - 2 \times 2x \times 3 + 3^2$
$= (2x-3)^2$

2.9

Worked Example 28

WE 28

Factorise the following using the difference of two squares rule.

(a) $x^2 - 4$ (b) $x^2 - 25$ (c) $16x^2 - 9$

Working

(a) $x^2 - 4$
$= x^2 - 2^2$
$= (x + 2)(x - 2)$

(b) $x^2 - 25$
$= x^2 - 5^2$
$= (x + 5)(x - 5)$

(c) $16x^2 - 9$
$= (4x)^2 - 3^2$
$= (4x + 3)(4x - 3)$

Answers page 430

2.9 Expanding and factorising special products

Navigator

Stage 5.3§

Q1 Column 1, Q2 Column 1, Q3, Q4, Q5, Q6, Q7, Q8, Q9, Q11 Column 1, Q13 (a), Q14, Q16	Q1 Column 2, Q2 Column 2, Q3, Q4, Q5, Q6, Q7, Q8, Q9, Q10, Q11 Column 1 & 2, Q13, Q14, Q16	Q1 Column 3, Q2 Column 3, Q3, Q4, Q5, Q6, Q7, Q8, Q9, Q10, Q11, Q13, Q14, Q15, Q16

Fluency and understanding

WE 25

1 Expand each of the following expressions using a perfect square rule.

(a) $(d - 1)^2$ (b) $(m + 5)^2$ (c) $(k + 2)^2$
(d) $(p - 3)^2$ (e) $(x - 8)^2$ (f) $(y - 12)^2$
(g) $(m + n)^2$ (h) $(w - k)^2$ (i) $(6 - c)^2$
(j) $(y + 6)(y + 6)$ (k) $(x + 9)(x + 9)$ (l) $(k - 7)(k - 7)$
(m) $(4y + 3)^2$ (n) $(2w + 9)^2$ (o) $(3x + 5)^2$
(p) $(6k - 1)^2$ (q) $(7 - 2a)^2$ (r) $(3 - 8d)^2$
(s) $(1 - 4c)(1 - 4c)$ (t) $(6 + 7y)(6 + 7y)$ (u) $(4g - h)(4g - h)$
(v) $(3a + d)^2$ (w) $(2a - 3b)^2$ (x) $(5m + 8n)^2$

WE 26

2 Use the difference of two squares rule to expand the following.

(a) $(c - 5)(c + 5)$ (b) $(a + 4)(a - 4)$ (c) $(k + 8)(k - 8)$
(d) $(9 + x)(9 - x)$ (e) $(3 - y)(3 + y)$ (f) $(1 - k)(1 + k)$
(g) $(m - n)(m + n)$ (h) $(x - y)(x + y)$ (i) $(c + d)(c - d)$
(j) $(v + w)(v - w)$ (k) $(p + q)(p - q)$ (l) $(h - g)(h + g)$
(m) $(2m - 1)(2m + 1)$ (n) $(3x + 4)(3x - 4)$ (o) $(6b - 7)(6b + 7)$
(p) $(1 + 5y)(1 - 5y)$ (q) $(8 - 5a)(8 + 5a)$ (r) $(3 - 2d)(3 + 2d)$
(s) $(4a + 5c)(4a - 5c)$ (t) $(3m - 2n)(3m + 2n)$ (u) $(2x - 7y)(2x + 7y)$
(v) $(4 + ab)(4 - ab)$ (w) $(3 - cd)(3 + cd)$ (x) $(5 + jk)(5 - jk)$

2.9

3 Factorise the following using a perfect square rule. WE27

(a) $x^2 + 4x + 4$ (b) $y^2 - 16y + 64$ (c) $9b^2 - 30b + 25$
(d) $t^2 - 24t + 144$ (e) $x^2 + 12x + 36$ (f) $4p^2 + 36p + 81$
(g) $m^2 + 20m + 100$ (h) $9s^2 + 24s + 16$ (i) $25h^2 - 30h + 9$
(j) $4f^2 - 20f + 25$ (k) $c^2 - 4cd + 4d^2$ (l) $49n^2 - 28kn + 4k^2$

4 Factorise the following using a perfect square rule. WE28

(a) $c^2 - 25$ (b) $k^2 - 100$ (c) $36m^2 - 49$
(d) $h^2 - 36$ (e) $m^2 - 81$ (f) $9y^2 - 121$
(g) $n^2 - 16$ (h) $v^2 - 144$ (i) $16c^2 - 25d^2$
(j) $4p^2 - 25$ (k) $36h^2 - k^2$ (l) $25m^2 - 9n^2$

5 $(a + 5)^2$ is the same as:

A $a^2 + 5a + 10$ B $a^2 + 10a + 25$ C $a^2 + 10a + 10$ D $a^2 + 25$

6 $(2k - 7)^2$ is the same as:

A $2k^2 - 28k + 49$ B $4k^2 + 28k + 49$ C $4k^2 - 49$ D $4k^2 - 28k + 49$

7 When expanded, $(2 - 4x)(2 + 4x)$ is:

A $4 - 16x^2$ B $4 + 8x$ C $4 - 8x^2$ D $4 - 8x$

8 When expanded, $(ab + 3cd)(ab - 3cd)$ is:

A $ab - 9cd$ B $a^2b^2 + 9c^2d^2$ C $a^2b^2 - 9c^2d^2$ D $a^2b^2 + 6c^2d^2$

9 (a) Find the area, in terms of x, of each of these squares. Expand your expressions.

(i)

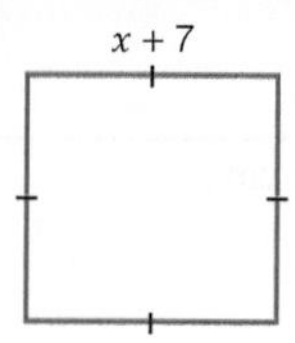

(ii)

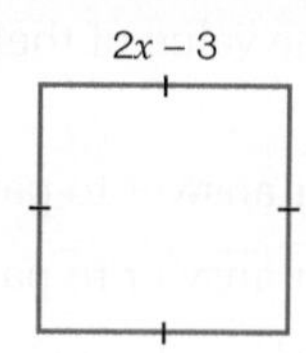

(b) Calculate the area of each when $x = 4$ cm.

10 A square playing field is to be changed to a rectangular shape by adding 2 m to the length and subtracting 2 m from the width.

(a) Write an expression for the area of the square playing field.
(b) Write an expression for the area of the new rectangular playing field.
(c) Which playing field has the larger area? How much more area is available?

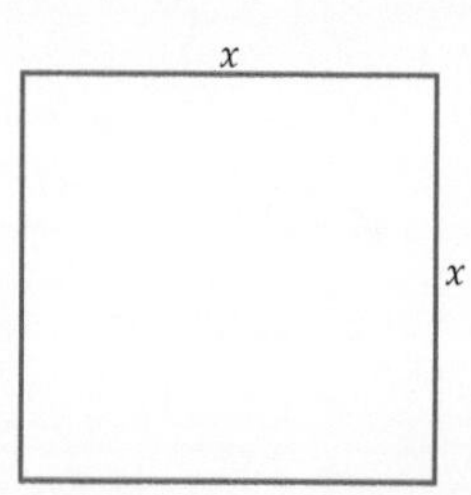

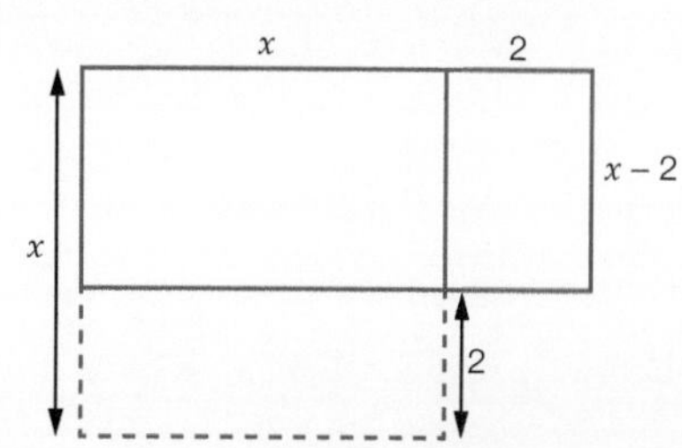

2.9

Reasoning, communicating and open-ended

11 Because 16 can be written as $(10 + 6)$ or $(20 - 4)$, 16^2 can be written as $(10 + 6)^2$ or $(20 - 4)^2$. Evaluate the following, without a calculator, using your knowledge of special products.

(a) 23^2 (b) 49^2 (c) 112^2
(d) 3.8^2 (e) 6.01^2 (f) 11.96^2
(g) 102×98 (h) 2.3×1.7 (i) 6.04×5.96

12 A steel washer has the dimensions shown.

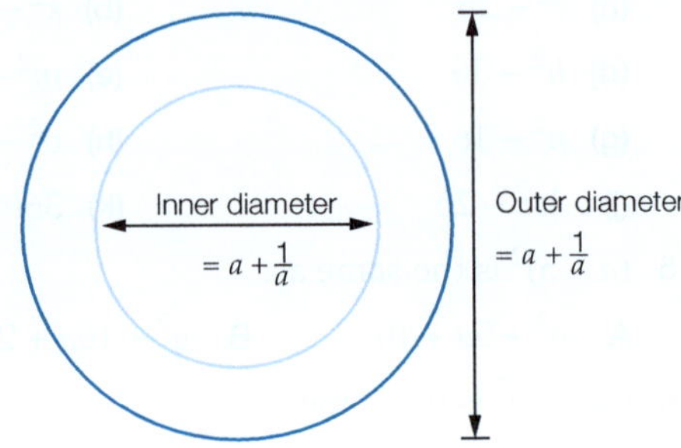

(a) Find the area of one side of the washer by using your knowledge of the expansion of special products.

(b) What can you say about the area of the washer for any value of a where $a > 0$?

13 (a) Find the shaded area using:

Area of a trapezium $= \frac{1}{2}(a + b)h$

(b) Find the shaded area using:

Area of a triangle $= \frac{1}{2}bh$

(c) Write an equation using part **(a)** and part **(b)**.

(d) If $L = 5l$, find the value of the shaded area:

(i) using your answer to part **(a)**

(ii) using your answer to part **(b)**.

(e) Show that the shaded area is 96% of the area of the large triangle when $L = 5l$.

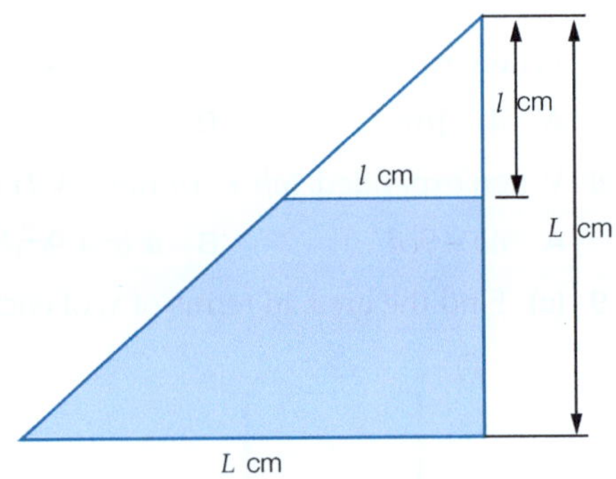

14 Find three values for a and b so that $a^2 - b^2$ produces a square number.

15 Add a middle term to $x^2 + 16$ so that it will factorise as a perfect square.

16 Explain why an expression that contains two terms where both are squares cannot be factorised unless the sign between the two squares is a subtraction sign.

2.9

Problem solving

Oranges and apples

The aim is to end up with the oranges and apples in opposite positions. You can make two types of moves:

- You can slide a piece of fruit one square into an empty space.
- You can hop a piece of fruit over a different type of fruit.

1 What is the minimum number of moves needed so that the oranges are on the left and the apples are on the right?

2 If there were four or five pieces of each fruit, how many moves would it take?

Strategy options

- Solve a simpler problem.
- Act it out.
- Look for a pattern.

Challenge 2

1 Write two numbers that have a sum of 27 and a product of 72.

2 Bridget once said, 'The day before yesterday I was 15 and next year I will be 18'. When is Bridget's birthday and on what day did she say this?

3 How many perfect cubes are there between 2000 and 10 000?

4 Part of a rectangle is shaded as shown in the diagram.

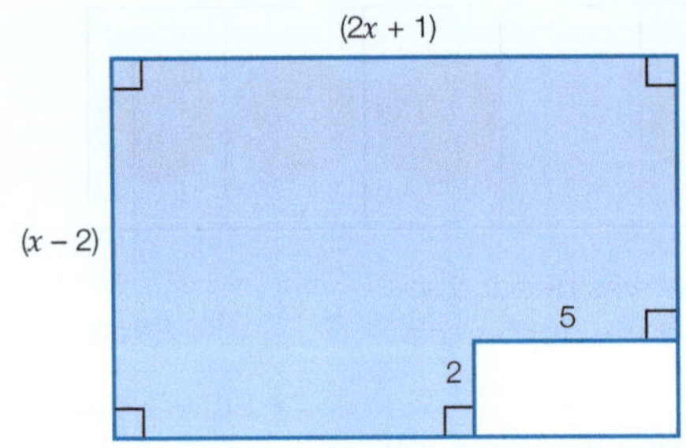

(*Note*: Diagram not to scale.)

Which expression gives the area of the shaded part?

A $2x^2 - x - 8$ **B** $2x^2 - x - 12$ **C** $2x^2 - 3x - 8$ **D** $2x^2 - 3x - 12$

5 If ♦ represents a digit, find the value of ♦ if $[3(170 + ♦)]^2 = 272\,48♦$.

6 a, b and c are positive whole numbers. Show that $(a - b)(b - c)(c - a)$ is divisible by 2.

7 Is the number $3^{404} + 4^{495}$ divisible by 5? Show why or why not.

8 Steven thinks of four different numbers. The sums Steven gets by pairing the numbers are 6, 7, 9, 9, 11 and 12. What are the four numbers?

9 A formula used in the science of optics is $\frac{1}{u} + \frac{1}{v} = \frac{1}{f}$.

(a) Make f the subject of the formula.

(b) Make v the subject of the formula.

10 (a) Show that the difference between any two-digit number and the number formed by reversing the digits is divisible by 9. (Hint: let the two-digit number be $10a + b$.)

(b) The difference between any three-digit number and the number formed by reversing its digits is obtained. By what number is it divisible?

(c) The difference between any four-digit number and the number formed by reversing its digits is obtained. By what number is it divisible?

Chapter review 2

D.I.Y. Summary

Key Words

Stage 5.1	expanded form	index form	scientific notation
	index	significant figures	variable
Stage 5.2	base	expand	highest common factor
	binomial expansion	expression	quadratic trinomial
	distributive law	factorise	reciprocal
Stage 5.3	difference of two squares	grouping in pairs	perfect square

Copy and complete the following using the words and phrases from the list where appropriate to write a summary for this chapter. A word or phrase may be used more than once.

Stage 5.1 Q1 **Stage 5.2 Q2, 3** **Stage 5.3 Q4**

1 If we write a number in __________, then it has a number between 1 and 10 multiplied by a positive or a negative power of 10.

2 If an expression is factorised to obtain the product of two brackets that are exactly the same, it is called a __________.

3 To __________ an expression is to remove brackets. The reverse operation is to __________ the expression and put it into brackets.

4 An expression of the form $x^2 - y^2$ is called the __________.

Stage 5.1 Q1, 2, 4, 5, 14, 15, 20 **Stage 5.2 Q3, 6–11, 17, 18, 19** **Stage 5.3 Q12, 13, 16, 21**

Fluency and understanding

1 Simplify each of the following. **Ex. 2.1**

(a) $a^3b^5 \times a^2b^2$ (b) $m^4n^7 \times m^3n$ (c) $\dfrac{p^4q^6}{p^3q^9}$

(d) $\dfrac{(x^5)^2 \times x^8}{(x^4)^4}$ (e) $\dfrac{(x^2)^3 \times (y^4)^2}{xy^3 \times x^5}$ (f) $\dfrac{c^2d^4}{c^4d^2} \times \dfrac{c^6d}{(cd)^2}$

2 $(2x^3y^2)^3$ is the same as: **Ex. 2.2**

A $2x^3y^6$ B $8x^3y^6$ C $2x^9y^6$ D $8x^9y^6$

3 Express each of the following with positive integers. **Ex. 2.3**

(a) $-5m^{-1}$ (b) a^2b^{-6} (c) $x^{-2}y^{-3}$ (d) $a^{-4}b^{-9}$

4 Write each of the following in scientific notation. **Ex. 2.4**

(a) 0.000 000 000 000 002 mm (b) 0.000 15 kg

(c) 320 000 000 L (d) 480 000 000 000 000 km

5 Write each of the following as a basic numeral. **Ex. 2.4**

(a) 5.4×10^7 (b) 8.273×10^{-4}

(c) 1.4×10^{-5} (d) 7.162×10^3

6 Rearrange each of the following to make x the subject. Ex 2.5

(a) $z = \frac{3 - xy}{2}$
(b) $x + 2a = 4x$
(c) $s = xt + 8t^2$
(d) $\frac{2 + 4z}{3x} = 10$

7 Expanded and simplified, $2c(3b - 4) - 3b(-5 + 2c)$ is: Ex 2.6

A $15b - 8c$
B $21b + 8c$
C $12bc - 12c$
D $7b + 14c$

8 Expand: Ex 2.6

(a) $4(x + 7)$
(b) $x(x - 3)$
(c) $-5x(2x - 1)$

9 (a) Expand and simplify: Ex 2.6

(i) $4(2x - 3) + 5(x + 1)$
(ii) $3x(6x + 5) - 2(3 - 2x)$

(b) Find the value of each expression when x has the value 4.

10 Expand and simplify: Ex 2.6

(a) $(x + 8)(x + 2)$
(b) $(2a + 5)(3a - 4)$

11 Factorise: Ex 2.7

(a) $8a + 12$
(b) $-15k - 20k^2$
(c) $5p^2q - 3pq + 2pq^2 - pqr$
(d) $-24a - 6ab - 72$

12 Expand and simplify: Ex 2.9

(a) $(x + 12)^2$
(b) $(5c - 2d)^2$
(c) $(4x - 3)(4x + 3)$

13 Factorise by grouping in pairs. Ex 2.8

(a) $ab + 4b + 3a + 12$
(b) $ab - 2b + a - 2$
(c) $ab - 6a - 5b + 30$
(d) $ab - 7 + 7b - a$

14 Express the following in scientific form.

(a) A 10-cent coin is about 0.0015 m thick.
(b) Pluto is approximately 5900 million km from the Sun.

15 Express each of the following as a basic numeral.

(a) The distance from the Sun to Jupiter is approximately 7.78×10^8 km.
(b) The temperature at the centre of the Sun is believed to be about 1.5×10^7°C.

16 The length of a greeting card is 20 cm and the width is $(5 + x)$ cm.

(a) Draw a diagram of the card.
(b) Write an expression for the perimeter.
(c) Factorise the expression obtained in part **(b)**.
(d) Write an expression for the area (in factorised form).
(e) Expand the expression obtained in part **(d)**.
(f) Check that the two expressions obtained for the area are equivalent by substituting a value for x.

17 If the area of a rectangular postage stamp is $4x^2 + 12x$, what is the width of the stamp if the height is $2x$?

18 The mass of an object can be determined by finding the product of its density and its volume. If the density can be represented by $(m - 2)$ and the volume by $(k + 8)$, write an expression for the mass.

Suggest possible values that m and k could take to give a mass of 48.

19 Find the height of this triangular support structure if the base is $14y$ and the area enclosed by the beams is $35y^2 - 21y$.

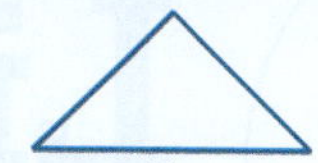

Reasoning and communicating

20 The heart pumps 7000 L of blood daily.

(a) Express this quantity in scientific notation.

(b) How many litres of blood are pumped by the heart:

- **(i)** each week
- **(ii)** each second, to one decimal place
- **(iii)** in 3 years (assuming no leap years)?

(c) Would this measurement be exact and the same for each person?

21 The Greenwood family had a rectangular block of land 2 km by 5 km for grazing sheep, as seen in the diagram. They then bought the land adjoining this leading down to the river. Depending on the time of year and the weather conditions, the width of the river can vary, so the land available for grazing sheep changes.

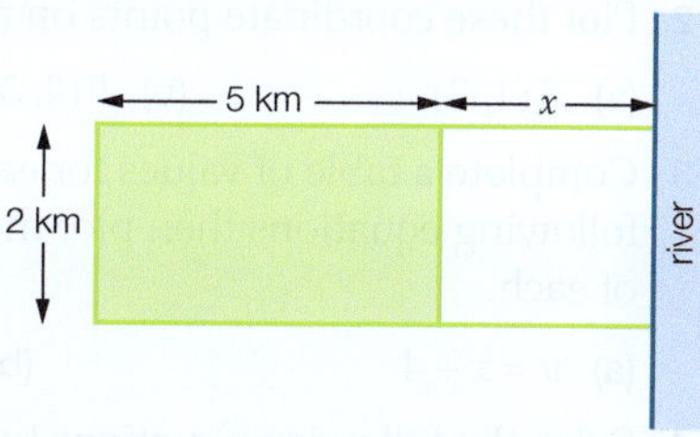

(a) Write an expression for the total area available for grazing sheep at any particular time. (Assume the land remains rectangular.)

(b) Suppose there are $\frac{1}{2}x$ sheep per km^2 of land. How many sheep are there altogether? Write your answer in expanded form.

(c) Fourteen sheep are moved to another location.

- **(i)** Write the expression for the number of sheep remaining.
- **(ii)** Expand $(x + 7)(x - 2)$. How is this related to the expression in **(i)**?
- **(iii)** Explain what happens when $x = 2$ km.

NAPLAN practice 2

Numeracy: Calculator allowed

1 Which of the following numbers is the largest?

A 3.2×10^{18} **B** 9.6×10^{3} **C** 3.2×10^{-18} **D** 9.6×10^{-3}

2 What is the value of $7(x - 9)$, when $x = 15$?

3 Factorise the expression $14x + 21$.

Numeracy: Non-calculator

4 Which one of the following expressions is equivalent to $3(5x - 2)$?

A $15x - 1$ **B** $15x - 6$ **C** $15x + 1$ **D** $15x + 6$

5 A rule for y in terms of x is $y = 12 - 3x$. Write an equivalent rule for x in terms of y.

6 Which one of the following expressions is a fully factorised form of $9xy + 12y$?

A $3(3xy + 4y)$ **B** $9y(x + 2)$ **C** $y(9x + 12)$ **D** $3y(3x + 4)$

7 When the expression $-3x^2y + 6xy$ is fully factorised, the result is:

A $3xy(2 - x)$ **B** $-3xy(2 - x)$ **C** $-3xy(x - y)$ **D** $-3xy(x + y)$

4 Linear relationships

Recall 4

Prepare for this chapter by attempting the following questions. If you have difficulty with a question, go to Pearson Places and download the Recall Worksheet.

1 **(a)** If $P = 2k + 5$, find P when $k = -9$. **(b)** If $C = 9 - 4d$, find C when $d = 4$.

2 Plot these coordinate points on a Cartesian plane and label them.

(a) $A(1, 2)$ **(b)** $B(0, 3)$ **(c)** $C(-2, 0)$ **(d)** $D(-4, 2)$

3 Complete a table of values for each of the following equations then plot the graph of each.

x	-2	-1	0	1	2
y					

(a) $y = x + 4$ **(b)** $y = x - 2$ **(c)** $y = 2x$

4 Solve the following equations to find the value of the variable.

(a) $x + 1 = 7$ **(b)** $3 + y = 18$ **(c)** $a + 4 = -6$ **(d)** $b - 5 = 16$

5 Solve the following equations to find the value of the variable.

(a) $2g = 38$ **(b)** $0.5x = 21$ **(c)** $-9y = -54$ **(d)** $\frac{x}{7} = 5$

6 Expand the brackets and simplify.

(a) $-3(5x + 2)$ **(b)** $2(7 - 3x) + 6$ **(c)** $3x + 1 + 6(2x + 7)$

7 Find the rule connecting x and y in the given table of values.

x	0	1	2	3	4	5
y	6	4	2	0	-2	-4

8 Use Pythagoras' Theorem to find the unknown sides in the following triangles.

(a)

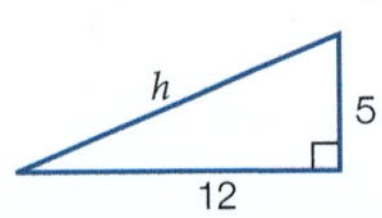

(b)

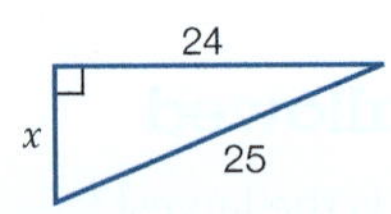

Syllabus references

Number and Algebra: Linear Relationships

- ✓ Find the midpoint and gradient of a line segment (interval) on the Cartesian plane using a range of strategies, including graphing software **(ACMNA294) (Stage 5.1)**
- ✓ Find the distance between two points located on the Cartesian plane using a range of strategies, including graphing software **(ACMNA214) (Stage 5.1)**
- ✓ Solve problems involving parallel lines **(ACMNA238) (Stage 5.1)**
- ✓ Sketch linear graphs using the coordinates of two points **(ACMNA215) (Stage 5.1)**
- ✓ Interpret and graph linear relationships using the gradient-intercept form of the equation of a straight line **(Stage 5.2)**
- ✓ Solve problems involving parallel and perpendicular lines **(ACMNA238) (Stage 5.2)**
- ✓ Find the midpoint and gradient of a line segment (interval) on the Cartesian plane **(ACMNA294) (Stage 5.3§)**
- ✓ Find the distance between two points located on the Cartesian plane **(ACMNA214) (Stage 5.3§)**
- ✓ Sketch linear graphs using the coordinates of two points **(ACMNA215) (Stage 5.3§)**
- ✓ Solve problems using various standard forms of the equation of a straight line **(Stage 5.3§)**

Number and Algebra: Equations

- ✓ Solve linear equations **(ACMNA215) (Stage 5.2)**
- ✓ Solve linear equations involving simple algebraic fractions **(ACMNA240) (Stage 5.2)**
- ✓ Solve problems involving linear equations, including those derived from formulas **(ACMNA235) (Stage 5.2)**
- ✓ Substitute values into formulas to determine an unknown **(ACMNA234) (Stage 5.2)**

Introducing coordinate geometry

4.1

Stage 5.1

Need to Know

- The horizontal axis is known as the x-axis and the vertical axis is known as the y-axis.
- The point of intersection of these axes is called the **origin**, which has the **coordinates** (0, 0).
- All points can be described by an ordered pair known as coordinates (x, y) as measured from the origin.
- An **interval** is the line segment formed when two points are joined.

The distance between two points

Pythagoras' Theorem can be used to find the distance between two points on a number plane.

For example, to find the distance between (1, 3) and (5, 6), plot the points on a Cartesian plane, join the points with a straight line and then draw a **right-angled triangle** by moving 4 units right and 3 units up from the point (1, 3) to reach (5, 6). Pythagoras' Theorem can be used to find the distance between (1, 3) and (5, 6).

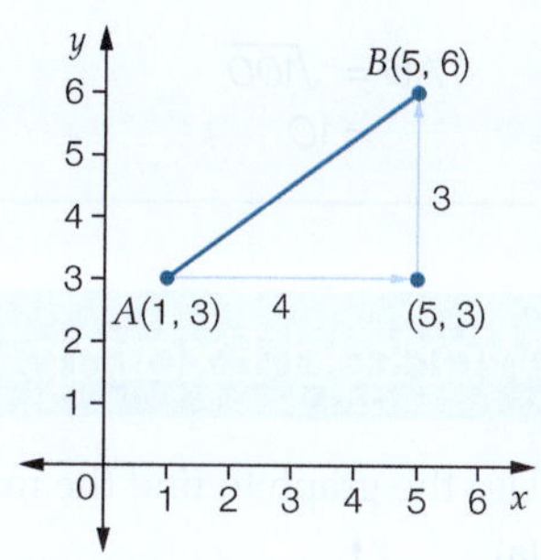

$$\text{Distance} = \sqrt{4^2 + 3^2}$$
$$= 5$$

The midpoint of an interval

The **midpoint** of an interval is halfway between the two points that form the interval. The average of the x-coordinates of the two points gives the x-coordinate of the midpoint and the average of the y-coordinates of the two points gives the y-coordinate of the midpoint.

For example, for points (2, 3) and (4, 7), plot the points on a Cartesian plane, join the points with a straight line then draw dotted lines parallel to the axes from the points as shown on the diagram. The coordinates of the midpoint lie halfway between the coordinates of the two points.

The x-coordinate of the midpoint is 3, the average of the x-coordinates 2 and 4. The y-coordinate of the midpoint is 5, the average of the y-coordinates 3 and 7. So the coordinates of the midpoint of the interval joining (2, 3) and (4, 7) are (3, 5).

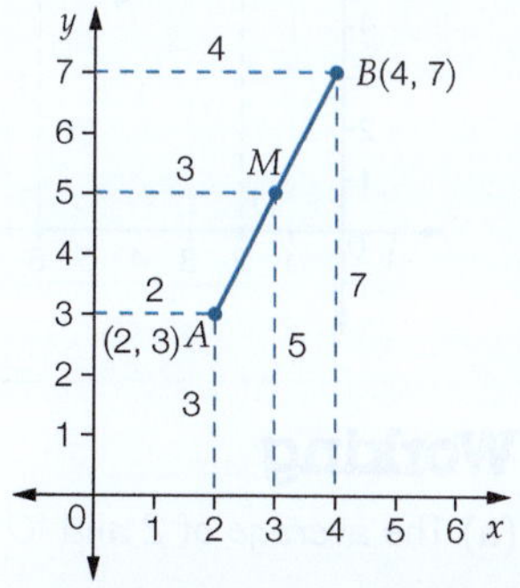

$$x\text{-coordinate of midpoint} = \frac{2+4}{2}$$
$$= 3$$

$$y\text{-coordinate of midpoint} = \frac{3+7}{2}$$
$$= 5$$

4.1

Worked Example 1 WE1

Use Pythagoras' Theorem to find the distance between points A and B for each of the following.

(a)

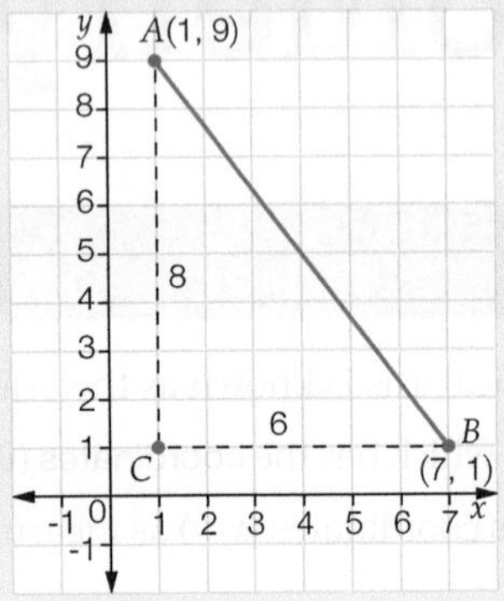

(b)

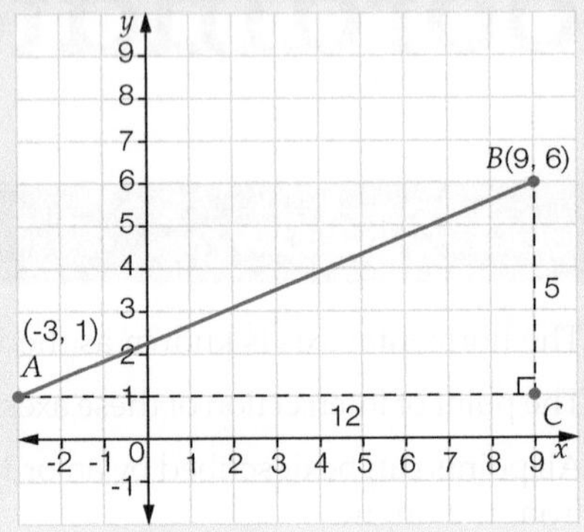

Working

(a) $AB^2 = 6^2 + 8^2$
$= 36 + 64$
$= 100$

$AB = \sqrt{100}$
$= 10$

(b) $AB^2 = 12^2 + 5^2$
$= 144 + 25$
$= 169$

$AB = \sqrt{169}$
$= 13$

Worked Example 2 WE2

Use the graph to find the midpoint of the interval AB for each of the following.

(a)

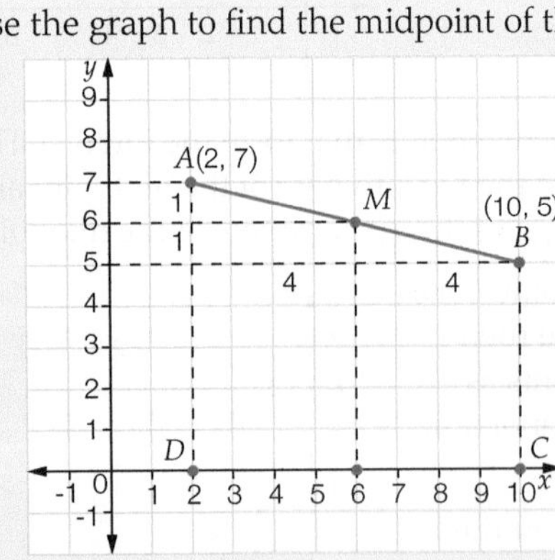

(b)

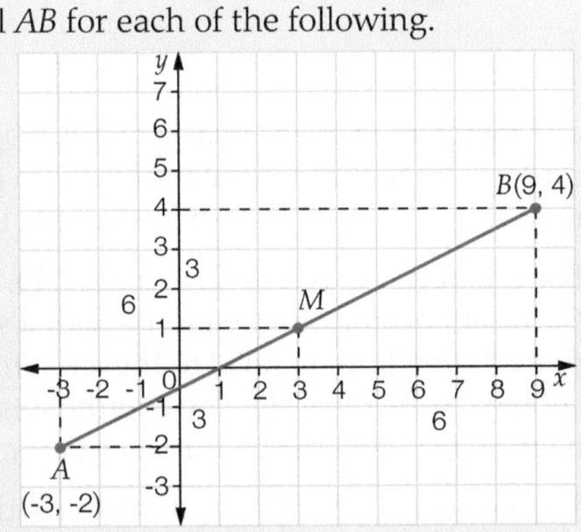

Working

(a) The average of 2 and 10 $= \dfrac{2+10}{2}$
$= 6$

The average of 7 and 5 $= \dfrac{7+5}{2}$
$= 6$
$M = (6, 6)$

(b) The average of -3 and 9 $= \dfrac{-3+9}{2}$
$= 3$

The average of -2 and 4 $= \dfrac{-2+4}{2}$
$= 1$
$M = (3, 1)$

4.1

Technology Exploration: GeoGebra

Finding the distance between two points and the midpoint of the line joining them

(a) Open GeoGebra and use the Interval between Two Points tool to draw an interval between the points $A(2, 1)$ and $B(6, 4)$. Use the Distance or Length tool. The length of the interval will appear on the screen. Select the Move tool to move either of the points around. The distance will change accordingly.

(b) Place a point C anywhere on the interval joining points A and B. Use the Distance or Length tool to measure the distances AC and BC. Use the Move tool to move the point along the line until the two distance measurements are equal. The coordinates of the point will be shown in the Algebra view. This will be the midpoint between the two points. Use the Move tool to move A and/or B around. The Point C will remain at the midpoint and its new coordinates will be shown in the Algebra view.

4.1 Introducing coordinate geometry

Answers page 437

Navigator

Q1, Q2, Q3, Q4, Q5, Q8, Q9	Q1, Q2, Q3, Q4, Q5, Q6, Q8, Q9	Q1, Q2, Q3, Q4, Q6, Q7, Q8, Q9

Stage 5.1

Equipment needed: Graph or grid paper

Fluency and understanding

1 Use Pythagoras' Theorem to find the distance between points A and B for each of the following.

WE1

(a)

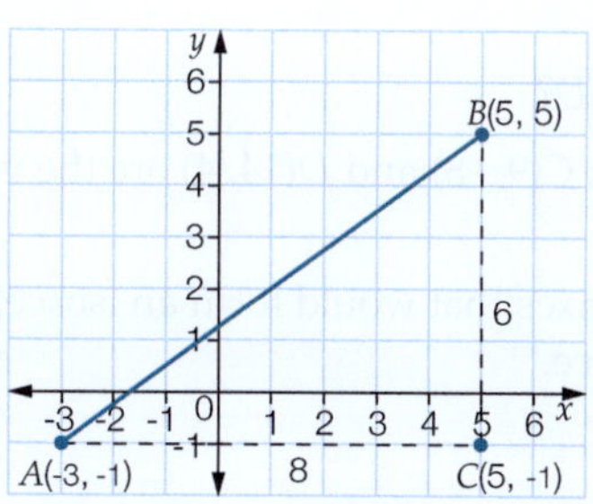

(b)

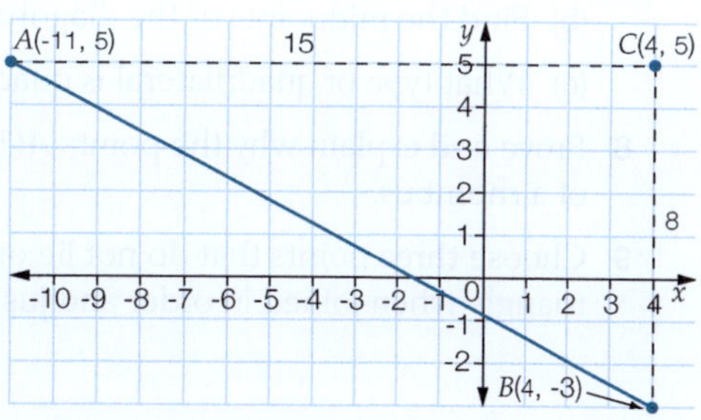

(c)

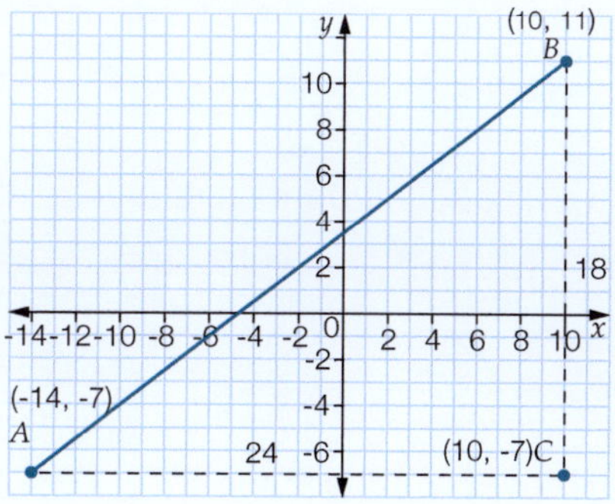

(d)

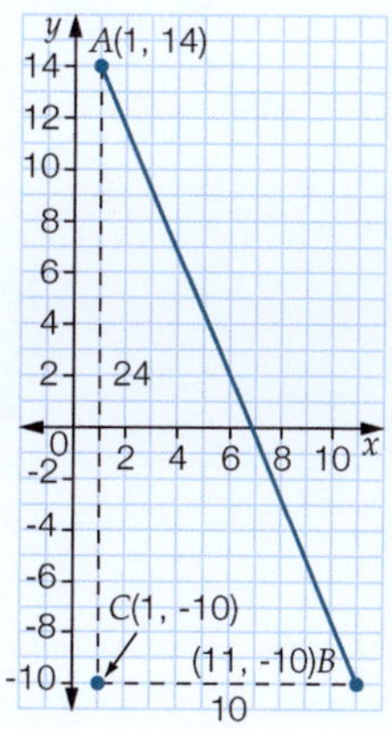

4.1

WE2

2 Use the graph to find the midpoint of the interval AB for each of the following.

(a)

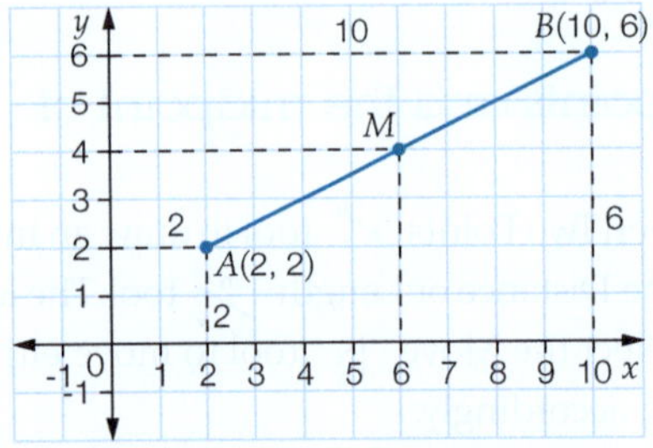

(b)

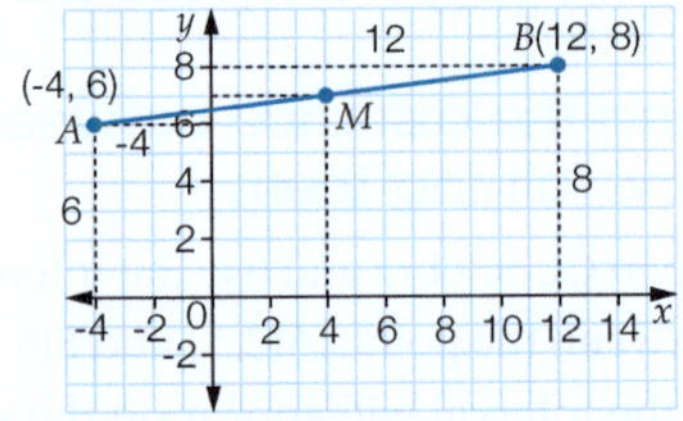

(c)

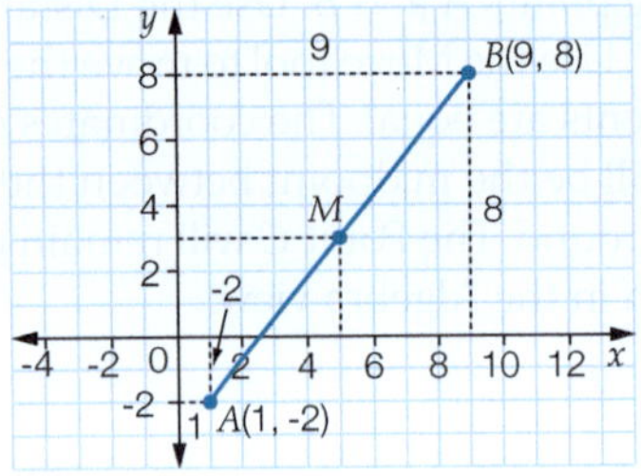

(d)

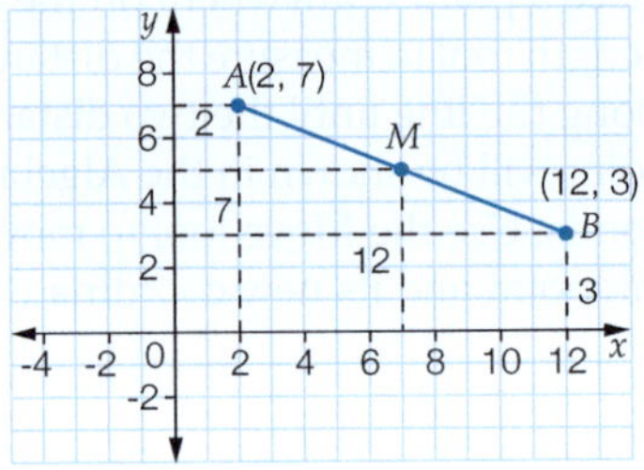

3 The distance between (1, -5) and (6, 7) is:

A 4 **B** 17 **C** 13 **D** 7

4 A line AB is drawn from Point A at (3, 14) to Point B (7, 10). Point C lies on the line AB halfway between points A and B. Find the coordinates of Point C.

5 Find the coordinates of the midpoint of the line joining (-4, 6) and (2, -2).

6 Use a diagram and Pythagoras' Theorem to show that the points $A(1, 2)$, $B(4, 6)$ and $C(8, 9)$ form an isosceles triangle.

Reasoning, communicating and open-ended

7 A quadrilateral has vertices $A(0, 2)$, $B(3, 6)$, $C(15, 1)$ and $D(12, -3)$.

(a) Find the length of each side.

(b) Find the midpoints of the diagonals.

(c) What type of quadrilateral is quadrilateral $ABCD$?

8 Prove and explain why the points $A(1, 4)$, $B(-4, -8)$, $C(9, -8)$ and $D(14, 4)$ are the vertices of a rhombus.

9 Choose three points that do not lie on the x- or y-axes that would form an isosceles triangle when joined in order and justify your choice.

Using coordinate geometry formulas

4.2

Stage 5.3 §

Need to Know

The distance between two points

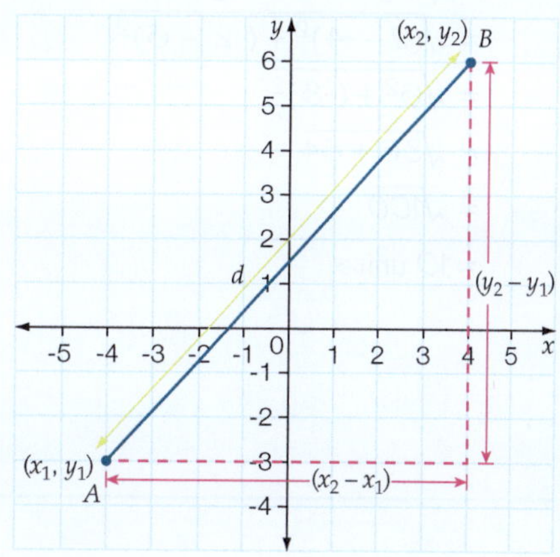

- The difference between the x-coordinates $(x_2 - x_1)$ gives the length of one side of a right-angled triangle.
- The difference between the y-coordinates $(y_2 - y_1)$ gives the length of another side.
- These two lines form two of the sides of a right-angled triangle, with the distance between the two points, d, the length of the hypotenuse.
- We use Pythagoras' Theorem to find the distance between the two points.

$$d^2 = (x_2 - x_1)^2 + (y_2 - y_1)^2$$

The distance, d, between any two Cartesian coordinates (x_1, y_1) and (x_2, y_2) is:

$$d = \sqrt{(x_2 - x_1)^2 + (y_2 - y_1)^2}$$

Exact and approximate answers

The exact distance between two points is often a surd, an irrational number.

For example, $\sqrt{35}$.

Unless an approximate answer is required, answers should always be left in exact form.

The midpoint of an interval

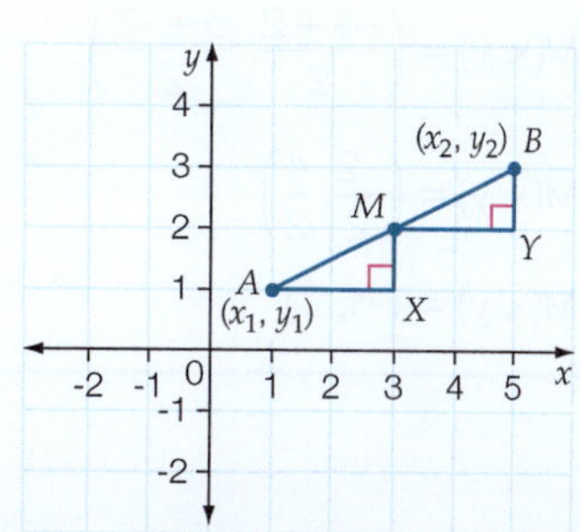

The midpoint of an interval is halfway between the two points that form the interval. The average of the x-coordinates of the two points gives the x-coordinate of the midpoint and the average of the y-coordinates of the two points gives the y-coordinate of the midpoint.

For points (x_1, y_1) and (x_2, y_2), the average of x_1 and x_2 is $\frac{x_1 + x_2}{2}$ and the average of y_1 and y_2 is $\frac{y_1 + y_2}{2}$.

The midpoint $M(x, y)$ between two points (x_1, y_1) and (x_2, y_2) is given by:

$$M(x, y) = \left(\frac{x_1 + x_2}{2}, \frac{y_1 + y_2}{2}\right)$$

4.2

Worked Example 3

WE3

Find the distance between the following Cartesian coordinates, giving your answers in exact simplest form.

(a) (-4, 6) and (2, -2) **(b)** (-2, 1) and (2, -4) **(c)** (3, 5) and (5, 2)

Working

(a) Let $(x_1, y_1) = (-4, 6)$ and $(x_2, y_2) = (2, -2)$

$$d = \sqrt{(x_2 - x_1)^2 + (y_2 - y_1)^2}$$
$$= \sqrt{(2 - -4)^2 + (-2 - 6)^2}$$
$$= \sqrt{6^2 + (-8)^2}$$
$$= \sqrt{36 + 64}$$
$$= \sqrt{100}$$
$$= 10 \text{ units}$$

(b) Let $(x_1, y_1) = (-2, 1)$ and $(x_2, y_2) = (2, -4)$

$$d = \sqrt{(x_2 - x_1)^2 + (y_2 - y_1)^2}$$
$$= \sqrt{(2 - -2)^2 + (-4 - 1)^2}$$
$$= \sqrt{4^2 + (-5)^2}$$
$$= \sqrt{41} \text{ units}$$

(c) Let $(x_1, y_1) = (3, 5)$ and $(x_2, y_2) = (5, 2)$

$$d = \sqrt{(x_2 - x_1)^2 + (y_2 - y_1)^2}$$
$$= \sqrt{(5 - 3)^2 + (2 - 5)^2}$$
$$= \sqrt{(2)^2 + (-3)^2}$$
$$= \sqrt{4 + 9}$$
$$= \sqrt{13} \text{ units}$$

Worked Example 4

WE4

Find the coordinates of the midpoint between (-4, 6) and (2, -2).

Working

Let $(x_1, y_1) = (-4, 6)$ and $(x_2, y_2) = (2, -2)$

$$M(x, y) = \left(\frac{x_1 + x_2}{2}, \frac{y_1 + y_2}{2}\right)$$

$$M(x, y) = \left(\frac{-4 + 2}{2}, \frac{6 + -2}{2}\right)$$

$$M(x, y) = \left(\frac{-2}{2}, \frac{4}{2}\right)$$

$$M(x, y) = (-1, 2)$$

4.2

4.2 Using coordinate geometry formulas

Answers page 437

Stage 5.3 §

Navigator

Q1 Column 1, Q2 Column 1, Q3, Q4, Q5, Q6, Q7, Q9, Q11, Q14	Q1 Column 2, Q2 Column 2, Q3, Q4, Q5, Q6, Q7, Q8, Q9, Q10, Q14	Q1 Column 3, Q2 Column 3, Q3, Q4, Q5, Q6, Q8, Q9, Q10, Q11, Q12, Q13

Fluency and understanding

WE3

1 Find the distance between the following Cartesian coordinates, giving your answers in exact simplest form.

(a) (2, 5) and (5, 9) **(b)** (-4, 2) and (8, -3) **(c)** (-4, 13) and (4, -2)

(d) (-6, 34) and (4, 10) **(e)** (-4, 5) and (4, -1) **(f)** (-11, -4) and (13, 3)

(g) (2, -5) and (3, -1) **(h)** (-1, 3) and (5, 2) **(i)** (4, -3) and (-2, 2)

(j) (-3, -5) and (-1, -7) **(k)** (3, -4) and (-1, 2) **(l)** (2, -7) and (6, 1)

WE4

2 Find the coordinates of the midpoints between the following Cartesian points.

(a) (-3, 4) and (0, 0) **(b)** (-3, 5) and (1, 3) **(c)** (-9, -11) and (7, -3)

(d) (-1, 1) and (2, 3) **(e)** (-5, 8) and (3, 7) **(f)** (-2, -4) and (5, 3)

(g) (1, 2) and (6, 7) **(h)** (-2, -5) and (7, 0) **(i)** (0, 0) and (9, 11)

3 The distance between (-4, -5) and (4, 1) is:

A 4 **B** 5 **C** 10 **D** 9

4 A line *AB* is drawn from Point *A* at (-2, 10) to Point *B* (4, 1). Point *C* lies on the line *AB* half-way between Points *A* and *B*. Find the coordinates of Point *C*.

5 **(a)** The midpoint of *CD* is (6, 5). If the coordinates of Point *C* are (3, 8), find the coordinates of Point *D*.

(b) The midpoint of *EF* is (-1, -2). If the coordinates of Point *E* are (-7, 6), find the coordinates of Point *F*.

(c) The midpoint of *AB* is (2, -4). If the coordinates of Point *A* are (-5, -8), what are the coordinates of Point *B*?

6 Use the distance formula and Pythagoras' Theorem to show that the points *A*(-3, -2), *B*(1, 1) and *C*(-2, 5) form an isosceles right-angled triangle.

7 A triangle has the vertices (0, 5), (3, 3) and (1, 0).

(a) Find the length of each side, correct to two decimal places.

(b) What type of triangle is it?

4.2

Reasoning, communicating and open-ended

Hint

Plot the known points on a Cartesian plane.

8 The proposed course for a yacht race has been mapped out with buoys positioned at $A(-3, -4)$, $B(-1, 3)$ and $C(9, 8)$, but the position of the last buoy $D(x, y)$ is yet to be finalised. The race will begin at position $O(0, 0)$, proceed through buoys A, B, C and D and finish back at the starting point O.

(a) If the coordinates indicate distances measured in kilometres, what is the distance from the starting position O to the first buoy A?

(b) The judge's boat must be placed in the middle of buoys A and C. What are the coordinates of the judge's boat?

(c) If the judge's boat is also in the middle of buoys B and D, as with A and C, what are the coordinates (x, y) of buoy D?

(d) What is the greatest distance that the yachts must travel between any two successive buoys?

(e) What geometric shape is formed by the buoys $ABCD$?

9 A quadrilateral has the vertices $A(0, -2)$, $B(-2, -6)$, $C(-6, -8)$ and $D(-4, -4)$.

(a) Find the midpoints of the diagonals.

(b) Find the length of each side.

(c) Find the length of each diagonal.

(d) What type of quadrilateral is it?

10 Prove and explain how the points $A(3, 5)$, $B(5, 1)$, $C(-1, -2)$ and $D(-3, 2)$ are the vertices of a rectangle.

11 If $E(1, 6)$, $F(-4, 3)$ and $G(-1, -2)$ are the vertices of a square, find the missing vertex.

12 Points $A(-6, 3)$, $B(2, 5)$, $C(0, -1)$ and $D(-2, 1)$ are joined to form a quadrilateral. The midpoints of AB, BC, CD, and DA are joined to form another quadrilateral $PQRS$. Show that $PQRS$ is a parallelogram.

13 Choose three points that would form an equilateral triangle when joined in order and justify your choice.

14 Choose four points that would form a kite when joined in order and justify your choice.

Solving linear equations

4.3

Stage 5.2

Need to Know

Linear relationships

A **linear relationship** between two variables exists if a uniform change in one variable, the **independent variable**, causes a uniform change in the other, the **dependent variable**.

A linear relationship produces a **linear graph** (or **straight-line graph**) and is represented by the equation $y = mx + b$.

Linear equations

A **linear equation** is a linear relationship where the dependent variable has a specific value.

We **solve** linear equations to find the value of the independent variable. A linear equation has only one solution. Linear graphs can be used to solve linear equations.

Solving linear equations

- Identify the operations used to construct the equation and use **inverse operations** to deconstruct the equation in the opposite order to the way it was constructed.
- If there is more than one term containing the unknown, collect those terms together on one side of the equation. Collect all numerical values on the other side.
- If fractions are involved, express each term with a common denominator and then eliminate the fractions by multiplying each term on both sides by the common denominator.
- If brackets are involved, expand the brackets.

Worked Example 5

WE5

Solve the following linear equations. Check your solutions by substitution.

(a) $3x - 5 = -11$ **(b)** $7 - 2x = 19$

Working

(a)

$$3x - 5 = -11$$
$$3x - 5 + 5 = -11 + 5$$
$$3x = -6$$
$$\frac{3x}{3} = \frac{-6}{3}$$
$$x = -2$$

Check: LHS $= 3 \times -2 - 5$
$= -6 - 5$
$= -11$
$=$ RHS

(b)

$$7 - 2x = 19$$
$$-2x + 7 = 19$$
$$-2x + 7 - 7 = 19 - 7$$
$$-2x = 12$$
$$\frac{-2x}{-2} = \frac{12}{-2}$$
$$x = -6$$

Check: LHS $= 7 - 2 \times -6$
$= 7 + 12$
$= 19$
$=$ RHS

4.3

Worked Example 6

WE6

Solve the following linear equations. Check your solutions by substitution.

(a) $\frac{2x}{5} - 1 = 7$ (b) $\frac{2x+4}{3} = 4$

Working

(a)

$$\frac{2x}{5} - 1 = 7$$
$$\frac{2x}{5} - 1 + 1 = 7 + 1$$
$$\frac{2x}{5} = 8$$
$$\frac{2x}{5} \times 5 = 8 \times 5$$
$$2x = 40$$
$$\frac{2x}{2} = \frac{40}{2}$$
$$x = 20$$

Check: $\text{LHS} = \frac{2 \times 20}{5} - 1$
$= \frac{40}{5} - 1$
$= 8 - 1$
$= 7$
$= \text{RHS}$

(b)

$$\frac{2x+4}{3} = 4$$
$$\frac{3 \times (2x+4)}{3} = 4 \times 3$$
$$2x + 4 = 12$$
$$2x + 4 - 4 = 12 - 4$$
$$2x = 8$$
$$\frac{2x}{2} = \frac{8}{2}$$
$$x = 4$$

Check: $\text{LHS} = \frac{2 \times 4 + 4}{3}$
$= \frac{8+4}{3}$
$= \frac{12}{3}$
$= 4$
$= \text{RHS}$

Worked Example 7

WE7

Solve the linear equation $5x - 2 = 3x + 8$. Check your solution by substitution.

Working

$$5x - 2 = 3x + 8$$
$$5x - 3x - 2 = 3x - 3x + 8$$
$$2x - 2 = 8$$
$$2x - 2 + 2 = 8 + 2$$
$$2x = 10$$
$$\frac{2x}{2} = \frac{10}{2}$$
$$x = 5$$

Check: $\text{LHS} = 5 \times 5 - 2$
$= 25 - 2$
$= 23$

$\text{RHS} = 3 \times 5 + 8$
$= 15 + 8$
$= 23$

$\text{LHS} = \text{RHS}$

4.3

Worked Example 8

WE8

Solve the linear equation $3(x + 4) = 2(1 - x)$. Check your solution by substitution.

Working

$$3(x+4) = 2(1-x)$$
$$3x + 12 = 2 - 2x$$
$$3x + 2x + 12 = 2 - 2x + 2x$$
$$5x + 12 = 2$$
$$5x + 12 - 12 = 2 - 12$$
$$5x = -10$$
$$\frac{5x}{5} = \frac{-10}{5}$$
$$x = -2$$

Check:
$$\text{LHS} = 3(-2 + 4) = 6$$
$$\text{RHS} = 2(1 - -2) = 6$$
$$\text{LHS} = \text{RHS}$$

Worked Example 9

WE9

Solve the linear equation $\frac{2a-1}{5} = \frac{a+7}{4}$. Check your solution by substitution.

Working

$$\frac{2a-1}{5} = \frac{a+7}{4}$$
$$\frac{4(2a-1)}{20} = \frac{5(a+7)}{20}$$
$$\frac{4(2a-1)}{20} \times 20 = \frac{5(a+7)}{20} \times 20$$
$$4(2a-1) = 5(a+7)$$
$$8a - 4 = 5a + 35$$
$$8a - 5a - 4 = 5a - 5a + 35$$
$$3a - 4 = 35$$
$$3a - 4 + 4 = 35 + 4$$
$$3a = 39$$
$$\frac{3a}{3} = \frac{39}{3}$$
$$a = 13$$

Check:
$$\text{LHS} = \frac{2 \times 13 - 1}{5} = \frac{26-1}{5} = \frac{25}{5} = 5$$
$$\text{RHS} = \frac{13+7}{4} = \frac{20}{4} = 5$$
$$\text{LHS} = \text{RHS}$$

4.3

Answers page 438

Stage 5.2

4.3 Solving linear equations

Navigator

Q1 Column 1, Q2 Column 1, Q3 Column 1, Q4 Column 1, Q5 Column 1, Q6 Column 1, Q7, Q8, Q9, Q10, Q11, Q12 Column 1, Q13 Column 1, Q14, Q17 (a), Q18, Q20	Q1 Column 2, Q2 Column 2, Q3 Column 2, Q4 Column 2, Q5 Column 2, Q6 Column 2, Q7, Q8, Q9, Q10, Q11, Q12 Column 2, Q13 Column 2, Q14, Q15, Q17 (a) & (b), Q18, Q20, Q21	Q1 Column 3, Q2 Column 3, Q3 Column 3, Q4 Column 3, Q5 Column 2, Q6 Column 3, Q7, Q8, Q9, Q10, Q12 Column 3, Q13 Column 3, Q14, Q15, Q16, Q17, Q18, Q19, Q20, Q21

Fluency and understanding

WE5

1 Solve the following linear equations. Check your solutions by substitution.

(a) $2h + 9 = 5$ (b) $5b + 3 = 28$ (c) $6x + 7 = 1$

(d) $6r - 3 = 2$ (e) $3f - 2 = -14$ (f) $5k - 3 = -1$

(g) $12 = 7x + 6$ (h) $-5 = 2x + 13$ (i) $-2 = 9y - 2$

(j) $-3b + 14 = -4$ (k) $-2a - 4 = 0$ (l) $-5b - 6 = 9$

(m) $-7 - p = 8$ (n) $15 - 4e = -1$ (o) $9 - 4g = -23$

(p) $4 = 17 - 5y$ (q) $-3 = 8 - 4p$ (r) $-5 = 11 - 8y$

WE6

2 Solve these linear equations. Check your solutions by substitution.

(a) $\frac{2x}{3} + 9 = 13$ (b) $\frac{3x}{7} + 5 = 8$ (c) $\frac{5x}{3} - 8 = 7$

(d) $4 - \frac{2x}{3} = 10$ (e) $8 - \frac{3x}{4} = -1$ (f) $12 - \frac{3x}{2} = 3$

(g) $\frac{2x + 3}{11} = 1$ (h) $\frac{3x + 1}{5} = -1$ (i) $\frac{7x + 4}{6} = 3$

(j) $\frac{1 - 7x}{4} = -5$ (k) $\frac{2 - 3x}{5} = 1$ (l) $\frac{9 - 4x}{3} = -5$

(m) $\frac{4x - 6}{7} = -2$ (n) $\frac{7x - 4}{5} = -12$ (o) $\frac{3x - 1}{8} = -2$

WE7

3 Solve the following linear equations. Check your solutions by substitution.

(a) $7c + 2 = 6c - 5$ (b) $4b + 1 = 2b - 15$ (c) $9f + 5 = 5f - 3$

(d) $5d - 4 = 2d + 17$ (e) $5e - 3 = 2e + 9$ (f) $3a - 5 = a + 7$

(g) $3g + 12 = 5 + 2g$ (h) $2h - 7 = 13 + h$ (i) $4x - 3 = 6 + x$

(j) $k + 6 = 9 - k$ (k) $10j + 1 = 25 - 2j$ (l) $3y - 22 = -4 - 3y$

(m) $2m + 5 = 4m - 3$ (n) $5 + 4n = 7n - 1$ (o) $7x - 4 = 9x - 6$

(p) $2 - y = 3 - 2y$ (q) $19 - p = 16 - 2p$ (r) $6 + 3x = 21 - 2x$

(s) $15a - 17 = 16a - 19$ (t) $14 - 3x = 5 - 2x$ (u) $30 - p = 5p + 6$

WE8

4 Solve the following linear equations. Check your solutions by substitution.

(a) $3(x - 2) = 2(x + 1)$ (b) $7(x + 4) = 3(x - 4)$ (c) $4(x - 5) = 5(3x + 7)$

(d) $5(2x + 3) = 9(x + 1)$ (e) $5(2x + 3) = 2(3x + 1)$ (f) $4(3x - 2) = 3(2x + 9)$

(g) $2(4x + 7) = 3(5 - x)$ (h) $3(5x + 4) = 4(2 - 3x)$ (i) $4(3x - 11) = 5(29 - 3x)$

4.3

5 Solve the following linear equations. Check your solutions by substitution. WE9

(a) $\frac{b+4}{3} = \frac{b-8}{5}$ **(b)** $\frac{a-3}{2} = \frac{a+6}{3}$

(c) $\frac{c+7}{4} = \frac{c+2}{2}$ **(d)** $\frac{d+5}{3} = \frac{d-1}{9}$

(e) $\frac{2x-3}{4} = \frac{x+6}{3}$ **(f)** $\frac{3y-2}{7} = \frac{y+6}{4}$

(g) $\frac{5g-1}{4} = \frac{7-2g}{5}$ **(h)** $\frac{7k-3}{4} = \frac{13-5k}{8}$

(i) $\frac{4-3m}{2} = \frac{1-5m}{3}$ **(j)** $\frac{2-3j}{5} = \frac{3-4j}{6}$

(k) $\frac{8x+3}{2} = 3x-4$ **(l)** $\frac{6p+4}{11} = p+9$

(m) $\frac{2(3x-5)}{3} = \frac{3(5x+1)}{4}$ **(n)** $\frac{4(2y+3)}{5} = \frac{5(2-y)}{2}$

6 Solve these equations. Check your solutions by substitution.

(a) $\frac{x}{2} - 3 = 0$ **(b)** $\frac{x}{8} - 6 = -5$ **(c)** $\frac{x}{9} + 3 = 3$

(d) $3 - \frac{x}{4} = 12$ **(e)** $5 - \frac{x}{7} = 11$ **(f)** $-4 - \frac{x}{3} = -8$

(g) $\frac{4x}{5} = 3$ **(h)** $\frac{7x}{13} = 0$ **(i)** $\frac{2x}{7} = -4$

(j) $\frac{x+3}{2} = 9$ **(k)** $\frac{x+7}{3} = 5$ **(l)** $\frac{x+5}{9} = 3$

(m) $\frac{x-6}{2} = 8$ **(n)** $\frac{x-9}{8} = 6$ **(o)** $\frac{x-4}{3} = 17$

7 **(a)** To obtain a from $2a + 7$ you would need to:

A add 7 then divide by 2 **B** subtract 7 then divide by 2
C divide by 2 then subtract 7 **D** multiply by 2 then add 7

(b) To obtain p from $5p - 4$ you would need to:

A add 4 then multiply by 5 **B** add 4 then divide by 5
C divide by 5 then add 4 **D** multiply by 5 then add 4

(c) To obtain x from $-6x - 11$ you would need to:

A add 11 then multiply by -6 **B** multiply by -6 then add 11
C add 11 then divide by -6 **D** subtract 11 then divide by -6

(d) To obtain n from $1 - 4n$ you would need to:

A add 4 then multiply by 1 **B** add 1 then divide by -4
C subtract 1 then divide by -4 **D** subtract 1 then multiply by -4

4.3

8 **(a)** The equation $2z + 1 = 11$ has the solution:

A $z = 4$ **B** $z = 5$ **C** $z = 6$ **D** $z = -5$

(b) The equation $4x - 5 = 3$ has the solution:

A $x = 3$ **B** $x = -1$ **C** $x = 2$ **D** $x = 0$

(c) The equation $7 - 3q = 22$ has the solution:

A $q = 4$ **B** $q = -3$ **C** $q = 5$ **D** $q = -5$

(d) The equation $5 - 8n = -11$ has the solution:

A $n = 2$ **B** $n = -3$ **C** $n = 0$ **D** $n = -2$

(e) The equation $\frac{r}{3} - 4 = 2$ has the solution:

A $r = 10$ **B** $r = -6$ **C** $r = 9$ **D** $r = 18$

(f) The equation $\frac{a+3}{4} = 6$ has the solution:

A $a = 5$ **B** $a = 7$ **C** $a = 21$ **D** $a = 27$

9 **(a)** The linear equation $\frac{2a}{5} - 3 = 1$ has the solution:

A $a = -10$ **B** $a = -5$ **C** $a = 4$ **D** $a = 10$

(b) The linear equation $\frac{4y}{3} + 7 = -1$ has the solution:

A $y = -9$ **B** $y = -6$ **C** $y = 6$ **D** $y = 9$

10 **(a)** $9x + 4 = 2x - 3$ has the solution:

A $x = 1$ **B** $x = -1$ **C** $x = 2$ **D** $x = 4$

(b) $2y + 7 = -8 - 3y$ has the solution:

A $y = 2$ **B** $y = 1$ **C** $y = -2$ **D** $y = -3$

(c) $4(3k - 2) = 5(k + 4)$ has the solution:

A $k = 3$ **B** $k = -2$ **C** $k = 1$ **D** $k = 4$

(d) $\frac{p-2}{2} = \frac{p+1}{3}$ has the solution:

A $p = 8$ **B** $p = 5$ **C** $p = -4$ **D** $p = 0$

11 Write an equation for each of the following and then solve the equation.

(a) 7 is added to x to give a result of 19.

(b) 5 is subtracted from y to give a result of 3.

(c) Three times a certain number is 14.

(d) One-quarter of a certain number is -5.

(e) Three more than two times a number is 13.

(f) 7 less than four times a number is -12.

12 Solve these equations. (First simplify each equation.)

(a) $3x + 2x + 7 = 32$ **(b)** $2x + x - 4 = -10$ **(c)** $6x + 3x - 4 = 14$

(d) $3(x + 5) + 8 = -4$ **(e)** $2(x + 3) + 1 = -5$ **(f)** $5(x - 2) - 3 = 7$

(g) $3(7 - x) - 6 = 0$ **(h)** $8(2 - x) + 1 = 9$ **(i)** $-4(3 - x) + 8 = -12$

(j) $-2(1 - 4x) - 3x = -2$ **(k)** $-7(5 - 2x) - 6x = 13$ **(l)** $-(8 - 5x) + 4x = -26$

4.3

13 Solve these linear equations.

(a) $\frac{4(2x+3)}{7} = 1$ **(b)** $\frac{3(4x+1)}{5} = -4$ **(c)** $\frac{5(3x+2)}{2} = 1$

(d) $\frac{2(x-1)}{3} = 5$ **(e)** $\frac{4(3x-2)}{5} = -1$ **(f)** $\frac{7(2x+1)}{2} = -2$

(g) $\frac{6(1-4x)}{15} = \frac{2}{3}$ **(h)** $\frac{3(2-5x)}{8} = -\frac{1}{2}$ **(i)** $\frac{3(7-2x)}{16} = \frac{1}{4}$

14 Subtracting 11 from four times a certain number gives the same result as subtracting two times the number from 7. Write an equation and solve it to find the unknown number.

15 A certain number when added to 36 gives the same result as adding 6 to this number and doubling the result. Write an equation and solve it to find the unknown number.

16 Half of the result of three times a certain number added to one is the same as one-fifth of the result when 4 is subtracted from this certain number. Write an equation and solve it to find the unknown number.

Reasoning, communicating and open-ended

17 Consecutive numbers, such as 6, 7 and 8, can be represented algebraically by n, $n + 1$ and $n + 2$, because to get from one number to the next consecutive number we simply add 1. Similarly, three consecutive odd numbers can be represented algebraically by n, $n + 2$ and $n + 4$ where n is an odd number. Using this information, write each of the following as an equation and solve to find the unknown numbers.

(a) Find the three consecutive odd numbers that sum to 195.

(b) Find the four consecutive even numbers that sum to 204.

(c) Find the six consecutive multiples of seven that sum to 609.

> **Hint**
> *n* does not have to be the first number. Three consecutive numbers could be $n - 1$, n and $n + 1$.

18 The ancient Egyptians solved equations using a method of guessing and checking. We call this method the 'rule of false position'.

For example, consider the equation $x + \frac{x}{9} = 80$, which the ancient Egyptians would have said as 'a number added to one-ninth of itself is 80'.

First, choose a possible value for x that is divisible by 9; for example, 9. Substitute this into the equation.

$$9 + \frac{9}{9} = 9 + 1 = 10$$

Because the answer we want is eight times the answer we found, try a number eight times the value of the number we tried.

$$9 \times 8 = 72$$

$$72 + \frac{72}{9} = 72 + 8 = 80$$

So, the solution to the equation 'a number added to one-ninth of itself is 80' is 72.

For the following questions, form an equation, then use the rule of false position to solve it.

(a) A number added to one-sixth of itself is 28.

(b) A number added to one-quarter of itself is 45.

(c) A number added to one-eighth of itself is 99.

19 On a photocopied test some numbers are not clear. One equation reads $2x + 7 =$ but the number on the right-hand side is unreadable. If we know that x is a whole negative number, give three examples of what the number on the right-hand side could be.

20 Use at least three mathematical operations to make up three different equations that have a solution of 4.

21 Make up three equations similar to those in Question **3**, where the solution is a positive whole number.

Problem solving

Burning down

Two different-sized candles are lit. Each candle burns at a different rate and one is 3 cm longer (or taller) than the other.

The longer candle was lit at 3.30 p.m. and the shorter one at 5 p.m.

At 7.30 p.m. they were both the same length.

The longer candle burned out at 9.30 p.m., whereas the shorter candle burned out at 9 p.m.

How long was each candle originally?

Strategy options

- Make a model.
- Draw a diagram.
- Guess and check.
- Break the problem into manageable parts.

Solving problems using linear equations

4.4

Stage 5.2

Need to Know

Problem-solving strategies

- Determine the unknown(s) you need to find and represent each with a variable.
- Form an equation, using the key words in the statement of the problem.
- Solve the equation using inverse operations.
- Answer the question in words.
- Check that your answers make sense and satisfy all conditions of the problem.

Worked Example 10

WE10

Tim is twice Sally's age. How old are Tim and Sally now, if in 10 years' time the sum of their ages will be 89? Use s to represent Sally's age.

Working

Sally's present age $= s$

Tim's present age $= 2s$

In 10 years' time:
Sally's age $= s + 10$
Tim's age $= 2s + 10$

$$3s + 20 = 89$$
$$3s = 69$$
$$s = 23$$

$$2s = 2 \times 23$$
$$= 46$$

Sally is 23 years old and Tim is 46 years old.

Checking the solution

In 10 years' time, they will be 33 and 56 years old respectively, $33 + 56 = 89$. Our answers solve the problem.

Answers page 438

4.4 Solving problems using linear equations

Stage 5.2

Navigator

Q1, Q2, Q3, Q4, Q5, Q6, Q7, Q8, Q9, Q10, Q14, Q15	Q1, Q2, Q3, Q4, Q6, Q7, Q8, Q9, Q10, Q11, Q12, Q14, Q15	Q1, Q2, Q3, Q4, Q6, Q7, Q8, Q9, Q10, Q11, Q12, Q13, Q14, Q15

Fluency and understanding

WE10

1 Kate's father is three times her age. How old are Kate and her father now if, in 4 years time, the sum of their ages is 68? Use K to represent Kate's age.

2 Andrew has x coins in his pocket. He tells Lori that she has five fewer coins than he has. If they have 17 coins between them, how many coins does each have?

3 Phil has $\$p$ in his piggy bank and Bill has $\$b$ in his. If Phil has three times the amount of money Bill has in his piggy bank and they save \$5 each per week for the next 10 weeks, the total amount of money in both piggy banks will be \$200. Which of the following equations represents this situation?

A $4p + 100 = 200$ **B** $4b + 100 = 200$
C $3b + 100 = 200$ **D** $3b + 1p + 100 = 150$

4 Choose the correct answer.

If $a = b + 2cx$, then:

A $x = \dfrac{a+b}{2c}$ **B** $x = \dfrac{a-b}{2c}$ **C** $x = \dfrac{a}{2c} - b$ **D** $x = a - \dfrac{2c}{b}$

5 If the perimeter of this rectangular block of land is 46 m, find the length and width of the block.

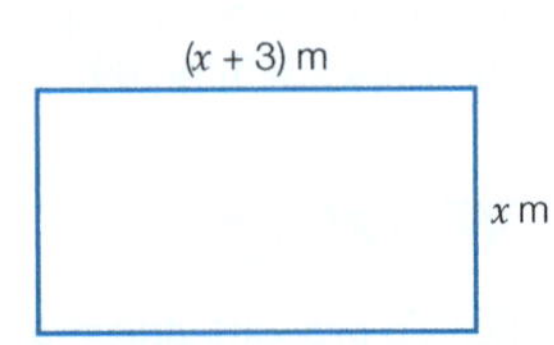

6 Three students take different routes home from school. Mark needs to walk twice as far as Annette, whereas Sue needs to walk four times as far as Annette. If the three cover a total distance of 2.1 km (or 2100 m), how far does each need to travel to get home from school?

7 Ben bought five ice-cream cones and received \$1.75 change from his \$10 note. How much did each ice-cream cone cost?

8 Peter is 7 years older than Sharon and John is twice as old as Peter. Find how old Peter is if the mean (arithmetic average) of their ages is 19.

9 Joseph has 24 cousins. He knows that he has six more female cousins than male cousins. How many female cousins does he have?

10 In 3 years' time Joanne will be twice as old as her daughter. If Joanne is 45 years old now, how old is her daughter?

4.4

Reasoning, communicating and open-ended

11 A triangle has angles of $(2x + 11)°$, $(3x - 2)°$ and $(4x)°$. Find each angle.

12 An isosceles triangle has a perimeter of 45 cm. Its equal sides each measure 6 cm longer than the base. How long is the base of the triangle?

13 Gabby is twice as old as Kate and Kate is 3 years older than Joe. Brett is four times as old as Joe. If the sum of the ages of Kate and Brett is 63, find how old each person is.

> **Hint**
> Let one of the ages be represented by a variable and express the other ages in terms of this variable.

14 A stock market analyst was investigating the trends of two independent stocks and was surprised to see that both stocks were at the same unit value after a 12-month period. He knew that both stocks had started at the same value but had undergone different trends over the past year. One stock had risen to three times its original unit value then lost \$8 a unit before finally finishing at $\frac{3}{4}$ of this reduced unit value. The other stock added \$3 a unit to its original unit value before crashing to $\frac{1}{5}$ of its improved unit value. It then showed a late improvement by increasing the unit value to seven times the reduced unit value. Express this information as an equation and solve it to find the original unit value of the stocks.

15 The sum of three consecutive whole numbers is a number between 70 and 100. Give three possible starting numbers for the group of three consecutive whole numbers.

Problem solving

Fractions in fractions in fractions

The fraction $\frac{2008}{1998}$ may be written in the form $a + \cfrac{1}{b + \cfrac{1}{c + \cfrac{1}{d}}}$ where a, b, c and d are all positive integers.

Can you find the value of d?

> **Hint**
> - Write the fraction $\frac{2008}{1998}$ as a mixed number.
>   This should help you find the value of a.
> - $\frac{x}{y} = \cfrac{1}{\frac{y}{x}}$ may be useful.

> **Strategy options**
> - Break the problem into manageable parts.
> - Solve a simpler problem.
> - Work backwards.

4.5 Plotting linear graphs

Stage 5.1

Need to Know

Linear relationships can be represented both graphically (using graphs) and algebraically (using equations).

Each axis on a graph should have a clearly marked scale and a label.

A linear relationship is a relationship between two variables that, when graphed, produces a straight line.

Plotting graphs accurately

Use graph paper. Each axis should have a clearly marked scale and a label. The scale may or may not be the same for each axis. On an axis, the scale divisions should be equally spaced with intervals representing equal amounts. An example is shown opposite.

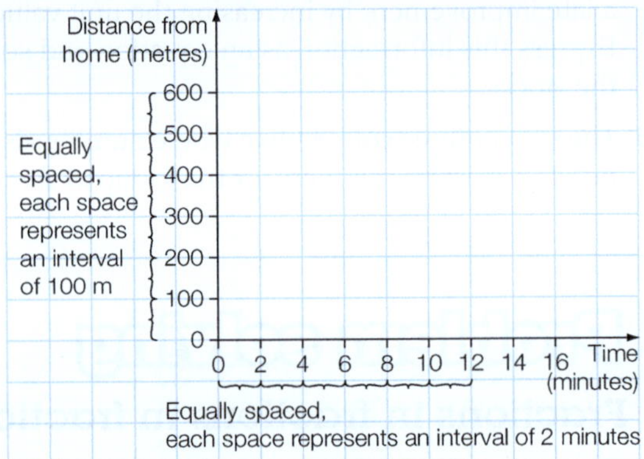

Worked Example 11

WE11

For each of the following table of values:

(i) plot a graph to represent the information

(ii) write an equation to represent the linear relationship.

(a)

f	0	1	2	3
g	0	5	10	15

(b)

v	0	1	2	3
w	7	5	3	1

Working

(a) (i)

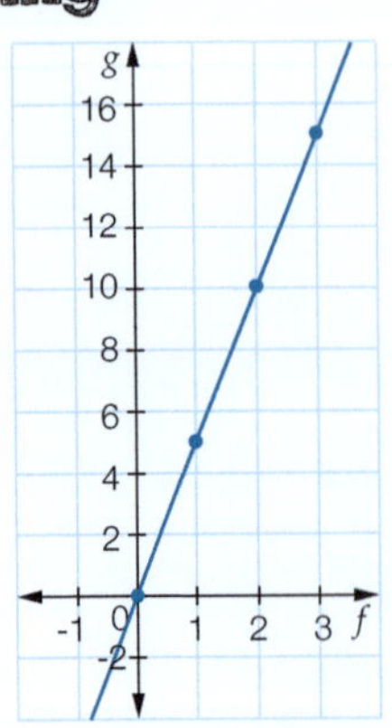

(ii) $g = 5f$

(b) (i)

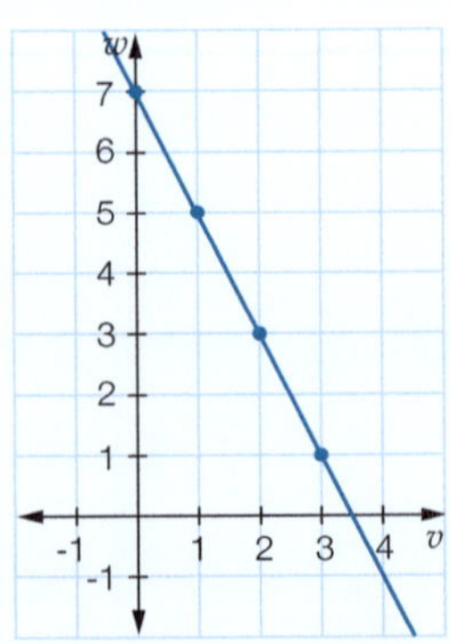

(ii) $-2v$ is in our equation

7 needs to be added.

$w = -2v + 7$

4.5

Worked Example 12

WE12

Use the equation given for each of the following to:

(i) copy and complete the table of values

(ii) plot the graph of each of the relationships.

(a) $M = 9n$

n	-3	0	3	6
M				

(b) $L = 5k - 8$

k	-4	0	4	8
L				

Working

(a) (i)

n	-3	0	3	6
M	-27	0	27	54

(ii)

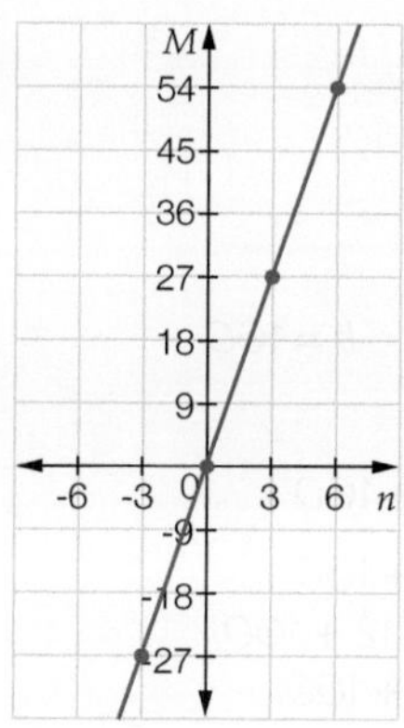

(b) (i)

k	-4	0	4	8
L	-28	-8	12	32

(ii)

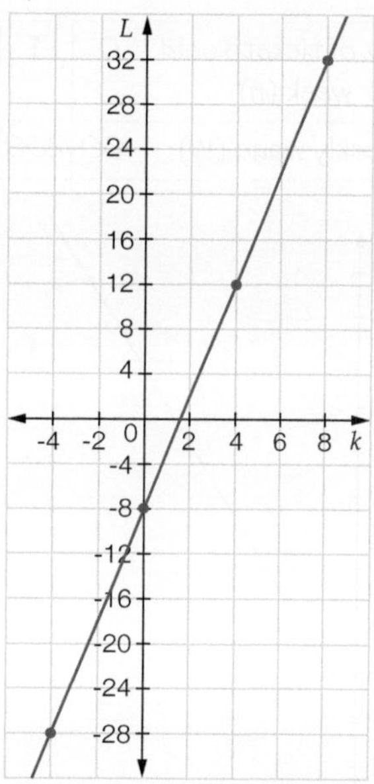

4.5

Worked Example 13

WE13

Lisa, a travel agent working for an airline, receives a base wage of $160 each week and a bonus of $40 for each flight ticket sold.

No. of tickets sold in a week (n)	0	1	2	3	4	5
Weekly wage (W)						

(a) Copy and complete the table above.

(b) Plot a graph to show the information contained in the table.

(c) Write an equation to represent the information.

(d) If Lisa sold 12 tickets one week, how much did she earn that week?

(e) Show, by substitution, that the point (8, 480) lies on the line.

Working

(a)

No. of tickets sold per week (n)	0	1	2	3	4	5
Weekly wage (W)	160	200	240	280	320	360

(b)

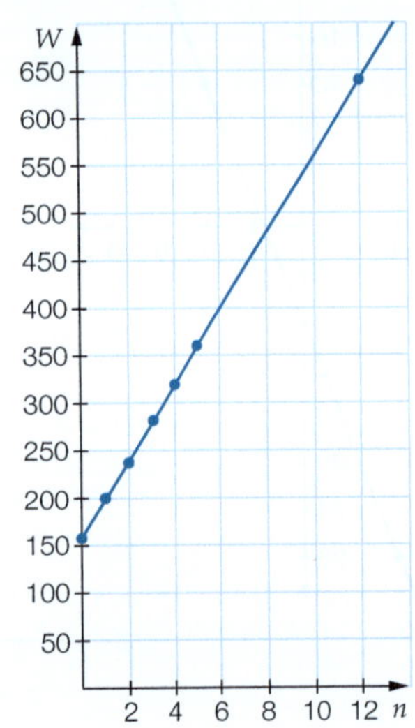

(c) $m = \frac{\text{rise}}{\text{run}}$ $\quad b = 160$

$= 40$

$W = 40n + 160$

(d) $W = 40n + 160$

$W = 40 \times 12 + 160$

$= 480 + 160$

$= 640$

Lisa would earn $640.

(e) $W = 40n + 160$

$= 40 \times 8 + 160$

$= 320 + 160$

$= 480$

(8, 480) lies on the line.

4.5 Plotting linear graphs

Answers page 438

Stage 5.1

Navigator

Q1, Q2 Column 1, Q3, Q4, Q5, Q7, Q9, Q10	Q1, Q2 Column 2, Q3, Q4, Q5, Q6, Q7, Q8, Q9, Q10	Q1, Q2 Column 2, Q3, Q4, Q5, Q6, Q7, Q8, Q9, Q10

Equipment required: graph or grid paper

Fluency and understanding

1 For each of the following tables of values: WE11

(i) plot a graph to represent the information

(ii) write an equation to represent the linear relationship.

(a)

a	0	1	2	3
b	0	2	4	6

(b)

c	0	3	6	9
d	5	2	-1	-4

2 Use the equation given for each of the following to: WE12

(i) copy and complete the table of values

(ii) plot the graph of each of the relationships.

(a) $A = 4L$

L	0	10	20	30
A				

(b) $Q = 5M$

M	0	2	4	6
Q				

(c) $J = 2G - 12$

G	0	4	8	12
J				

(d) $d = -6e + 5$

e	0	3	6	9
d				

(e) $B = \frac{3R}{5}$

R	-20	0	20	40
B				

(f) $C = \frac{5F}{6} - 4$

F	-6	0	6	12
C				

3 Eddy, a television salesperson, receives a base wage of \$80 each week and a bonus of \$100 for each television he sells. WE13

Number of televisions sold	0	1	2	3	4	5
Weekly wage (dollars)						

(a) Copy and complete the table above.

(b) Plot a graph to show the information contained in the table. Use the horizontal axis for the number of televisions sold and the vertical axis for weekly wage (dollars).

(c) Write an equation to represent the information.

(d) If Eddy sold eight televisions in one week, how much did he earn that week?

(e) Show by substitution, that the point (7, 780) lies on the graph.

4.5

4 (a) In the linear relationship $V = 3W - 14$, when $W = 7$ the value of V will be:

A -14 **B** -8 **C** -2 **D** 7

(b) In the linear relationship $P = 5R$, P has a value of -30 when R is equal to:

A -10 **B** -6 **C** 6 **D** 10

(c) In the linear relationship $C = \frac{5}{9}F - 32$, the value of C when F is equal to 63 is:

A -32 **B** -3 **C** 3 **D** 18

5 A no-name brand detergent can be bought wholesale for 95c a litre. To supply information to customers quickly, the wholesaler has prepared a graph representing this information.

(a) Copy and complete this table.

Number of litres	0	1	2	3	4	5	6	7
Cost (dollars)								

(b) Plot the cost of the detergent against the number of litres of detergent bought. This is the wholesaler's graph. Use the horizontal axis for the number of litres and the vertical axis for the cost (dollars).

(c) Is the relationship linear? How can you tell?

(d) From your graph, how much would it cost to buy:

(i) 3 L of detergent **(ii)** 2.4 L of detergent **(iii)** 5.7 L of detergent?

(e) From your graph, how many litres of detergent (to the nearest tenth of a litre) could be bought for:

(i) \$3.80 **(ii)** \$2.50 **(iii)** \$5.20?

(f) Write a relationship connecting the cost in dollars, C, to the number of litres of detergent, d, bought.

6 Steve is cycling at a constant speed of 185 metres per minute.

(a) Copy and complete the following table.

Time (minutes)	0	1	2	3	4	5	6	7
Distance cycled (metres)								

(b) Plot the distance against the time. Use the horizontal axis for time (minutes) and the vertical axis for distance cycled (metres).

(c) From the graph, how far does Steve cycle in 2.7 minutes?

(d) How long does it take Steve to cycle 700 metres?

(e) Extend your graph to find how long it takes Steve to cycle 2 km.

(f) Write the rule (or equation) connecting the distance cycled to the time taken.

(g) If Steve continued cycling, what distance would he have covered in 40 minutes?

Reasoning, communicating and open-ended

7 Phuong works at the stock exchange buying and selling shares for different international and local companies on behalf of her client. She needs to be able to convert quickly from one currency to another, and uses either a calculator or a linear graph to obtain her information.

(a) If \$20 in Australian currency converts to \$21 in American (US) currency, draw up a graph showing the relationship between the two currencies. Plot Australian currency on the horizontal axis and American currency on the vertical axis.

Hint

What is \$A1.00 equivalent to in \$US?

Use the graph to convert from one currency to the other for each of these amounts (round off to the nearest 10c).

(b) A$7.00 to US$

(c) A$17.60 to US$

(d) US$5.50 to A$

(e) US$9.60 to A$

8 Wally, Sarah and Cleo are collecting money for an upcoming charity walk. Wally is asking for $1 for each kilometre he walks. Sarah, on the other hand, is asking for a $4 donation plus $0.50 per kilometre walked. Finally, Cleo is asking for a donation of $8 regardless of how many kilometres she walks.

(a) Draw up and complete a table of values to show the amount donated for each kilometre walked for Wally, Sarah and Cleo for distances from 0 to 10 kilometres.

(b) Use your table of values to plot graphs showing the amount donated for each kilometre walked for Wally, Sarah and Cleo. You should plot each graph on the same set of axes. Plot distance on the horizontal axis and the amount donated on the vertical axis.

(c) From the graph and table of values, identify the distance at which all three collect the same amount of money.

(d) Identify the distances for which:

(i) Wally raises the most money

(ii) Sarah raises the most money

(iii) Cleo raises the most money.

9 Frank and Bruno are two emerging football players about to sign contracts with FC Mathematica. In the first year, Frank earns $120 000, which will increase by $60 000 for each successive year that he stays with the club. Bruno earns $80 000 in his first year, which increases by $80 000 for each successive year that he stays with the club. Explain in which years Frank will earn more than Bruno and in which years Bruno will earn more than Frank, assuming each stays with the club.

10 Draw three different linear relationships that show the same amount of distance travelled over different periods of time. Explain what happens to the graph as the time increases.

Problem solving

How many edges?

Equipment required: 1–2 brains, scissors, ruler and several pieces of A4 paper

Instructions: Divide an A4 rectangular sheet of paper into two pieces using a single straight cut. Then divide one of those pieces into two, also using a single straight cut. What is the maximum number of edges that the resulting three pieces can have in total? What would the minimum number of edges of the three pieces be?

Strategy options

- Act it out.
- Look for a pattern.
- Test all possible combinations.

Half-time 4

Stage 5.1 Q1, 2, 3 **Stage 5.2 Q4–7** **Stage 5.3 Q8**

Ex.4.5

1 Decide which, if any, of the following tables represent linear relationships.

(a)

x	-1	0	1	2
y	-5	-3	-1	1

(b)

x	-2	0	2	4
y	5	2	-1	-3

(c)

x	0	1	2	3
y	0	1	1	2

(d)

x	-4	0	4	8
y	0	1	2	3

Ex.4.1

2 Find the distance between the points (-1, 7) and (4, -5) by plotting the points on a Cartesian plane and using Pythagoras' Theorem.

Ex.4.5

3 Use the equation $P = 3 - 4R$ to:

(a) copy and complete the following table of values

R	0	2	4	6
P				

(b) plot the graph of the relationship.

Ex.4.3

4 Solve the following linear equations.

(a) $2x - 5 = 13$ **(b)** $8 - 3y = -7$ **(c)** $\frac{3x}{4} - 2 = 10$

(d) $\frac{4x+1}{5} = -7$ **(e)** $3x - 2 = 5x - 8$ **(f)** $5(x - 1) = 2(8 - x)$

(g) $4(3p - 5) = 5(7 - 2p)$ **(h)** $\frac{a+11}{3} = \frac{3-a}{4}$ **(i)** $\frac{3m-5}{2} = \frac{2m+25}{3}$

Ex.4.4

5 Kelvin and Rob are training to be weightlifters. Rob weighs 6 kg more than Kelvin. Kelvin can lift one-and-a-half times his weight less 6 kg. Rob can lift a quarter more than his weight. They found they can both lift the same weight. How much do Kelvin and Rob each weigh?

Ex.4.3

6 The equation $3x - 4 = 2(x - 5)$ has the solution:

A $x = -2$ **B** $x = -1$ **C** $x = -6$ **D** $x = 6$

Ex.4.4

7 The total perimeter of this shape is 36 m. Find x.

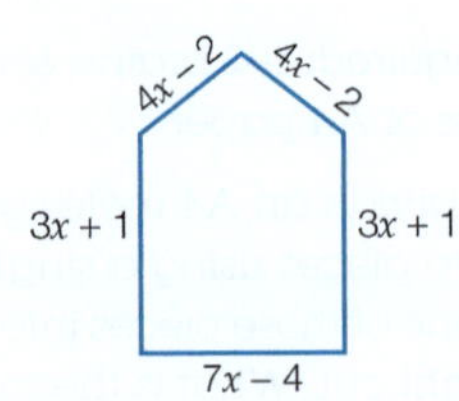

Ex.4.2

8 Find the midpoint of the line joining (6, 2) and (3, 8) using the midpoint formula.

Gradient

4.6

Stage 5.1

Need to Know

The measure of a line's steepness is called the **gradient** or **slope** of the line. The gradient of a line describes the change in the y values as the x value increases by 1.

Line				
Magnitude of gradient	Small	Large	Large	Small
Sign of gradient	Negative	Negative	Positive	Positive

The gradient of a line can be found by considering the ratio of the vertical **rise** to the horizontal **run** between any two points on the line.

$$\text{Gradient } AB = \frac{\text{vertical rise}}{\text{horizontal run}}$$

We use the letter m to represent the gradient.

$$m = \frac{\text{rise}}{\text{run}}$$

$$= \frac{\text{change in } y\text{-value}}{\text{change in } x\text{-value}}$$

$$= \frac{y_2 - y_1}{x_2 - x_1}$$

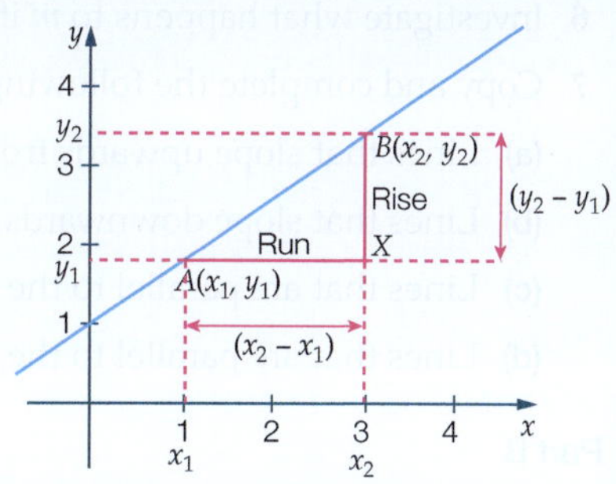

Special cases for the gradient

The gradient of any horizontal line is zero.

The gradient of any vertical line is **undefined**.

Technology Exploration: GeoGebra

Finding the gradients of lines and exploring the gradients of parallel lines

Part A

1. Open GeoGebra and use the Line through Two Points tool to draw a line through the points $A(2, 1)$ and $B(6, 3)$. Plot the point $C(6, 1)$ and use the Interval between Two Points tool to join AC and AB. This will form a right-angled triangle ACB with the right angle at C.
2. Use the Distance or Length tool to measure the lengths of AC (the run) and CB (the rise). Calculate $\frac{rise}{run}$ to get 0.5.
3. Now select the Slope tool. Click on the line AB and a small shaded triangle will appear with $m = 0.5$. This is the gradient of the line.
4. Use the Move tool to move point A or B around. Move point C to form a right-angled triangle ACB with the right angle at C. Recalculate $\frac{rise}{run}$. The value obtained should be the same as the value of m on the screen. The direction of the interval must be used in the calculations, so note that movements to the right and up are positive while movements to the left and down are negative.
5. Investigate what happens to m if the line AB is a horizontal line.
6. Investigate what happens to m if the line AB is a vertical line. (See the Algebra view.)
7. Copy and complete the following sentences.
   (a) Lines that slope upwards from left to right have a .... gradient.
   (b) Lines that slope downwards from left to right have a .... gradient.
   (c) Lines that are parallel to the x-axis have a .... gradient.
   (d) Lines that are parallel to the y-axis have a .... gradient.

Part B

1. Plot a point anywhere in the Graphics view. Select the Parallel Line tool and click on the line AB. A line parallel to AB will be drawn through the new point. Use the Slope tool to find the gradient of the new line.
2. Move the line AB around, finding a new gradient each time. Find the gradient of the parallel line. What do you notice about the gradients of the parallel lines?
3. Copy and complete the following sentence:
   Lines that are parallel have the .... gradient.

4.6

Worked Example 14

WE14

Find the gradient of:

(a) the line AB

(b) the line CD.

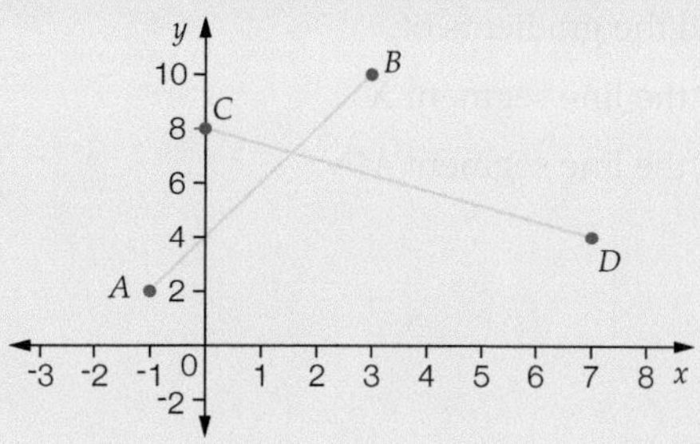

Working

(a)

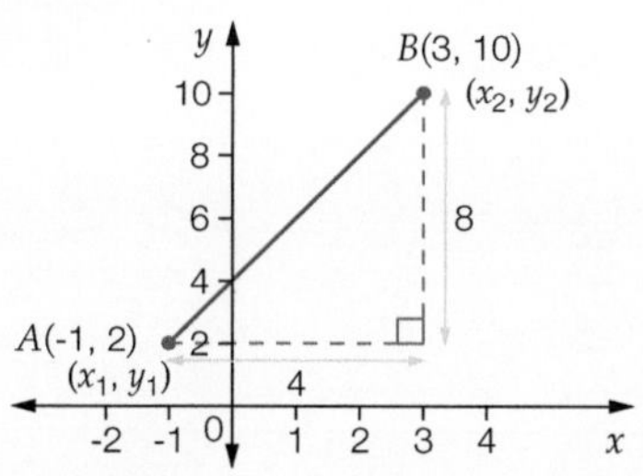

$$m = \frac{\text{rise}}{\text{run}}$$

$$= \frac{y_2 - y_1}{x_2 - x_1}$$

$$= \frac{10 - 2}{3 - -1}$$

$$= \frac{8}{4}$$

$$= 2$$

(b)

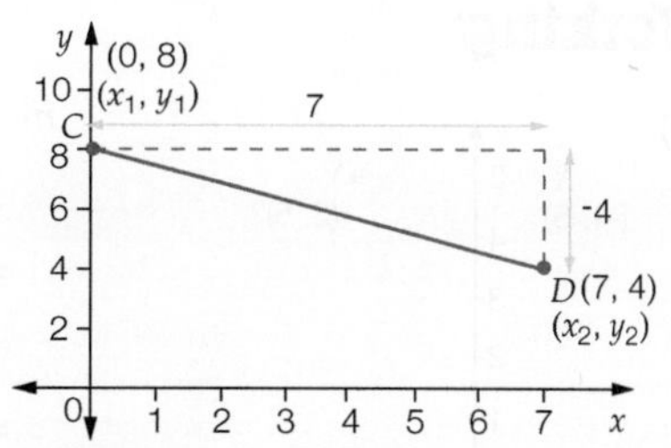

$$m = \frac{\text{rise}}{\text{run}}$$

$$= \frac{y_2 - y_1}{x_2 - x_1}$$

$$= \frac{4 - 8}{7 - 0}$$

$$= \frac{-4}{7}$$

$$= -\frac{4}{7}$$

4.6

Worked Example 15

WE 15

Find the gradients of:

(a) the line segment XY

(b) the line segment MN.

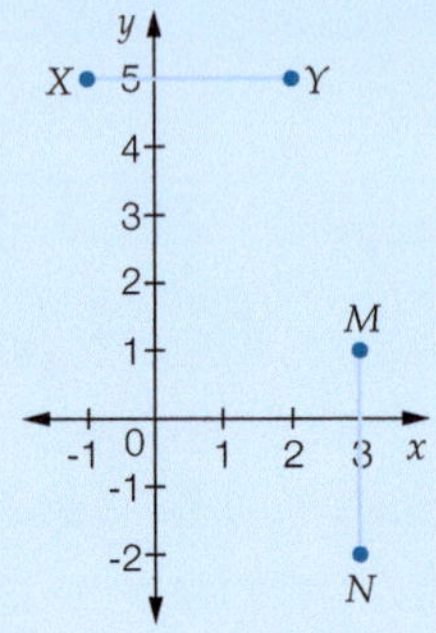

Working

(a)

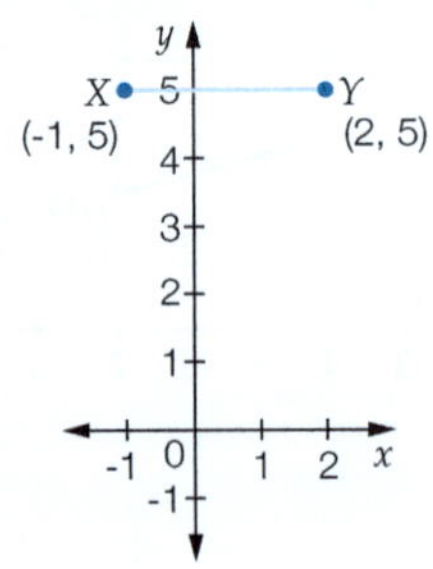

$$m = \frac{\text{rise}}{\text{run}}$$

$$= \frac{y_2 - y_1}{x_2 - x_1}$$

$$= \frac{5 - 5}{2 - 1}$$

$$= \frac{0}{1}$$

$$= 0$$

The gradient is equal to zero.

(b)

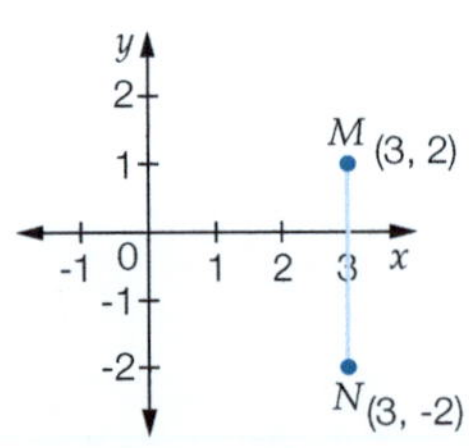

$$m = \frac{\text{rise}}{\text{run}}$$

$$= \frac{y_2 - y_1}{x_2 - x_1}$$

$$= \frac{-2 - 1}{3 - 3}$$

$$= \frac{-3}{0}$$

This cannot be evaluated, so the gradient is undefined.

Worked Example 16

WE16

Find the gradient of the line joining (5, -2) and (-1, 4).

Working

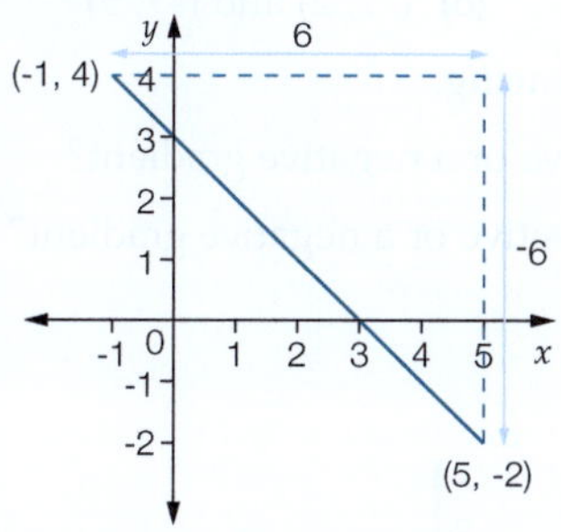

$$m = \frac{\text{rise}}{\text{run}}$$

$$= \frac{y_2 - y_1}{x_2 - x_1}$$

$$m = \frac{-2 - 4}{5 - -1}$$

$$= \frac{-6}{6}$$

$$= -1$$

4.6 Gradient

Answers page 440

Stage 5.1

Navigator

Q1, Q2, Q3 Column 1, Q4, Q5, Q6, Q7, Q10, Q11, Q15	Q1, Q2, Q3 Column 2, Q4, Q5, Q6, Q7, Q10, Q11, Q12, Q15	Q1, Q2, Q3 Column 3, Q4, Q5, Q6, Q7, Q9, Q10, Q12, Q13, Q14, Q15

Fluency and understanding

1 Find the gradient of each of the lines *a*–*d* using the points and triangles indicated.

WE14

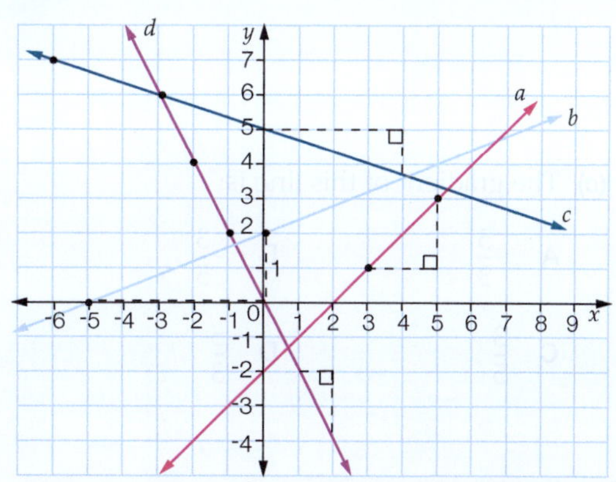

2 Find the gradient of each of the lines *a*–*d* using the points on the lines as indicated.

WE15

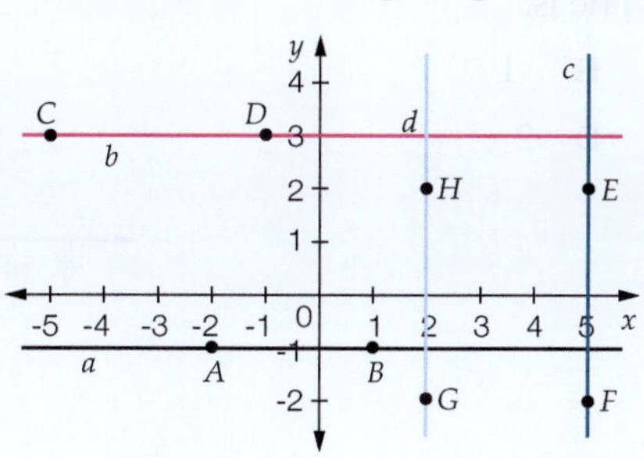

4.6

WE16

3 Find the gradient of the line joining each of the following pairs of points.

(a) (2, 3) and (4, 7) **(b)** (1, 5) and (3, 11) **(c)** (5, 8) and (6, 4)

(d) (7, 0) and (-1, 8) **(e)** (0, -2) and (3, 4) **(f)** (1, -4) and (3, -2)

(g) (-7, 3) and (5, -4) **(h)** (-6, 2) and (-2, -2) **(i)** (-1, -3) and (-4, -2)

(j) (2, -1) and (-7, -1) **(k)** (4, 2) and (-3, 2) **(l)** (-3, -5) and (1, -5)

(m) (-1, -5) and (-1, 5) **(n)** (3, -4) and (3, 8) **(o)** (-7, 2) and (-7, 5)

4 Use your answers to Questions **1** and **2** to answer the following.

(a) If a line is sloping up to the right, does it have a positive or a negative gradient?

(b) If a line is sloping down to the right, does it have a positive or a negative gradient?

(c) What is the gradient of a horizontal line?

(d) What is the gradient of a vertical line?

5 **(a)** The gradient of this line is:

A $-\frac{3}{2}$ B $-\frac{2}{3}$

C $\frac{2}{3}$ D $\frac{3}{2}$

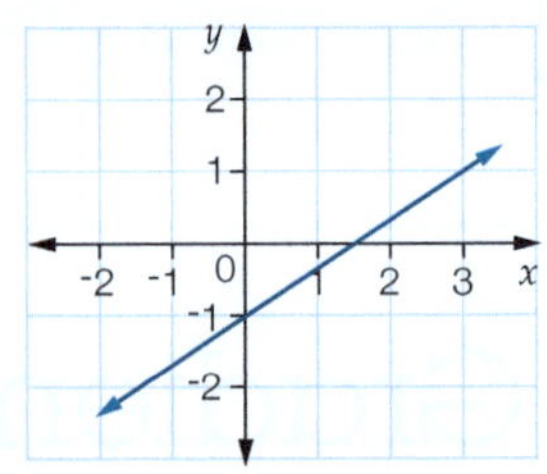

(b) The gradient of this line is:

A -2 B $-\frac{1}{2}$

C $\frac{1}{2}$ D 2

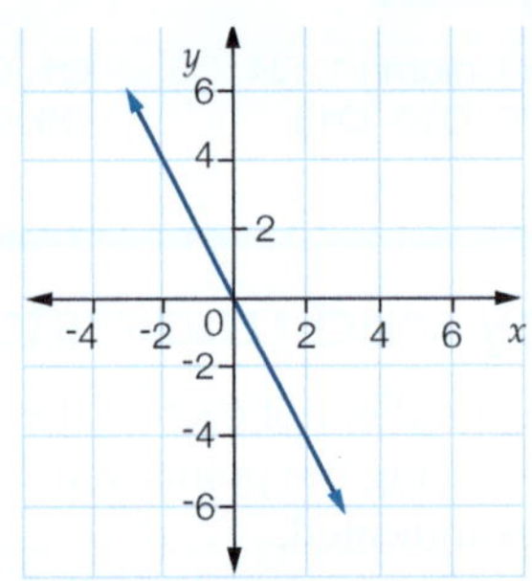

(c) The gradient of this line is:

A $-\frac{5}{3}$ B $-\frac{3}{5}$

C $\frac{3}{5}$ D $\frac{5}{3}$

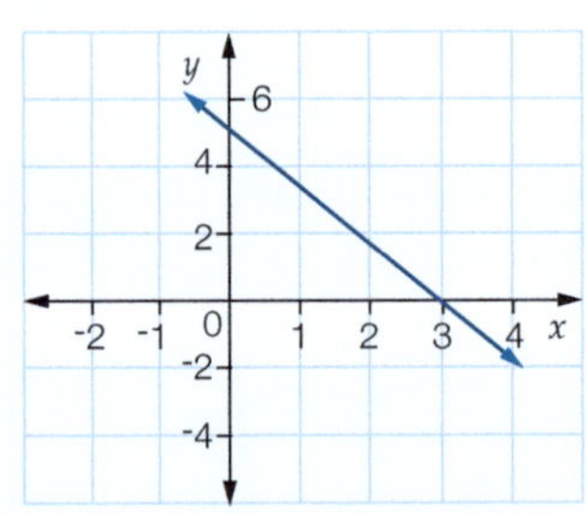

(d) The gradient of this line is:

A -2 B -1

C 0 D 2

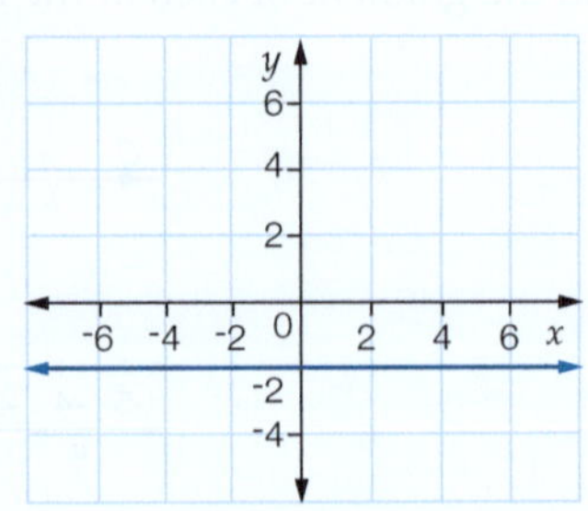

6 For each line *a*–*e* decide whether the gradient is:

A positive

B negative

C zero

D undefined.

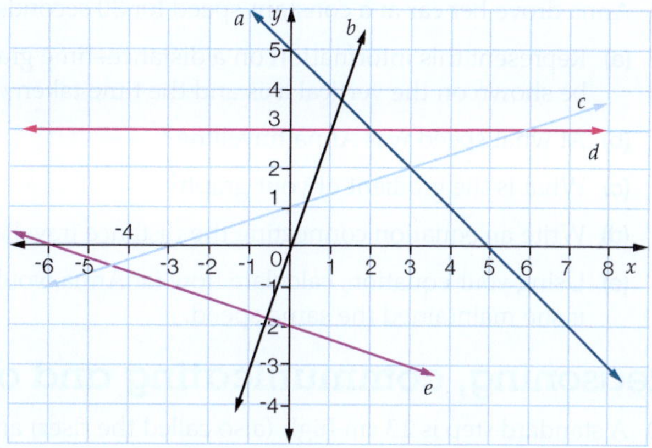

7 **(a)** Find the gradient of each of the lines in Question **6**.

(b) Which line has the steepest positive gradient?

(c) Which line has the steepest negative gradient?

8 Thanh decided to walk to school rather than take the bus. This graph shows his distance from home at any particular time on the walk. The school is located at point *P*.

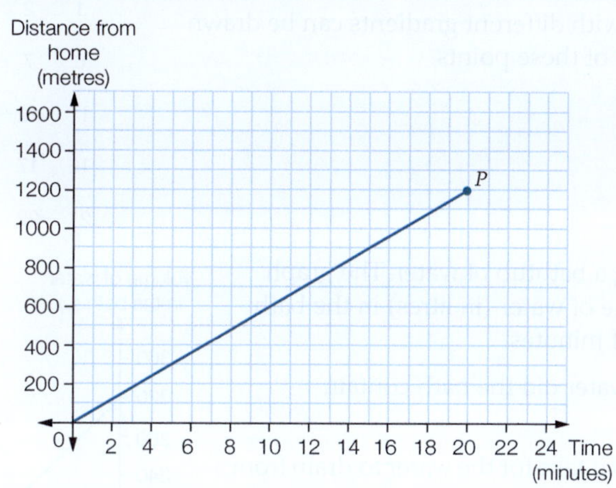

(a) How far is the school from Thanh's home?

(b) How long did it take Thanh to walk to school?

(c) How far did Thanh walk every minute?

(d) What is the gradient of this graph?

(e) What was Thanh's speed in m/min on the walk?

(f) How far from home was Thanh after walking for:

(i) 6 minutes **(ii)** 12 minutes **(iii)** 18 minutes **(iv)** *t* minutes?

(g) Write the equation to find the distance from home, *D* (in metres), after Thanh has walked *t* minutes.

4.6

9 Anna drove her car at a constant speed for 30 seconds, covering a distance of 450 m.

(a) Represent this information on a distance–time graph. (The distance travelled should be shown on the vertical axis and the time taken on the horizontal axis.)

(b) At what speed was Anna travelling?

(c) What is the gradient of your graph?

(d) Write an equation connecting the distance travelled to the time taken.

(e) Using your equation, calculate how far Anna would have travelled after 53 seconds if she maintained the same speed.

Reasoning, communicating and open-ended

10 A standard step is 13 cm high (also called the riser) and has a width of 23 cm (also called the tread). A wheelchair ramp is to be constructed to reach a platform that is four steps above ground level. If wheelchair access ramps have a gradient of 1:12, (this means a rise of 1 for a run of 12) find how far from the bottom of the stairs the ramp would need to start if it is to reach the top of the platform.

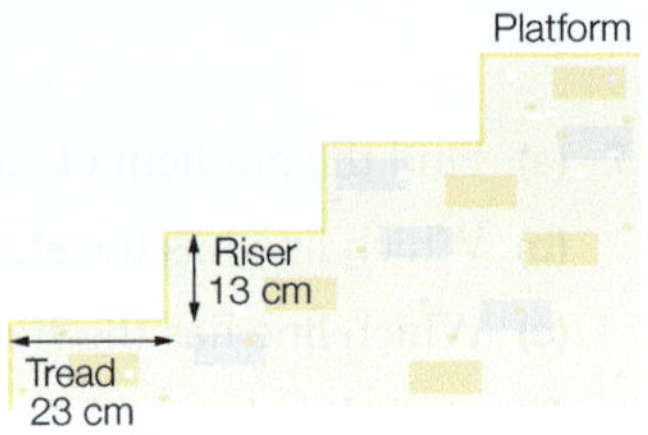

Hint

The gradients can be whole numbers, fractions, decimals positive and negative.

11 Using the 5×5 grid labelled with the numbers 1–25, show how many lines with different gradients can be drawn between any two of these points.

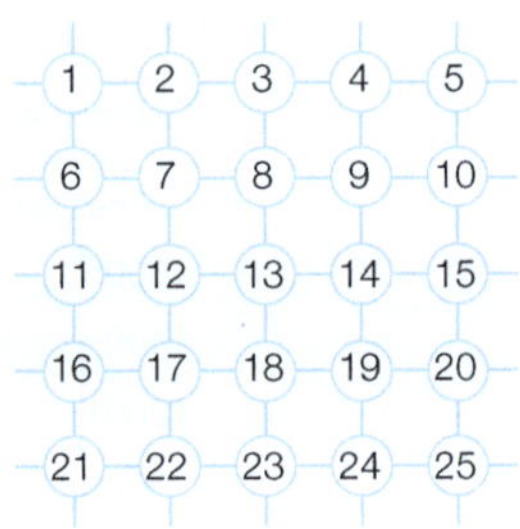

12 Sylvia is draining a bathtub of water. The graph shows the volume of water (in litres) in the bath after a number of minutes.

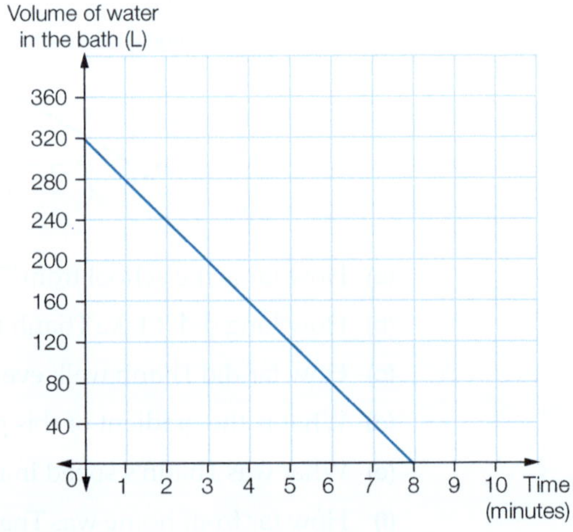

(a) How much water did the bath contain originally?

(b) How long did it take for the water to drain from the bath completely?

(c) What is the gradient of this graph?

(d) At what rate was the water being drained away? (That is, how many litres of water drained away each minute?)

(e) What does the negative represent in your answer to part **(c)**?

(f) How much water was left in the bath after:

(i) 4 minutes **(ii)** 6 minutes **(iii)** 2 minutes **(iv)** t minutes?

(g) Write the equation for the volume of water left in the bath, V (in litres), after t minutes.

(h) Why is this graph only drawn in the first quadrant?

13 Monique works for a courier company and had to make a long-distance delivery one very rainy day. This graph represents her journey. The rain had cleared for Monique's return journey, so she could travel twice as fast as she did on her way to make the delivery. It took her 4 hours and 10 minutes to make the delivery.

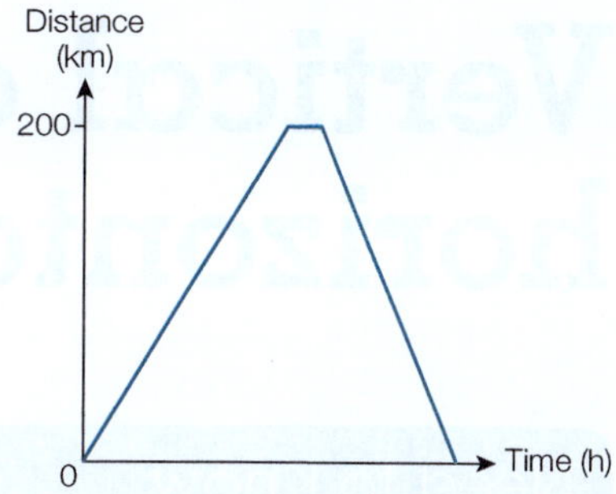

(a) At what speed was Monique travelling on her way to make the delivery?

(b) At what speed was Monique travelling on her return journey?

(c) How long did the entire trip take if Monique stopped for 45 minutes for lunch?

14 The gradient of the line joining two points A and B is $-\frac{2}{3}$. If A is the point (1, 6), write three possible coordinates for B.

15 Kate and Sally were looking at a diagram the teacher had drawn on the whiteboard and the question written underneath, 'What is the gradient of this line?'

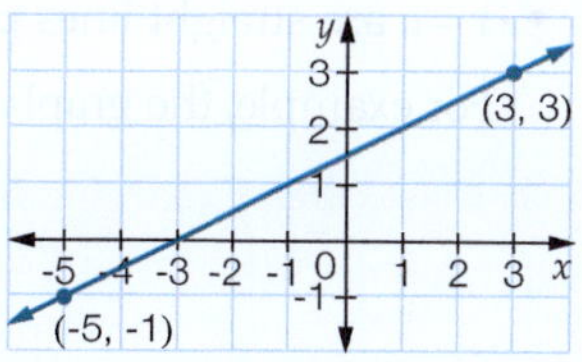

Kate wrote $m = \frac{\text{rise}}{\text{run}} = \frac{3 - -1}{3 - -5} = \frac{4}{8} = \frac{1}{2}$

Sally wrote $m = \frac{\text{rise}}{\text{run}} = \frac{3 - 1}{3 - 5} = \frac{2}{-2} = -1$

The teacher tells Kate she is correct. Explain Sally's mistake.

Problem solving

Lost in the desert

An aeroplane has crashed in the desert at a point that could be labelled (-15, 12) on a Cartesian plane. A search party sets out by 4WD from the point (18, 3) and travels in a straight line passing through the point (3, 9). A light plane sets out from point (3, -15) and travels in a straight line passing over the point (-3, -6).

1 Will either of the search parties pass the point where the aeroplane lies if they continue on these paths? Explain how you obtained your answer.

2 If either (or both) of the original routes does not pass the point where the aeroplane lies, give a different point (but with the same x-coordinate) that the searchers should pass through (or over) to be on course to locate the aeroplane.

Strategy options
- Draw a diagram.
- Break the problem into manageable parts.

4.7 Vertical and horizontal graphs

Stage 5.1

Need to Know

Linear graphs of the form:

- $y = b$ are straight lines parallel to the x-axis (i.e. horizontal lines).
  For example, the graph of $y = 3$ is a line parallel to the x-axis through the point (0, 3).
- $x = a$ are straight lines parallel to the y-axis (i.e. vertical lines).
  For example, the graph of $x = -4$ is a line parallel to the y-axis through the point (-4, 0).

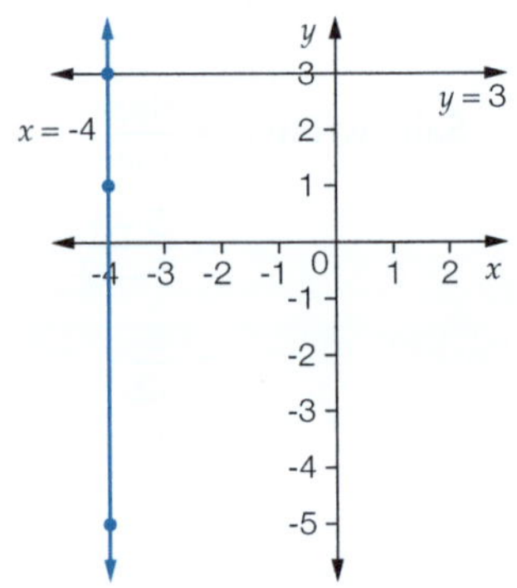

Worked Example 17

WE 17

Plot the following graphs, showing any intercepts on the axes.

(a) $y = -2$ (b) $x = 5$

Working

(a)

x	-4	-3	-2	-1	0	1	2	3	4
y	-2	-2	-2	-2	-2	-2	-2	-2	-2

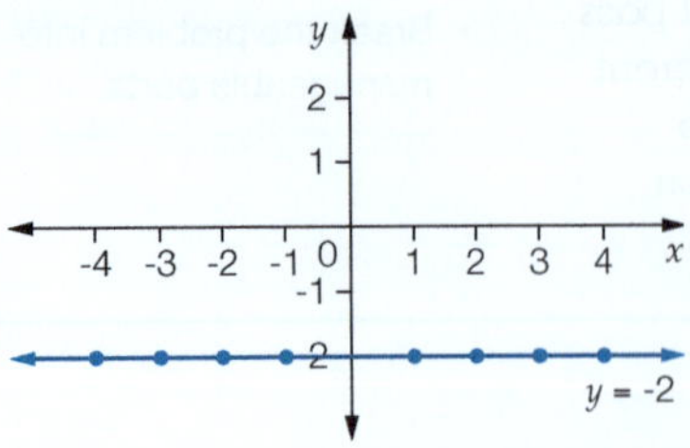

(b)

x	5	5	5	5	5
y	-2	-1	0	1	2

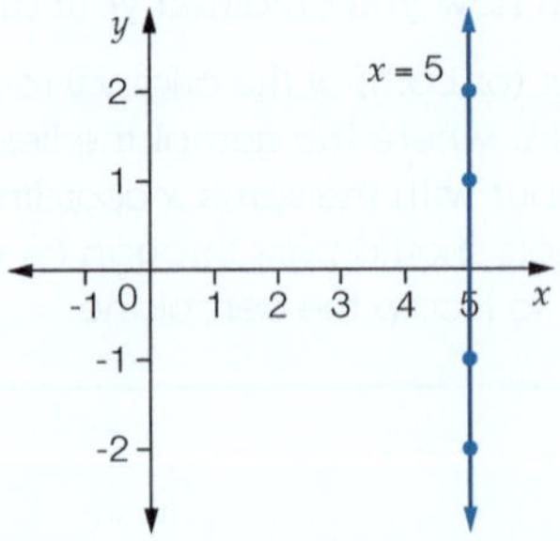

Hint

The scales on the x- and y-axes do not need to be the same.

4.7

4.7 Vertical and horizontal graphs

Answers page 441

Stage 5.1

Navigator

Q1 Column 1, Q2, Q3, Q4 Column 1, Q5 Column 1, Q6, Q7, Q8	Q1 Column 2, Q2, Q3, Q4, Q5 Column 2, Q6, Q7, Q8	Q1 Column 3, Q2, Q3, Q4 Column 3, Q5 Column 3, Q6, Q7, Q8, Q9

Fluency and understanding

WE17

1 Plot the graphs of the following, showing any intercepts on the axes.

(a) $y = 2$ **(b)** $y = 4$ **(c)** $x = -3$

(d) $x = -1$ **(e)** $x = 6$ **(f)** $x = \frac{5}{2}$

(g) $y = -3$ **(h)** $x = -2$ **(i)** $y = -\frac{3}{4}$

(j) $x = 4$ **(k)** $y = -\frac{7}{2}$ **(l)** $y = 5$

2 Write the equation of the straight line given.

(a) $m = 0$ and $b = -6$ **(b)** $m = 0$ and $b = 2$

(c) gradient 0 and y-intercept $\frac{2}{3}$ **(d)** gradient 0 and y-intercept $\frac{3}{4}$

3 (a) A coordinate point that *does not* lie on the line $y = 5$ is:

A (0, 5) **B** (-1, -5) **C** (-5, 5) **D** (5, 5)

(b) A coordinate point that *does not* lie on the line $x = 1$ is:

A (1, 0) **B** (-1, 1) **C** (1, -1) **D** (1, 11)

(c) A coordinate point that *does not* lie on the line $y = -8$ is:

A (8, -8) **B** (-8, -8) **C** (-8, 0) **D** (0, -8)

4 Sketch each pair of linear graphs on the same number plane. What coordinate point is common to both graphs in each pair?

(a) $x = 2$ and $y = 8$ **(b)** $y = 4$ and $x = -3$ **(c)** $y = 0$ and $x = -5$

(d) $y = 2$ and $x = 3$ **(e)** $y = -5$ and $x = 10$ **(f)** $x = 0$ and $y = 7$

5 Sketch the graphs of the following linear relationships using the most appropriate method, showing any relevant points.

(a) $2y = -6$ **(b)** $y = -6$ **(c)** $2x = 8$

(d) $x + 1 = 0$ **(e)** $x + 4 = 0$ **(f)** $y - 3 = 0$

(g) $5x = 4$ **(h)** $3y = 2$ **(i)** $2y = 5$

(j) $2y = 7$ **(k)** $3x = -5$ **(l)** $3y = -4$

(m) $5x = 0$ **(n)** $6x = 0$ **(o)** $7y = 0$

(p) $3y - 6 = 0$ **(q)** $2y - 8 = 0$ **(r)** $4x + 12 = 0$

(s) $4y - 3 = 0$ **(t)** $3x - 8 = 0$ **(u)** $2x + 7 = 0$

4.7

Reasoning, communicating and open-ended

6 Five streets form the boundary of an undeveloped outer suburban block of land. Taking the southernmost point of the block as the coordinate point (0, 0), each of the roads can be plotted on a number plane. The graphs that represent the roads are $y = 2x$, $x = 3$, $y = 10$, $x + 2 = 0$ and $2y + 3x = 0$.

The local council plans to subdivide this land by putting in new roads along the lines $x = 0$, $y = -5x$ and $3y - 10x = 0$.

- **(a)** Sketch each of the linear graphs that represent the boundary roads on the same number plane.
- **(b)** From your sketch, determine the coordinate points that are common to each adjacent pair of graphs.
- **(c)** Sketch each of the linear graphs that represent the new roads.
- **(d)** Will an old shed located at the coordinate point (-1, 5) be affected by the new road?

7 Write three equations whose graphs are parallel to the x-axis. Sketch the graphs for each.

8 Write three equations whose graphs are parallel to the y-axis. Sketch the graphs for each.

9 Draw three different pairs of linear graphs on the same number plane. Each pair of graphs must have a graph parallel to the x-axis and the other parallel to the y-axis. From these graphs, determine the coordinate point which is common to each pair.

Sketching linear graphs using the gradient and *y*-intercept

4.8

Stage 5.2

Need to Know

The gradient–intercept form

$y = mx + b$, where m is the gradient and b is the y coordinate of the y-**intercept**.

Graphs of equations in the form of $y = mx + b$ are linear graphs.

For example: $y = 2x - 6$ is a linear graph.

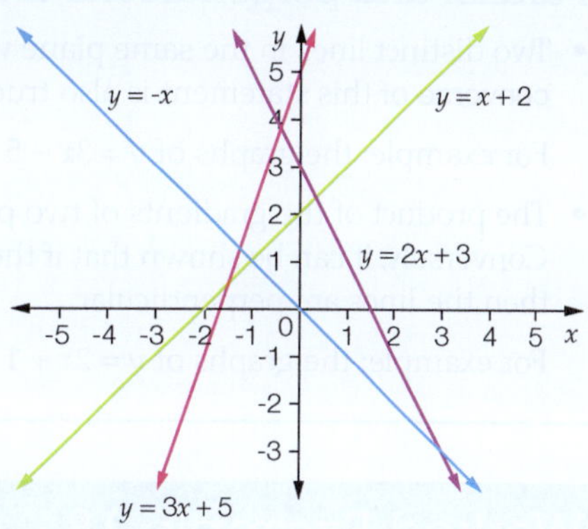

The effect of m

As m increases and b stays constant, the gradient of the graph increases. The coefficient m gives the value of the gradient, $m = \frac{rise}{run}$.

For example: $y = 3x + 1$ has a larger gradient than $y = 2x + 1$.

The effect of b

As b increases and m stays constant, the y-intercept increases. The value of the constant term b is the value of the y coordinate of the y-intercept.

For example: $y = 3x + 5$ has a y-intercept of (0, 5).

Sketching linear graphs that pass through the origin

Graphs of equations in the form of $y = mx$ are graphs through the origin. The origin is the y-intercept. This is one point on the graph. Use the gradient to find a second point. The line through the origin and this second point is the graph of the equation.

For example: the graphs of $y = 2x$, $y = 6x$, $y = -9x$ all pass through the origin.

Sketching other linear graphs using gradient and y-intercept

Identify the y-intercept using the value of b in the equation $y = mx + b$. This is one point on the graph. Using $m = \frac{rise}{run}$, identify a second point on the graph. The line through these two points is the graph of the equation.

4.8

The general form

$ax + by + c = 0$, where a, b and c are constants and a and $b \neq 0$.

In this equation, the gradient $m = \frac{-a}{b}$ and the y coordinate of the y-intercept $= \frac{-c}{b}$ can be found by rearranging the equation to make y the subject.

For example: $3x + 2y - 8 = 0$.

$m = -\frac{3}{2}$

y coordinate of the y-intercept $= \frac{8}{2}$
$= 4$

If $3x + 2y - 8 = 0$ is rearranged to give $y = -\frac{3}{2}x + 4$, the answers are the same.

Parallel and perpendicular lines

- Two distinct lines in the same plane with the same gradient are **parallel** lines. The **converse** of this statement is also true: Parallel lines have the same gradient, $m_1 = m_2$.

  For example: the graphs of $y = 3x - 5$ and $y = 3x + 1$ are parallel lines.
- The product of the gradients of two **perpendicular** lines is equal to -1, $m_1m_2 = -1$. Conversely, it can be shown that if the product of the gradients of two lines $m_1m_2 = -1$, then the lines are perpendicular.

  For example: the graphs of $y = 2x + 1$ and $y = 1 - \frac{x}{2}$ are perpendicular lines.

Worked Example 18

WE18

State the gradient and coordinates of the y-intercept of the graphs of each of the following linear equations.

(a) $y = -3x + 2$
(b) $y = x - 1$
(c) $2y = 5x - 4$
(d) $y + 2x = 6$

Working

(a) $y = -3x + 2$

gradient = -3

y-intercept: (0, 2)

(b) $y = x - 1$
$= 1x + (-1)$

gradient = 1

y-intercept: (0, -1)

(c) $2y = 5x - 4$
$y = \frac{5}{2}x - 2$

gradient $= \frac{5}{2}$

y-intercept: (0, -2)

(d) $y + 2x = 6$
$y = 6 - 2x$
$= -2x + 6$

gradient = -2

y-intercept: (0, 6)

Worked Example 19

WE19

Sketch the graph of the following relationships.

(a) $y = 4x$ (b) $5x - 2y = 0$

Working

(a) y-intercept:

When $x = 0$, $y = 4 \times 0$

$y = 0$

y-intercept is (0, 0)

$m = 4$

$= \frac{4}{1}$

rise = 4; run = 1

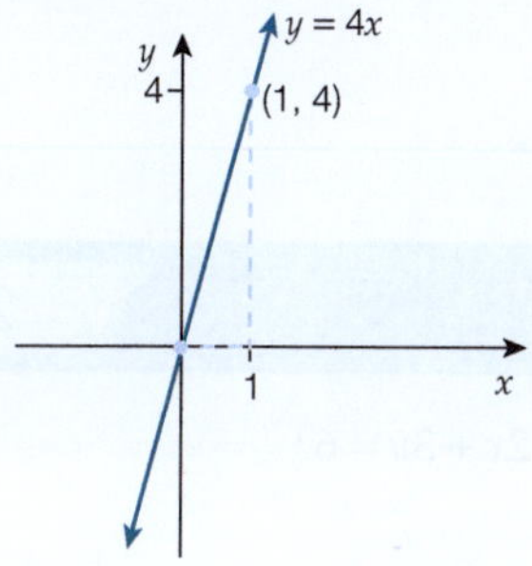

(b) y-intercept: When $x = 0$, $-2y = 0$

$y = 0$

y-intercept is (0, 0)

$$5x - 2y = 0$$

$$5x - 5x - 2y = 0 - 5x$$

$$-2y = -5x$$

$$\frac{-2y}{-2} = \frac{-5x}{-2}$$

$$y = \frac{5}{2}x$$

$$m = \frac{5}{2}$$

rise = 5; run = 2

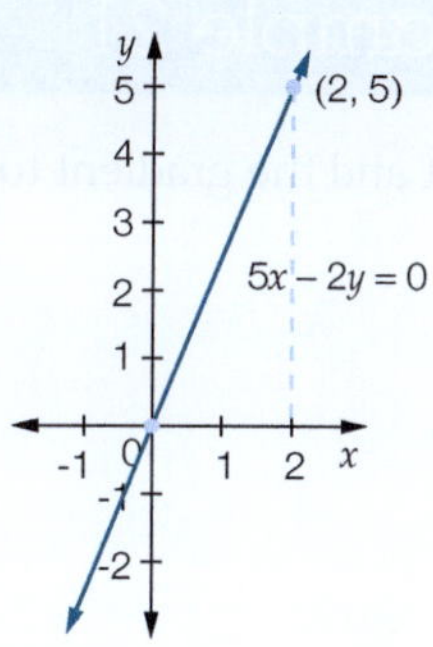

4.8

Worked Example 20

WE 20

Use the y-intercept and the gradient to sketch the graph of $y = 2x + 1$.

Working

y-intercept: $(0, 1)$

$$\text{Gradient} = 2 = \frac{2}{1} = \frac{\text{rise}}{\text{run}}$$

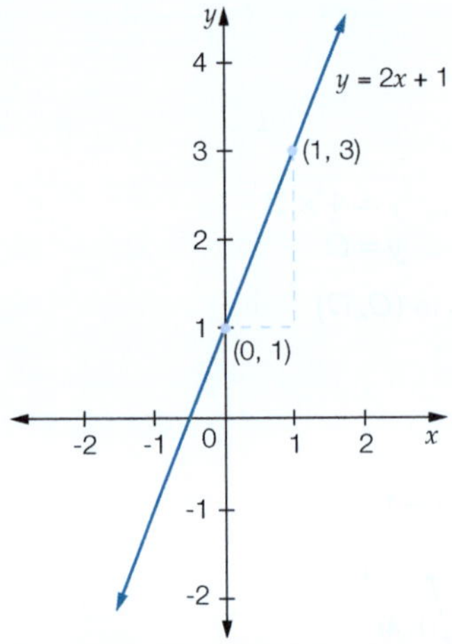

Worked Example 21

WE 21

Use the y-intercept and the gradient to sketch the graph of $2x + 3y = 6$.

Working

$$2x + 3y = 6$$
$$3y = 6 - 2x$$
$$y = \frac{6 - 2x}{3}$$
$$= 2 - \frac{2x}{3}$$
$$= -\frac{2x}{3} + 2$$

y-intercept: $(0, 2)$

$$\text{Gradient} = -\frac{2}{3} = \frac{\text{rise}}{\text{run}}$$

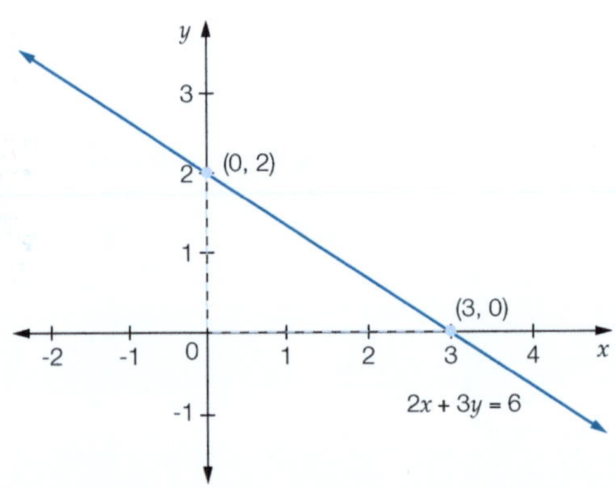

4.8

Worked Example 22

WE 22

Find the gradient of a line parallel to each of the following.

(a) $y = 2x - 5$ (b) $2x + 3y = 6$

Working

(a) $y = 2x - 5$

$m = 2$

Any parallel line will have a gradient of 2.

(b) $2x + 3y = 6$

$2x - 2x + 3y = 6 - 2x$

$\frac{3y}{3} = -\frac{2}{3}x + \frac{6}{3}$

$y = -\frac{2}{3}x + 2$

$m_1 = -\frac{2}{3}$

Any parallel line will have a gradient of $-\frac{2}{3}$.

Worked Example 23

WE 23

Find the gradient of a line perpendicular to each of the following.

(a) $y = 4x + 3$ (b) $3x - 5y = 10$

Working

(a) $y = 4x + 3$

$m_1 = 4$

$m_2 = -\frac{1}{m_1}$

$= -\frac{1}{4}$

(b) $3x - 5y = 10$

$3x - 3x - 5y = 10 - 3x$

$-5y = -3x + 10$

$\frac{-5y}{-5} = \frac{-3x}{-5} + \frac{10}{-5}$

$y = \frac{3}{5}x - 2$

$m_1 = \frac{3}{5}$

$m_2 = -\frac{1}{m_1}$

$= -\frac{5}{3}$

4.8

Answers page 445

4.8 Sketching linear graphs using the gradient and *y*-intercept

Stage 5.2

Navigator

Q1 Column 1, Q2 Column 2, Q3 Column 3, Q4 Column 1, Q5, Q6 Column 1, Q7 Column 1, Q8, Q9, Q10, Q11, Q13, Q14, Q15, Q18, Q20	Q1 Column 2, Q2 Column 2, Q3 Column 2, Q4 Column 2, Q5, Q6 Column 2, Q7 Column 2, Q8, Q9, Q10, Q12, Q13, Q14, Q15, Q17, Q18, Q20	Q1 Column 3, Q2 Column 3, Q3 Column 3, Q4 Column 3, Q5, Q6 Column 3, Q7 Column 3, Q8, Q9, Q10, Q12, Q13, Q15, Q16, Q17, Q18, Q19, Q20

Fluency and understanding

WE18

1 State the gradient and the coordinates of the y-intercept of the graphs of each of the following linear equations.

(a) $y = x + 3$ **(b)** $y = 2x + 1$ **(c)** $y = 7x + 3$

(d) $y = 3x - 2$ **(e)** $y = x - 4$ **(f)** $y = -4x + 3$

(g) $y = -x + 4$ **(h)** $y = 6 - 2x$ **(i)** $y = -5 + 2x$

WE19

2 Sketch the graphs of the following relationships.

(a) $y = 2x$ **(b)** $y = 3x$ **(c)** $y = -x$

(d) $y = -2x$ **(e)** $y = \frac{2}{3}x$ **(f)** $y = \frac{4}{5}x$

(g) $4x + y = 0$ **(h)** $6x + y = 0$ **(i)** $4x + 3y = 0$

WE20

3 Use the y-intercept and the gradient to sketch the graph of each of the equations in Question **1**.

WE21

4 Use the y-intercept and the gradient to sketch the graph of each of the following.

(a) $2x + y = 4$ **(b)** $-x + y = -3$ **(c)** $-3x + y = -2$

(d) $4x - y = 1$ **(e)** $2x - y = -3$ **(f)** $3x + 2y = 4$

(g) $2x + 3y = 3$ **(h)** $2x - 5y = 10$ **(i)** $7x + 3y - 6 = 0$

WE22

5 Find the gradient of lines parallel to each of the following.

(a) $y = 3x + 5$ **(b)** $y = 7 - 4x$ **(c)** $2y = 6x - 5$

(d) $6x - 2y = 3$ **(e)** $2x + 5y = 10$ **(f)** $4x - 3y = 8$

WE23

6 Find the gradient of lines perpendicular to each of the equations given in Question **5**.

7 Write the equation of the straight line given:

(a) $m = 3$ and $b = 4$ **(b)** $m = 1$ and $b = 5$ **(c)** $m = 2$ and $b = 0$

(d) gradient is 4 and y-intercept is 0 **(e)** gradient is -5 and y-intercept is 1 **(f)** gradient is 3 and y-intercept is -4

8 **(a)** The equation of a linear graph with y-intercept 3 and gradient of -2 is:

A $y = 2x + 3$ **B** $2x - y = 3$ **C** $3x + 6 = 2$ **D** $2x + y = 3$

(b) The equation of a linear graph with gradient $\frac{1}{2}$ and y-intercept 1 is:

A $y = 2x + 1$ **B** $y = x + 2$ **C** $x - 2y = -2$ **D** $x + 2y = -2$

4.8

9 When using the y-intercept and gradient method of graphing, we first mark in the y-intercept and then use the gradient to locate a second coordinate point.

(a) A line has a gradient of $\frac{3}{4}$. From the y-intercept, we move:

A 3 units right and 4 units up　**B** 4 units right and 3 units up
C 3 units right and 4 units down　**D** 5 units right and 3 units down

(b) A line has a gradient of $-\frac{2}{5}$. From the y-intercept, we move:

A 2 units right and 5 units down　**B** 5 units right and 2 units down
C 2 units left and 5 units up　**D** 5 units left and 2 units down

(c) A line has a gradient of 5. From the y-intercept, we move:

A 1 unit right and 5 units up　**B** 5 units right and 5 units up
C 1 unit right and 5 units down　**D** 5 units right and 5 units down

(d) A line has a gradient of -3. From the y-intercept, we move:

A 3 units right and 3 units up　**B** 1 unit right and 3 units down
C 0 units right and 3 units down　**D** 1 unit right and 3 units up

10 (a) A coordinate point that lies on the line $y = x + 7$ is:

A (2, 10)　**B** (-3, -4)　**C** (0, -7)　**D** (-4, 3)

(b) A coordinate point that lies on the line $3x + 2y = 11$ is:

A (5, -2)　**B** (2, 3)　**C** (-3, -1)　**D** (4, 0)

(c) A coordinate point that lies on the line $3y - x + 6 = 0$ is:

A (0, 2)　**B** (2, 0)　**C** (3, -1)　**D** (-3, 1)

11 Find the gradient of the line perpendicular to:

(a) a line with gradient 2
(b) a line with gradient 0
(c) the line passing through (4, 5) and (3, -2)
(d) the line passing through (-2, 6) and (3, 8).

12 Match the following.

(a) A line parallel to the line joining the points (2, 3) and (4, 7)　**A** A line joining (-6, 2) and (-2, -2)
(b) A line perpendicular to the line joining the points (2, 3) and (4, 7)　**B** A line joining (0, -2) and (3, 4)
(c) A line parallel to the line joining the points (7, 0) and (-1, 8)　**C** A line joining (-3, -5) and (1, -5)
(d) A line perpendicular to the line joining the points (7, 0) and (-1, 8)　**D** A line joining (7, 3) and (5, 4)
(e) A line perpendicular to the line joining the points (-1, -3) and (-4, -2)　**E** A line joining (-1, -5) and (-1, 5)
(f) A line parallel to the line joining the points (4, 2) and (-3, 2)　**F** A line joining (1, -4) and (3, -2)
(g) A line parallel to the line joining the points (3, -4) and (3, 8)　**G** A line joining (1, 5) and (3, 11)

4.8

13 What is special about the following lines?

(a) Lines parallel to the line joining (4, 2) and (-3, 2).

(b) Lines parallel to the line joining (3, -4) and (3, 8).

14 Consider this distance–time graph for Cassia as she walks to the shops.

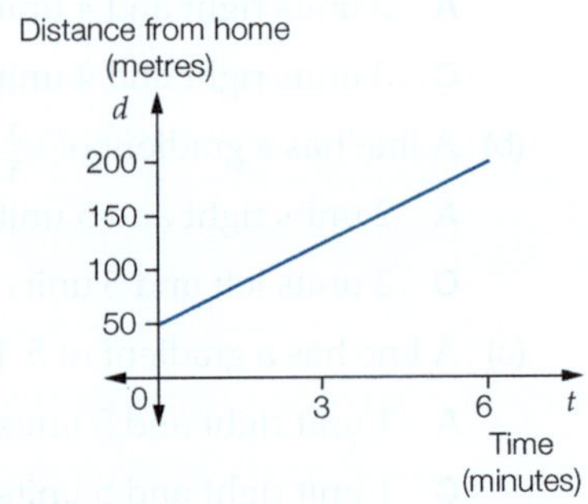

(a) Write the intercept with the vertical axis.

(b) Find the gradient of this graph.

(c) Use this information to write an equation linking the distance from home (d) to the time (t).

(d) At what speed was she walking? (Refer to the equation.)

(e) If Cassia continued walking at the same speed, how far from home would she be after 20 minutes?

15 A distance–time graph is represented by the equation $d = 30t + 20$, where d is the distance from Newcastle in kilometres and t is the time in hours.

(a) Write the gradient of the line.

(b) What is the speed of travel?

(c) How far from Newcastle did the journey start?

(d) If the journey had started 90 km from Newcastle and a constant speed of 100 km/h was maintained, what equation could now be written to represent the relationship between d and t?

Reasoning, communicating and open-ended

16 An oil-tanker is being filled with crude oil. The oil-tanker initially contained 10 000 tonnes of oil and, over an 8-hour period, it received a further 96 000 tonnes of crude oil. Three full storage tanks, each of 48 000 tonnes capacity, were being used to transfer the crude oil to the tanker. Each tank was being emptied simultaneously at the same rate of 4000 tonnes per hour.

(a) Sketch a graph showing the amount of crude oil (in tonnes) in the oil-tanker during the 8-hour period.

(b) Find the gradient of this graph.

(c) Use this information to write an equation relating the amount of crude oil (m) in the tanker to the time (t).

(d) Sketch a graph showing the amount of crude oil (m) left in one of the storage tanks over the 8-hour period as oil is pumped out of it.

(e) Write an equation relating the amount of crude oil (m) left in one of the storage tanks to the time (t).

(f) How much crude oil was left in each of the storage tanks after the 8-hour period?

(g) If the pumps from one of the storage tanks were under repair, how long would it take to transfer the same amount of crude oil from the remaining storage tanks?

17 This graph shows the journeys two friends made between Towns A and B, leaving at midday and arriving at their respective destinations 2 hours later.

Yvonne travels at a constant speed straight from Town A to Town B. Cameron travels from Town B towards Town A at a constant speed for the first hour. He rests for 15 minutes, then continues to Town A at the same speed that Yvonne is travelling.

Find how far each of them was from Town A after travelling for half an hour.

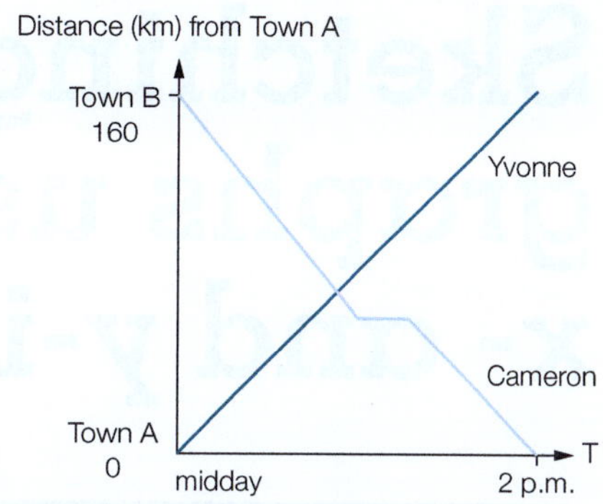

18 State at least three different coordinate points that lie on each of the following lines.

(a) $y = x - 4$ **(b)** $2x - 5y = -10$ **(c)** $4y - x + 2 = 0$

19 When the equations $y = 4x + 3$, $y = 4x - 2$, $y = -\frac{1}{4}x - 2$ and $y = -\frac{1}{4}x - \frac{3}{4}$ are plotted on a Cartesian plane, the area enclosed by the points of their intersection is a square.

Can you find another set of sloping lines that enclose a square?

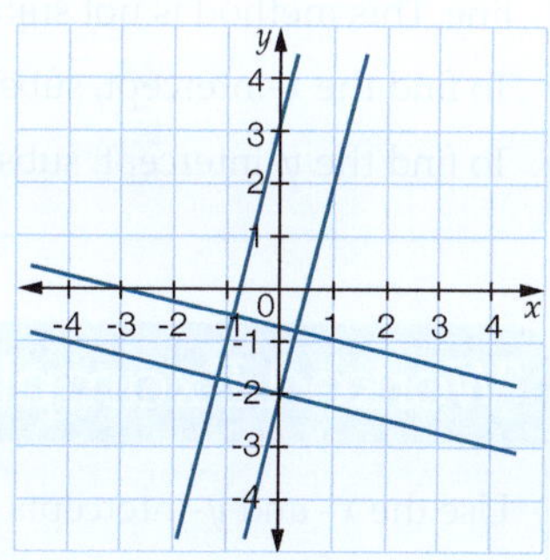

Hint

Remember the properties of parallel and perpendicular lines.

20 **(a)** Write an equation for a line parallel to $y = 4x - 3$.

(b) Write an equation for a line perpendicular to $y = 4x - 3$.

Problem solving

Area of a rectangle

In the diagram, $ABCD$ is a rectangle with A on the line $y = x + 10$, B on the line $y = -2x + 10$ and C and D on the x-axis. If $AD = 4$, what is the area of rectangle $ABCD$?

Strategy options

- Solve a simpler problem.
- Break the problem into manageable parts.

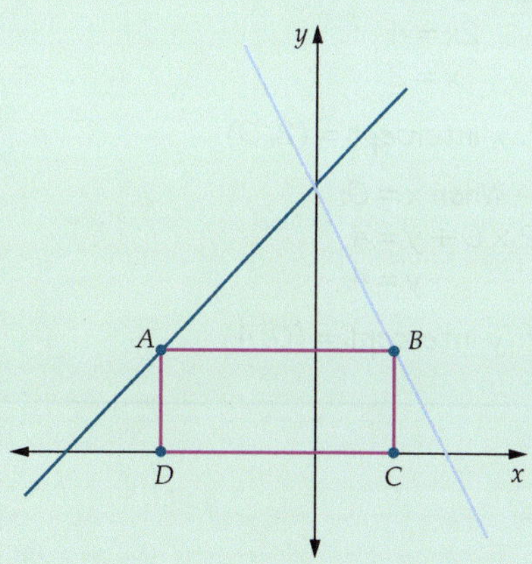

4.9 Sketching linear graphs using the x- and y-intercepts

Stage 5.3 §

Need to Know

A linear equation can be sketched by using the x- and y-intercepts as two points on the line. This method is not suitable for graphs in the form of $y = mx$.

To find the x-intercept, substitute $y = 0$ into the equation.

To find the y-intercept, substitute $x = 0$ into the equation.

Worked Example 24

WE 24

Use the x- and y-intercepts to sketch the graph of $2x + y = 4$.

Working

$2x + y = 4$

When $y = 0$:

$2x + 0 = 4$

$2x = 4$

$x = 2$

$\therefore$ x-intercept $= (2, 0)$

When $x = 0$:

$2 \times 0 + y = 4$

$y = 4$

$\therefore$ y-intercept $= (0, 4)$

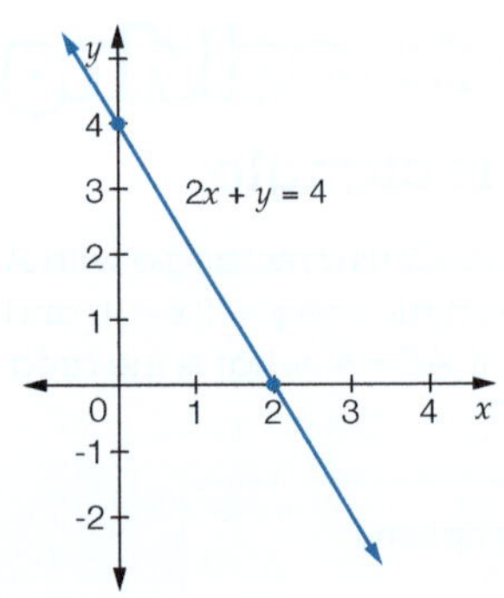

Worked Example 25

WE25

Use the x- and y-intercepts to sketch the graph of $y = 6x - 2$.

Working

$y = 6x - 2$

When $y = 0$:

$0 = 6x - 2$

$6x = 2$

$x = \frac{2}{6}$

$= \frac{1}{3}$

$\therefore$ x-intercept $= \left(\frac{1}{3}, 0\right)$

When $x = 0$:

$y = 6 \times 0 - 2$

$= ^-2$

$\therefore$ y-intercept $= (0, ^-2)$

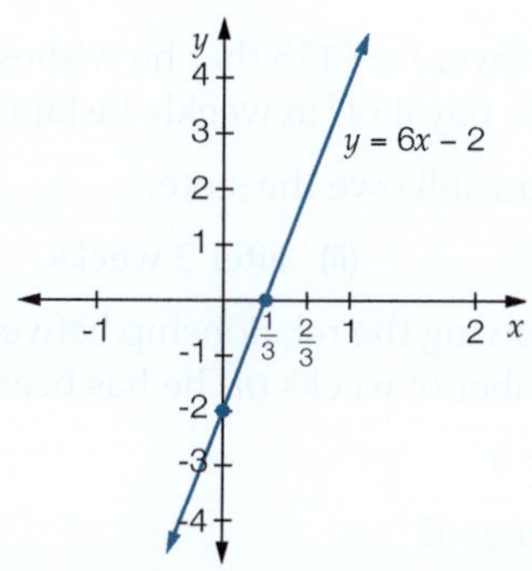

4.9 Sketching linear graphs using the x- and y-intercepts

Answers page 449

Navigator

Stage 5.3 §

Q1 Column 1, Q2 Column 1, Q3, Q4, Q6, Q7	Q1 Column 2, Q2 Column 2, Q3, Q4, Q5, Q6, Q7	Q1 Column 3, Q2 Column 3, Q3, Q4, Q5, Q6, Q7

Fluency and understanding

1 Use the x- and y-intercepts to sketch a graph for each of the following. WE24

(a) $x + 2y = 4$ (b) $3x + y = 6$ (c) $3x - y = 6$

(d) $x - y = 3$ (e) $4x - 2y = 12$ (f) $2x - 5y = 20$

(g) $2x + 5y = 10$ (h) $4x + 3y = 12$ (i) $3x + 4y = 2$

2 Use the x- and y-intercepts to sketch a graph for each of the following. WE25

(a) $y = 2x - 8$ (b) $y = 3x - 9$ (c) $y = x + 5$

(d) $y = x + 2$ (e) $y = 6 - 3x$ (f) $y = 2 - 4x$

(g) $2y = 3x + 4$ (h) $5y = 2x - 5$ (i) $3y = 6 - 2x$

4.9

3 (a) The coordinates of two points that lie on the line $2x + 3y = 6$ are:

A (2, 0) and (0, 3) B (-2, 0) and (0, -3)

C (3, 0) and (0, 2) D (-3, 0) and (0, -2)

(b) The coordinates of two points that lie on the line $3y = 2x - 6$ are:

A (-3, 0) and (0, 2) B (3, 0) and (0, -2)

C (-3, 0) and (0, -2) D (2, 0) and (0, -3)

4 Todd has seen a Blu-ray player for $175 that he wishes to buy. The store will allow him to pay it off in weekly instalments of $25.

(a) How much would Todd still owe the store:

(i) after 1 week (ii) after 3 weeks (iii) after n weeks?

(b) Write an equation showing the relationship between the amount (A) that Todd owes the store and the number of weeks (n) he has been paying off the Blu-ray player.

(c) (i) Find A when $n = 0$.

(ii) What does this mean?

(d) (i) Find n when $A = 0$.

(ii) What does this mean?

(e) Sketch the graph of this relationship. Write A on the vertical axis and n on the horizontal axis.

5 A milk vat is being drained so that regular cleaning and maintenance can take place. The volume, V, of milk remaining in the vat after a given time, t, is given by the relationship $V = 630 - 15t$, where V is in litres and t is in minutes.

(a) How much milk is left in the vat after 5 minutes?

(b) How much milk did the vat originally contain?

(c) How long will it take for the vat to be completely empty?

(d) Use this information to graph this relationship on a number plane.

Reasoning, communicating and open-ended

6 Chantal has borrowed $7200 from her parents to buy a car. She agrees to repay the amount, plus an extra $300 in interest, to her parents at the rate of $300 per month.

(a) What is the total amount she owes to her parents?

(b) How long will it take to pay off the loan?

(c) Sketch a graph of this relationship. (Put amount owing A, in $, on the vertical axis and time, t, in months on the horizontal axis.)

(d) Write an equation describing the amount, A, Chantal owes her parents after t months.

(e) How much money does Chantal owe after 6 months?

(f) How much money has Chantal repaid after 14 months?

7 Find three different pairs of x- and y-intercepts when the gradient of the graph is equal to 3.

Challenge 4

1 It takes Vinny 30 minutes to walk around the block, but Jesse can ride around on his bike in 15 minutes. If Vinny starts walking at 5:35 p.m. and Jesse follows at 5:48 p.m., when will Jesse catch up to Vinny?

2 Jakki is travelling through country Victoria and is trying to decide on her mode of travel for the next 5 days. If she hires a bicycle for 2 days and a car for the other 3 days, she will pay $150. If she hires a bicycle for 4 days and a car for the remaining 1 day, she will pay $90. What is the cost of hiring a bicycle for one day?

3 Nina uses the following table to produce a straight line graph.

x	0	2	4
y	3	7	11

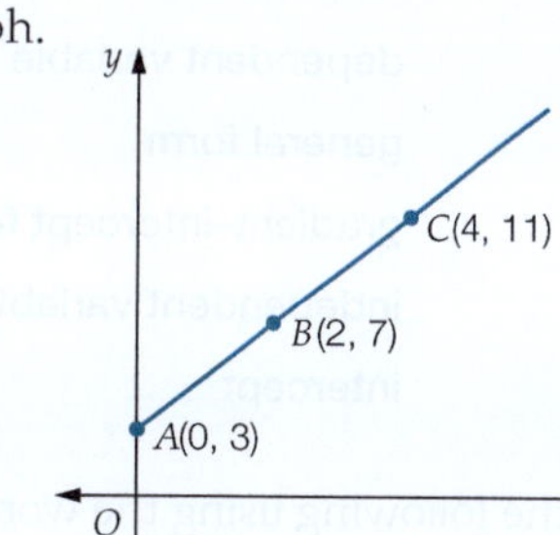

The point $D(7, w)$ also lies on the line. Find the value of w.

4 Which line has a gradient of $-\frac{3}{4}$? Diagrams are not to scale.

A
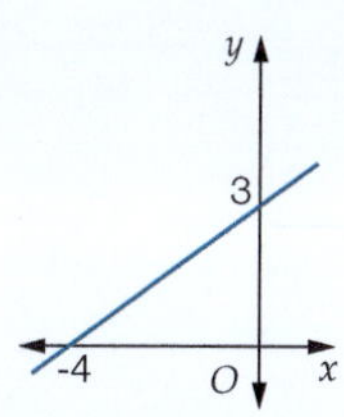

B
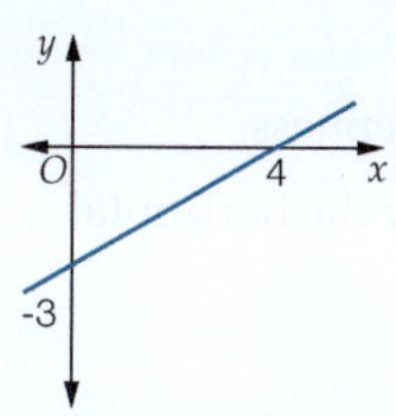

C
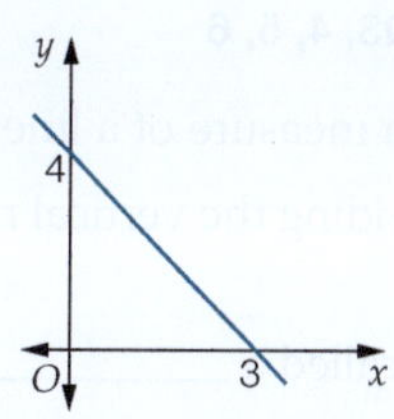

D
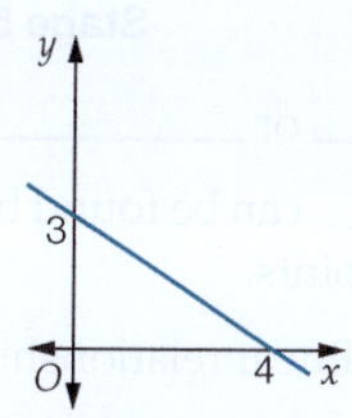

5 **(a)** The square $ABCD$ is to be plotted on a number plane. Given $A(1, 1)$, $B(3, 2)$ and $C(2, 4)$, find the coordinates of D.

(b) Write the gradients of AB and BC. What do you notice about your answers?

(c) Write the gradients of AD and DC. What do you notice about your answers?

(d) What can you say about the gradients of the parallel sides of the square?

(e) What can you say about the gradients of the perpendicular sides of the square?

6 The square $KLMN$ is to be constructed on the side KL. Given $K(-2, -1)$ and $L(1, 3)$, find two pairs of values for the coordinates of M and N.

7 The diagonals of this rectangle intersect at a point. What are the coordinates of this point?

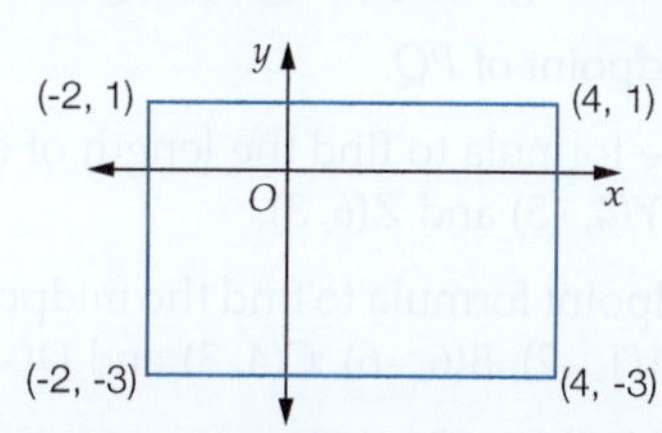

8 Solve each equation for x.

(a) $\frac{x-a}{x-b} = 1 - \frac{x}{x-b}; x \neq b$

(b) $\frac{a}{bx} - \frac{b}{ax} = a^2 - b^2; a \neq 0, b \neq 0$

(c) $\frac{a-b}{x-c} = \frac{a+b}{x+c}; x \neq \pm c$

Chapter review 4

D.I.Y. Summary

Key Words

Stage 5.1	coordinates	midpoint	run
	gradient	origin	slope
	interval	rise	undefined
Stage 5.2	dependent variable	inverse operation	perpendicular
	general form	linear equation	solve
	gradient–intercept form	linear graph	straight line
	independent variable	linear relationship	
	intercept	parallel	

Copy and complete the following using the words and phrases from the list, where appropriate, to write a summary for this chapter. A word or phrase may be used more than once.

Stage 5.1 Q1, 2 **Stage 5.2 Q3, 4, 5, 6**

1 The __________ or __________ is a measure of a line's steepness.

2 The __________ can be found by dividing the vertical rise by the horizontal __________ between two points.

3 The graph of a linear relationship is called a __________.

4 To solve an equation, we perform the __________ in the reverse order.

5 A __________ has only one solution.

6 The point where a linear graph cuts the y-axis is called the y-__________.

Stage 5.1 Q1, 9–12, 18, 21, 22, 24 **Stage 5.2 Q4–8, 13, 14** **Stage 5.3 Q2, 3, 15–17, 19–20, 23**

Fluency and understanding

1 **(a)** Plot the points $P(-3, 5)$ and $Q(6, 11)$ on a Cartesian plane and join the points PQ. **Ex 4.1**

(b) Find the exact length PQ using Pythagoras' Theorem.

(c) Find the midpoint of PQ.

2 Use the distance formula to find the length of each side of a triangle formed by the points $X(1, -9)$, $Y(4, -5)$ and $Z(6, 3)$. **Ex 4.2**

3 **(a)** Use the midpoint formula to find the midpoints of each side of the polygon formed by the points $A(1, -2)$, $B(6, -6)$, $C(4, 3)$ and $D(-4, 2)$. **Ex 4.2**

(b) What shape is this polygon?

4 Solve the following equations. **Ex 4.3**

(a) $3x - 7 = -22$ **(b)** $7x + 4 = -3$ **(c)** $\frac{x}{5} + 9 = 0$

(d) $5 - \frac{x}{6} = -7$ **(e)** $6(x - 2) + 7 = -5$ **(f)** $4(2x - 5) = 0$

5 Solve these equations. **Ex 4.3**

(a) $\frac{4x}{3} - 20 = -8$

(b) $\frac{2x-3}{5} = 7$

(c) $\frac{3-9x}{2} = -4$

(d) $\frac{4(7x+1)}{11} = 8$

6 Solve these equations. **Ex 4.3**

(a) $4x - 5 = 8 - 2x$

(b) $8x + 5 = 3(2x - 11)$

(c) $5x + 18 = 3x$

(d) $2x = 9 - 7x$

(e) $\frac{1-4x}{8} = \frac{3x-7}{4}$

(f) $\frac{2x-5}{3} = \frac{x+7}{4}$

7 (a) The linear equation $\frac{x}{8} - 5 = 4$ has the solution: **Ex 4.3**

A $x = 8$ B $x = 72$ C $x = 37$ D $x = 13$

(b) The linear equation $6x + 21 = 4x + 1$ has the solution:

A $x = 2$ B $x = 10$ C $x = -10$ D $x = -2$

8 Pete keeps pet mice. He has quite a lot of mice. He has just been given four more. His friend Mal also keeps pet mice, but he has 15 fewer than Pete has now. **Ex 4.4**

(a) Write an expression to show how many pet mice Pete has now.

(b) Write an expression to show how many pet mice Mal has now.

(c) If Mal has 10 mice now, use an equation to find out how many mice Pete had originally.

9 A 68 cm screen television set is on sale for \$1850. Terry arranges to pay \$500 as a deposit and then weekly payments of \$90. **Ex 4.5**

(a) Copy and complete this table.

Number of weeks	0	1	2	3	4	5	6
Amount paid (dollars)							

(b) Plot the amount paid for the television set against the number of weeks. Extend your graph to show the amount paid up to 20 weeks. (Plot number of weeks along the horizontal axis.)

(c) From your graph, how much would Terry have paid after 11 weeks?

(d) How long would it take Terry to pay for the television set in full?

(e) Write a relationship connecting the amount paid in dollars, A, to the number of weeks, n.

(f) Use this relationship to calculate the amount paid after 9 weeks. Check to see whether you obtained the same result from your graph.

10 Find the gradient of the line joining (3, 2) and (4, 7). **Ex 4.6**

11 (a) The gradient of the line joining the points (-5, 3) and (2, -4) is: **Ex 4.6**

A -1 B $\frac{1}{7}$ C 1 D $\frac{7}{3}$

(b) The gradient of the line joining the points (-2, 3) and (3, 3) is:

A 0 B $\frac{1}{5}$ C 1 D 5

12 Sketch these graphs on the one number plane: **Ex 4.7**

(a) $y = -5$ (b) $y = 3$ (c) $x = -2$

13 Plot the graph of $y = 2x + 4$ by first completing a table of values for x and y. Find the gradient, x-intercept and y-intercept for this line. **Ex 4.8**

14 Write four points that lie on each of the following lines and draw the graph. **Ex 4.8**

(a) $y = 0$ **(b)** $x = -2$ **(c)** $y = 4$ **(d)** $x = 0$

15 **(a)** The y-intercept of the graph of the linear equation $y + 2x = 3$ is: **Ex 4.9**

A -3 **B** -2 **C** 2 **D** 3

(b) The x-intercept of the graph of the linear equation $3x + 5y = 45$ is:

A 3 **B** 5 **C** 15 **D** 45

(c) The gradient of the graph of the linear equation $7x - 4y = 28$ is:

A $-\frac{7}{4}$ **B** $-\frac{4}{7}$ **C** $\frac{4}{7}$ **D** $\frac{7}{4}$

16 Find the y-intercept and gradient for each of the following, and sketch the graph of each. **Ex 4.9**

(a) $y = 2x - 4$ **(b)** $2y = 3x - 4$ **(c)** $x + y = 5$

(d) $3x + 2y = 4$ **(e)** $y = 3x$ **(f)** $y = -2x$

17 Find the x- and y-intercepts for each of the following and hence sketch graphs of each, showing the intercepts on the axes. **Ex 4.9**

(a) $4x - y = 4$ **(b)** $2x + 5y = 20$ **(c)** $y - 3x = 6$ **(d)** $2x + y = 5$

18 Polly walked at a constant speed for 5 minutes, covering 350 metres.

(a) To construct a distance–time graph, what two points can you use? (Do not draw the graph.)

(b) What is the gradient between these two points?

(c) What is Polly's average speed during these 5 minutes?

(d) Write an equation relating distance covered, d metres, to time taken, t minutes.

(e) Use your equation to find the distance Polly would travel in 8 minutes if she maintained her walking speed.

19 **(a)** The sum of three consecutive whole numbers is 66. What are the numbers?

(b) The sum of four consecutive even whole numbers is 100. What are the numbers?

20 The length of a rectangular paddock is 7 m less than twice the width. If the perimeter is 58 m, write an equation and solve it to find the length and width of the paddock.

Reasoning and communicating

21 Two lines are given by $kx + 6y - 5 = 0$ and $3x + 2y + 4 = 0$. Find the value of k such that the lines are parallel.

22 A bicycle manufacturing company needs to make a certain component to increase its range of products. $3000 needs to be spent on set-up costs and each unit that is produced costs an additional $5.75.

(a) Construct a cost function that describes the total cost of producing x components.

(b) Find the cost of producing 500 components.

(c) How many components could be produced on a budget of $9000?

(d) Draw a graph of the cost function to find the cost of 1000 components.

23 An electrician charges a fixed fee plus an hourly rate. A job that takes 2 hours to complete costs $280 and a job that takes 3.5 hours to complete costs $437.50.

(a) Find the linear model to express the charge $C for t hours of work.

(b) Find the amount she earns if she spends 8 hours on one job.

(c) Find the time taken to complete the job if the charge was $306.25.

24 Mrs Thomas, who works in Lithgow, has arranged to meet her husband at home in exactly 1 hour. Mr Thomas works closer to home but on the same route. He leaves work at the same time as Mrs Thomas but needs to make a stop along the way, as seen in the graph below; however, he still arrives home on time.

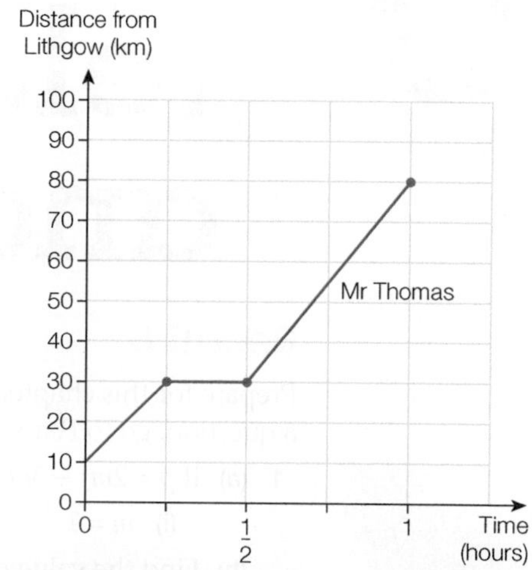

(a) Copy this graph onto graph paper.

(b) How far from Lithgow do they live?

(c) How fast was Mr Thomas travelling in the first 15 minutes?

(d) How long did he stop on his trip home?

(e) How fast did he travel in the last leg of his trip?

(f) How far from Lithgow does Mr Thomas work?

(g) If Mrs Thomas drives at a constant speed from Lithgow, rule a line on your graph that represents her travel.

(h) At what speed does Mrs Thomas need to travel to arrive home 1 hour later?

(i) Write a linear equation that represents her travel, giving the distance from Lithgow at any particular time.

(j) At what time does Mrs Thomas pass Mr Thomas on the way home? (Use your graph.)

(k) At what distance from home does she pass him?

NAPLAN practice 4

Numeracy: Non-calculator

1 Which expression is equivalent to $5 - 7x$?

A $7x - 5$ **B** $-7x - 5$ **C** $-7x + 5$ **D** $7x + 5$

2 What is the gradient of the straight line connecting the points (-3, -4) and (-1, 8)?

A 8 **B** 4 **C** 2 **D** 6

3 What is the value of b in the equation $5b - 4 = 2b + 17$?

4 If $y = -x + 4$ and $y = 2x - 8$, which point lies on both lines?

A (-3, 7) **B** (4, 0) **C** (1, -6) **D** (3, 1)

5 Which of these points lies on the straight line joining (4, 4) and (20, 12)?

A (5, 5) **B** (6, 6) **C** (10, 7) **D** (14, 8)

Numeracy: Calculator allowed

6 What number comes next in the following sequence?

25, 50, 75, 100, …

7 $y = 2x - 1$

$y = 3x + 2$

Which value of x satisfies both these equations?

A $x = -3$ **B** $x = -1$ **C** $x = 1$ **D** $x = 3$

8 Find the value of x that makes the equation $\frac{3x - 5}{2} = 5$ true.

9 The sum of two consecutive numbers is 55. What are the numbers?

A 33 and 22 **B** 22 and 23 **C** 27 and 28 **D** 30 and 25

10 An isosceles triangle has two equal sides and a third side that is 15 cm longer. Find the length of the longest side if the perimeter is 84 cm.

8 Non-linear relationships and proportion

Recall 8

Prepare for this chapter by attempting the following questions. If you have difficulty with a question, go to Pearson Places and download the Recall Worksheet.

1 **(a)** If $p = 2m^2 - 3m + 5$, find the value of p when:

(i) $m = 1$ **(ii)** $m = 0$ **(iii)** $m = -1$

(b) Find the value of $x(x - 3)$ when:

(i) $x = 5$ **(ii)** $x = -3$ **(iii)** $x = 0$ **(iv)** $x = 3$

2 Consider this graph showing the cost charged to dryclean carpet squares of different lengths.

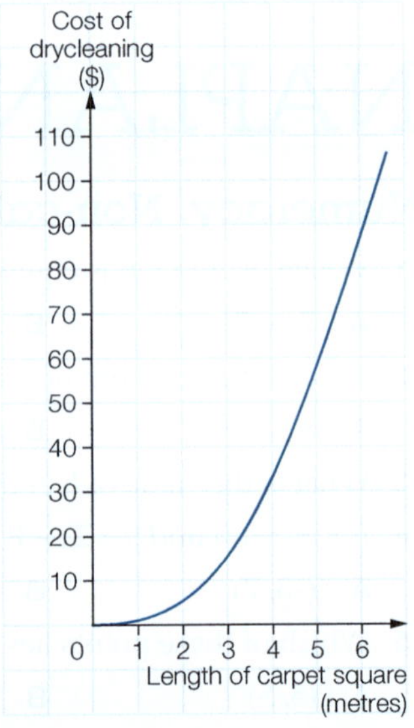

(a) Find the approximate cost to dryclean a carpet square with a length of:

(i) 3 m **(ii)** 5.5 m

(b) Find the approximate length of the carpet square if the drycleaning cost is:

(i) \$48.00 **(ii)** \$18.75

3 Expand the following.

(a) $3x(x + 4)$ **(b)** $(x + 2)(x + 5)$

(c) $(x - 7)(x + 3)$ **(d)** $(x - 5)(x + 5)$

4 Factorise the following.

(a) $x^2 + 2x$ **(b)** $x^2 - 16$ **(c)** $x^2 + 12x + 36$

5 Factorise the following.

(a) $x^2 + 2x + 5x + 10$ **(b)** $x^2 - 2x - 3x + 6$

(c) $x^2 - 14x + x - 14$ **(d)** $x^2 - 7x - 4x + 28$

6 Solve for x.

(a) $3x + 5 = 7$ **(b)** $5x - 2 = 11$ **(c)** $6x - 5 = -11$

Syllabus references

Number and Algebra: Non-linear Relationships

✓ Graph simple non-linear relations, with and without the use of digital technologies **(ACMNA296) (Stage 5.1)**

✓ Explore the connection between algebraic and graphical representations of relations such as simple quadratics, circles and exponentials using digital technologies as appropriate **(ACMNA239) (Stage 5.1) (Stage 5.2◊)**

✓ Graph simple non-linear relationships, with and without the use of digital technologies, and solve simple related equations **(ACMNA296) (Stage 5.2◊)**

✓ Describe, interpret and sketch parabolas, hyperbolas, circles and exponential functions and their transformations **(ACMNA267) (Stage 5.3§)**

Number and Algebra: Equations

✓ Solve simple quadratic equations using a range of strategies **(ACMNA241) (Stage 5.2)**

Number and Algebra: Ratios and Rates

✓ Solve problems involving direct proportion; explore the relationship between graphs and equations corresponding to simple rate problems **(ACMNA208) (Stage 5.2)**

Non-linear graphs

8.1

Stage 5.1

Need to Know

A non-linear graph is any graph connecting variables that do not graph as a straight line. Common non-linear graphs are **parabolas, circles** and **exponential graphs.**

Parabolas

A parabola is a U-shaped graph and can be drawn as  or . Quadratic equations graph as parabolas.

Circles

Circles map points that remain a fixed distance, called the radius, from a given point, the centre of the circle.

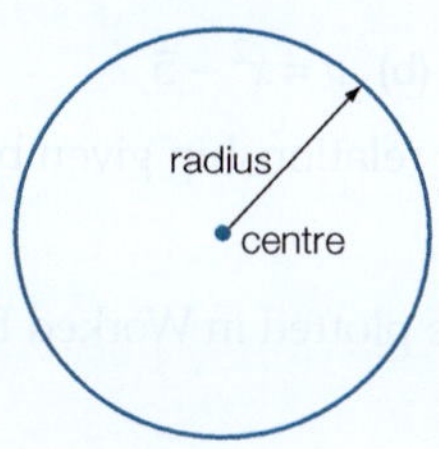

Exponential graphs

Exponential graphs are used to show rapid growth and decay and are drawn as shown below.

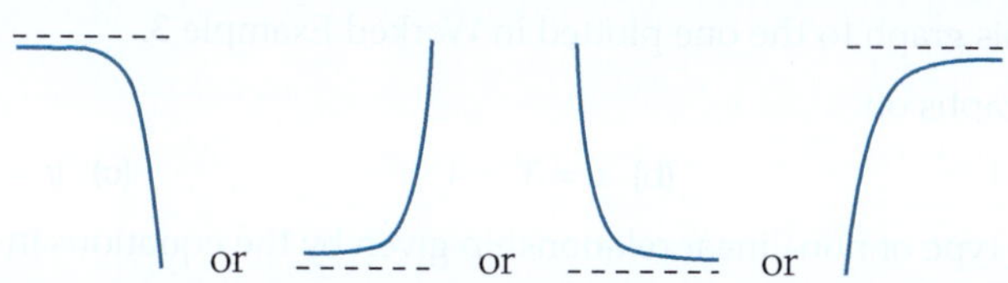

All exponential graphs approach lines that they do not touch or cross. These lines are called **asymptotes**.

Technology Exploration: GeoGebra

Non-linear graphs

Open GeoGebra. Use the Input bar at the bottom of the window to enter the equations to be graphed.

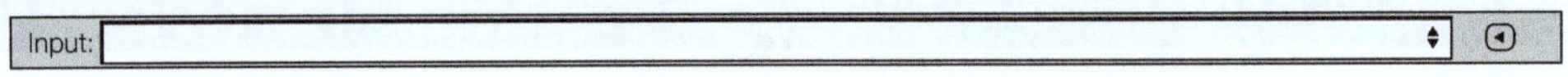

To see the graph clearly, select from the menu View > Layout… and then in the pop-up window go to Preferences - Layout > Basic. You can also adjust the preferences by right-clicking in the Graphics view. The tools Move Graphics View, Zoom Out and Zoom In may also be used.

In the Input bar, use ^ to enter powers and * to enter multiplication.

1 Draw the graph of $y = x^2 + 1$.

2 Compare this graph to the one plotted in Worked Example 1.

3 Draw the graphs of:

(a) $y = 2x^2$ (b) $y = x^2 - 5$ (c) $y = -x^2$

4 Identify the type of non-linear relationship given by the equations in **1** and **3** above.

5 Draw the graph of $x^2 + y^2 = 25$.

6 Compare this graph to the one plotted in Worked Example 2.

7 Draw the graphs of:

(a) $x^2 + y^2 = 64$ (b) $x^2 + y^2 = 100$ (c) $x^2 + y^2 = 12$

8 Identify the type of non-linear relationship given by the equations in **5** and **7** above.

9 Draw the graph of $y = 2^x - 3$.

10 Compare this graph to the one plotted in Worked Example 3.

11 Draw the graphs of:

(a) $y = 2^x + 1$ (b) $y = 3^x - 4$ (c) $y = 1 - 2^x$

12 Identify the type of non-linear relationship given by the equations in **9** and **11** above.

8.1

Worked Example 1

WE1

(a) Copy and complete the following table of values for the equation $y = x^2 + 1$.

x	-5	-4	-3	-2	-1	0	1	2	3	4	5
y											

(b) Plot the graph of the equation given above using the table of values from part **(a)**.

(c) Use the graph to identify the type of non-linear relationship given by the equation.

Working

(a)

x	-5	-4	-3	-2	-1	0	1	2	3	4	5
y	26	17	10	5	2	1	2	5	10	17	26

(b)

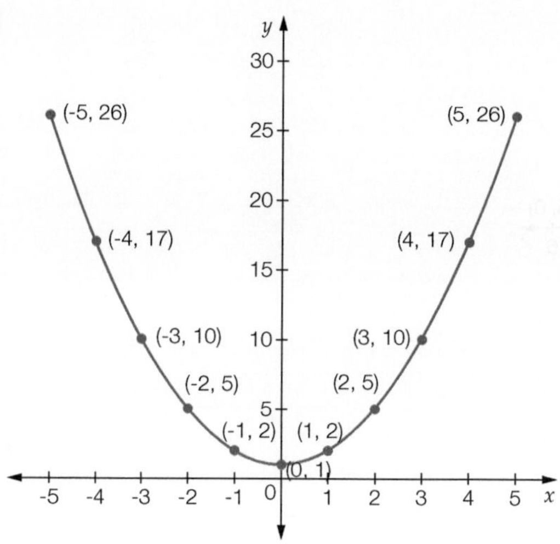

(c) *Quadratic equation.*

8.1

Worked Example 2

WE2

(a) Copy and complete the following table of values for the equation $x^2 + y^2 = 25$.

x	-5	0	5
y			

(b) Plot the graph of the equation given above using the table of values from part **(a)**.

(c) Use the graph to identify the type of non-linear relationship given by the equation.

Working

(a)

x	-5	0	5
y	0	±5	0

(b)

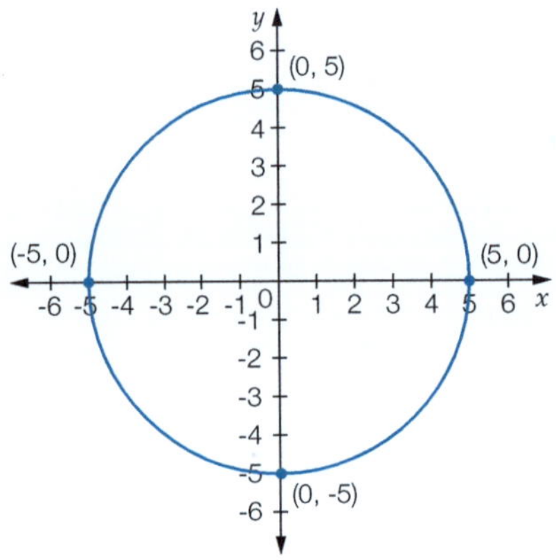

(c) Circle equation.

8.1

Worked Example 3

WE3

(a) Copy and complete the following table of values for the equation $y = 2^x - 3$.

x	-2	-1	0	1	2	3	4	5
y								

(b) Plot the graph of the equation given above using the table of values in part **(a)**.

(c) Use the graph to identify the type of non-linear relationship given by the equation.

Working

(a)

x	-2	-1	0	1	2	3	4	5
y	-2.75	-2.5	-2	-1	1	5	13	29

(b)

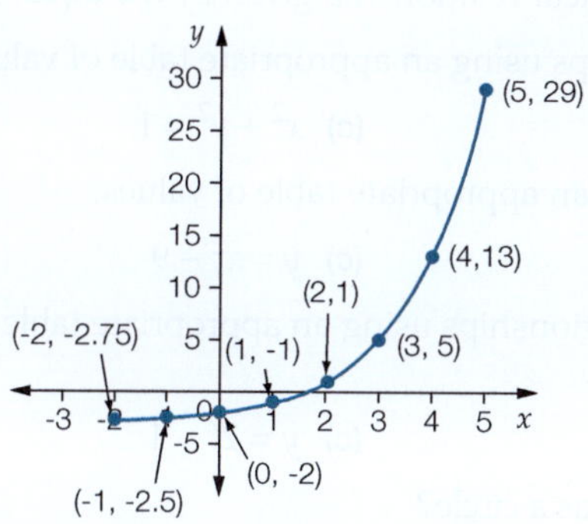

(c) Exponential equation.

8.1 Non-linear graphs

Answers page 475

Stage 5.1

Navigator

Q1, Q2, Q3, Q4, Q5, Q6, Q7, Q10	Q1, Q2, Q3, Q4, Q5, Q6, Q7, Q8, Q10	Q1, Q2, Q3, Q4, Q5, Q6, Q7, Q8, Q9, Q10

Fluency and understanding

WE1

1 **(a)** Copy and complete the table of values for each of the following equations.

(i) $y = 3x^2$ **(ii)** $y = x^2 - 4x + 3$

x	-5	-4	-3	-2	-1	0	1	2	3	4	5
y											

(b) Plot the graph of each of the equations using the table of values in part **(a)**.

(c) Use the graph to identify the type of non-linear relationship given by the equation.

8.1

WE2

2 **(a)** Copy and complete the table of values for each of the following equations.

(i) $x^2 + y^2 = 36$

x	-6	0	6
y			

(ii) $x^2 + y^2 = 4$

x	-2	0	2
y			

(b) Plot the graph of each of the equations given above using the tables of values in part **(a)**.

(c) Use the graph to identify the type of non-linear relationship given by the equation.

WE3

3 **(a)** Copy and complete the table of values for each of the following equations.

(i) $y = 2^x$ **(ii)** $y = 3^x$

x	-2	-1	0	1	2	3	4	5
y								

(b) Plot the graph of each of the equations given above using the table of values in part **(a)**.

(c) Use the graph to identify the type of non-linear relationship given by the equation.

4 Plot the graph of the following circle relationships using an appropriate table of values.

(a) $x^2 + y^2 = 16$ **(b)** $x^2 + y^2 = 9$ **(c)** $x^2 + y^2 = 1$

5 Plot the graph of the following parabolas using an appropriate table of values.

(a) $y = x^2 + 6$ **(b)** $y = 3 - x^2$ **(c)** $y = x^2 - 9$

6 Plot the graph of the following exponential relationships using an appropriate table of values.

(a) $y = -2^x$ **(b)** $y = 3^x + 1$ **(c)** $y = 2^x - 1$

7 Which of the following equations would graph as a circle?

A $y - x^2 = 4$ **B** $y = 5^x - 9$ **C** $x^2 + y^2 = 81$ **D** $y = 5x^2 - 3$

Reasoning, communicating and open-ended

8 Match the following graphs with the equations.

(a) $x^2 + y^2 = 49$ **(b)** $y = 9 + x^2$ **(c)** $y = 3^x - 1$

A

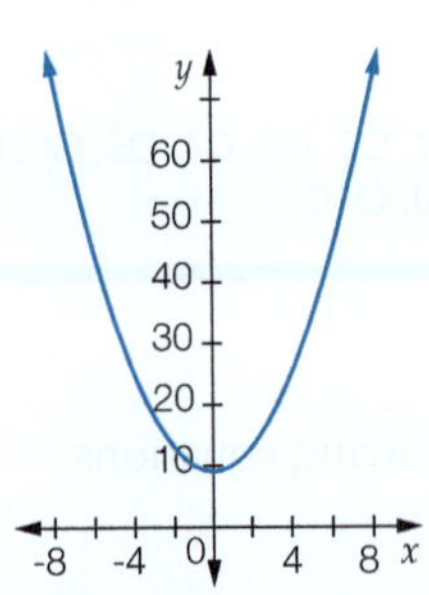

B

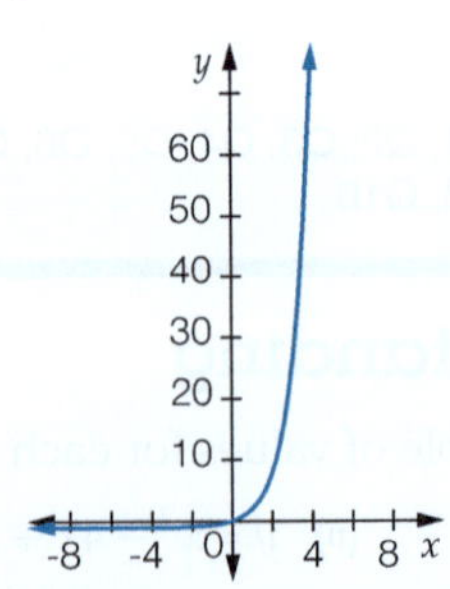

C

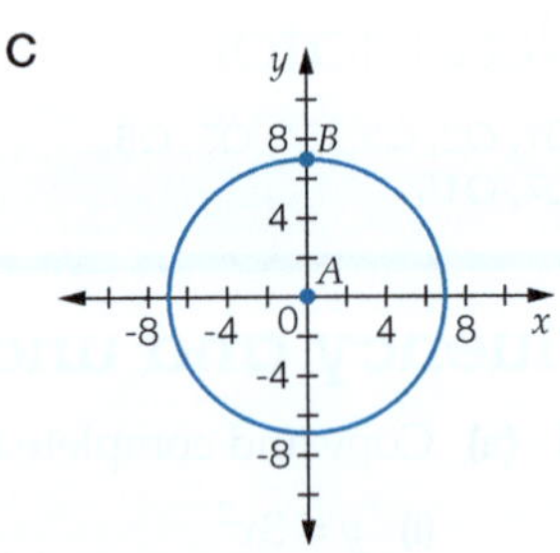

9 **(a)** Explain how you would recognise the equation of a circle relationship.

(b) Explain how you would recognise the equation of an exponential relationship.

(c) Explain how you would recognise the equation of a parabola.

10 Write an equation that would graph as a:

(a) circle **(b)** exponential **(c)** parabola

Quadratic relationships

8.2

Stage 5.2◊

Need to Know

Quadratic relationships

A **quadratic relationship** is a very common type of non-linear relationship. The graph of a quadratic relationship is a special shape called a parabola.

The general **quadratic equation** is given by $y = ax^2 + bx + c$, where $a \neq 0$.

All quadratic equations have a term where the highest power of the independent variable is 2, the dependent variable has a highest power of 1, and all powers are integers.

$y = x^2 + 3x - 4$ and $y = -2x^2 + 9$ are examples of quadratic equations where x is the independent variable and y is the dependent variable.

Axis of symmetry and turning point

A parabola is the graph of a quadratic relationship. Each parabola has an **axis of symmetry** and a **turning point**. An axis of symmetry is an imaginary vertical line that divides the parabola into two mirror-image or symmetrical halves passing through the turning point. The turning point or **vertex** is where the graph changes direction.

A minimum turning point occurs where the graph changes direction at the lowest point (the lowest y-value) of the graph. A maximum turning point occurs where the graph changes direction at the highest point (the highest y-value) of the graph.

The parabola shown on the right has a minimum turning point at (1, -3) and an axis of symmetry at $x = 1$.

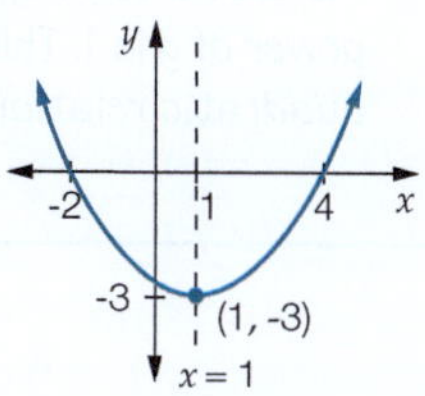

8.2

Graphing quadratic equations

1 Complete a table of values.
2 Draw a set of appropriately sized and scaled axes.
3 Plot the points from the table.
4 Join the points with a smooth curve.
5 Label the graph, the turning point and the axis of symmetry.

Features of the parabola:

- Every parabola is symmetrical about the axis of symmetry.
- Every parabola has a turning point.
- A parabola can have a minimum turning point or a maximum turning point.

x- and y-intercepts

- All parabolas have a y-intercept.
- A parabola can have zero, one or two x-intercepts.

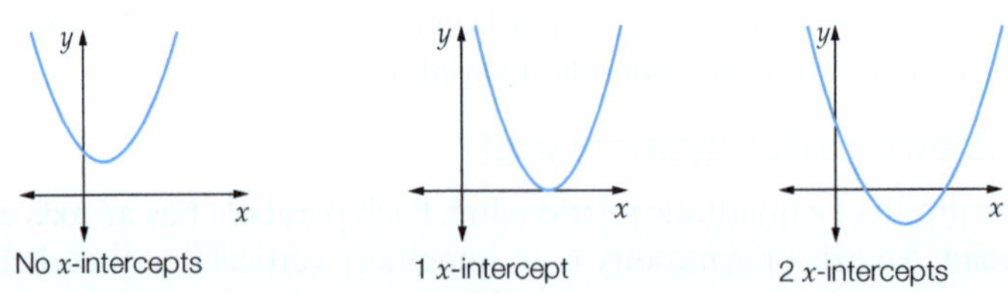

Worked Example 4

WE4

State which of the following are quadratic relationships.

(a) $g = 8f + 6f^2$ (b) $p^2 = r - 3r^2$ (c) $d = 4b^3 - b^2 + 5$

Working

(a) The highest power of f is 2 and the highest power of g is 1. This is a quadratic relationship.

(b) The highest power of r is 2 and the highest power of p is 2. This is not a quadratic relationship.

(c) The highest power of b is 3 and the highest power of d is 1. This is not a quadratic relationship.

8.2

Worked Example 5

WE 5

Plot the graph of the following quadratic equations for integer values of x between -3 and 3 inclusive.

(a) $y = x^2$ **(b)** $y = 2x^2$ **(c)** $y = -x^2$

Working

(a) $y = x^2$

x	-3	-2	-1	0	1	2	3
y	9	4	1	0	1	4	9

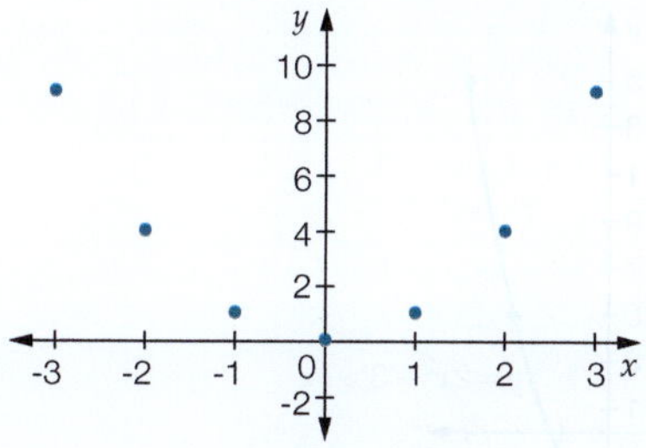

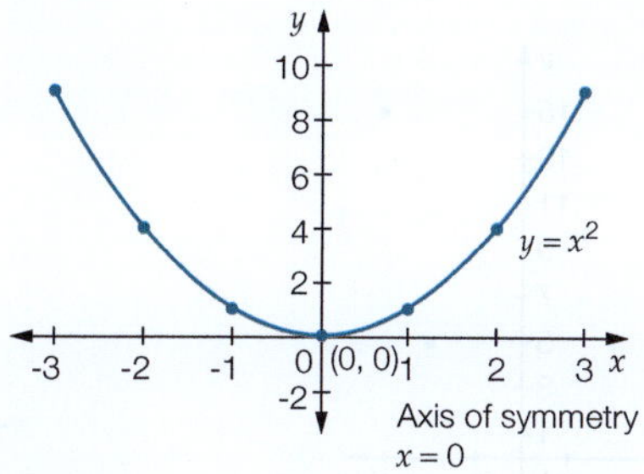

(b) $y = 2x^2$

x	-3	-2	-1	0	1	2	3
y	18	8	2	0	2	8	18

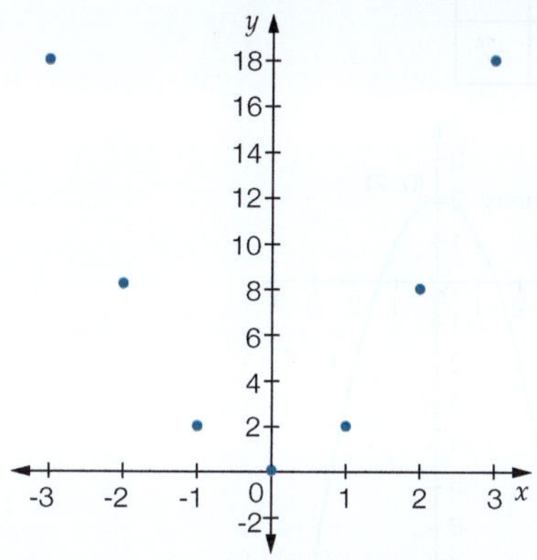

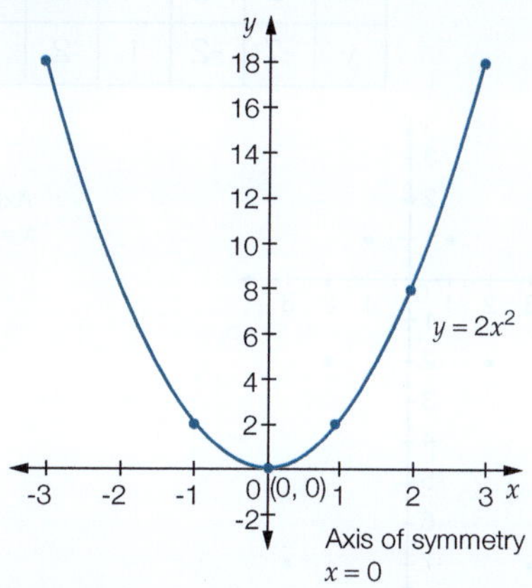

(c) $y = -x^2$

x	-3	-2	-1	0	1	2	3
y	-9	-4	-1	0	-1	-4	-9

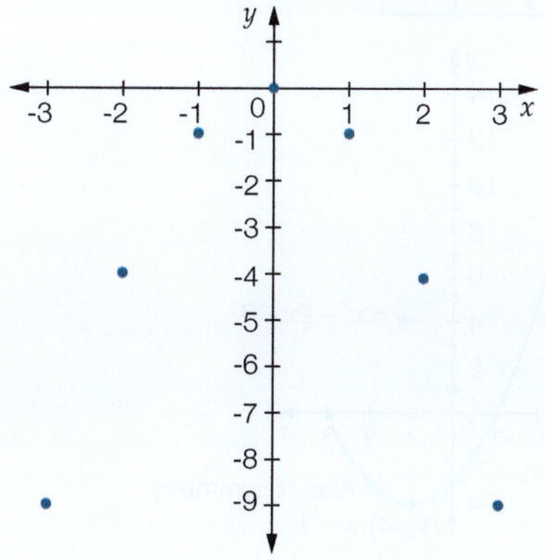

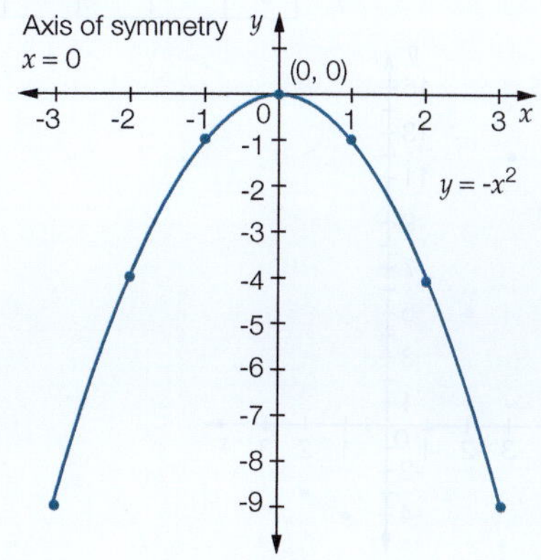

8.2

Worked Example 6

WE6

Plot the graph of the following quadratic equations for integer values of x between -3 and 3 inclusive.

(a) $y = 2x^2 - 3$ **(b)** $y = 2 - x^2$ **(c)** $y = x^2 - 2x - 3$

Working

(a) $y = 2x^2 - 3$

x	-3	-2	-1	0	1	2	3
y	15	5	-1	-3	-1	5	15

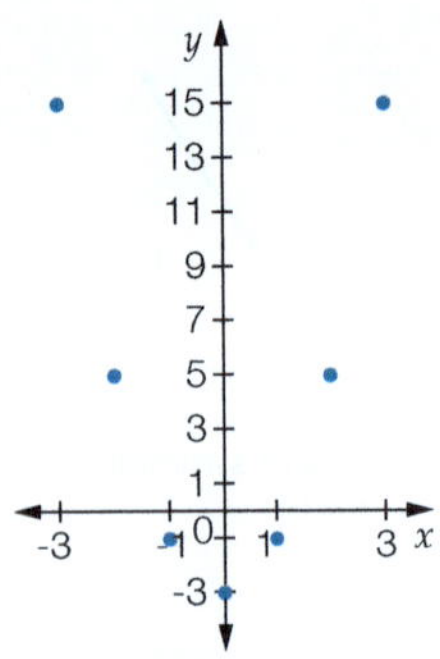

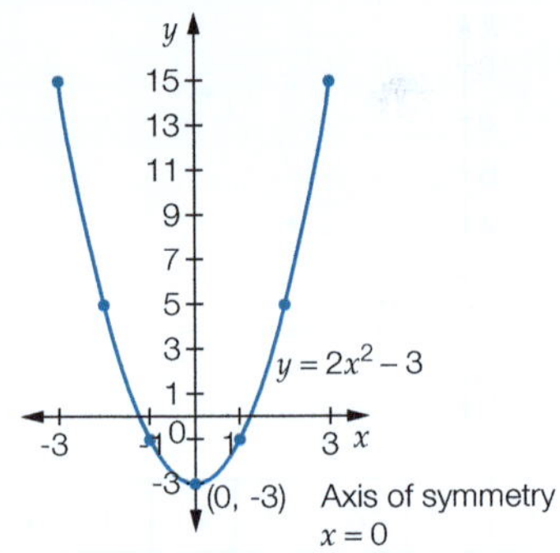

(b) $y = 2 - x^2$

x	-3	-2	-1	0	1	2	3
y	-7	-2	1	2	1	-2	-7

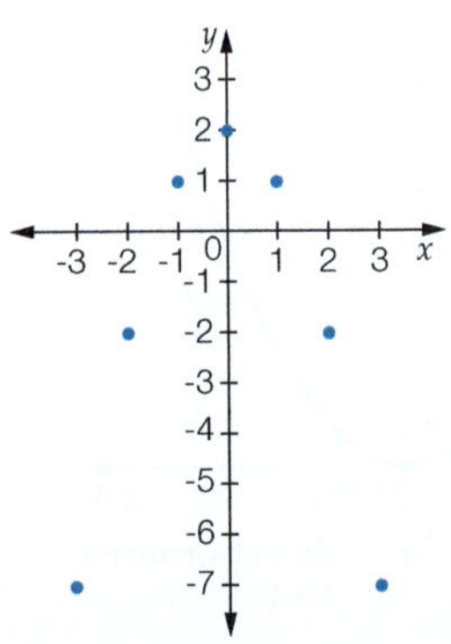

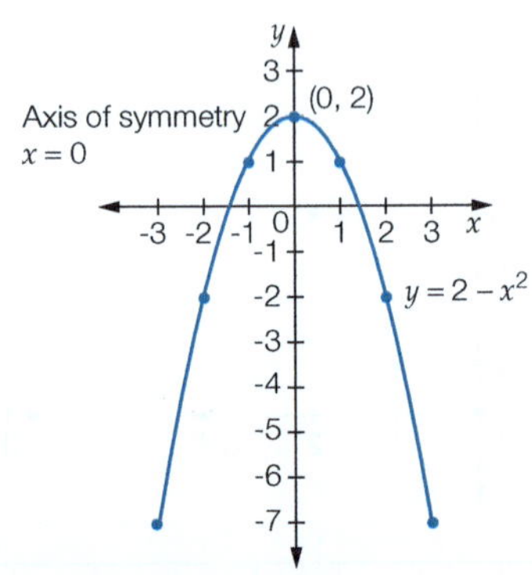

(c) $y = x^2 - 2x - 3$

x	-3	-2	-1	0	1	2	3
y	12	5	0	-3	-4	-3	0

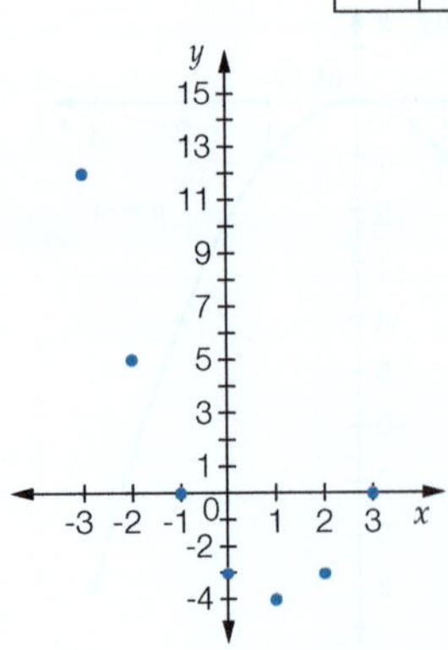

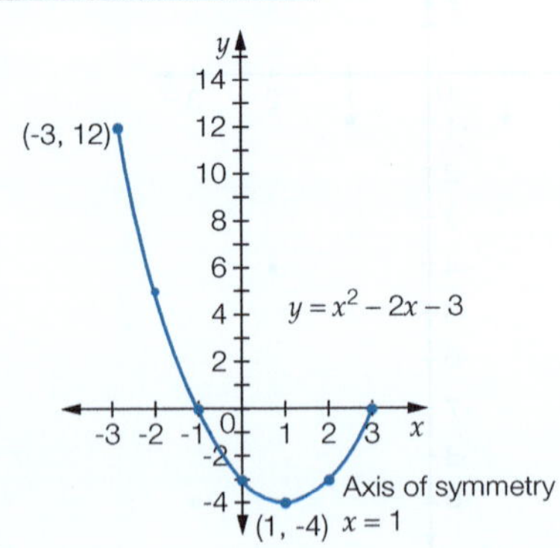

8.2

Worked Example 7

WE 7

For the graph shown, write down:

(a) the coordinates of the turning point
(b) the type of turning point
(c) the equation of the axis of symmetry
(d) the y-intercept
(e) the x-intercepts.

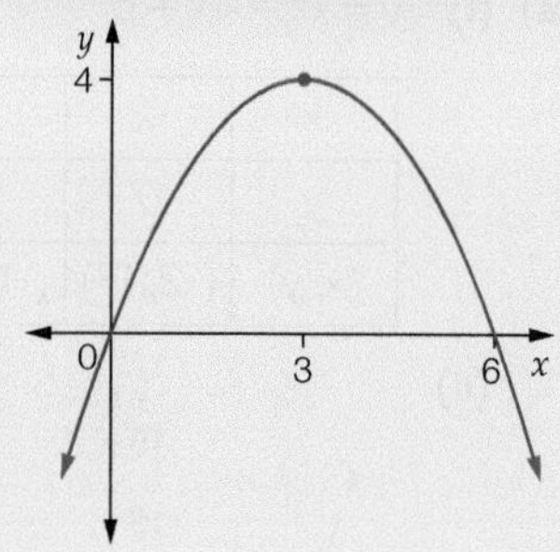

Working

(a) There is a turning point at (3, 4).
(b) Because $y = 4$ is the maximum value of y, this is a maximum turning point.
(c) The axis of symmetry is the line $x = 3$.
(d) The y-intercept is (0, 0).
(e) The x-intercepts are (0, 0) and (6, 0).

Worked Example 8

WE 8

For each of the following equations:

(a) $y = x^2 - 4x + 5$ (b) $y = x^2 - 4x + 4$ (c) $y = x^2 - 4x + 3$

(i) Complete the table of values.

x	-2	-1	0	1	2	3	4	5
y								
(x, y)								

(ii) Plot the points on a Cartesian plane for the values in the table and join with a smooth curve.
(iii) Write down the coordinates of the turning point.
(iv) Does the parabola have a maximum or a minimum turning point?
(v) Write down the equation of the axis of symmetry of the parabola.
(vi) Write down the coordinates of the y-intercept.
(vii) Write down the coordinates of the x-intercepts.
(viii) From the graph, find the values of x when $y = 8$. Hence, state the values of x when the given equation is equal to 8. (Answer correct to one decimal place.)

Working

(a) (i) $y = x^2 - 4x + 5$

x	-2	-1	0	1	2	3	4	5
y	17	10	5	2	1	2	5	10
(x, y)	(-2, 17)	(-1, 10)	(0, 5)	(1, 2)	(2, 1)	(3, 2)	(4, 5)	(5, 10)

(ii)

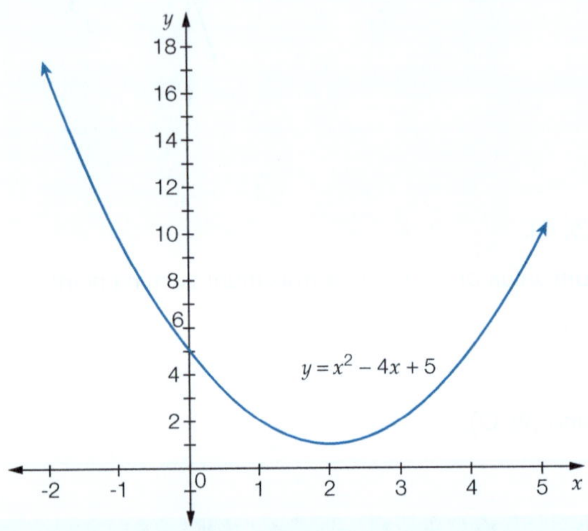

(iii) The turning point is (2, 1).

(iv) $y = 1$ is the minimum value of y, so this is a minimum turning point.

(v) The axis of symmetry is $x = 2$.

(vi) The y-intercept is (0, 5).

(vii) There are no x-intercepts.

(viii)

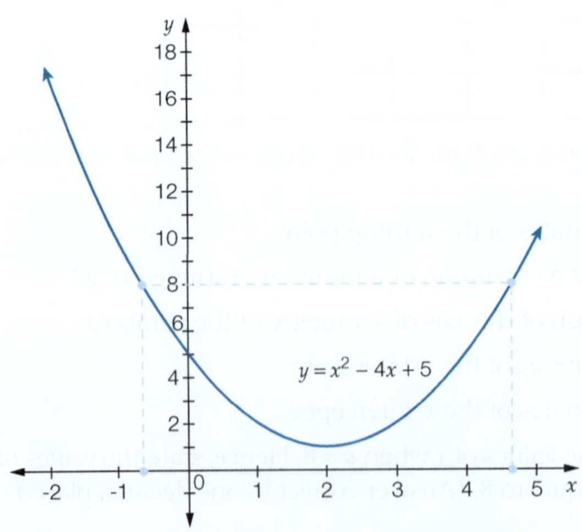

The graph shows $x \approx 4.6$ and -0.6 when $y = 8$.

So, $x^2 - 4x + 5 = 8$ when $x \approx 4.6$ and -0.6.

(b) **(i)** $y = x^2 - 4x + 4$

x	-2	-1	0	1	2	3	4	5
y	16	9	4	2	01	1	4	9
(x, y)	(-2, 16)	(-1, 9)	(0, 4)	(1, 1)	(2, 0)	(3, 1)	(4, 4)	(5, 9)

(ii)

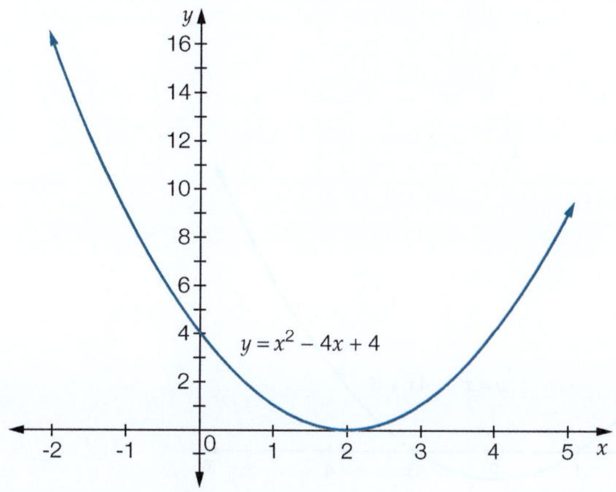

(iii) The turning point is (2, 0).

(iv) $y = 0$ is the minimum value of y, so this is a minimum turning point.

(v) The axis of symmetry is $x = 2$.

(vi) The y-intercept is (0, 4).

(vii) There is one x-intercept at (2, 0).

(viii)

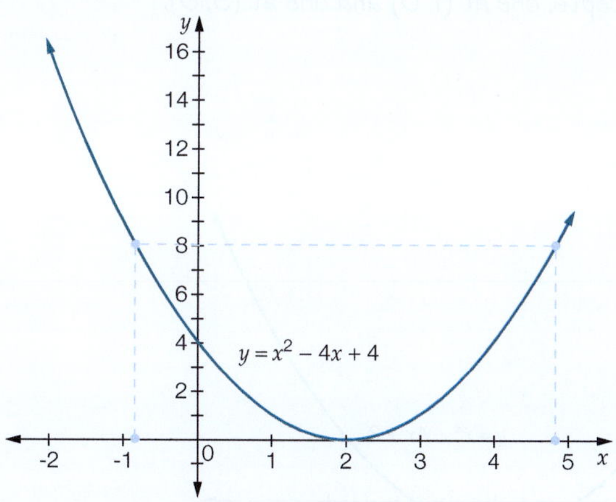

The graph shows $x \approx 4.8$ and -0.8 when $y = 8$.

So, $x^2 - 4x + 4 = 8$ when $x \approx 4.8$ and -0.8.

8.2

(c) (i) $y = x^2 - 4x + 3$

x	-2	-1	0	1	2	3	4	5
y	15	8	3	0	-1	0	3	8
(x, y)	(-2, 15)	(-1, 8)	(0, 3)	(1, 0)	(2, -1)	(3, 0)	(4, 3)	(5, 8)

(ii)

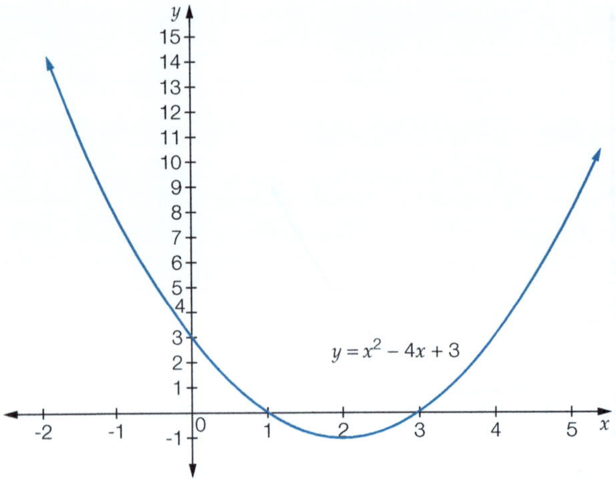

(iii) The turning point is (2, -1).

(iv) y = -1 is the minimum value of y, so this is a minimum turning point.

(v) The axis of symmetry is x = 2.

(vi) The y-intercept is (0, 3).

(vii) There are two x-intercepts, one at (1, 0) and one at (3, 0).

(viii)

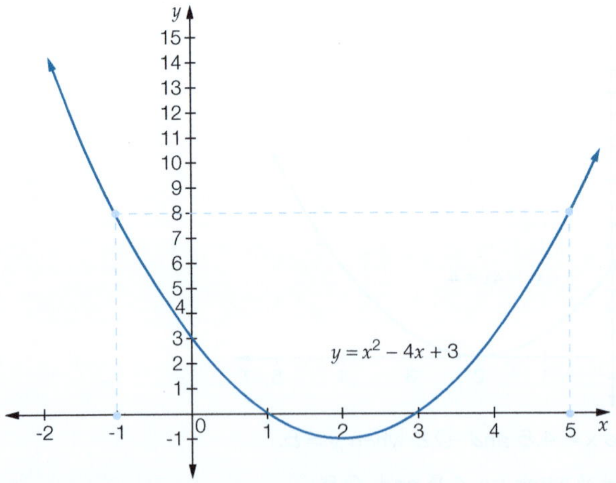

The graph shows x = -1 and 5 when y = 8.

So, $x^2 - 4x + 3 = 8$ when x = -1 and 5.

8.2

8.2 Quadratic relationships

Answers page 478

Stage 5.2◊

Navigator

Q1, Q2, Q3, Q4, Q5, Q6, Q7, Q8, Q9, Q10, Q11, Q12, Q13, Q14, Q16, Q21, Q22, Q24	Q1, Q2, Q3, Q4, Q5, Q6, Q7, Q8, Q9, Q10, Q11, Q12, Q13, Q15, Q16, Q17, Q19, Q20, Q22, Q23	Q1, Q2, Q3, Q4, Q5, Q6, Q7, Q8, Q9, Q10, Q11, Q12, Q14, Q15, Q17, Q18, Q19, Q20, Q21, Q22, Q23

Equipment required: graph paper for most questions

Fluency and understanding

WE4

1 State which of the following are quadratic relationships.

(a) $m = 4k^2 - 6$ (b) $a = -b^2$ (c) $s = 4t^4 - 5t^2 + 7$

(d) $r = 4w - 3$ (e) $l = 2 - 3n^2$ (f) $a^2 = 9 - x^2$

WE5

2 Plot the graph of the following quadratic equations for integer values of x between -3 and 3 inclusive.

(a) $y = 3x^2$ (b) $y = -2x^2$ (c) $y = \frac{x^2}{2}$

WE6

3 Plot the graph of the following quadratic equations for integer values of x between -4 and 4 inclusive.

(a) $y = 3x^2 - 3$ (b) $y = -2x^2 + 3$ (c) $y = \frac{x^2}{2} - 2$

(d) $y = 3x^2 - 6x$ (e) $y = -2x^2 + 5x + 3$ (f) $y = \frac{x^2}{2} - \frac{3x}{2} - 2$

WE7

4 For each of the following graphs, write down:

(i) the coordinates of the turning point (ii) the type of turning point

(iii) the equation of the axis of symmetry (iv) the y-intercept

(v) the x-intercepts.

(a)

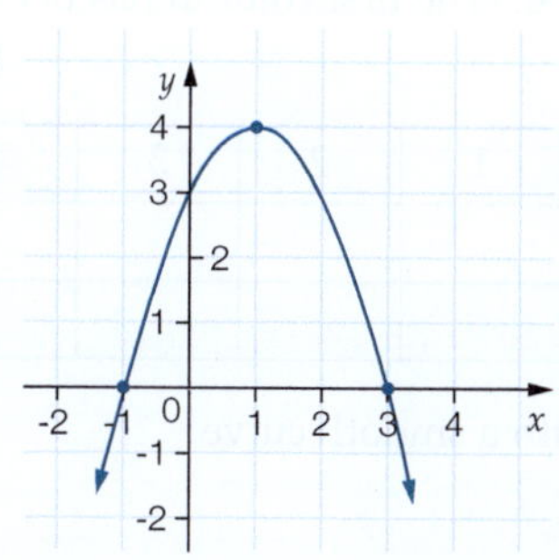

(b)

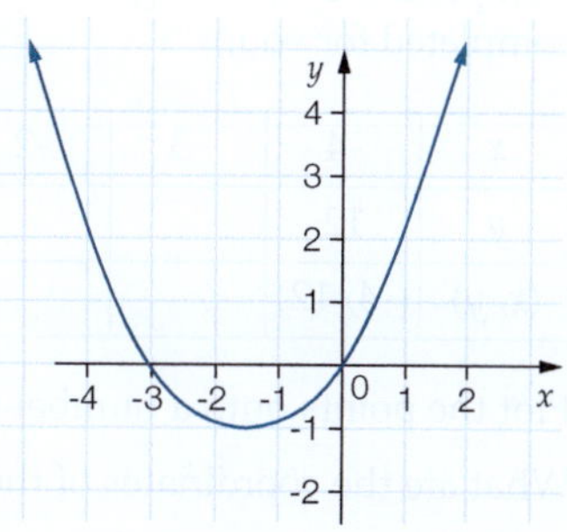

WE8

5 For each of the following equations:

(a) $y = x^2 - 2x + 2$ (b) $y = x^2 - 2x - 3$ (c) $y = x^2 - 2x + 1$

x	-2	-1	0	1	2	3	4	5
y								

(i) copy and complete the table of values.

(ii) Plot the points on a Cartesian plane for the values in the table and join with a smooth curve.

8.2

(iii) Write down the coordinates of the turning point.

(iv) Does the parabola have a maximum or a minimum turning point?

(v) Write down the equation of the axis of symmetry of the parabola.

(vi) Write down the coordinates of the y-intercept.

(vii) Write down the coordinates of the x-intercepts.

(viii) From the graph, find the values of x when $y = 5$. Hence, state the values of x when the given equation is equal to 5.

6 (a) Which of the following does not show a quadratic relationship?

A $v = w - 4w^2$ B $f^2 = 3d^2 - 1$ C $p = 6r^2 - 7$ D $k^2 = t$

(b) If $p = 3m^2 + 4$, then if m equals 5, p will be equal to:

A 19 B 34 C 75 D 79

7 The graph of $y = 2x^2 - 4x + 1$ is shown.

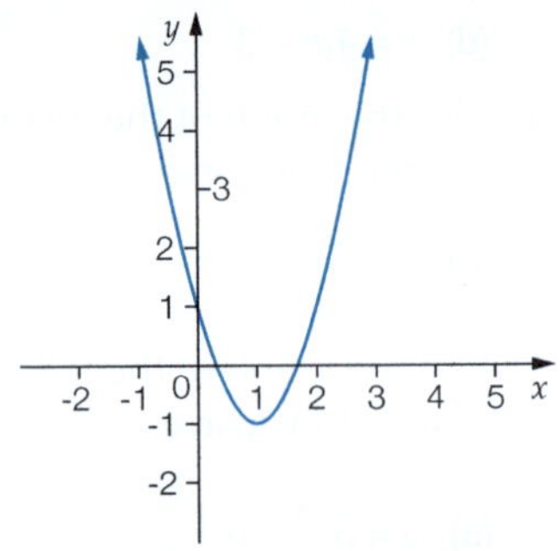

(a) For this graph the turning point would be:

A (1, 0) B (1, 1)

C (1, -1) D (-1, 1)

(b) The value(s) of x for $2x^2 - 4x + 1 = 0$ would be approximately:

A -0.8 and 1.8 B 1 and 0

C 0.3 and 1.7 D 0 and 2

(c) The value(s) of x for $2x^2 - 4x + 1 = 1$ would be approximately:

A 0 and 2 B 0.3 and 1.7 C 0 and 1 D -0.8 and 1.8

8 Compare your graphs in Question **2 (a)–(c)** to your graphs in Question **3 (a)–(c)**.

Comment on any similarities and differences.

9 Compare your graphs in Question **2 (a)–(c)** to your graphs in Question **3 (d)–(f)**.

Comment on any similarities and differences.

10 (a) Complete the following table of values for $y = x^2 - 4$. (The first column has been completed for you.)

x	-4	-3	-2	-1	0	1	2	3	4
y	12								
(x, y)	(-4, 12)								

(b) Plot the points onto a number plane. Join them with a smooth curve.

(c) What are the coordinates of the turning point?

(d) Does the parabola have a maximum or a minimum turning point?

(e) Write down the equation of the axis of symmetry of the parabola.

(f) Write down the coordinates of the y-intercept.

(g) Write down the coordinates of the x-intercepts.

(h) From the graph, find the value of y when $x = 3.5$ (correct to one decimal place).

(i) From the graph, find the values of x when $y = -2$ (correct to one decimal place).

(j) State the values of x when $x^2 - 4 = 5$. Hence, find the values of x when $x^2 - 9 = 0$.

8.2

11 **(a)** Complete the following table of values for $y = -2x^2 - 6x$.

x	-5	-4	-3	-2	-1	0	1	2
x^2								

(b) Plot the graph of $y = -2x^2 - 6x$ using graph paper.

(c) Write down the coordinates of the turning point. State whether the parabola has a maximum or a minimum turning point.

(d) Write down the coordinates of any x- and y-intercepts.

(e) What are the x-values when $y = 4$?

(f) If $y = 4$, what does $-2x^2 - 6x$ equal? Hence, state the x-values when $-2x^2 - 6x = 4$. These x-values are the solution to the equation $-2x^2 - 6x = 4$.

(g) Use the graph to find the x-values that solve the equation $-2x^2 - 6x = -8$.

(h) Find the x-values that solve $-2x^2 - 6x = 0$.

12 **(a)** Complete a table of values for $y = x^2 - x - 2$ for integer values of x between -3 and 5.

(b) Plot the graph of $y = x^2 - x - 2$ for the points in your table.

(c) Write down the coordinates of the turning point. Does the parabola have a maximum or a minimum turning point?

(d) Write down the coordinates of any x- and y-intercepts.

(e) Use the parabola to solve x in each of the following equations.

(i) $x^2 - x - 2 = 0$ **(ii)** $x^2 - x - 2 = 4$ **(iii)** $x^2 - x - 2 = 2$

(iv) $x^2 - x - 2 = -2.25$ **(v)** $x^2 - x - 2 = -4$ **(vi)** $x^2 - x - 2 = 1\frac{1}{2}$

13 **(a)** Plot the graph of $y = x^2 + 6x + 10$. Use integer values of x between -7 and 1.

(b) What are the coordinates of the turning point?

(c) Write down the coordinates of any x- and y-intercepts.

(d) Use the graph to solve each of the following equations.

(i) $x^2 + 6x + 10 = 1$ **(ii)** $x^2 + 6x + 10 = 10$

(iii) $x^2 + 6x + 10 = 0$ **(iv)** $x^2 + 6x + 10 = 15$

14 **(a)** Plot the graphs of the following parabolas on separate axes on graph paper. (Remember to complete a table of values first.)

(i) $y = -x^2 + 4x$ for $-1 \le x \le 5$

(ii) $y = x^2 + 3x - 10$ for $-6 \le x \le 3$

(iii) $y = -x^2 + 2x + 3$ for $-2 \le x \le 4$

(b) Use your graphs to solve the following equations to find x.

(i) $-x^2 + 4x = 0$ **(ii)** $x^2 + 3x - 10 = 0$ **(iii)** $-x^2 + 2x + 3 = 4$

(iv) $-x^2 + 4x = 4$ **(v)** $x^2 + 3x - 10 = 8$ **(vi)** $-x^2 + 2x + 3 = 0$

(vii) $-x^2 + 4x = 2$ **(viii)** $x^2 + 3x - 10 = 4$ **(ix)** $-x^2 + 2x + 3 = -2$

Reasoning, communicating and open-ended

Hint

Use only realistic values of x for your table of values.

15 A postage stamp has a length of $(x + 2)$ cm and a width of $(x + 1)$ cm.

(a) Write an equation for the area, h, of the postage stamp in expanded form.

(b) Complete a table of values for x and A.

(c) Plot a graph of your equation showing x on the horizontal axis (use 1 cm = 1 cm) and A on the vertical axis (use 1 cm = 1 cm^2).

(d) Use the graph to find the value x represents if the area of the stamp is 8 cm^2. Did you obtain one or two values for x in this problem? Explain.

(e) What is the length and width of the stamp if its area is 8 cm^2? (Give your answers correct to one decimal place.)

16 Karrie, a keen golfer, hits a golf ball along the fairway. The height of the golf ball, h, in metres, at any time, t, in seconds, during its flight is given by the equation $h = -5t^2 + 20t$.

(a) Complete a table of values for t and h.

(b) Plot the graph of $h = -5t^2 + 20t$. Show the values of t on the horizontal axis (use 1 cm = 0.25 s) and values for h on the vertical axis (use 1 cm = 2 m).

Use the graph to answer the following questions.

(c) How high does the ball get above the ground?

(d) How long does it take to reach this maximum height?

(e) When does the ball first reach a height of 12 m?

(f) How long does it take before the ball hits the ground?

(g) For how long was the ball at least 8 m above the ground?

17 A worker's vacuum flask was accidentally dropped from the top of a multistorey building. The relationship between the height, h, above the ground of the flask (in metres) and the time, t, for it to fall to that height (in seconds) was found to be:

$h = 125 - 5t^2$

(a) Complete a table of values for t and h.

(b) Plot the graph of $h = 125 - 5t^2$. Show values of t on the horizontal axis (use 1 cm = 1 s) and values for the height h on the vertical axis (use 1 cm = 20 m).

(c) What is the height of the building?

Hint

The first second occurs between $t = 0$ and $t = 1$.

(d) How long did it take for the flask to hit the ground?

(e) How high above the ground was the flask after 2.5 s? (Answer correct to the nearest metre.)

(f) What distance, to the nearest metre, had the flask fallen after 2.5 s?

(g) How long did it take for the flask to fall 50 m? (Answer correct to one decimal place.)

(h) How long did it take for the flask to fall 100 m? (Answer correct to one decimal place.)

(i) What distance did the flask fall during the fourth second?

(j) During which second did the flask fall the greatest distance?

18 The speed of a car was tracked by radar in a police helicopter over a period of 10 minutes. The police found that the car's speed could be represented by the equation, $s = 3t^2 - 30t + 135$, where s is the speed of the car in km/h after t minutes have passed.

(a) Plot a graph representing this information, showing values of t on the horizontal axis. (Use 1 cm = 2 minutes) and values of s on the vertical axis (use 1 cm = 20 km/h).

Use the graph to answer the following questions.

(b) What was the maximum speed recorded for the car?

(c) What was the minimum speed recorded for the car?

(d) What was the speed of the car after 3 minutes?

(e) What was the speed of the car after 7 minutes?

(f) When was the car travelling at 80 km/h?

(g) When was the car travelling at 120 km/h?

(h) The speed limit along this section of road is 100 km/h. In which time intervals was the car driven above the speed limit?

19 The entrance to a one-way railway tunnel forms an arch. At any particular distance, d, across the base of the tunnel, starting from the left, the height, h, of the arch can be found using the equation $h = -d^2 + 6d$. Both h and d are in metres.

(a) Plot the graph of $h = -d^2 + 6d$. (Show d on the horizontal axis.)

Use the graph to answer the following questions.

(b) How wide is the base of the tunnel?

(c) What is the maximum height of the tunnel?

(d) How high is the tunnel 1 m along the base from either side of the tunnel?

For a train to pass safely through the tunnel there must be a minimum vertical clearance of 1 m between the train and the arch. Assume the train has a rectangular cross-section. Use this information to answer the following questions.

(e) If a particular train happened to be 4 m wide, what would be its maximum allowable height?

(f) What is the maximum allowable height of a train that is 3 m wide?

(g) Find the maximum width of a train if it just fits through the tunnel and is 4.8 m high.

(h) Decide whether trains of the following dimensions would fit through the tunnel.

(i) width 2.6 m and height 6 m **(ii)** width 3.8 m and height 4.5 m

20 The volume of liquid in a container can be found using the equation $V = 3t^2 + 4t$, where V is in cm^3 and t is the number of seconds after pouring commences.

(a) Plot the graph of $V = 3t^2 + 4t$. (Use values of t from 0 to 7 and show t on the horizontal axis.)

Use your graph to answer the following questions.

(b) Find the amount of liquid in the container at the start.

(c) Find the amount of liquid in the container after 4 minutes.

(d) The container can hold only 132 cm^3. How long, in seconds, does it take to fill the container?

(e) How much liquid is added in the first 2 seconds?

(f) What are the feasible values for time and for volume?

21 **(a)** Write down at least three expressions that are quadratic expressions.

(b) Write down at least three expressions that are not quadratic expressions.

22 Make up at least two quadratic equations. For each equation:

(a) draw up a table of values **(b)** plot the graph

(c) find the turning point **(d)** find the value of any x- and y-intercepts.

23 **(a)** If the length of a rectangle is 3 cm longer than its width, write down the area of the rectangle.

(b) Draw a graph of this relationship.

(c) Choose a value for the area of the rectangle and use your graph to find the dimensions of your rectangle, correct to two decimal places.

24 Josh is having some trouble filling in a table of values for a quadratic equation. Explain what he is doing wrong.

(a) $y = -x^2 + 2x$

x	-3	-2	-1	0
y	3	0	-1	0

(b) $y = 2x^2 + 1$

x	-2	-1	0	1	2
y	17	5	1	5	17

Solving quadratic equations

8.3

Stage 5.2

Need to Know

The Null Factor Law

If the product of two or more numbers is zero, then at least one of the numbers must be zero.

If $ab = 0$, then $a = 0$, $b = 0$ or both a and $b = 0$.

To use the **Null Factor Law** to solve quadratic equations, two requirements need to be met.

1. One side of the equation must equal zero.
2. The other side must be the product of factors. (Factorisation may be necessary to achieve this.)

If the quadratic expression is not in a factorised form, we must try to factorise it first. Not all quadratics will factorise to give real number solutions.

Solving non-factorised quadratics

For equations in the form of $ax^2 + bx = 0$:

- identify the highest common factor (HCF) and factorise
- use the Null Factor Law to find solutions.

For equations in the form of $x^2 - a^2 = 0$:

- insert $+ ax - ax$ to make four terms
- factorise by grouping in pairs to get $(x + a)(x - a) = 0$
- use the Null Factor Law to find solutions.

For equations in the form of $x^2 + 2ax + a^2 = 0$ or $x^2 - 2ax + a^2 = 0$:

- split $2ax$ into ax and ax to make four terms
- factorise by grouping in pairs to get $(x + a) = 0$ or $(x - a) = 0$
- use the Null Factor Law to find solutions.

Notice that there is only one solution for quadratics in this form. Remember that some parabolas only touch the x-axis at one point. The equation to find the x-intercept of these parabolas will have only one solution.

8.3

Worked Example 9

WE9

Solve the following equations using the Null Factor Law.

(a) $x(x+3)=0$ **(b)** $(x+3)(x-2)=0$ **(c)** $(x-1)^2=0$

Working

(a) $x(x+3)=0$

$x=0$ or $x+3=0$

$x=0$ or $x=-3$

(b) $(x+3)(x-2)=0$

$x+3=0$ or $x-2=0$

$x=-3$ or $x=2$

(c) $(x-1)^2=0$

$(x-1)(x-1)=0$

$x-1=0$

$x=1$

Worked Example 10

WE10

Solve the following equations using the Null Factor Law.

(a) $2x^2+4x=0$ **(b)** $7x^2=x$

Working

(a) $2x^2+4x=0$

$2x(x+2)=0$ (2x is the HCF)

$2x=0$ or $x+2=0$

$x=0$ or $x=-2$

(b) $7x^2=x$

$7x^2-x=0$

$x(7x-1)=0$ (x is the HCF)

$x=0$ or $7x-1=0$

$x=0$ or $7x=1$

$x=\frac{1}{7}$

Worked Example 11

WE11

Solve the following equations using the Null Factor Law.

(a) $x^2+10x+21=0$ **(b)** $x^2-12x+27=0$

Working

(a) $x^2+10x+21=0$

$x^2+3x+7x+21=0$

$x(x+3)+7(x+3)=0$

$(x+3)(x+7)=0$

$x+3=0$ or $x+7=0$

$x=-3$ or $x=-7$

(b) $x^2-12x+27=0$

$x^2-9x-3x+27=0$

$x(x-9)-3(x-9)=0$

$(x-9)(x-3)=0$

$x-9=0$ or $x-3=0$

$x=9$ or $x=3$

8.3

Worked Example 12

WE 12

Solve the following equations using the Null Factor Law.

(a) $x^2 - 49 = 0$ (b) $x^2 - 7 = 0$

Working

(a)

$$x^2 - 49 = 0$$
$$x^2 - 7x + 7x - 49 = 0$$
$$x(x - 7) + 7(x - 7) = 0$$
$$(x - 7)(x + 7) = 0$$
$$x - 7 = 0 \quad \text{or} \quad x + 7 = 0$$
$$x = 7 \quad \text{or} \quad x = -7$$

(b)

$$x^2 - 7 = 0$$
$$x^2 - \sqrt{7}x + \sqrt{7}x - 7 = 0$$
$$x(x - \sqrt{7}) + \sqrt{7}(x - \sqrt{7}) = 0$$
$$(x - \sqrt{7})(x + \sqrt{7}) = 0$$
$$(x - \sqrt{7}) = 0 \quad \text{or} \quad (x + \sqrt{7}) = 0$$
$$x = \sqrt{7} \quad \text{or} \quad x = -\sqrt{7}$$

Worked Example 13

WE 13

Solve the following equations using the Null Factor Law.

(a) $x^2 - 6x + 9 = 0$ (b) $2x^2 + 48x = -288$

Working

(a)

$$x^2 - 6x + 9 = 0$$
$$x^2 - 3x - 3x + 9 = 0$$
$$x(x - 3) - 3(x - 3) = 0$$
$$(x - 3)(x - 3) = 0$$
$$x - 3 = 0$$
$$x = 3$$

(b)

$$2x^2 + 48x = -288$$
$$2x^2 + 48x + 288 = 0$$
$$2(x^2 + 24x + 144) = 0$$
$$2(x^2 + 12x + 12x + 144) = 0$$
$$2(x(x + 12) + 12(x + 12)) = 0$$
$$2(x + 12)(x + 12) = 0$$
$$(x + 12)^2 = 0$$
$$(x + 12)(x + 12) = 0$$
$$x + 12 = 0$$
$$x = -12$$

8.3

Answers page 481

8.3 Solving quadratic equations

Stage 5.2

Navigator

Q1 Column 1, Q2 Column 1, Q3 Column 1, Q4 Column 1, Q5 Column 1, Q6, Q7, Q8, Q9, Q15	Q1 Column 2, Q2 Column 2, Q3 Column 2, Q4 Column 2, Q5 Column 2, Q6, Q7, Q8, Q9, Q10, Q11	Q1 Column 3, Q2 Column 3, Q3 Column 3, Q4 Column 3, Q5 Column 3, Q6, Q7, Q8, Q9, Q10, Q11, Q12, Q13, Q14

Fluency and understanding

WE9

1 Solve the following equations using the Null Factor Law.

(a) $x(x-1)=0$ (b) $x(x+8)=0$ (c) $x(x+5)=0$

(d) $(x-4)(x-1)=0$ (e) $(x-2)(x-6)=0$ (f) $(x-3)(x-9)=0$

(g) $(x-4)(x+1)=0$ (h) $(x-3)(x+7)=0$ (i) $(x-5)(x+2)=0$

(j) $(x+4)(x+6)=0$ (k) $(x+7)(x+2)=0$ (l) $(x+3)(x+10)=0$

(m) $(x-4)^2=0$ (n) $(x+7)^2=0$ (o) $(x-9)^2=0$

WE10

2 Solve the following equations using the Null Factor Law.

(a) $x^2-6x=0$ (b) $x^2-x=0$ (c) $x^2-8x=0$

(d) $x^2+7x=0$ (e) $x^2+5x=0$ (f) $x^2+x=0$

(g) $2x^2-8x=0$ (h) $3x^2-9x=0$ (i) $7x^2-28x=0$

(j) $3x^2+27x=0$ (k) $5x^2+45x=0$ (l) $12x^2+4x=0$

WE11

3 Solve the following equations using the Null Factor Law.

(a) $x^2+6x+8=0$ (b) $x^2+9x+20=0$ (c) $x^2+12x+32=0$

(d) $x^2-11x+28=0$ (e) $x^2-12x+35=0$ (f) $x^2-15x+56=0$

(g) $x^2+2x-8=0$ (h) $x^2+2x-15=0$ (i) $x^2-9x-22=0$

WE12

4 Solve the following equations using the Null Factor Law.

(a) $x^2-9=0$ (b) $x^2-1=0$ (c) $x^2-16=0$

(d) $2x^2-98=0$ (e) $3x^2-27=0$ (f) $7x^2-28=0$

(g) $x^2-11=0$ (h) $x^2-29=0$ (i) $x^2-47=0$

WE13

5 Solve the following equations using the Null Factor Law.

(a) $x^2-6x+9=0$ (b) $x^2-2x+1=0$ (c) $x^2-8x+16=0$

(d) $x^2+10x+25=0$ (e) $x^2+12x+36=0$ (f) $x^2+18x+81=0$

(g) $2x^2-28x+98=0$ (h) $3x^2+18x+27=0$ (i) $7x^2-28x+28=0$

6 (a) The solution(s) to $x(x+1)=0$ is/are:

A $x=-1$ B $x=1$ C $x=0, x=-1$ D $x=0, x=1$

(b) The solution(s) to $(x+2)(x-3)=0$ is/are:

A $x=2, x=3$ B $x=-2, x=3$ C $x=2, x=-3$ D $x=-2, x=-3$

8.3

7 When three times a number is added to the square of that number, the result is seven times the number. Use an equation to find the number if it is greater than zero.

8 The product of $(2x + 1)$ and $(2x - 3)$ is equal to the product of $(x + 2)$ and $(5x - 6)$. Find x.

9 The height of a golf ball can be represented by the equation $h = 5t - t^2$, where t represents the time in seconds.

The golf ball will be 4 m above the ground when t equals:

A 0 seconds **B** 4 seconds **C** 1 second **D** Never

Reasoning, communicating and open-ended

10 A square vegetable garden has one side extended by 4 m and the other side reduced by 4 m. The area is now 65 m^2. What was the side length of the original vegetable garden?

> **Hint**
> Write an equation and solve it to find the original length of the vegetable garden.

11 A graph has equation $y = (x + 1)(x - 4)$. Find the x-intercepts of this graph.

12 Determine the x-intercepts of the graphs of the following quadratic relationships.

(a) $y = x^2 - 144$ **(b)** $y = 6x^2 + 24x$ **(c)** $y = x^2 - 12x + 36$

13 **(a)** Expand $(x - 6)(x + 2)$.

(b) Use your answer in part **(a)** to find the solution to $x^2 - 4x - 12 = 0$.

(c) How are your solutions in part **(b)** related to the constant term in $y = x^2 - 4x - 12$?

(d) Expand $(x - 5)(x + 1)$.

(e) Use your answer in part **(d)** to find the solution to $x^2 - 4x - 5 = 0$.

(f) How are your solutions for part **(e)** related to the constant term in $y = x^2 - 4x - 5$?

(g) Solve the quadratic equation $x^2 - 4x = 0$.

(h) What is the value of the constant term in $x^2 - 4x$?

(i) How are your solutions for part **(g)** related to the constant term in $y = x^2 - 4x$?

(j) Expand $(x - 1)(x - 3)$.

(k) Use your answer in part **(j)** to find the solution to $x^2 - 4x + 3 = 0$.

(l) How are your solutions for part **(k)** related to the constant term in $y = x^2 - 4x + 3$?

(m) Solve the quadratic equation $x^2 - 4x + 4 = 0$.

(n) How are your solutions in part **(m)** related to the constant term in $y = x^2 - 4x + 4$?

(o) What does the constant term in the equations in parts **(c)**, **(f)**, **(i)** and **(n)** represent graphically?

(p) What do the equations in parts **(b)**, **(e)**, **(g)**, **(k)** and **(m)** all have in common?

(q) What information does the solution to these five equations give you about the graphs of each of these equations?

(r) What happens to the x-intercepts of the graphs of these equations as the constant term increases?

(s) Write a statement about the solutions to $x^2 - 4x + c = 0$ where $c > 4$. Hence, make a statement about the x-intercepts of graphs in the form of $y = x^2 - 4x + c$ where $c > 4$.

14 Decide on two values of x that will be the x-intercepts of a quadratic graph. Write down at least two possible equations for the graph.

15 Louise is solving the equation $8x - x^2 = 16$. She writes:

$8x - x^2 = 16$
$x(8 - x) = 16$
$x = 16$ or $8 - x = 16$
$x = 16$ or $x = -8$

Explain to Louise what she has done incorrectly and write the correct solution.

Problem solving

Stopping distances

The distance it takes for a car to stop could be described by the equation $y = 0.005x^2 + 0.12x$, where y is the distance (m) it takes to stop and x is the speed (km/h) of the car.

(a) What extra distance will it take a car to stop if it is travelling at 60 km/h instead of 40 km/h through a school zone?

(b) What is the extra distance required for a car to stop if it is exceeding a 100 km/h speed limit by 10 km/h?

(c) Show that the speed of the car must be less than 12 km/h to ensure it does not hit a stationary car 2 m in front of it.

(d) What is the effect on the stopping distance of wiping 5 km/h off a 100 km/h speed?

Strategy options

- Have I seen a similar problem?
- Break the problem into manageable parts.

Sketching parabolas using transformations

8.4

Stage 5.3 §

Need to Know

A **transformation** is any change that can be made to a graph or shape using a **dilation**, **reflection**, **translation** or rotation.

Dilations and reflections

A dilation is when the graph of x^2 is multiplied by a constant, a, to form graphs in the form of $y = ax^2$. The **coefficient** of x^2, a, is referred to as the **dilation factor**.

For equations of the form $y = ax^2$:

- When the coefficient of x^2 is negative ($a < 0$) the parabola is inverted. These graphs are reflected in the x-axis.
- When the coefficient of x^2 is greater than 1 ($a > 1$), the parabola is narrower than $y = x^2$.
- When the coefficient of x^2 is greater than zero but less than 1 ($0 < a < 1$), the parabola is wider than $y = x^2$.
- When the coefficient of x^2 is greater than -1 but less than zero ($-1 < a < 0$) the parabola is also wider than $y = x^2$.

Translations

Horizontal translations:

The graph of $y = (x - h)^2$ is the graph of $y = x^2$ translated h units to the right.
If the value of h is negative, this will mean a translation to the left.

Vertical translations:

The graph of $y = x^2 + k$ is the graph of $y = x^2$ translated k units upwards. If the value of k is negative, this will mean a translation downwards.

Combined horizontal and vertical translations:

The graph of $y = (x - h)^2 + k$ is the graph of $y = x^2$ translated h units to the right and k units upwards. If the value of h is negative, this will mean a translation to the left. If the value of k is negative, this will mean a translation downwards.

Locating x- and y-intercepts

The y-intercept occurs where the parabola intersects the y-axis (i.e. when $x = 0$). The y-value of the y-intercept can therefore be found by substituting $x = 0$ into the quadratic equation.

If $y = x^2 - 5x + 6$,

$y = 0 - 5 \times 0 + 6$ (substituting $x = 0$)

$= 6$

The coordinates of the y-intercept would be (0, 6).

The x-intercept occurs where the parabola intersects with the x-axis (when $y = 0$). The x-value of the x-intercept can therefore be found by substituting $y = 0$ into the quadratic equation.

If $y = (x - 2)(x - 3)$, (this quadratic, when expanded, gives $y = x^2 - 5x + 6$)

$0 = (x - 2)(x - 3)$ (substituting $y = 0$)

$x - 2 = 0$ or $x - 3 = 0$ (using the Null Factor Law)

$x = 2$ or $x = 3$

The coordinates of the x-intercepts would be (2, 0) or (3, 0).

Many quadratics will have two x-intercepts.

Parabolas with equations in the form of $y = (x - h)^2$ will have only one x-intercept.

Some parabolas will have no x-intercepts.

Intercepts and the turning point

All quadratic graphs have a vertex or turning point. This can be a minimum value or a maximum value. When a graph has two x-intercepts, the x-coordinate of the turning point is halfway between the two x-intercepts.

For the quadratic equation $y = (x - 3)(x + 5)$, the x-intercepts are (3, 0) and (-5, 0) and $x = -1$ is the x-coordinate of the turning point.

The corresponding y-coordinate can be found by substituting the x-value into the quadratic equation.

$y = (-1 - 3)(-1 + 5)$

$y = -4 \times 4$

$= -16$

For the quadratic equation $y = (x - h)^2$, the x-intercept and turning point is $(h, 0)$.

When a graph has only one x-intercept, the turning point is the x-intercept, i.e. for $y = (x - 4)^2$, the x-intercept and turning point is (4, 0).

For the quadratic equation $y = x^2 + k$, the turning point is $(0, k)$. The turning point is also the y-intercept.

If k is a positive integer there will be no x-intercepts, i.e. for $y = x^2 + 3$, the coordinates of the turning point are (0, 3) and there are no x-intercepts.

For the quadratic equation $y = (x - h)^2 + k$, the turning point is (h, k), i.e. for $y = (x - 2)^2 + 7$, the coordinates of the turning point are (2, 7).

In a quadratic equation in the form of $y = ax^2 + bx + c$, the x-value of the turning point can be found using $x = -\frac{b}{2a}$.

Technology Exploration: GeoGebra

Investigating non-linear equations

1 Open GeoGebra. From the menu, select Options > Advanced... and in the pop-up window go to/Preferences-Graphics > Grid and select Show Grid. In the same window, go to the Basic tab and in Dimensions set the x-axis values from -10 to 10 and set the y-axis values from -8 to 8. In the same window, go to the xAxis tab and set Label to x, then go to the yAxis tab and set Label to y. In both these tabs, select 'Distance' and set it to 1. Finally, from the main menu select Options > Point Capturing > Automatic and then select Options > Labelling > No New Objects.

2 If a larger font is required, select from the menu Options > Font Size. Choose an appropriate size from the list provided.

Investigating parabolas

All parabolas can be represented by the equation $y = ax^2 + bx + c$, where a represents the dilation and c represents the y-coordinate of the y-intercept (where the line cuts the y-axis). We will now create parabolas using different values of a, b and c to investigate how the graphs change when we change these values.

3 Select the Slider 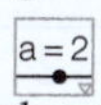

tool. Click in the right-hand side of the Graphics view to create a slider there. In the pop-up window that appears, in Name, type a. Change Increment to 0.2. Click on the Slider tab and change Width to 200. Click Apply. A line like this will appear on the screen.

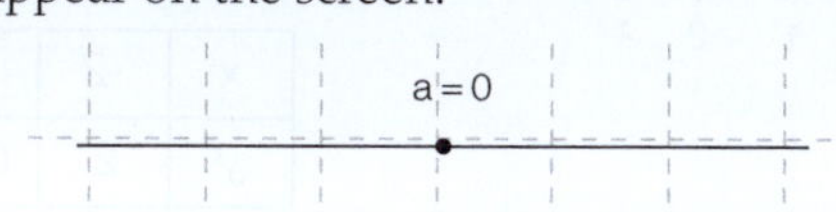

4 Construct a second slider by repeating Step **3**, naming it b.

5 Construct a third slider by repeating Step **3**, naming it c.

6 Set the slider values for a, b and c to 0.

7 In the Input bar at the bottom of the window, type *y=x^2* (the ^ is used to show indices) and press enter. A parabola should appear on the screen.

8 In the Input bar, enter *y=a*x^2+b*x+c* (the * is used to show multiplication). Increase the slider value for a from -5 to 5. A parabola should appear on the screen, and it should change as the slider value changes.

9 Write what you see as the slider values for a increase from -5 to 5. What happens when $a = 0$? Explain. What happens when $a = 1$? Explain. What happens when $a = -1$? Explain.

10 Set the value of a to 1 and increase the value of c from -5 to 5. Describe what happens to the parabola as the c value changes. What do you notice about the shape of each graph? What do you notice about the y-intercept for each graph? What do you notice about the x-intercepts? (Hint: $c = -4$ and -1 should help you.)

11 Leave the a value at 1 and set the c value to 0. Change the b value from -5 to 5. Write about any connection between the graphs and the b value. Use algebra to explain this connection. You should also notice that an equation appears in the Algebra view (on the left-hand side of the window).

12 Find a value for each of b and c so that the parabola moves to the right. Explain why you chose these values.

13 Find a value for each of b and c so that the parabola moves to the left. Explain why you chose these values.

Worked Example 14

WE14

(a) Plot the graph of $y = x^2$ for $-2 \le x \le 2$.

(b) Plot the graphs of **(i)** $y = -x^2$, **(ii)** $y = 2x^2$ and **(iii)** $y = \frac{1}{2}x^2$ on the same set of axes as your graph in part **(a)** using the same values for x.

(c) Compare each of the graphs in part **(b)** to $y = x^2$ using the words narrower, wider or inverted (upside down).

(d) State the dilation factor for each of the graphs in part **(b)**.

Working

(a)

x	-2	-1	0	1	2
y	4	-1	0	1	4

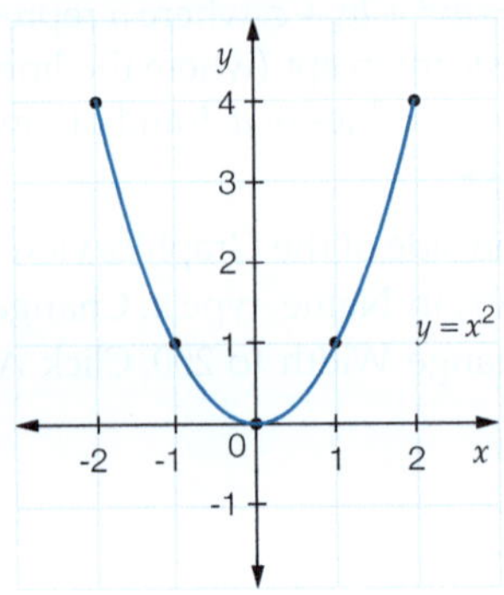

(b) (i) $y = -x^2$

x	-2	-1	0	1	2
y	-4	-1	0	-1	-4

(ii) $y = 2x^2$

x	-2	-1	0	1	2
y	8	2	0	2	8

(iii) $y = \frac{1}{2}x^2$

x	-2	-1	0	1	2
y	2	0.5	0	0.5	2

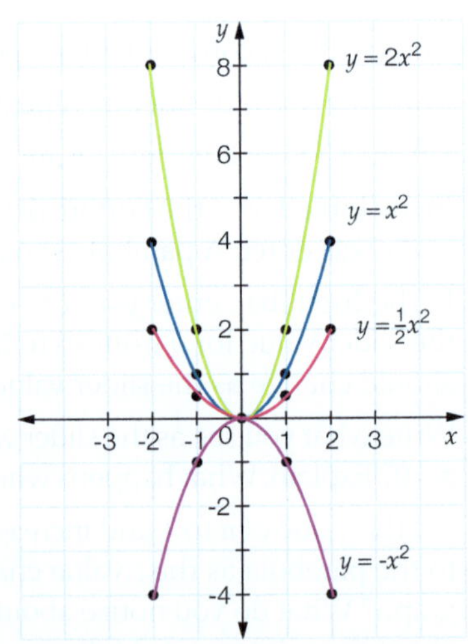

(c) (i) $y = -x^2$ is inverted, but the same width as $y = x^2$.

(ii) $y = 2x^2$ is narrower than $y = x^2$.

(iii) $y = \frac{1}{2}x^2$ is wider than $y = x^2$.

(d) (i) The dilation factor, a, is -1.

(ii) The dilation factor, a, is 2.

(iii) The dilation factor, a, is $\frac{1}{2}$.

Worked Example 15

WE 15

(a) Plot the graph of $y = x^2$ for $-4 \leq x \leq 4$.

(b) Plot the graphs of **(i)** $y = (x - 1)^2$ and **(ii)** $y = (x + 2)^2$ on the same set of axes as your graph in part **(a)** using the same values for x.

(c) Compare each of the graphs in part **(b)** to $y = x^2$.

(d) State the x-intercept for each graph in part **(b)** and identify the x-value of the intercept from the equation.

Working

(a) $y = x^2$

x	-4	-3	-2	-1	0	1	2	3	4
y	16	9	4	1	0	1	4	9	16

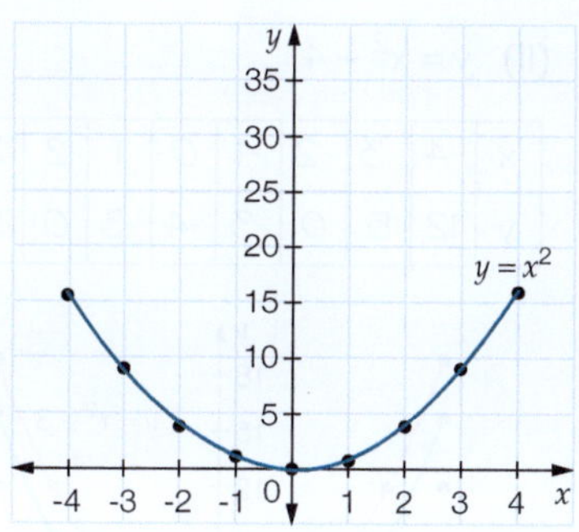

(b) (i) $y = (x - 1)^2$

x	-4	-3	-2	-1	0	1	2	3	4
y	25	16	9	4	1	0	1	4	9

(ii) $y = (x + 2)^2$

x	-4	-3	-2	-1	0	1	2	3	4
y	4	1	0	1	4	9	16	25	36

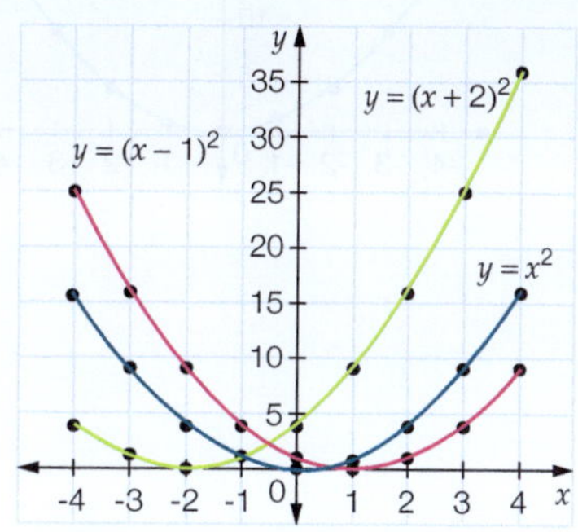

(c) (i) The graph of $y = (x - 1)^2$ has been translated 1 unit to the right.

(ii) The graph of $y = (x + 2)^2$ has been translated 2 units to the left.

(d) (i) The x-intercept of $y = (x - 1)^2$ is the point (1, 0). The x-value of the intercept is $x = 1$. This is the number subtracted from x before squaring in the equation.

(ii) The x-intercept of $y = (x + 2)^2$ is the point (-2, 0). The x-value of the intercept is $x = -2$. This is the number subtracted from x before squaring in the equation.

8.4

Worked Example 16

WE16

(a) Plot the graph of $y = x^2$ for $-4 \le x \le 4$.

(b) Plot the graphs of **(i)** $y = x^2 + 3$ and **(ii)** $y = x^2 - 4$ on the same set of axes as your graph in part **(a)** using the same values for x.

(c) Compare each of the graphs in part **(b)** to $y = x^2$.

(d) State the y-intercept for each graph in part **(b)** and identify the y-intercept from the equation.

Working

(a)

x	-4	-3	-2	-1	0	1	2	3	4
y	16	9	4	1	0	1	4	9	16

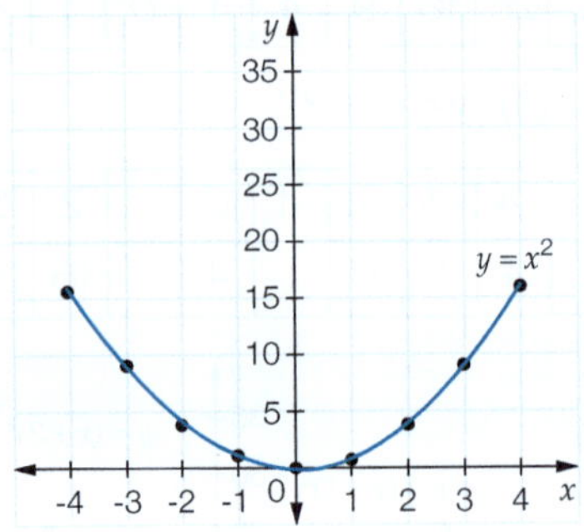

(b) (i) $y = x^2 + 3$

x	-4	-3	-2	-1	0	1	2	3	4
y	19	12	7	4	3	4	7	12	19

(ii) $y = x^2 - 4$

x	-4	-3	-2	-1	0	1	2	3	4
y	12	5	0	-3	-4	-3	0	5	12

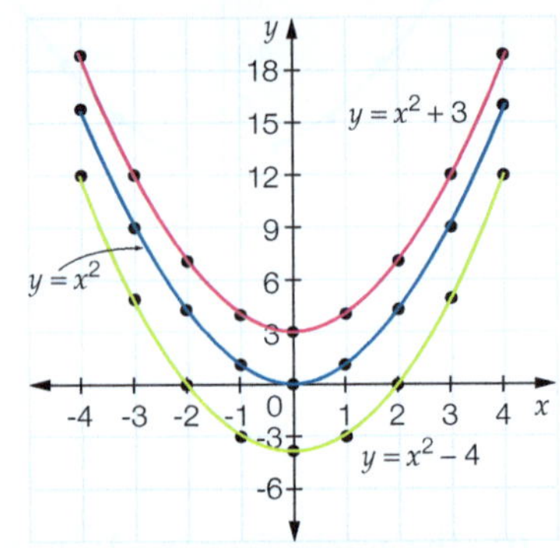

(c) (i) The graph of $y = x^2$ has been translated 3 units upwards.

(ii) The graph of $y = x^2$ has been translated 4 units downwards.

(d) (i) The y-intercept of $y = x^2 + 3$ is the point (0, 3). The y-value of the intercept is $y = 3$. This is the constant in the equation.

(ii) The y-intercept of $y = x^2 - 4$ is the point (0, -4). The y-value of the intercept is $y = -4$. This is the constant in the equation.

Worked Example 17

WE17

(a) Plot the graph of $y = x^2$ for $-5 \le x \le 5$.

(b) Plot the graphs of **(i)** $y = (x-3)^2 + 1$ and **(ii)** $y = (x+1)^2 - 5$ on the same set of axes as your graph in part **(a)** using the same values for x.

(c) Compare each of the graphs in part **(b)** to $y = x^2$.

(d) Find the y-intercept for each graph in part **(b)**.

(e) For each graph, identify whether there are zero, one or two x-intercepts.

Working

(a) $y = x^2$

x	-5	-4	-3	-2	-1	0	1	2	3	4	5
y	25	16	9	4	1	0	1	4	9	16	25

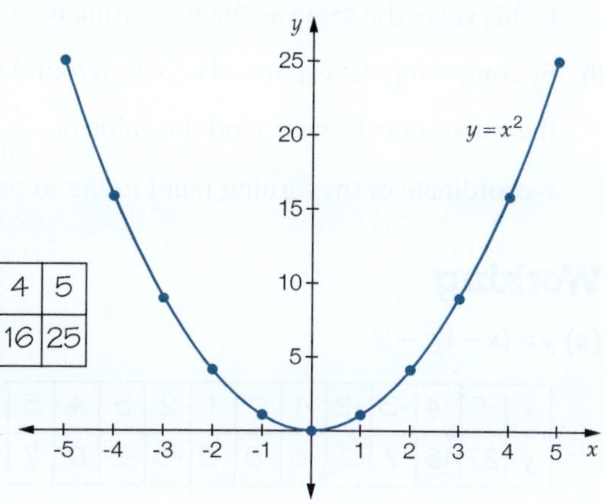

(b) (i) $y = (x-3)^2 + 1$

x	-5	-4	-3	-2	-1	0	1	2	3	4	5
y	65	50	37	26	17	10	5	2	1	2	5

(ii) $y = (x+1)^2 - 5$

x	-5	-4	-3	-2	-1	0	1	2	3	4	5
y	11	4	-1	-4	-5	-4	-1	4	11	20	31

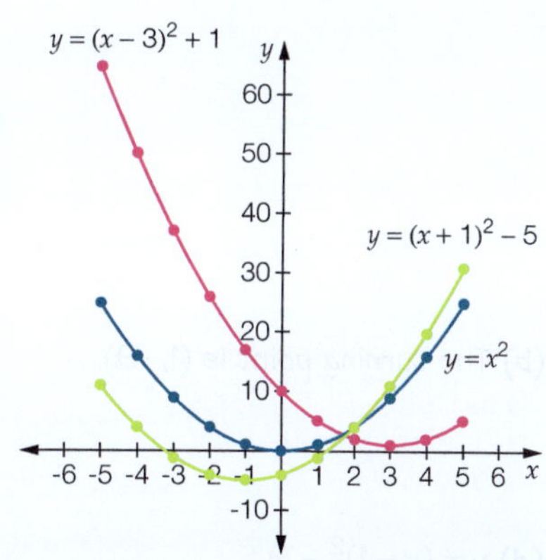

(c) (i) The graph of $y = x^2$ has been translated 3 units to the right and 1 unit upwards.

(ii) The graph of $y = x^2$ has been translated 1 unit to the left and 5 units down.

(d) (i) The y-intercept is (0, 10).

(ii) The y-intercept is (0, -4).

(e) (i) The graph has no x-intercepts.

(ii) The graph has two x-intercepts.

Worked Example 18

WE18

(a) Plot the graph of $y = (x - 1)^2 - 9$ for $-5 \le x \le 5$.

(b) From your graph, identify the turning point.

(c) By comparing the equation in part **(a)** to $y = (x - h)^2 + k$, write down the values of h and k. Is the point (h, k) the same point as the turning point found in part **(b)**?

(d) Expand the given equation and expand $y = (x - 4)(x + 2)$ to show that they are equivalent expressions. Use this information to find the x-coordinates of the x-intercepts. Find the average of these two values. Is this value the same as the x-coordinate of the turning point found in part **(b)**?

(e) Substitute this value into the equation in part **(a)** to find the corresponding y-value. Is this value the same as the y-coordinate of the turning point found in part **(b)**?

(f) By comparing the expanded equation found in part **(d)** to $y = ax^2 + bx + c$, write down the values of a, b and c. Find the value of $-\frac{b}{2a}$. Is this value the same as the x-coordinate of the turning point found in part **(b)**?

Working

(a) $y = (x - 1)^2 - 9$

x	-5	-4	-3	-2	-1	0	1	2	3	4	5
y	27	16	7	0	-5	-8	-9	-8	-5	0	7

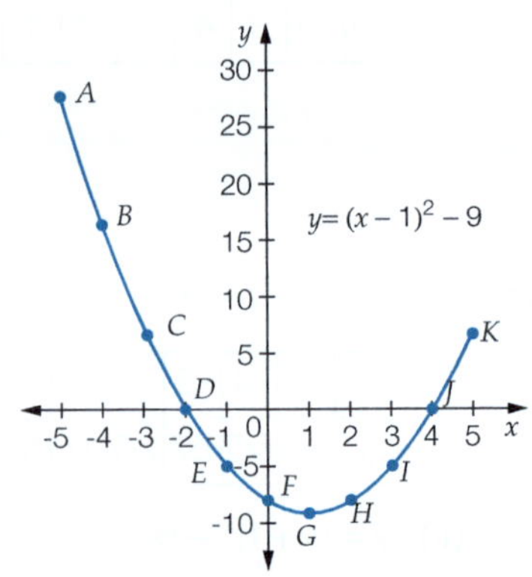

(b) The turning point is (1, -9).

(c) $h = 1$ and $k = -9$

$(h, k) = (1, -9)$

Yes, it is the same point.

(d) $y = (x - 1)^2 - 9$
$= x^2 - 2x + 1 - 9$
$= x^2 - 2x - 8$

$(x - 4)(x + 2)$
$= x^2 - 4x + 2x - 8$
$= x^2 - 2x - 8$

$(x - 1)^2 - 9 = (x - 4)(x + 2)$

$(x - 4)(x + 2) = 0$

$x = 4$ and $x = -2$

Average of 4 and -2 $= \frac{4 + -2}{2} = 1$

Yes, this is the same value.

(e) $y = (x - 1)^2 - 9 = (1 - 1)^2 - 9 = -9$

Yes, this is the same value.

(f) $a = 1$, $b = -2$, $c = -8$

$-\frac{b}{2a} = -\frac{-2}{2} = 1$

Yes, this is the same value.

8.4

8.4 Sketching parabolas using transformations

Answers page 482

Stage 5.3 §

Navigator

Q1, Q2, Q3, Q4, Q5, Q6, Q7 Columns 1 & 2, Q8 Columns 1 & 2, Q9, Q10, Q11, Q12, Q13 (a)–(h), Q14, Q17, Q19, Q20	Q1, Q2, Q3, Q4, Q5, Q6, Q7 Columns 2 & 3, Q8 Columns 2 & 3, Q9, Q10, Q11, Q12, Q13 (a)–(h), Q14, Q17, Q18, Q19, Q20	Q1, Q2, Q3, Q4, Q5, Q6, Q7 Columns 3 & 4, Q8 Columns 3 & 4, Q9, Q10, Q11, Q12, Q13 (i)–(p), Q14, Q15, Q16, Q17, Q18, Q20

Fluency and understanding

1 **(a)** Plot the graph of $y = x^2$ for $-3 \le x \le 3$. WE14

(b) Plot the graphs of **(i)** $y = -2x^2$, **(ii)** $y = 3x^2$ and **(iii)** $y = \frac{x^2}{4}$ on the same set of axes as your graph in part **(a)** using the same values for x.

(c) Compare each of the graphs in part **(b)** to $y = x^2$ using the words narrower, wider or inverted (upside down).

(d) State the dilation factor for each of the graphs in part **(b)**.

2 **(a)** Plot the graph of $y = x^2$ for $-3 \le x \le 3$. WE15

(b) Plot the graphs of **(i)** $y = (x - 2)^2$ and **(ii)** $y = (x + 1)^2$ on the same set of axes as your graph in part **(a)** using the same values for x.

(c) Compare each of the graphs in part **(b)** to $y = x^2$.

(d) State the x-intercept for each graph in part **(b)** and identify the x-value of the intercept from the equation.

3 **(a)** Plot the graph of $y = x^2$ for $-3 \le x \le 3$. WE16

(b) Plot the graphs of **(i)** $y = x^2 + 4$ and **(ii)** $y = x^2 - 2$ on the same set of axes as your graph in part **(a)** using the same values for x.

(c) Compare each of the graphs in part **(b)** to $y = x^2$.

(d) State the y-intercept for each graph in part **(b)** and identify the y-intercept from the equation.

4 **(a)** Plot the graph of $y = x^2$ for $-3 \le x \le 3$. WE17

(b) Plot the graphs of **(i)** $y = (x - 1)^2 - 2$ and **(ii)** $y = (x + 2)^2 + 1$ on the same set of axes as your graph in part **(a)** using the same values for x.

(c) Compare each of the graphs in part **(b)** to $y = x^2$.

(d) Find the y-intercept for each graph in part **(b)**.

(e) For each graph, identify whether there are zero, one or two x-intercepts.

5 **(a)** Plot the graph of $y = (x - 2)^2 - 1$ for $-5 \le x \le 5$. WE18

(b) From your graph, identify the turning point.

(c) By comparing the equation in part **(a)** to $y = (x - h)^2 + k$, write down the values of h and k. Is the point (h, k) the same point as the turning point found in part **(b)**?

(d) Expand the given equation and expand $y = (x - 1)(x - 3)$ to show that they are equivalent expressions. Use this information to find the x-coordinates of the x-intercepts. Find the average of these two values. Is this value the same as the x-coordinate of the turning point found in part **(b)**?

(e) Substitute this value into the equation in part **(a)** to find the corresponding y-value. Is this value the same as the y-coordinate of the turning point found in part **(b)**?

8.4

(f) By comparing the expanded equation found in part (d) to $y = ax^2 + bx + c$, write down the values of a, b and c. Find the value of $-\frac{b}{2a}$. Is this value the same as the x-coordinate of the turning point found in part (b)?

6 (a) State whether the graph of each of these equations will be narrower or wider than the parabola $y = x^2$.

(i) $y = 5x^2$ (ii) $y = \frac{7}{8}x^2$ (iii) $y = -\frac{1}{6}x^2$ (iv) $y = -14x^2$

(b) Which of the graphs of the equations in part (a) would be upside down (inverted) compared with $y = x^2$?

(c) State the dilation factor of each of the following.

(i) $y = 4x^2$ (ii) $y = \frac{1}{5}x^2$ (iii) $y = 7x^2 + 2x$ (iv) $y = \frac{2}{3}x^2 - 3x + 1$

(d) Will a parabola be narrower or wider than $y = x^2$ if it has a dilation factor of:

(i) 3 (ii) $\frac{4}{5}$ (iii) 16 (iv) $\frac{1}{7}$?

7 How is the graph of $y = x^2$ translated (moved) to obtain each of the following parabolas?

(a) $y = x^2 + 4$ (b) $y = x^2 - 7$ (c) $y = x^2 - 1$ (d) $y = x^2 + 2$

(e) $y = x^2 - 8$ (f) $y = x^2 + 5$ (g) $y = x^2 + \frac{1}{2}$ (h) $y = x^2 - \frac{3}{4}$

8 How is the graph of $y = x^2$ translated (moved) to obtain each of the following parabolas?

(a) $y = (x - 5)^2$ (b) $y = (x + 4)^2$ (c) $y = (x + 1)^2$ (d) $y = (x - 7)^2$

(e) $y = (x + 6)^2$ (f) $y = (x - 3)^2$ (g) $y = (x - \frac{2}{3})^2$ (h) $y = (x + \frac{3}{2})$

9 For each of the following quadratic relationships:

(i) find the transformations of the graph $y = x^2$ that have taken place; and

(ii) find the turning point.

(a) $y = 2(x - 1)^2$ (b) $y = -(x + 2)^2$ (c) $y = (x + 3)^2 - 2$

(d) $y = (x + 5)^2 + 3$ (e) $y = (x + 4)^2 - 3$ (f) $y = -2(x + 3)^2$

10 (a) To obtain the graph of $y = x^2 + 6$ we need to move the graph of $y = x^2$:

A 6 units to the right B 6 units to the left

C 6 units up D 6 units down

(b) To obtain the graph of $y = (x + 6)^2$ we need to move the graph of $y = x^2$:

A 6 units to the right B 6 units to the left

C 6 units up D 6 units down

(c) To obtain the graph of $y = (x - 2)^2$ we need to move the graph of $y = x^2$:

A 2 units to the right B 2 units to the left

C 2 units up D 2 units down

(d) To obtain the graph of $y = x^2 - 4$ we need to move the graph of $y = x^2$:

A 4 units to the right B 4 units to the left

C 4 units up D 4 units down

11 To obtain the graph of $y = (x - 1)^2 - 3$, we need to move the graph of $y = x^2$:

A 1 unit to the right and 3 units down

B 3 units to the right and 1 unit down

C 1 unit to the left and 3 units up

D 3 units to the right and 1 unit down

8.4

12 **(a)** The equation of the unlabelled parabola shown could be:

A $y = x^2 + 2$

B $y = x^2 - 2$

C $y = (x - 2)^2$

D $y = (x + 2)^2$

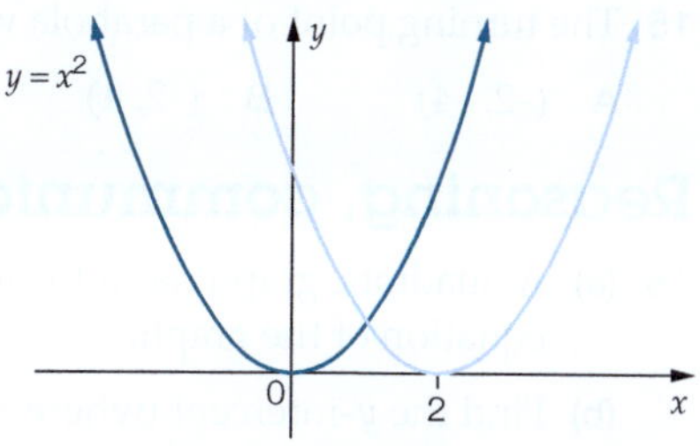

(b) The equation of the unlabelled parabola shown could be:

A $y = x^2 + 1$

B $y = (x + 3)^2 + 1$

C $y = 3x^2 + 1$

D $y = (x - 3)^2 + 1$

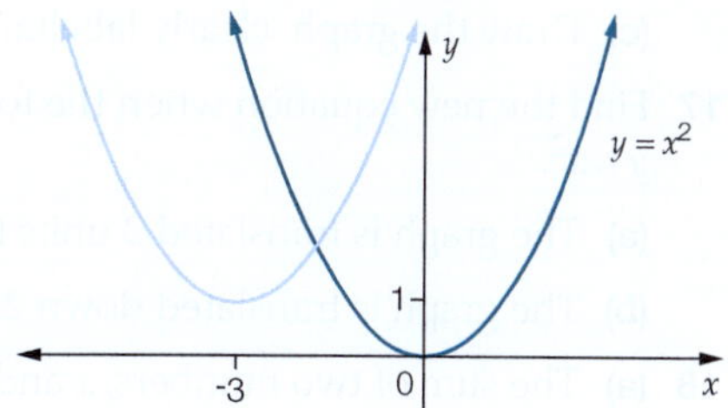

(c) The equation of the unlabelled parabola shown could be:

A $y = (x + 4)^2 - 1$

B $y = -(x + 4)^2 - 1$

C $y = (x - 4)^2 - 1$

D $y = 4(x - 4)^2 - 1$

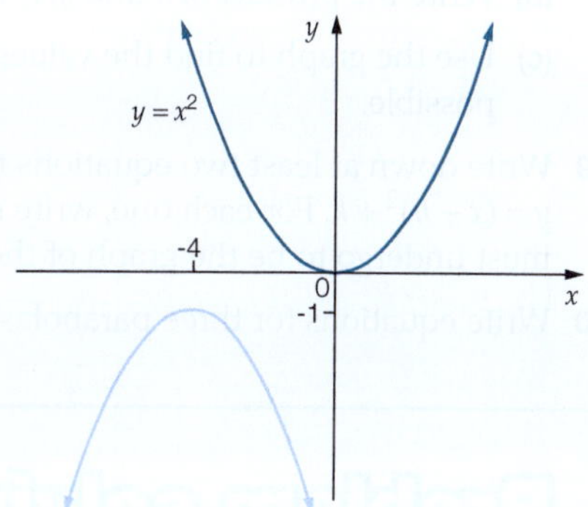

13 State the changes that should be made to the graph of $y = x^2$ to obtain the graph of each of the following.

		Dilation	Reflection in *x*-axis	Translation in *x* direction	Translation in *y* direction
(a)	$y = (x + 1)^2 + 2$				
(b)	$y = (x - 3)^2 + 1$				
(c)	$y = 5x^2 + 4$				
(d)	$y = -x^2 - 6$				
(e)	$y = -(x - 2)^2 + 3$				
(f)	$y = (x - 4)^2 - 5$				
(g)	$y = (x + 2)^2 - 7$				
(h)	$y = 3x^2 - 5$				
(i)	$y = -x^2 - 2$				
(j)	$y = -(x - 5)^2 + 2$				

14 When graphed, how many of the following equations would have a shape like that of the parabola shown here?

Identify them.

(a) $y = x^2 + 2x - 5$

(b) $y = -x^2 - 2x + 4$

(c) $y = 1 - x^2$

(d) $y = 3 - 4x + x^2$

(e) $y = 6 - 2x - x^2$

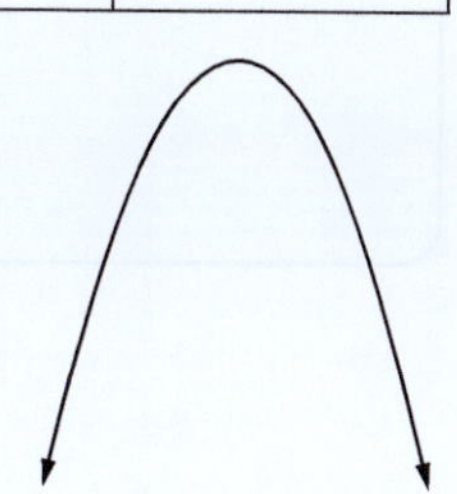

15 The turning point of a parabola with the equation $y = (x - 2)^2 + 4$ is:

A (-2, -4) B (-2, 4) C (2, 4) D (2, -4)

Reasoning, communicating and open-ended

16 (a) A quadratic graph has a turning point at (-2, 5) and a dilation factor of 1. Find the equation of the graph.

(b) Find the y-intercept (where $x = 0$).

(c) Draw the graph, clearly labelling the turning point and the y-intercept.

17 Find the new equation when the following transformations are completed on the graph $y = x^2$.

(a) The graph is translated 3 units to the right and 2 units up.

(b) The graph is translated down 2 units and to the right 4 units.

18 (a) The sum of two numbers, x and y, is 20. Express y in terms of x.

(b) Write the product of x and y, P, in terms of x and sketch a graph of P.

(c) Use the graph to find the values of x and y that will make the product as large as possible.

19 Write down at least two equations that show a quadratic relationship in the form $y = (x - h)^2 + k$. For each one, write down the transformations that the graph $y = x^2$ must undergo to be the graph of the relationship you have written.

20 Write equations for three parabolas that have no x-intercepts.

Problem solving

Let's go camping

You decide to go camping at a camping ground by a lake. The manager of the camping ground has designed a novel way of determining individual campsites. She gives you a 60-m long piece of rope and four pegs. The only condition is that your campsite must be rectangular. You decide to pitch your tent by the lake and only need to use the rope on three sides. Find the largest campsite that you could make.

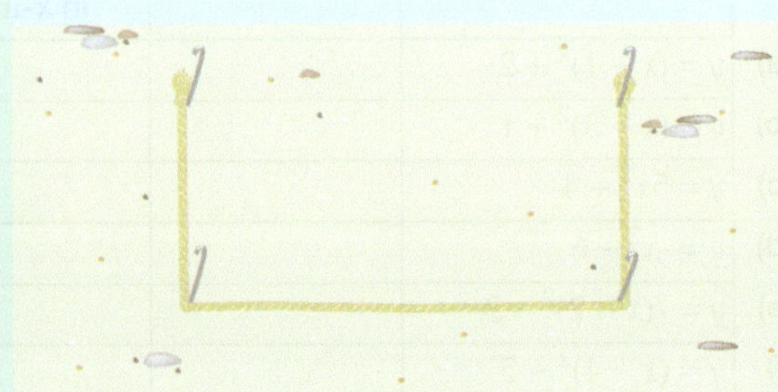

Strategy options

- Test all possible combinations.
- Make a table.
- Make a model.
- Draw a diagram.

Half-time 8

Stage 5.1 Q1 **Stage 5.2 Q2–7** **Stage 5.3 Q8–10**

1 Plot the graphs of each of the following equations using an appropriate table of values. **Ex 8.1**

(a) $x^2 + y^2 = 144$ **(b)** $y = x^2 - 7$ **(c)** $y = 2x + 3$

2 Which of the following is not a quadratic expression? **Ex 8.2**

A $2x^2 + 3x + 5$ **B** $-x + 3x^2$ **C** $3x + 2$ **D** $4x^2 - 3x - 1$

3 Solve each of the following. **Ex 8.3**

(a) $x(x + 4) = 0$ **(b)** $2x(x - 3) = 0$ **(c)** $(x + 2)(x - 3) = 0$

4 Lisa throws a ball upwards. The equation of the flight of the ball is $h = -5t^2 + 8t + 1.3$, where h is the height of the ball in metres and t is the time in seconds. **Ex 8.2**

(a) Complete the table of values below.

t	0	0.2	0.4	0.6	0.8	1	1.2	1.4	1.6
h									

(b) Draw a graph showing this relationship.

(c) From the graph, find when the ball first reaches a height of 3 m.

(d) How high is the ball after 1.2 seconds?

5 Solve each of the following. **Ex 8.3**

(a) $x^2 + 4x + 4 = 0$ **(b)** $x^2 + 12x + 36 = 0$ **(c)** $x^2 - 18x + 81 = 0$

6 Solve each of the following. **Ex 8.3**

(a) $x^2 - 6x = 0$ **(b)** $2x^2 + 20x + 50 = 0$ **(c)** $3x^2 = 27x$

7 Solve each of the following. **Ex 8.3**

(a) $x^2 - 64 = 0$ **(b)** $(x + 1)^2 = 0$ **(c)** $(x - 2)^2 = 4$

8 **(a)** Complete a table of values for $y = 2x^2 + 4x$ from $x = -3$ to $x = 3$ inclusive. **Ex 8.4**

(b) Plot the points on a number plane and join them using a smooth curve.

(c) Write down the coordinates of the turning point.

(d) State whether the parabola has a maximum or a minimum turning point.

(e) Write down the coordinates of the x- and y-intercepts.

9 **(a)** Plot the graph of $y = x^2 - 2x + 3$. Use integer values of x between -3 and 5 inclusive. **Ex 8.4**

(b) What are the coordinates of the turning point?

(c) Write down the coordinates of any x- and y-intercepts.

(d) Use the graph to solve the following equations.

(i) $x^2 - 2x + 3 = 3$ **(ii)** $x^2 - 2x + 3 = 2$

(iii) $x^2 - 2x + 3 = 6$ **(iv)** $x^2 - 2x + 3 = 0$

10 Consider the equation $y = -x^2 + 1$. **Ex 8.4**

(a) Set up a table of values using integer values of x between -3 and 3 inclusive.

(b) Plot the points on a number plane and join them with a smooth curve.

(c) Find the coordinates of the turning point.

(d) Find the coordinates of the y-intercept.

(e) Find the coordinates of the x-intercepts.

8.5 Circle relationships

Stage 5.2◊

Need to Know

Circle equations are important non-linear relationships.

Circles with centre at the origin

All points on the circle are the same distance from the centre of the circle, the origin, so the graph of a circle and Pythagoras' Theorem can be used to derive the equation of a circle with the centre at the origin.

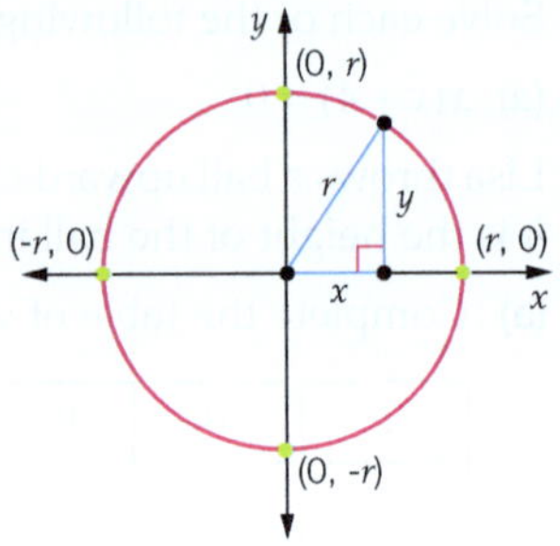

The equation of a circle with centre (0, 0) and radius r is:

$$x^2 + y^2 = r^2$$

In circle equations, both the x and y variables are squared.

Graphing a circle from its equation

Find the radius of the circle, then find four points on the circle circumference to give maximum and minimum values by adding and subtracting the radius from the centre coordinates (0, 0).

For a circle with centre (0, 0) and radius r, the four points are $(r, 0)$, $(0, r)$ $(-r, 0)$ and $(0, -r)$.

Worked Example 19

WE19

Determine the centre and radius of each of the following, and hence the equation of the circle.

(a)

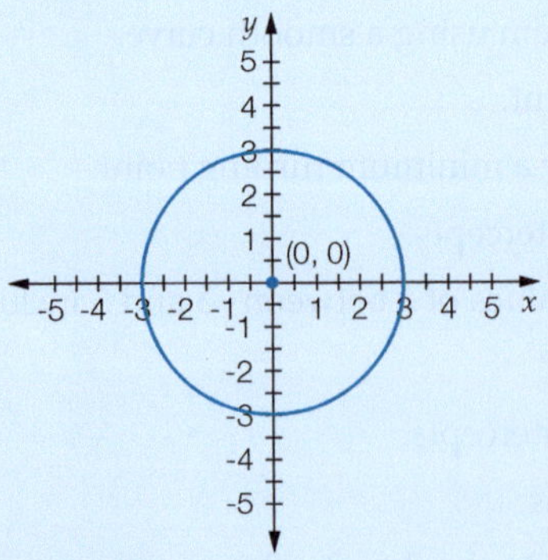

(b)

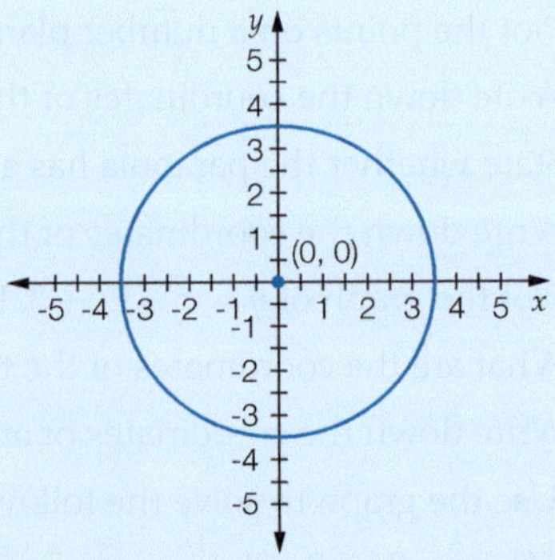

Working

(a) Centre = (0, 0), Radius = 3

Equation

$x^2 + y^2 = r^2$

$x^2 + y^2 = 3^2$

$x^2 + y^2 = 9$

(b) Centre = (0, 0), Radius = 3.5

Equation

$x^2 + y^2 = r^2$

$x^2 + y^2 = 3.5^2$

$x^2 + y^2 = 12.25$

8.5

8.5 Circle relationships

Answers page 485

Stage 5.2◊

Navigator

Q1, Q2, Q3, Q4 (a) & (b), Q5, Q6, Q9	Q1, Q2, Q3, Q4, Q5, Q6, Q7, Q9	Q1, Q2, Q3, Q4, Q5, Q6, Q7, Q8, Q9

Fluency and understanding

WE19

1 Determine the centre and radius of each of the following, and hence the equation of the circle.

(a)

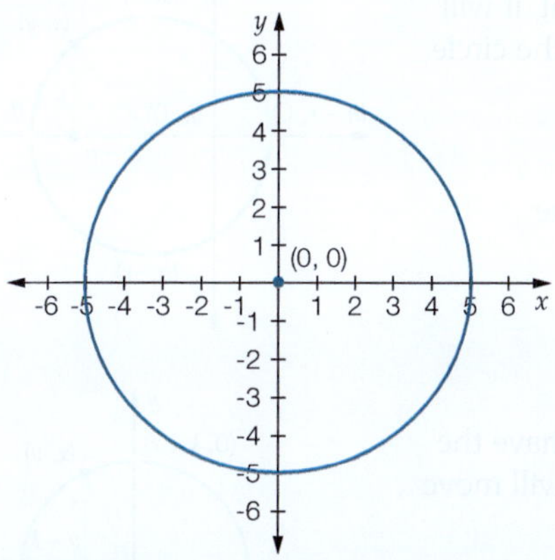

(b)

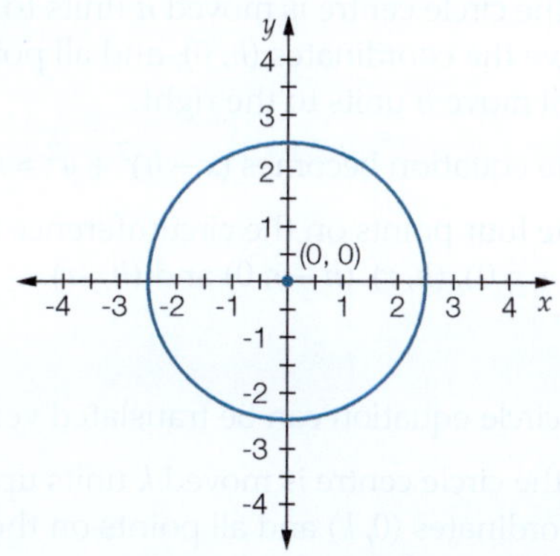

2 Find the radius of the following circles.

(a) $x^2 + y^2 = 4$ **(b)** $x^2 + y^2 = 25$

(c) $x^2 + y^2 = 49$ **(d)** $x^2 + y^2 = 81$

(e) $x^2 + y^2 = 6.25$ **(f)** $x^2 + y^2 = 1.96$

3 Graph each of the circles in question **2** on separate number planes.

4 Find the equation of a circle with its centre at the origin if its radius is:

(a) 2 units **(b)** 8 units **(c)** 0.6 units

(d) 1.2 units **(e)** $\frac{1}{2}$ unit **(f)** $1\frac{3}{5}$ units

5 Which point lies on the circle with equation $x^2 + y^2 = 25$?

A (2, 3) **B** (3, -4) **C** (2, -3) **D** (1, 4)

6 Decide whether the equation would graph as a circle in each of the following. Answer Yes or No.

(a) $y - x^2 = 4$ **(b)** $y^2 = x^2 - 9$ **(c)** $x^2 = 1 - y^2$

(d) $y = 25 - x^2$ **(e)** $x^2 + y^2 = 7$ **(f)** $y^2 = x + 4$

Reasoning, communicating and open-ended

7 **(a)** Explain how to determine whether a point lies inside, on or outside a circle.

(b) Use this reasoning to decide whether the following points lie inside, on or outside the circle with equation $x^2 + y^2 = 13^2$.

(i) (3, 10) **(ii)** (-2, 3) **(iii)** (5, -12) **(iv)** (15, -2)

8 Find the equation of a circle with its centre the origin that passes through the point (-8, -15).

9 List two points that lie on the circle $x^2 + y^2 = 100$ but do not lie on the x- or y-axes.

8.6 Circle transformations

Stage 5.3 §

Need to Know

Translated circles

A circle equation can be translated horizontally.

If the circle centre is moved h units to the right, it will have the coordinates $(h, 0)$, and all points on the circle will move h units to the right.

The equation becomes $(x-h)^2 + y^2 = r^2$.

The four points on the circumference would be $(h+r, 0)$, (h, r), $(h-r, 0)$ and $(h, -r)$.

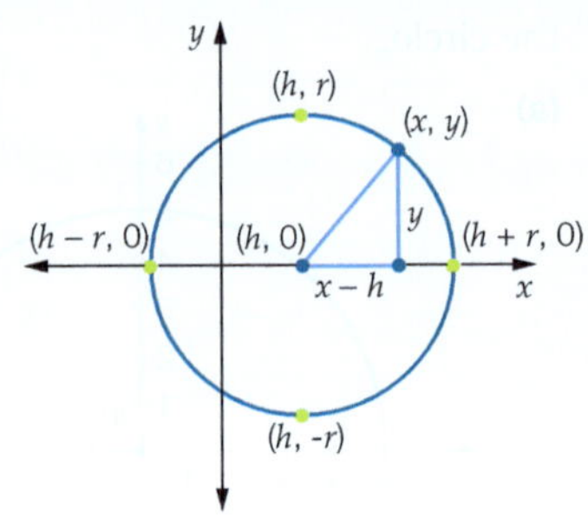

A circle equation can be translated vertically.

If the circle centre is moved k units up, it will have the coordinates $(0, k)$ and all points on the circle will move k units up.

The equation becomes $x^2 + (y-k)^2 = r^2$.

The four points on the circumference would be (r, k), $(0, k+r)$, $(-r, k)$ and $(0, k-r)$.

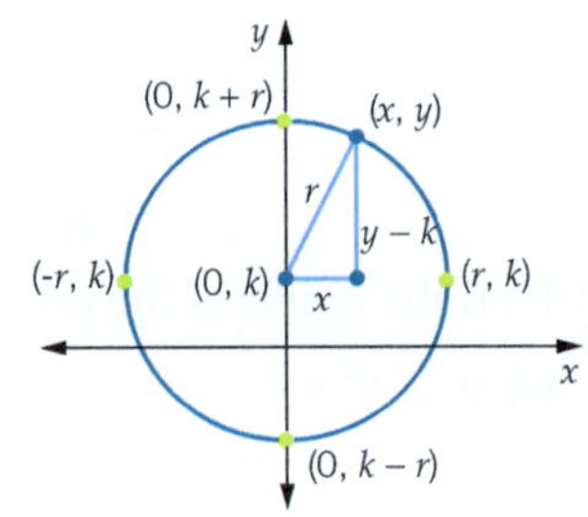

A circle can be translated both horizontally and vertically.

If the circle centre is moved h units to the right and k units up, it will have the coordinates (h, k) and all points on the circle will move h units to the right and k units up.

The equation becomes $(x-h)^2 + (y-k)^2 = r^2$.

The four points on the circumference would be $(h+r, k)$, $(h, k+r)$, $(h-r, k)$ and $(h, k-r)$.

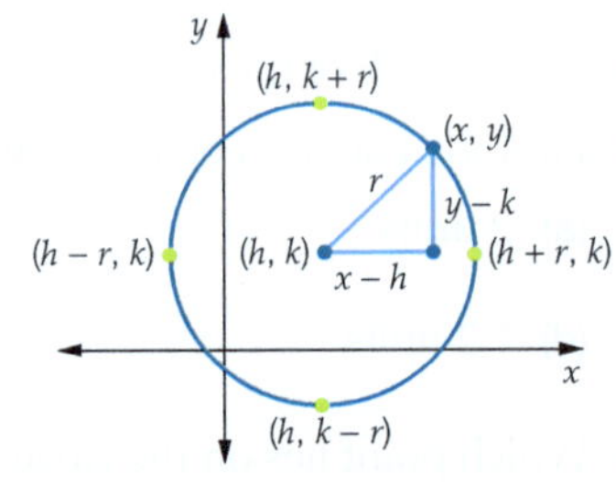

The important features of a circle are:

- the centre with coordinates (h, k)
- the radius, r.

Circles with centre the origin can be translated by h units to the right and k units up. The new centre has coordinates (h, k) but the radius remains the same.

The general equation of a circle is given by $(x-h)^2 + (y-k)^2 = r^2$.

8.6

Technology Exploration: GeoGebra

Investigating circles

1 Open a new GeoGebra window and set it up as for the Technology Exploration in section 8.4. Select the Circle with Centre and Radius tool, and use this tool to draw a circle with radius 4 and centre at the origin. Use the New Point tool to insert a point on the circle in each quadrant. Use the coordinates of each point, as given in the Algebra view, to show that $x^2 + y^2 = 16$. If you do not get 16 exactly, explain why.

2 Use the Input bar to enter the equation (*x*–*h*)^2 + (*y*–*k*)^2 =*r*^2. Set up three new sliders and call them *h*, *k* and *r*, and set the *r* values from 0 to 5. Set both *h* and *k* values to zero and *r* to 4. What do you see? Explain.

3 Leave the *k* value on 0 and the *r* value on 4. Change the slider value of *h* from -5 to 5. Write what you see.

4 Now set the value of *h* to 0 and change the slider value of *k* from -5 to 5. Write what you see.

5 What happens when neither *h* nor *k* is zero? Choose integer values for each of *h* and *k*. Write these values down and write the centre of the circle that is drawn. What do you see?

6 Use the New Point tool to insert a point on this circle in each quadrant. Use the coordinates of each point (given in the Algebra view) to show that $(x - h)^2 + (y - k)^2 = 16$. If you do not get 16 exactly, explain why.

7 Set the values of *h* and *k* to zero and change the value of *r* from 0 to 5. Describe what happens.

Worked Example 20

WE 20

Determine the centre and radius of each of the following, and hence the equation of the circle.

(a)

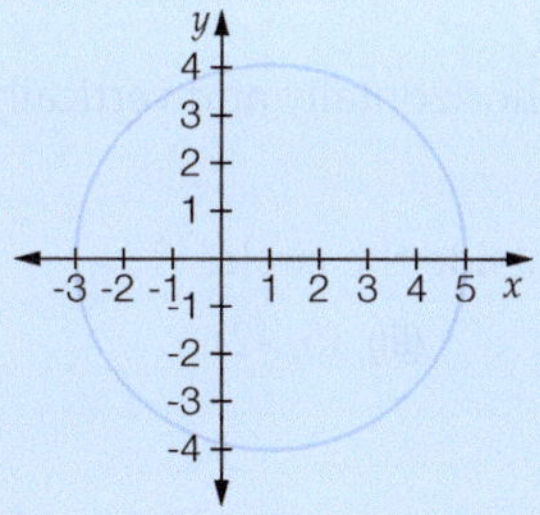

(b)

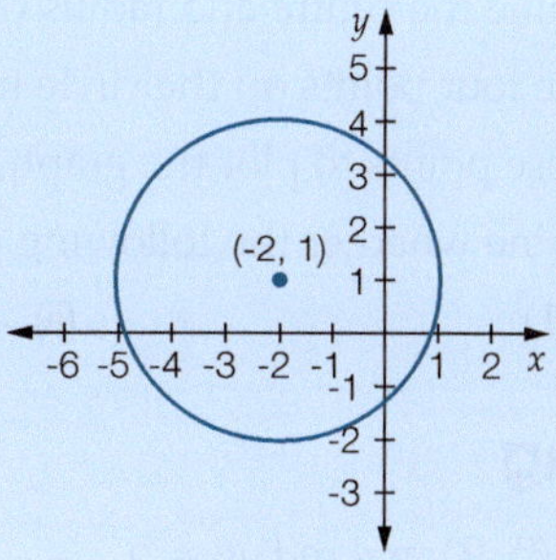

Working

(a) Centre = (*h*, *k*)

Centre = (1, 0)

Radius = 5 – 1

= 4

$(x - h)^2 + (y - k)^2 = r^2$

$(x - 1)^2 + y^2 = 16$

(b) Centre = (*h*, *k*)

Centre = (-2, 1)

Radius = (1 – -2)

= 3

$(x - h)^2 + (y - k)^2 = r^2$

$(x + 2)^2 + (y - 1)^2 = 9$

Worked Example 21

WE 21

Sketch each of the following graphs of circles.

(a) $(x + 1)^2 + (y + 2)^2 = 9$ **(b)** $x^2 + (y - 2)^2 = 6$

Working

(a) $(x + 1)^2 + (y + 2)^2 = 9$

Centre: (-1, -2)

Radius: $\sqrt{9} = 3$

Points (2, -2) (-1 ,1) (-4, -2) (-1, -5)

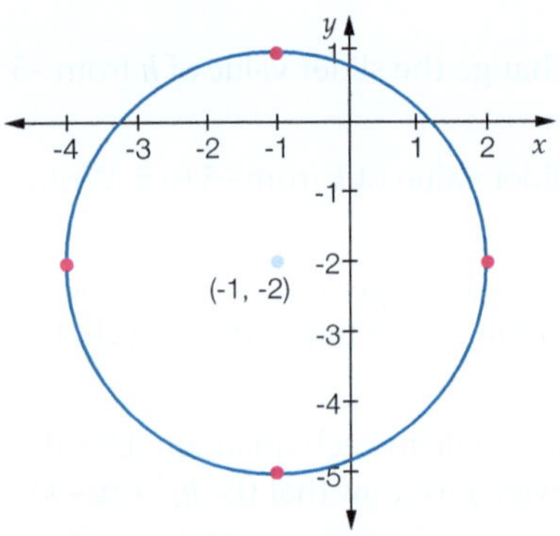

(b) $x^2 + (y - 2)^2 = 6$

Centre: (0, 2)

Radius = $\sqrt{6}$

Points: $(\sqrt{6}, 2)$ $(0, \sqrt{6} + 2)$ $(-\sqrt{6}, 2)$, $(0, \sqrt{6} - 2)$

$\approx$ (2.45, 2), (0, 4.45), (-2.45, 2), (0, -0.45)

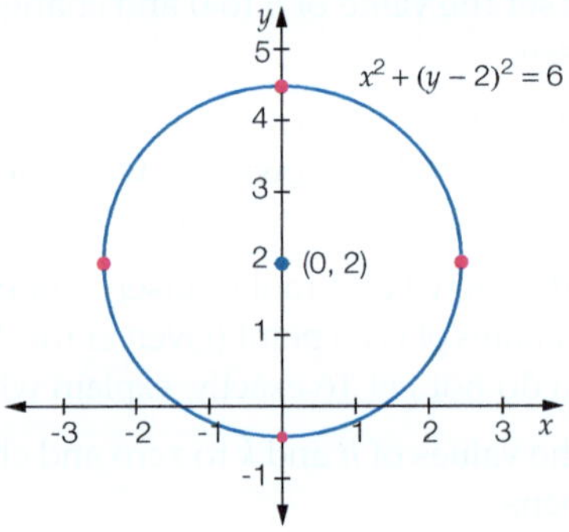

Worked Example 22

WE 22

A path around an area on a coordinate map is described by the equation $(x - 3)^2 + (y - 2)^2 = 9$.

(a) Determine the centre and radius of the circle.

(b) Find the four points on the circle in line with the centre horizontally and vertically.

(c) Use these points to plot the graph.

(d) Determine whether the following points are inside or outside the circle.

(i) (0, 1) **(ii)** (2, 4) **(iii)** (5, -2)

Working

(a) Centre (3, 2) and radius = 3

(b) (0, 2), (6, 2) (3, -1) (3, 5)

(c)

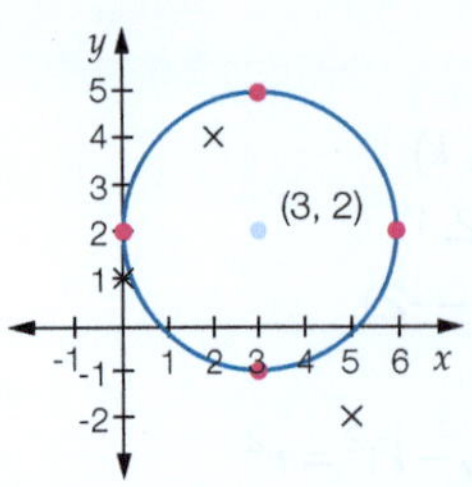

(d) **(i)** (0, 1) is outside the circle.

(ii) (2, 4) is inside the circle.

(iii) (5, -2) is outside the circle.

8.6

8.6 Circle transformations

Answers page 487

Stage 5.3 §

Navigator

Q1, Q2, Q3, Q4, Q5, Q6, Q9, Q10	Q1, Q2, Q3, Q4, Q5, Q6, Q7, Q9, Q10, Q11	Q1, Q2, Q3, Q4, Q5, Q6, Q7, Q8, Q9, Q10, Q11

Fluency and understanding

WE20

1 Determine the centre and radius of each of the following, and hence the equation of the circle.

(a)

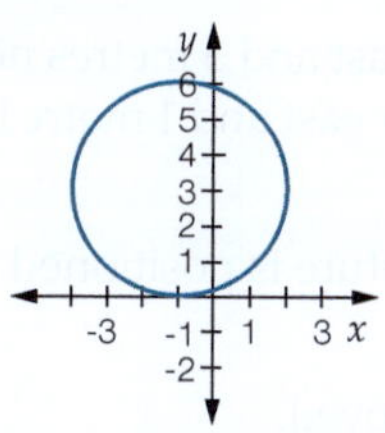

(b)

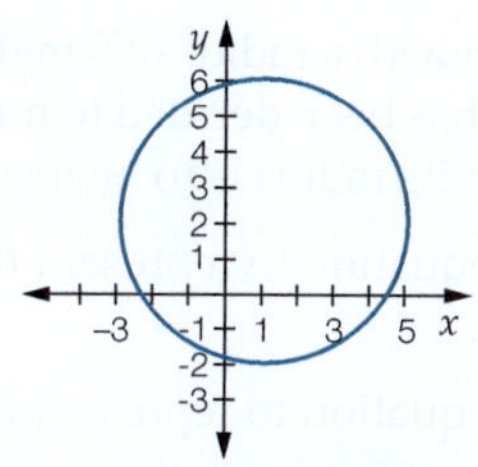

(c)

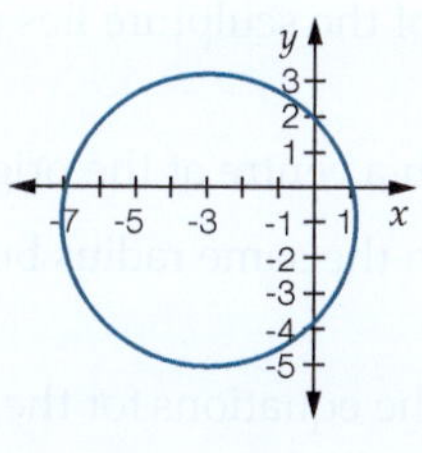

(d)

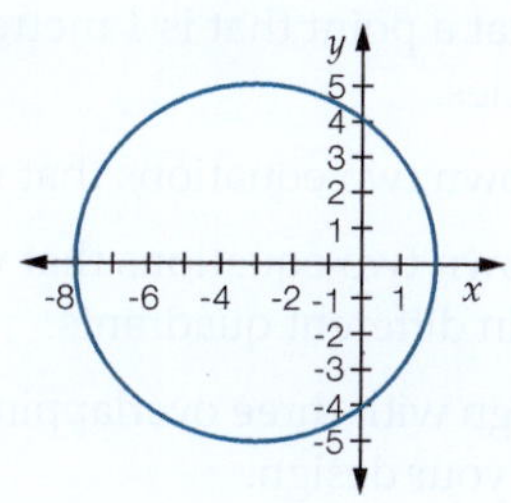

WE21

2 Sketch each of the following graphs of circles.

(a) $(x+2)^2+(y+3)^2=25$

(b) $(x-2)^2+(y-1)^2=16$

(c) $(x+1)^2+(y-1)^2=4$

(d) $x^2+(y-1)^2=9$

(e) $(x+2)^2+y^2=7$

(f) $(x-3)^2+(y+1)^2=1$

WE22

3 A path around an area on a coordinate map is described by the equation $(x+1)^2+(y-3)^2=16$.

(a) Determine the centre and radius of the circle.

(b) Find the four points on the circle in line with the centre horizontally and vertically.

(c) Use these points to plot the graph.

(d) Determine whether the following points are inside or outside the circle.

(i) (1, 2) (ii) (3, 6) (iii) (-2, 0)

4 Which of the following equations would graph as a circle?

A $x^2+y=4$ B $x^2+(y-3)^2=5$ C $y^2=x^2+9$ D $y^2=x-3$

5 Write down the equation if the circle with equation $x^2+y^2=36$ undergoes the following transformations.

(a) a move of 4 units to the right and 3 units up

(b) a move of 6 units to the left and 4 units up

(c) a move of 7 units to the right

(d) a move of 5 units down

(e) a dilation to a radius of 7

8.6

6 What are the centre coordinates and radius of a circle with the equation $(x-5)^2+(y+3)^2=81$?

A (5, 3), 9 B (-5, 3), 9 C (5, -3), 81 D (5, -3), 9

Reasoning, communicating and open-ended

7 A circle has an equation $(x+4)^2+(y+2)^2=25$. What would be the equation of this circle if it is translated 2 units left and 6 units up and the radius is decreased by 1 unit?

8 (a) Show, by substitution, that the equations $(x-8)^2+(y+8)^2=169$, $(x+9)^2+(y-9)^2=169$ and $x^2+y^2=25$ are three circles that intersect at (3, 4) and (-4, -3).

(b) Draw a graph of each of these three circles on the same set of axes.

9 A circular path with a radius of 5 metres has its centre 2 metres east and 3 metres north of a sculpture. It has been decided to move the path 9 metres further east and 1 metre further north and that its radius is to be increased by 8 metres.

(a) Write an equation to represent the original circle if the sculpture is positioned at the origin.

(b) Write an equation to represent the circle after it has been moved.

(c) Show that a point that is 1 metre west and 1 metre south of the sculpture lies on both circles.

10 (a) Write down two equations that would graph as circles with a centre at the origin.

(b) Write down two equations that would graph as circles with the same radius but with centres in different quadrants.

11 Make a design with three overlapping circles and write down the equations for the circles you used in your design.

Problem solving

Circles and circles

Find the equations for the eight full circles in the pattern shown.

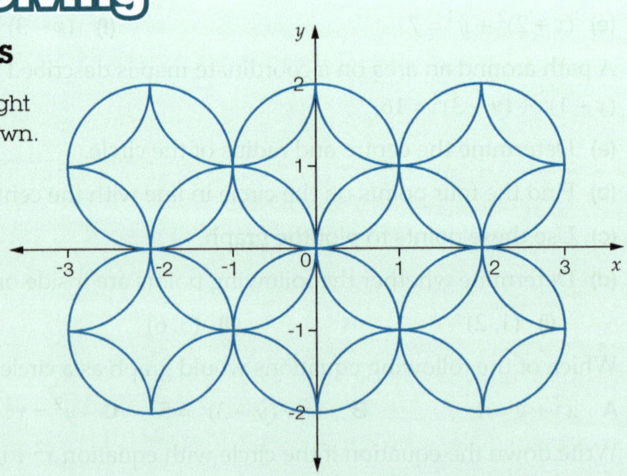

Strategy options

- Draw a diagram.
- Solve a simpler problem.
- Look for a pattern.
- Guess and check.

Exponential relationships

8.7

Stage 5.2◊

Need to Know

Exponential relationships

An exponential relationship is a relationship between two quantities that involves an **exponent** (or **index**), where the exponent is a variable. Exponential graphs always go in the same direction, either continually getting bigger or continually getting smaller, but, unlike straight lines, there is a limiting feature to this. Exponential graphs will always have a value that they will get closer and closer to but never reach. This value is called the **asymptote** of the graph.

The simplest exponential equations are in the form $y = a^x$ where a is the base and is any non-zero number and x is the index or exponent.

Because $a^0 = 1$ for all values of a except 0, all these exponential graphs will have a y-intercept of (0, 1).

As x gets smaller (i.e. negatively large), the value of y gets closer and closer to 0, but it never reaches it. The line with the equation $y = 0$ (the x-axis) is the asymptote.

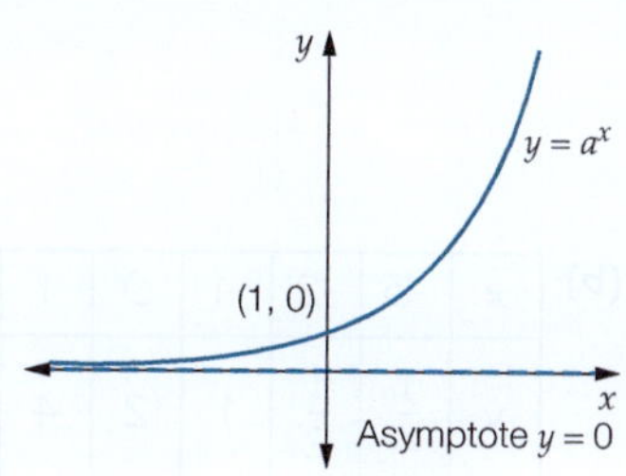

Translations of exponential graphs

An exponential graph with equation $y = a^x$ can be translated k units to the right to give the equation $y = a^{x-h}$. The asymptote remains $y = 0$.

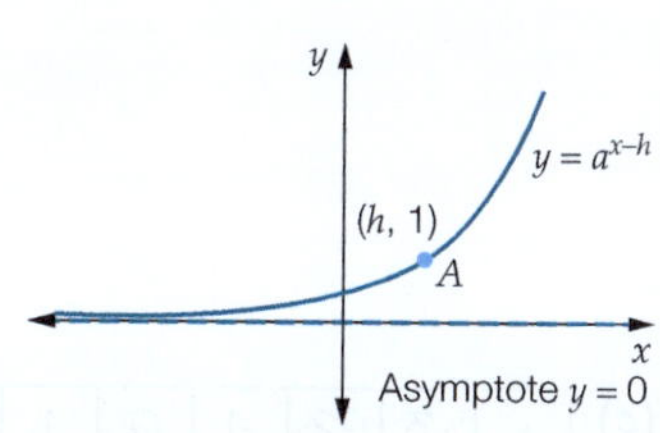

An exponential graph with an equation $y = a^x$ can be translated k units up to give the equation $y = a^x + k$. Here the asymptote becomes $y = k$.

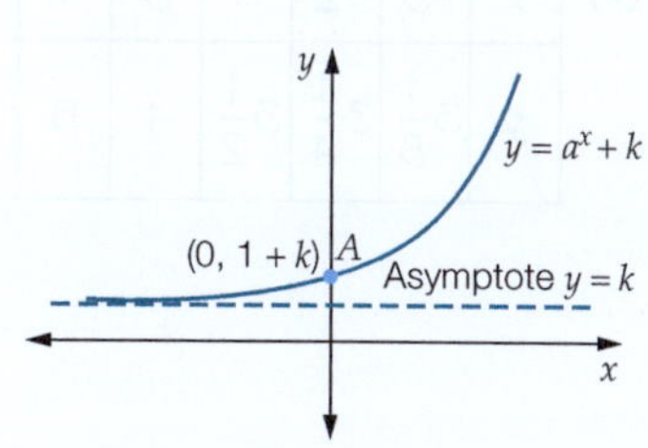

The general equation of an exponential relationship is $y = a^{x-h} + k$, where

- a is the base
- x is the exponent
- h indicates the translation to the right
- k indicates the translation upwards
- $y = k$ is the asymptote.

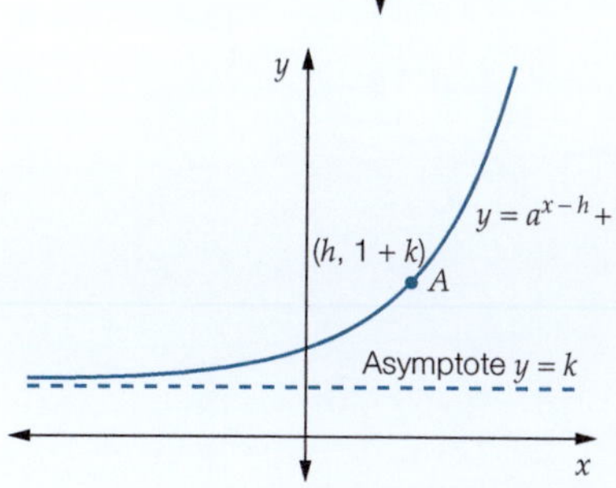

8.7

Worked Example 23

WE 23

Plot the following graphs by setting up tables of values from -3 to 3 inclusive.

(a) $y = 2^x$ (b) $y = 2^{x+1}$ (c) $y = 2^x + 3$

Working

(a)

x	-3	-2	-1	0	1	2	3
y	$\frac{1}{8}$	$\frac{1}{4}$	$\frac{1}{2}$	1	2	4	8

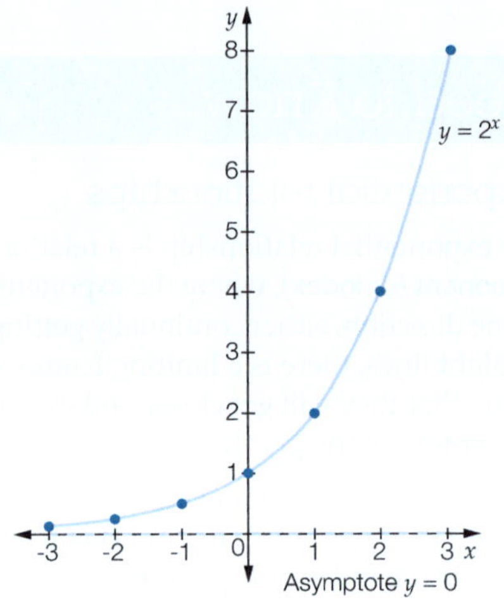

(b)

x	-3	-2	-1	0	1	2	3
y	$\frac{1}{4}$	$\frac{1}{2}$	1	2	4	8	16

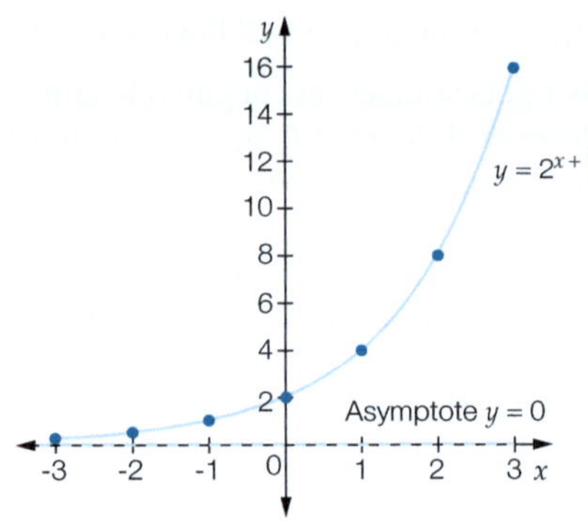

(c)

x	-3	-2	-1	0	1	2	3
y	$3\frac{1}{8}$	$3\frac{1}{4}$	$3\frac{1}{2}$	4	5	7	11

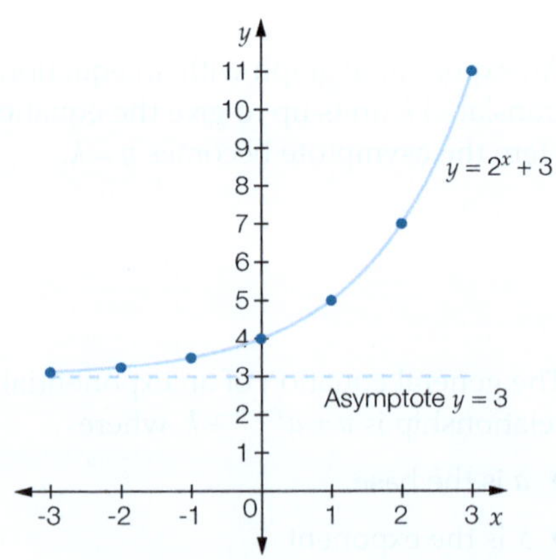

8.7

Worked Example 24

WE 24

Bacteria in a culture are multiplying rapidly. Each bacterial cell divides into two, then each of these cells divides into two and so on. This can be expressed in the form $N = 2^t$ where N is the number of bacteria and t is the time in hours.

(a) For the relationship $N = 2^t$, copy and complete the table of values for t and N.

t	0	1	2	3	4	5	6
N							

(b) Use this table of values to plot a graph of $N = 2^t$. Join the points with a smooth curve.

(c) Use your graph to determine how long it takes for the number of bacteria to exceed 40.

(d) Evaluate N for $t = 5.3$ (give your answer correct to the nearest integer) and explain why this value is meaningless in this situation.

(e) Determine how many bacteria there are after 3 hours 20 minutes.

Working

(a)

t	0	1	2	3	4	5	6
N	1	2	4	8	16	32	64

(b)

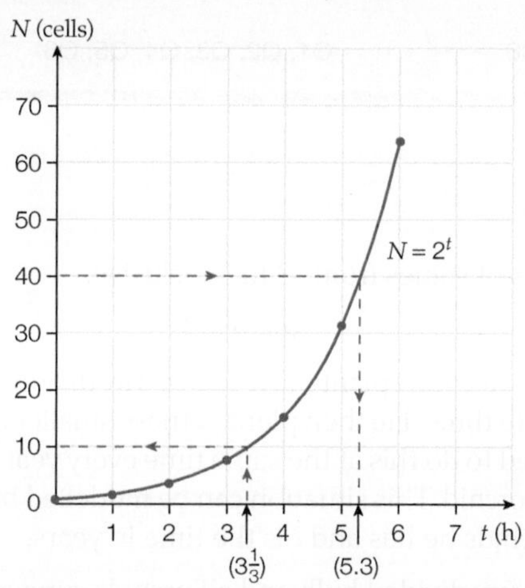

(c) From the graph, when $N = 40$, $t = 5.3$ hours. However, it takes a full hour for the 32 cells present at $t = 5$ to become 64 cells, so $t = 6$.

(d) The answer using a calculator is ≈ 39.4, but cell division only takes place at the end of each hour, so this value is meaningless in this situation.

(e) At 3 hours 20 minutes there will still only be 8 cells, as the next cell division does not occur until $t = 4$.

8.7

Technology Exploration: GeoGebra

Investigating exponentials

1 Open a new GeoGebra window and set it up as for the GeoGebra activity in section 8.4. Use the Input bar to enter the equation $y = a^\wedge(h) + k$. Set up three new sliders as you did in previous GeoGebra activities and call them h, k and a, and set the a values from 0 to 5. Set both h and k values to 0 and set a to 1. What do you see?

2 Leave the k value on 0 and the a value on 1. Change the slider value of h from -5 to 5. Write what you see.

3 Now set the value of h to 0 and change the slider value of k from -5 to 5. Write what you see.

4 What happens when neither h nor k is zero? Choose integer values for each of h and k. Write these values down and write the equation of the asymptote formed. What do you see?

5 Set the values of h and k to 0 and change the value of a from 0 to 5. Describe what happens.

Answers page 488

8.7 Exponential relationships

Stage 5.2◊

Navigator

Q1, Q2, Q3, Q4, Q6	Q1, Q2, Q3, Q4, Q6	Q1, Q2, Q3, Q4, Q5, Q6

Equipment required: graph paper for Question **6**

Fluency and understanding

WE23

1 Plot the following graphs by setting up tables of values from -3 to 3 inclusive.

(a) $y = 3^x$ (b) $y = 3^{x+1}$ (c) $y = 3^x - 2$

WE24

2 A gardener decided to propagate his orchid (make new plants). He did this by dividing the bulb on his plant after it finished flowering into three. He then planted these smaller bulbs giving him three orchids in total. He continued to do this at the same time every year with every orchid he had grown from the original orchid. This situation can be modelled by the equation $P = 3^t$ where P is the number of orchids he has and t is the time in years.

(Assume that a healthy orchid grows from every divided bulb and all orchids survive for at least 6 years.)

(a) For the relationship of orchids to the time in years where $P = 3^t$, copy and complete the table of values below.

t (years)	0	1	2	3	4	5	6
P (orchids)							

(b) Use this table of values to plot a graph of $P = 3^t$. Join the points with a smooth curve.

(c) Use your graph to determine how many years it takes for the number of orchids to exceed 125.

(d) Evaluate P for $t = 3.8$ (give your answer correct to the nearest integer) and explain why this value is meaningless in this situation.

(e) Determine how many orchids there are after $5\frac{1}{2}$ years.

3 (a) Plot the graphs of $y = 2^x$ and $y = 3^x$ on the same set of axes.

(b) By comparing asymptotes, y-intercepts and shape, describe the similarities and differences between the two graphs.

4 An exponential graph of $y = 5^x$ is shifted 4 units left and 3 units down. Which of the following is the equation of the new graph?

A $y = 5^{(x-4)} + 3$ B $y = 5^{(x-4)} - 3$ C $y = 5^{(x+4)} + 3$ D $y = 5^{(x+4)} - 3$

Reasoning, communicating and open-ended

5 The population of a town has been increasing at a rate of 10% each year since the beginning of 2005. This can be modelled by the equation $P = 2000 \times 1.1^t$ where P is the number of people in the town t years after the beginning of 2005. Initially, when $t = 0$, the population numbered 2000.

(a) Set up a table to show the population of the town for the first 10 years after 2005.

(b) Draw a graph showing this information.

(c) From the graph, approximately how long did it take for the town to double in size?

(d) Jason said it would take 10 years for the population to double. Explain why this isn't the case.

6 Substitute your own values for h and k into the equation of an exponential in the form of $y = 2^{(x-h)} + k$ and draw a graph of this relationship.

Problem solving

How many triangles?

Six equilateral triangles are joined to make a hexagon.

Six hexagons are joined to make the hexagon pattern as shown.

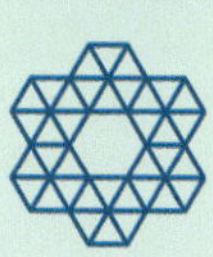

Six of these hexagon patterns are joined to make a larger pattern.

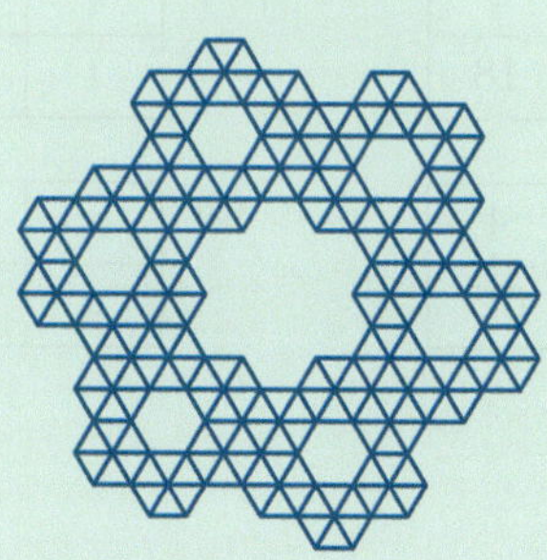

Six of these larger hexagon patterns are joined in the same way to make an even larger pattern. How many triangles would there be in this pattern?

Strategy options

- Look for a pattern.
- Make a table.

8.8 Direct proportion

Stage 5.2

Need to Know

The simplest type of relationship between two continuous quantities is one in which any change in one quantity results in an identical change to the other quantity. For example, if we double one quantity, the other will double also; halve one quantity and the other will halve also. In this case we say that one quantity is in direct proportion to the other quantity.

The relationship between two quantities, say a and b, is one of **direct proportion** when a increases as b increases and the graph of the relationship passes through the origin.

$a \propto b$ means 'a is directly proportional to b'.

$a \propto b$ is not the same as $a = b$.

In general, there is direct proportion between two variables, x and y, when

$y \propto x^n$ where n is any positive number;

or, as an equation, when

$y = kx^n$ where k is the **constant of proportionality**

Two variables are in direct proportion if the graph of y plotted against x^n is a linear graph passing through the origin, (0, 0).

As x increases, y increases.

Worked Example 25

WE25

Which of the following could be examples of direct proportion?

(a)

x	1	3	6	8
y	2	6	12	18

(b)

s	1	2	3	5
t	2	8	18	50

(c)

v	1	2	3	4
w	4	2	1	0.5

(d)

x	1	2	3	4
y	5	7	7	3

Working

(a) Could be direct proportion. (y is increasing as x is increasing.)

(b) Could be direct proportion. (t is increasing as s is increasing.)

(c) Not direct proportion. (w is decreasing as v is increasing.)

(d) Not direct proportion. (q is fluctuating as p is increasing.)

8.8

Worked Example 26

WE 26

Find the equation for each of the following tables of values.

(a)

x	1	3	6	8
y	2	6	12	16

(b)

s	0	1	2	3	5
t	0	2	8	18	50

(c)

a	0	1	4	9
b	0	3	6	9

Working

(a) $\frac{y}{x} = \frac{2}{1} = \frac{6}{3} = \frac{12}{6} = \frac{16}{8} = 2$

$\therefore k = 2$

$y \propto x$

$y = kx$

$y = 2x$

(b) $\frac{t}{s} = \frac{2}{1} = 2$, $\frac{t}{s} = \frac{8}{2} = 4$, $\frac{t}{s} = \frac{18}{3} = 6$,

$\frac{t}{s} = \frac{50}{5} = 10$

s^2	0	1	4	9	25
t	0	2	8	18	50

$\frac{t}{s^2} = \frac{2}{1} = \frac{8}{4} = \frac{18}{9} = \frac{50}{25} = 2$

$\therefore k = 2$

$t \propto s^2$

$t = ks^2$

$t = 2s^2$

(c) $\frac{b}{a} = \frac{3}{1} = 3$, $\frac{b}{a} = \frac{6}{4} = \frac{3}{2}$, $\frac{b}{a} = \frac{9}{9} = 1$

$\sqrt{a}$	0	1	2	3
b	0	3	6	9

$\frac{b}{\sqrt{a}} = \frac{3}{1} = \frac{6}{2} = \frac{9}{3} = 3$

$\therefore k = 3$

$b \propto \sqrt{a}$

$b = k\sqrt{a}$

$b = 3\sqrt{a}$

8.8

Answers page 489

Stage 5.2

8.8 Direct proportion

Navigator

Q1, Q2 Column 1, Q3, Q4, Q5, Q6, Q7, Q8, Q9, Q10, Q13	Q1, Q2 Column 2, Q3, Q4, Q5, Q6, Q7, Q8, Q9, Q10, Q11, Q13	Q1, Q2 Column 2, Q3, Q4, Q5, Q6, Q7, Q8, Q9, Q10, Q11, Q12, Q13

Fluency and understanding

WE25

1 Which of the following could be examples of direct proportion?

(a)

v	1	2	4	5
w	7	3	2	1

(b)

p	1	2	3	4
q	3	4	5	0

(c)

x	1	2	3	4
w	0	1	2	4

(d)

s	1	2	6	5
t	5	10	30	25

WE26

2 Find the equation for each of the following tables of values.

(a)

a	0	1	2	3
b	0	4	8	12

(b)

w	0	1	2	3
v	0	5	10	15

(c)

c	0	1	4	6
d	0	0.5	2	3

(d)

g	0	4	6	8
h	0	1	1.5	2

(e)

e	0	1	2	3
h	0	3	6	9

(f)

p	0	2	4	6
q	0	3	6	9

(g)

i	1	2	3	0
j	2	8	18	0

(h)

a	0	1	2	3
d	0	3	12	27

(i)

m	0	1	2	3
n	0	4	32	108

(j)

g	1	2	3	0
h	2	16	54	0

(k)

p	0	1	2	3
q	0	0.5	2	4.5

(l)

b	0	1	2	3
f	0	0.2	0.8	1.8

(m)

r	0	1	2	3
s	0	3	24	81

(n)

p	0	1	2	3
m	0	0.25	2	6.75

(o)

t	0	1	2	3
u	0	0.5	4	13.5

(p)

v	0	1	4	9
w	0	2	4	6

(q)

u	0	1	4	9
t	6	0.1	0.2	0.3

(r)

d	0	1	4	9
l	0	$\frac{1}{3}$	$\frac{2}{3}$	1

8.8

3 State whether or not each graph could be an example of direct proportion.

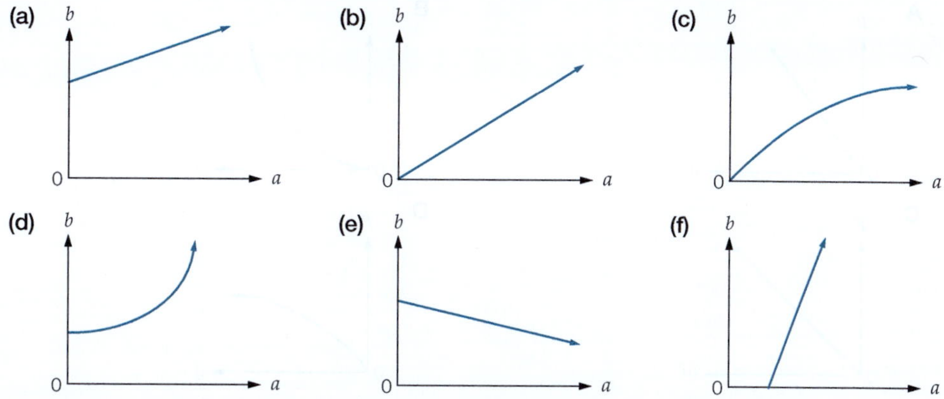

4 Express each of these relationships involving direct proportion as an equation, using k as the constant.

(a) $m \propto n$

(b) $a \propto b^2$

(c) $g \propto h^3$

(d) t is proportional to w.

(e) N is proportional to p^2.

(f) y is proportional to the cube of x.

(g) V is proportional to the square of r.

(h) L is proportional to the square root of d.

5 **(a)** The volume of a sphere (V) is directly proportional to the cube of its radius (r). This may be written as:

A $V \propto r^2$ **B** $V \propto r^3$ **C** $V \propto r$ **D** $V \propto \sqrt{r}$

(b) The radius of a cylinder (r) is directly proportional to the square root of the volume (V). This may be written as:

A $r \propto \sqrt{V}$ **B** $r \propto V^2$ **C** $r \propto V^3$ **D** $r \propto V$

6 State the value of the constant of proportionality in each of the following.

(a) $p = 3q$ **(b)** $a = 0.2b^2$ **(c)** $M = \frac{1}{4}n$ **(d)** $y = 200\sqrt{x}$

(e) $c = n^5$ **(f)** $F = \frac{a}{6}$ **(g)** $x = \frac{w^3}{20}$ **(h)** $P = \frac{r^4}{35}$

7 In each of the following table of values find the constant of proportionality and hence find the missing values in the table.

(a) $p \propto q$

q	5	7	9	
p	15	21		33

(b) $m \propto n^2$

n	1	3	5	
m	7		175	252

(c) $d \propto c^3$

c	1		4	5
d	1	8		125

(d) $w \propto \sqrt{x}$

x	1		16	49
w	3	6	12	

8 Which graph shows direct proportion of the form $p \propto q^2$?

A

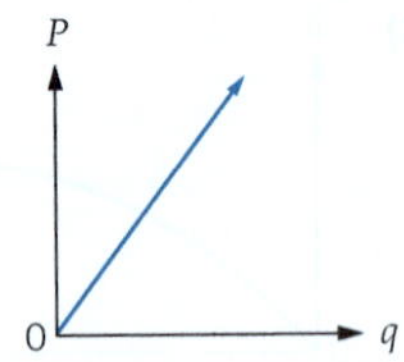

B

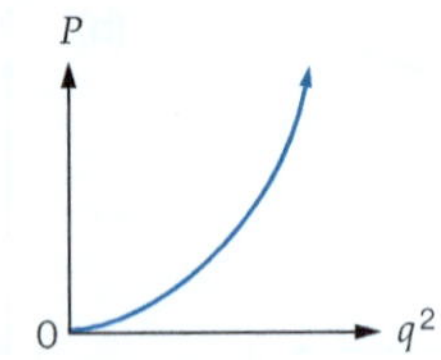

C

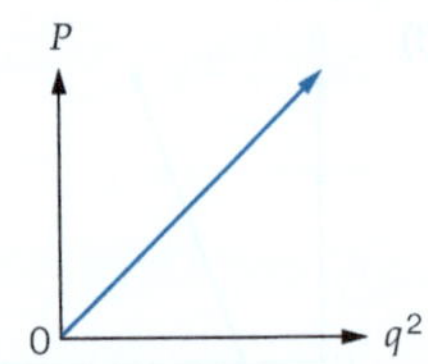

D

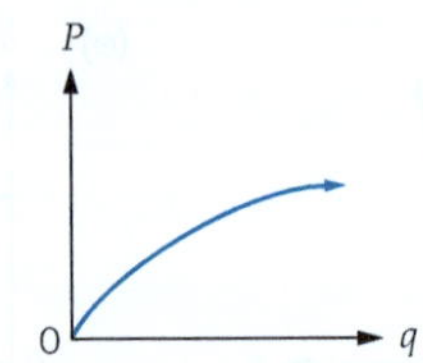

9 A student measures the current and voltage of a simple electrical circuit to determine the value of a resistor within the circuit. The results obtained are shown in the table.

Current, *I* (amps)	0.001	0.002	0.003	0.004
Voltage, *V* (volts)	3	6	9	12

(a) Graph the relationship between *I* amps and *V* volts. Show *I* on the horizontal axis and *V* on the vertical axis.

(b) Calculate the constant of proportionality, in this case called the resistance *R* (ohms).

(c) Calculate the voltage if 0.01 amps of current is flowing through the circuit.

(d) What is the current flowing through the circuit when a voltage of 45 volts is measured in the circuit?

10 The following table shows the relationship between the surface area of a sphere and its radius.

Radius, *r* (cm)	2	7	15	20	50
Surface area, *A* (cm^2)	50.24	615.44	2826	5024	31 400

(a) Determine the constant of variation for this relationship.

(b) Express this relationship as an equation using $\pi = 3.14$.

Reasoning, communicating and open-ended

11 The distance to the horizon is proportional to the square root of the height above sea level. A man standing on a cliff 32 m above sea level can see a distance of 18 km.

(a) How far could he see if he was standing on the viewing platform of a lighthouse 18 m above the cliff top?

(b) What height above sea level would he need to be to see a distance of 100 km?

12 The following table shows the circumference, surface area and volume of spheres of various radii, in similar units.

Radius, *r*	**Circumference, *C***	**Surface area, *A***	**Volume, *V***
2	12.56	50.24	33.49
4	25.12	200.96	267.95
6	37.68	452.16	904.32

8	50.24	803.84	2143.57
10	62.80	1256.00	4186.67
12	75.36	1808.64	7234.56

(a) Find the ratio $\frac{C}{r}$, $\frac{A}{r}$ and $\frac{V}{r}$ for each pair of values.

(b) Which of the ratios is constant? Describe your results.

(c) Plot graphs of C against r, A against r and V against r.

(d) Plot a graph of A against r^2. Describe the graph.

(e) Plot a graph of V against r^3. Describe the graph.

(f) Which of the graphs show direct proportion between the quantities? Write a relationship for each of these.

(g) Calculate the constant of proportionality, k, for each of these relationships. Express each as an equation in terms of π (where $\pi = 3.14$).

13 **(a)** Write two different relationships of the type $y = kx^n$, where $n > 0$ and $0 < k < 1$. Use the same value of n for both of your relationships.

(b) Sketch the graph of $y = x^n$ for your chosen value of n and, on the same set of axes, sketch the two relationships you have written. Describe what happens as k decreases from 1 to 0.

Problem solving

Relationships between variables

Determine the relationship between x and y for each of these tables.

1

x	0	1	2	3	4
y	0	3	12	27	48

2

x	0	1	2	3	4
y	1	5	25	125	625

Strategy options

- Look for a pattern.
- Guess and check.

Challenge 8

1 The solutions of the equation $x^2 - 13x - 48 = 0$ are:

A -3, 16 B 3, -16 C 4, -12 D -6, 8

2 How many positive integers less than or equal to 3000 can be written without using the digits 4, 5 or 6?

3 If $(x - 2)(2x + 3) = 0$, then possible values for $2x + 3$ are:

A 0 only B 0 and 3 C 0 and 7 D $-\frac{3}{2}$ and $-\frac{7}{2}$

4 The perimeter of the rectangle in the diagram is 112 cm. Its total area is:

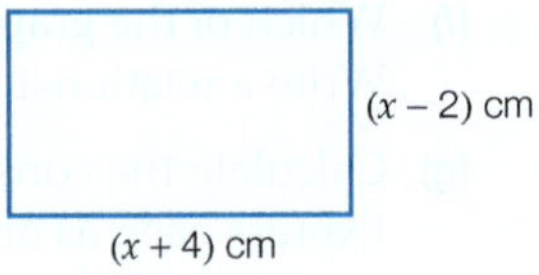

A 44 cm^2 B 187 cm^2

C 775 cm^2 D 3127 cm^2

5 If $u \# v$ means $2u - v$, what is the value of $6 \# (2 \# 5)$?

6 The mean of the solutions of the quadratic equation $x^2 + 10x - 11 = 0$ is:

A -10 B -5 C 5 D 10

7 A 6-m wide strip is mown around the inside edges of a square lawn. If eight-ninths of the lawn has been mowed (shaded area), calculate the area, in square metres, that still has to be mown (unshaded area).

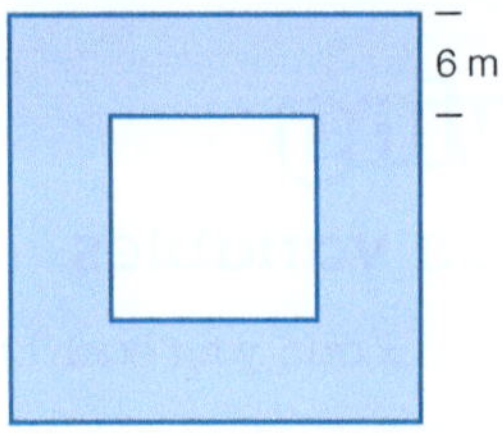

8 Two three-digit numbers, *abc* and *def*, have the following property:

$$\begin{array}{r} a\ b\ c \\ +\quad d\ e\ f \\ \hline 1\ 0\ 0\ 0 \end{array}$$

None of a, b, c, d, e and f is zero. Therefore, the value of $a + b + c + d + e + f$ is:

A 10 B 19 C 21 D 28

9 If $x + \frac{1}{x} = 5$, then $x^2 + \frac{1}{x^2}$ equals:

A 25 B 23 C 27 D 3

10 Five identical rectangles are arranged to form a larger rectangle *KLMN*, as shown. The area of *KLMN* is 4000 cm^2. The length, x cm, of each of the identical rectangles is closest to:

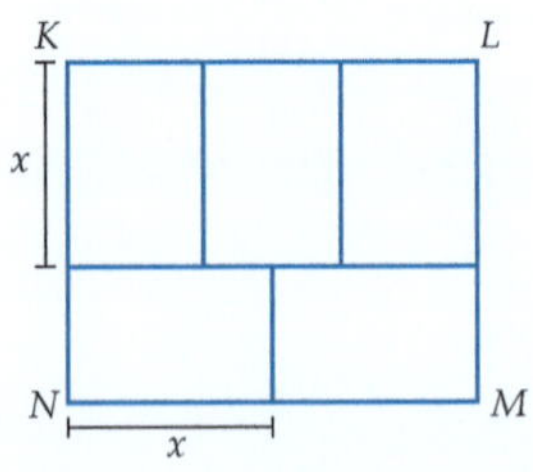

A 33 cm B 35 cm

C 37 cm D 39 cm

Chapter review 8

D.I.Y. Summary

Key Words

Stage 5.1	circle	exponential	parabola
Stage 5.2	asymptote	direct proportion	quadratic relationship
	axis of symmetry	Null Factor Law	turning point
	constant of proportionality	quadratic equation	vertex
Stage 5.3	coefficient	dilation factor	transformation
	dilation	reflection	translation

Copy and complete the following using the words and phrases from the list where appropriate to write a summary for this chapter. A word or phrase may be used more than once

Stage 5.1 Q1, 2 **Stage 5.2 Q3–6** **Stage 5.3 Q7–9**

1 A ___________ is an expression where the highest power of x is 2 and the highest power of y is 1.

2 A ___________ is the type of graph produced by the equation $y = x^2$.

3 The ___________ passes through the ___________ of a quadratic graph.

4 Another name for the turning point is the ___________.

5 Using the ___________ it is possible to find the solutions for x when an equation equals 0.

6 The equation $x^2 + y^2 = 9$ graphs as a ___________ with a radius of 3.

7 The graph $y = x^2$ is a ___________ of the graph $y = -x^2$.

8 The graph $y = x^2$ can be changed into $y = 2(x-1)^2 + 3$ by undergoing a number of ___________s.

9 The main features of a sketch curve of a quadratic relationship are: y-intercept, x-intercepts and the ___________.

Equipment required: graph paper for Question **14**.

Stage 5.1 Q1 **Stage 5.2 Q2–4, 9, 11–15, 17–21, 24** **Stage 5.3 Q5–8, 10, 16, 22, 23**

Fluency and understanding

1 Which of the following equations would NOT graph as a parabola. **Ex 8.1**

A $y = 4 - x^2$ **B** $y = 4x^2$ **C** $y^2 = 4 - x^2$ **D** $y = \frac{x^2}{4}$

2 Plot the graph of $y = x^2 - 2x - 8$. **Ex 8.2**

(a) Write down the coordinates of the x- and y-intercepts.

(b) Write down the coordinates of the turning point. Does the parabola have a maximum or a minimum turning point?

(c) Write down the equation of the axis of symmetry.

(d) From the graph, find the value of y when $x = 3$.

(e) What are the values of x when $y = 7$?

(f) Hence state the values of x when $x^2 - 2x - 8 = 7$.

(g) Write down the values of x when $x^2 - 2x - 8 = 0$.

3 Solve the following using the Null Factor Law. **Ex 8.3**

(a) $2x(x+4)=0$ (b) $(x-7)(x+1)=0$ (c) $x^2-16=0$

(d) $x^2+6x+9=0$ (e) $x^2=8x$ (f) $x^2+36=12x$

4 Solve these quadratic equations. **Ex 8.3**

(a) $x^2-4x=0$ (b) $x^2-6=0$ (c) $x^2+12=0$

(d) $3x^2=27$ (e) $(x-3)(x+7)=11$ (f) $(x-3)(x+2)=6$

5 (a) The equation of the unlabelled parabola shown could be: **Ex 8.4**

A $y=x^2+2$

B $y=x^2-2$

C $y=(x-2)^2$

D $y=(x+2)^2$

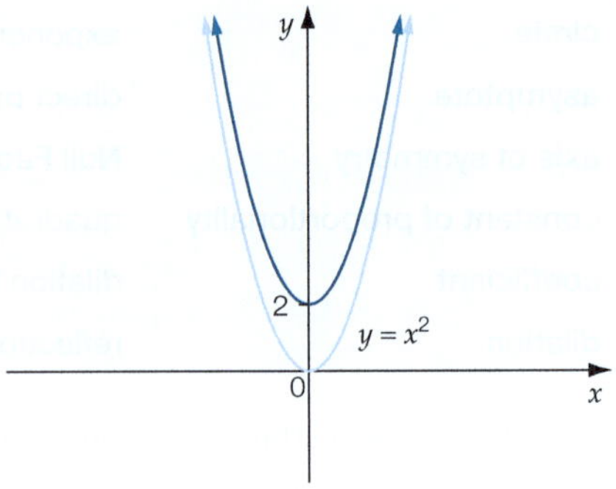

(b) The equation of the unlabelled parabola shown could be:

A $y=(x-1)^2-2$

B $y=(x-1)^2+2$

C $y=(x+1)^2+2$

D $y=(x+1)^2-2$

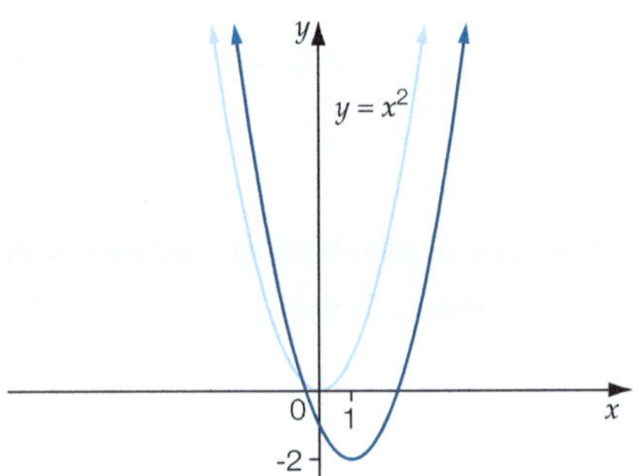

6 State the changes that should be made to the graph of $y=x^2$ to obtain the graph of each of the following. **Ex 8.4**

		Dilation	Reflection in *x*-axis	Translation in *x* direction	Translation in *y* direction
(a)	$y=(x-1)^2+3$				
(b)	$y=-x^2+4$				
(c)	$y=-(x+2)^2-1$				

7 Sketch the graphs of: **Ex 8.4**

(a) $y=x^2+8x+16$ (b) $y=-x^2+2x-1$

(c) $y=3x^2+9x$ (d) $y=x^2-4x+4$

8 (a) The graph of the quadratic relationship $y=x^2-64$ crosses the x-axis at: **Ex 8.4**

A $x=-8$ and $x=8$ B $x=8$ C $x=-8$ D $x=64$

(b) The graph of the quadratic relationship $y=x^2-8x+16$ crosses the x-axis at:

A $x=2$ and $x=8$ B $x=-4$ and $x=4$ C $x=4$ D $x=-4$

9 Identify the centre and radius for each of the following circle equations, then sketch the graphs. **Ex 8.5**

(a) $x^2+y^2=9$ (b) $x^2+y^2=5$

10 Sketch graphs for each of the following circle equations. Ex 8.6

(a) $x^2 + (y - 2)^2 = 25$ (b) $(x + 3)^2 + (y - 1)^2 = 16$

11 Sketch graphs for each of the following equations, showing the asymptotes. Ex 8.7

(a) $y = 2^{-x}$ (b) $y = 2^{x-1} + 3$

12 Find the equation for each of the following tables of values. Ex 8.8

(a)

a	1	3	4
b	3	9	12

(b)

g	0	4	9
h	0	6	9

13 (a) The volume of a cube is directly proportional to the cube of its height. This may be written as: Ex 8.8

A $V \propto H^3$ B $V = H^3 + 1$ C $V \propto H^{1/3}$ D $V = H^{1/3}$

(b) The surface area of a cylinder is directly proportional to the square of its radius. This may be written as:

A $A = r^2 + 3r$ B $A \propto r^2$ C $A \propto \sqrt{r}$ D $A \propto \frac{1}{r^2}$

(c) The profit of a company is directly proportional to the square root of its number of employees. This may be written as:

A $p \propto n^2$ B $p = n^2$ C $p = \sqrt{n^2}$ D $p \propto \sqrt{n}$

14 A softball thrown by an outfielder to the pitcher follows a parabolic path where its height, h metres, at any time, t seconds, is given by the relationship:

$$h = -5t^2 + 10t + 1$$

(a) Complete a table of values for this relationship.

(b) Plot the graph of this relationship on graph paper.

(c) What was the height of the softball after 0.5 seconds?

(d) When was the ball first at a height of 5 m above the ground?

(e) During which time interval was the ball above a height of 2 m?

(f) What was the maximum height of the softball during its flight?

(g) How long was the ball in flight if it was caught at a height of 1.6 m above the ground on its downward path?

15 A jet of water from an outlet level with the surface of a pond follows a parabolic path. The height of the water at any point can be described by the equation:

$$h = -d^2 + 6d$$

where h is the height of the water a horizontal distance d from the water outlet. Both h and d are in metres.

(a) Sketch the graph of $h = -d^2 + 6d$.

(b) From your sketch, state the maximum height of the water jet above the surface of the pond.

(c) What horizontal distance does the jet of water cover from its outlet back to the surface of the pond?

16 Describe the transformation that has been performed when $(x + 2)^2 + (y - 1)^2 = 25$ has been changed to $(x - 4)^2 + (y - 3)^2 = 16$.

17 (a) Show that the graphs of $y = 2^{x+1} + 5$ and $y = 3^{x+2} - 2$ have the same y-intercept.

(b) Write down the equation of the asymptotes for each graph.

Reasoning and communicating

18 Di and Sam found the following table of values, but some values were missing. Di could not see a pattern, but Sam thought she could see one.

x	1	2	3	4	5	
y	0.1		2.7		12.5	21.6

(a) Is there a relationship between the two values? If so, write the relationship.

(b) Use the relationship to fill in the missing values.

19 Leonard and Faye want to develop a vegetable garden in their backyard. They have 20 m of chicken wire and enough posts to erect a fence around the four sides of the rectangular garden.

(a) If the length of the rectangular garden is x m, find an expression (in terms of x) for the width of the garden.

(b) Show that the area of the garden, A m^2, is given by the quadratic expression $A = -x^2 + 10x$.

(c) Sketch the graph of the area of the vegetable garden, A m^2, against the length of the garden, x m. (Show length on the horizontal axis.)

(d) What should the dimensions of the garden be if the area is to be as large as possible?

(e) What is the area of the largest vegetable garden that Leonard and Faye can make?

20 A book cover has an area of 64 cm^2. If the length is x cm and the length and width add to 16 cm, find the dimensions of the book cover.

21 The sum of two numbers is 50. What are the numbers if their product is to be as large as possible?

22 A parabola has a minimum turning point at (3, 5) and a dilation factor of 1. What is the equation of the curve?

23 A circle has an equation $(x + 1)^2 + (y - 3)^2 = 121$. What would be the equation of this circle if it is translated 4 units right and 3 units down and the radius is decreased by 2 units?

24 $v^2 = u^2 + 2as$ is a formula used in motion. If $u = 0$, the formula becomes $v^2 = 2as$. If a is a constant:

(a) what happens to v when s is doubled?

(b) what happens to s when v is reduced to $\frac{3}{4}$ of its original value?

NAPLAN practice 8

Numeracy: Non-calculator

1 Find the value of x in the following rectangle given that the area is 8 m^2.

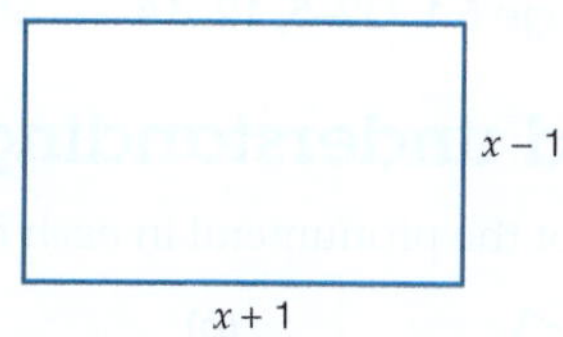

2 Here is a table of values for x and y.

x	-3	-2	-1	0	1
y	18	8	2	0	2

Which of the following is a correct rule for y in terms of x?

A $y = 6x$ **B** $y = 4x$ **C** $y = x^2 + 3$ **D** $y = 2x^2$

3 The solution/s to $x^2 - 4 = 0$ is/are:

A $x = -2$ **B** $x = 0$ **C** $x = 2$ and -2 **D** $x = 2$

Numeracy: Calculator allowed

4 Jane drew this plan for her square block of land. (All measurements are in metres.)

The shaded area shows the garage. Apart from the garage, what is the area of the land?

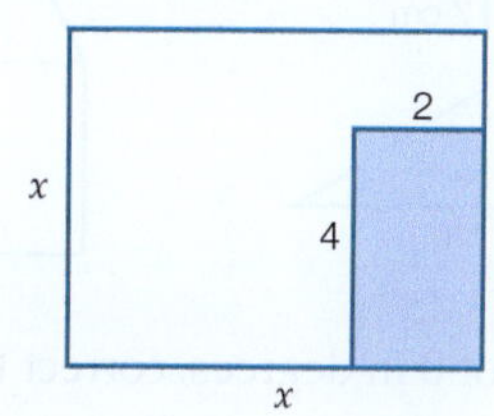

A $x^2 + 4x$ **B** $(x + 4)(x - 2)$

C $(x - 4)(x - 2)$ **D** $x^2 - 8$

5 Complete the table of values below that would be used to plot points for the relationship $y = x^2 - 3x + 2$.

x	-3	-2	-1	0	1	2	3
y							

6 If $x = 2$, then which of these expressions will be equal to 6?

A $x^2 + 5x - 3$ **B** $(x + 1)(x - 4)$ **C** $4x^2 - 2x + 3$ **D** $-x^2 + 4x + 2$

INTRODUCTION TO MATHEMATICAL THINKING
A publication from Pearson Custom Publishing

Mixed review C

Stage 4 Q1 Stage 5.1 Q2–5, 12, 14 Stage 5.2 Q6–10, 13, 15 Stage 5.3 Q11, 16

Fluency and understanding

1 Find the value of the pronumeral in each figure. **Ex 5.1**

(a)

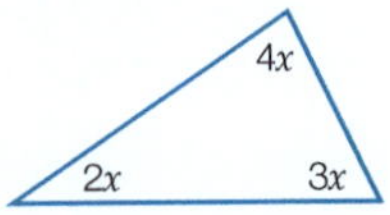

(b)

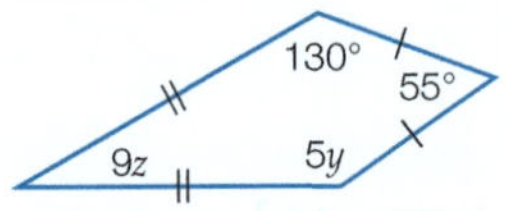

(c)

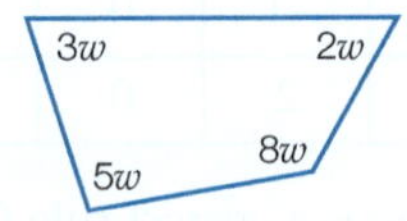

2 An investment of $5000 is made at 3.5% p.a. for 6 years. Find the difference in interest earned if interest is compounded monthly compared with annually. **Ex 1.5**

3 Find the area of each of the following shapes. **Ex 3.2**

(a)

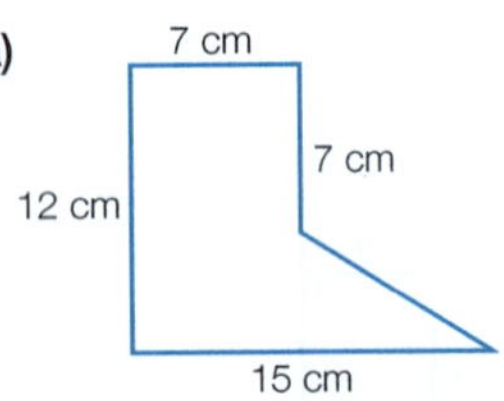

(b)

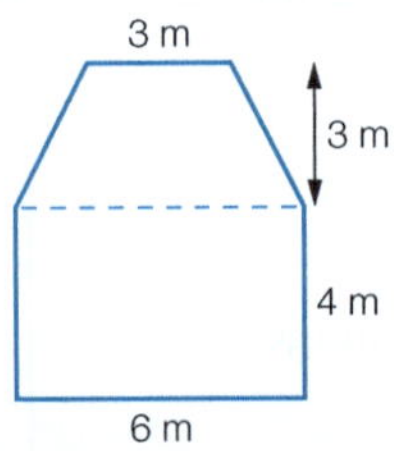

(c)

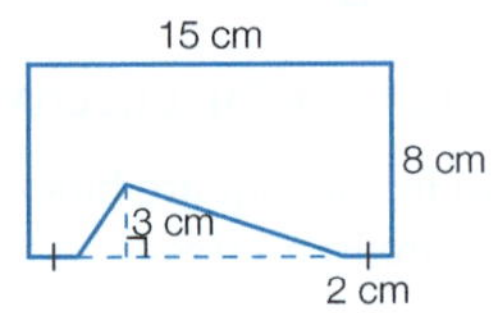

4 Find the value of θ in degrees, correct to one decimal place in each of the following. **Ex 6.4**

(a)

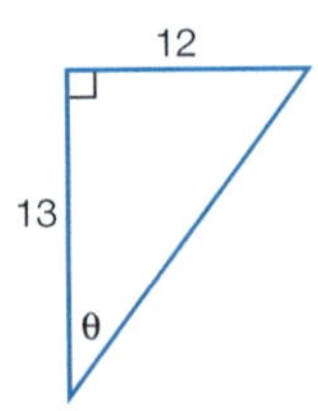

(b)

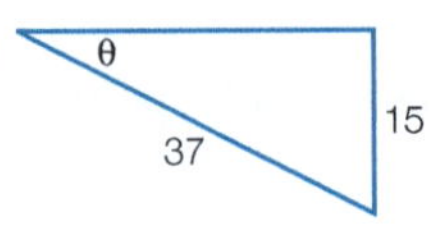

(c)

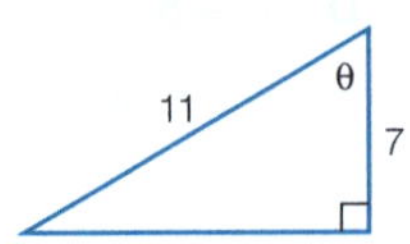

5 Find the mean for each of the following grouped data frequency tables. Give your answers correct to two decimal places. **Ex 7.3**

(a)

Class interval	Frequency
20–<25	8
25–<30	6
30–<35	3
35–<40	2
40–<45	5
45–<50	5

(b)

Class interval	Frequency
111–<115	12
116–<120	15
121–<125	17
126–<130	19
131–<135	13
136–<140	11

6 Solve the following equations. **Ex 4.3**

(a) $\frac{3a+4}{7} = -2$ (b) $\frac{4a+3}{2} = \frac{2a}{5}$ (c) $\frac{3a-2}{3} = \frac{4a-5}{2}$

7 Solve each of the following for x. **Ex 8.3**

(a) $2x^2 - 50 = 0$ (b) $4x^2 + 2x = 0$ (c) $x^2 - 6x + 9 = 0$

8 Sketch the graphs of each of the following, clearly showing the turning point. Ex 8.2

(a) $y = (x + 3)(x - 2)$ (b) $y = x^2 + 2x + 5$ (c) $y = 2x^2 - 4x$

9 Sketch the graph of each of the following quadratics, showing all important features. Ex 8.2

(a) $y = x^2 + 6x - 4$ (b) $y = x^2 - 4x + 4$ (c) $y = -2x^2 + 3x$

10 Consider the equation $y = 2x^2 + x$. Ex 8.2

(a) Draw up and fill in a table of values with x-values from -3 to 3.

(b) Use the table of values to draw the graph of the equation.

(c) Find the turning point of the graph.

(d) Determine the x- and y-intercepts for the graph.

11 Factorise the following quadratic expressions. Ex 2.8

(a) $6a^2 - 24$ (b) $3a^2 + 6a + 3$ (c) $x^2 + 5x - 7x - 35$

12 The 26 letters of the alphabet are written on 26 identical cards and placed in a bag. A card is chosen at random. What is the probability that the card chosen is:

(a) a vowel

(b) one of the letters in the word 'card'

(c) not a letter in the word 'letter'?

13 Jodie is standing on the top of a sand dune and spots a dolphin in the water. If the distance from Jodie to the dolphin along her line of sight is 100 m, find the angle of depression from Jodie to the dolphin if she is 4.3 m above sea level. Answer to one decimal place.

Reasoning and communicating

14 To meet Australian standards, ramps for wheelchair access that are longer than 152 cm must only rise 1 cm vertically for every 14 cm horizontal distance.

(a) What is the largest angle, θ, between the ramp and the horizontal that can be used to build a wheelchair ramp 2 m long? Answer to the nearest degree.

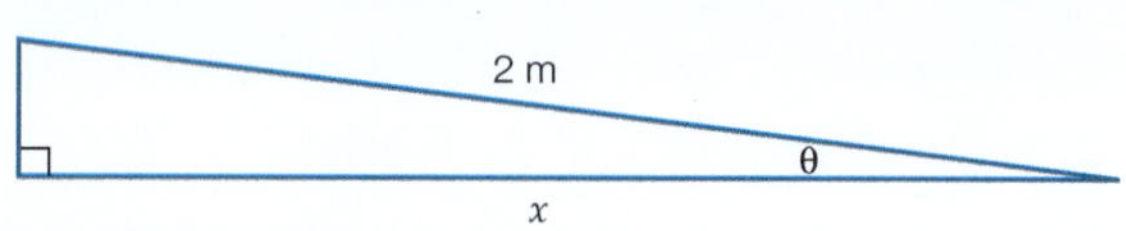

(b) Use your answer to (a) to find:

(i) the horizontal length, x

(ii) the height of the top of the ramp. Answer to the nearest mm.

15 A book cover has an area of 60 cm^2. If the width is 7 cm less than the length, find the dimensions of the book cover.

16 A parabola has x-intercepts at $x = 4$ and $x = 6$.

(a) Determine the x-coordinate of the turning point.

(b) If the coefficient of x^2 is 1, determine the equation of the graph.

SECTION 3

Probability

11 Probability

P Recall 11

Prepare for this chapter by attempting the following questions. If you have difficulty with a question, go to Pearson Places and download the Recall Worksheet.

1 Express in simplest form.

(a) $\frac{3}{15}$ (b) $\frac{16}{52}$

2 Evaluate the following.

(a) $\frac{1}{8} + \frac{3}{4}$ (b) $\frac{2}{5} + \frac{1}{3}$ (c) $1 - \frac{3}{5}$

(d) $\frac{2}{5} \times \frac{3}{4}$ (e) 0.8×0.5 (f) 0.05×0.2

3 (a) Two students are to be chosen from a class of girls and boys. Complete a list of the gender pairings possible, in the order first student, second student.

(b) What are the two possible outcomes when a coin is tossed?

4 (a) A container has 25 numbered tokens: 6 are red, 10 are blue and the rest are yellow. A token is chosen at random from the container.

(i) How many different tokens can be chosen?

(ii) How many possible outcomes are there?

(iii) How many of these outcomes result in a red token being chosen?

(iv) How many of these outcomes result in a yellow token being chosen?

(v) How many of these outcomes result in a token that is not blue being chosen?

(b) A fair six-sided *die* (singular of *dice*) is rolled.

(i) List all the possible outcomes.

(ii) How many of these outcomes are an even number?

Syllabus references

Statistics and Probability: Probability

- ✓ Calculate relative frequencies from given or collected data to estimate probabilities of events involving 'and' or 'or' **(ACMSP226) (Stage 5.1)**
- ✓ List all outcomes for two-step chance experiments, with and without replacement, using tree diagrams or arrays; assign probabilities to outcomes and determine probabilities for events **(ACMSP225) (Stage 5.2)**
- ✓ Describe the results of two- and three-step chance experiments, with and without replacement, assign probabilities to outcomes, and determine probabilities of events; investigate the concept of independence **(ACMSP246) (Stage 5.2)**
- ✓ Use the language of 'if ... then', 'given', 'of', 'knowing that' to investigate conditional statements and to identify common mistakes in interpreting such language **(ACMSP247) (Stage 5.2)**

Probability review

11.1

Stage 5.2

Need to Know

Probability is the study of the chance, or likelihood, of a particular **event** occurring.

The probability of an event lies between 0 and 1 with 0 being impossible and 1 being certain.

$$0 \le P(\text{event}) \le 1$$

Estimating probability

An estimate of the probability of an outcome can sometimes be found by examining past results. The long-term pattern of results can be used to indicate a similar event occurring in the future. This process is known as estimating probability and assumes that conditions will remain the same for future events. The greater the number of trials, the more accurate this estimate will be.

An estimate, based on **long-run proportion**, of $P(\text{success}) = \dfrac{\text{number of past successes}}{\text{total number of past trials}}$.

For example:

Greg is playing tennis and has served 15 aces in the last 80 points. An estimate of the probability of Greg serving an ace is:

$$P(\text{Greg serving an ace}) = \frac{15}{80} = \frac{3}{16}$$

Assuming this will always be the proportion of aces Greg will serve, then $\frac{3}{16}$ can be used as the long-run proportion. However, it is important to understand that this is an estimate.

Theoretical probability

Probabilities can be deduced in situations where all the possible outcomes can be listed, if the outcomes can be assumed to be equally likely.

For example:

If a coin is flipped, it would be reasonable to assume that heads and tails are equally likely, but if two different footballers were lined up to take a shot at goal from the same position it would not be reasonable to assume that they had the same probability of scoring a goal.

The full list of possible outcomes in a probability experiment is called the **sample space**.

$$P(\text{success}) = \frac{\text{number of equally likely successful outcomes}}{\text{total number of equally likely possible outcomes}}$$

Complementary events

A **complementary event** is the opposite of the event under consideration. The sum of the probability of an event and its complementary event is 1.

$$P(A) + P(\text{not } A) = 1$$

Multiple events

Consider flipping a coin once, twice or three times.

Flipping a coin once has two possible outcomes, a head H or a tail T.

Flipping a coin twice has four possible outcomes, HH, HT, TH, TT.

Flipping a coin three times has eight possible outcomes, HHH, HHT, HTH, HTT, TTH, THT, THH, TTT.

Multiple events can comprise two events with different numbers of equally likely outcomes.

For example:

Flipping a coin (2 outcomes: H or T) and rolling a die (6 outcomes 1, 2, 3, 4, 5, 6). There are 12 possible equally likely outcomes, H1, H2, H3, H4, H5, H6, T1, T2, T3, T4, T5, T6.

11.1

Two-way tables and tree diagrams

Two-way tables and tree diagrams are used to list the sample space.

Consider spinning a spinner with three possible outcomes and rolling a die.

All possible outcomes can be listed using a two-way table or using a tree diagram. From this, we can find the probability of obtaining a 2 with the spinner and rolling a 5 on the die.

Possible outcomes		**Die**					
		1	**2**	**3**	**4**	**5**	**6**
Spinner	**1**	(1, 1)	(1, 2)	(1, 3)	(1, 4)	(1, 5)	(1, 6)
	2	(2, 1)	(2, 2)	(2, 3)	(2, 4)	(2, 5)	(2, 6)
	3	(3, 1)	(3, 2)	(3, 3)	(3, 4)	(3, 5)	(3, 6)

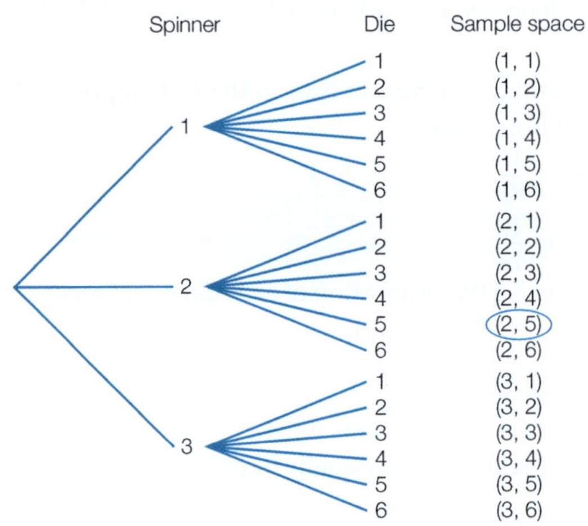

The probability of obtaining a 2 on the spinner and a 5 on the die is $\frac{1}{18}$.

Worked Example 1

WE1

Over the past 6 months Jake has been late to school on 12 occasions in 100 school days.

(a) What is the estimated probability he will be late next Wednesday?

(b) What is the estimated *P*(Jake being on time next Friday)?

Working

(a) Jake has been late 12 out of 100 days.

$$P(\text{late}) = \frac{12}{100}$$

$$= \frac{3}{25}$$

(b) $P(\text{late}) + P(\text{on time}) = 1$

$$P(\text{on time}) = 1 - P(\text{late})$$

$$= 1 - \frac{3}{25}$$

$$= \frac{22}{25}$$

11.1

Worked Example 2

WE2

Robin loves to play football with his mates after school but he also loves to play his guitar. He can't decide what to do, so he flips a coin. A head means he will play football and a tail means he will play his guitar. Find the probability that he:

(a) plays football 3 afternoons in a row

(b) plays football at least twice in 3 afternoons

(c) plays his guitar 2 afternoons in a row

(d) doesn't play football 3 afternoons in a row.

Working

HHH, HHT, HTH, HTT, THH, TTH, THT, TTT are all equally likely as long as the coin is fair.

(a) P(football 3 afternoons in a row)

$= P(\text{HHH})$

$= \dfrac{\text{number of successful outcomes}}{\text{number of equally likely outcomes}}$

$= \dfrac{1}{8}$

(b) P(football at least twice in 3 afternoons)

$= P(\text{HHT or HTH or THH or HHH})$

$= \dfrac{\text{number of successful outcomes}}{\text{number of equally likely outcomes}}$

$= \dfrac{4}{8}$

$= \dfrac{1}{2}$

(c) P(plays guitar 2 afternoons in a row)

$= P(\text{TTH or HTT or TTT})$

$= \dfrac{\text{number of successful outcomes}}{\text{number of equally likely outcomes}}$

$= \dfrac{3}{8}$

(d) P(doesn't play football 3 afternoons in a row)

$= 1 - P(\text{plays football 3 afternoons in a row})$

$= 1 - \dfrac{1}{8}$

$= \dfrac{7}{8}$

11.1

Worked Example 3

WE3

Jemma has a dilemma. She has six favourite outfits: a black, a red, a blue, a green, a purple and a multi-coloured outfit. She has two favourite pairs of shoes; one has a pattern in the leather and one is plain. She can never decide what to wear so she rolls a die to choose her outfit and flips a coin to choose her shoes. On the die: 1 represents the black outfit, 2 the red, 3 the blue, 4 the green, 5 the purple and 6 the multi-coloured outfit. On the coin: heads represents the shoes with pattern, tails represents the plain pair. Find the probability that she will wear:

(a) the black outfit with the plain shoes

(b) the green, purple or red outfit

(c) the multi-coloured outfit and the patterned shoes

(d) the red or green outfit with plain shoes

(e) the red outfit

(f) the plain pair of shoes.

Working

A two-way table can be used to find the sample space.

		Die					
		1	2	3	4	5	6
Coin	H	H1	H2	H3	H4	H5	H6
	T	T1	T2	T3	T4	T5	T6

H1, H2, H3, H4, H5, H6, T1, T2, T3, T4, T5, T6

There are 12 equally likely outcomes.

(a) P(black outfit, plain shoes)
$= P(T1)$
$= \frac{1}{12}$

(b) P(green or purple or red outfit)
$= P(H2, H4, H5, T2, T4, T5)$
$= \frac{6}{12}$
$= \frac{1}{2}$

(c) P(multi-coloured outfit and shoes with pattern)
$= P(H6)$
$= \frac{1}{12}$

(d) P(red or green outfit with plain shoes)
$= P(T2, T4)$
$= \frac{2}{12}$
$= \frac{1}{6}$

(e) P(red outfit)
$= P(T2, H2)$
$= \frac{2}{12}$
$= \frac{1}{6}$

(f) P(plain shoes)
$= P(T1, T2, T3, T4, T5, T6)$
$= \frac{6}{12}$
$= \frac{1}{2}$

11.1

Worked Example 4

WE4

Libby and Chris wish to start a family. They plan to have three children. The chance of any particular child being a boy or a girl is considered to be equally likely.

(a) Represent all the possible outcomes using a tree diagram.

(b) List the sample space.

(c) Find the probability that they have:

- **(i)** a girl, a boy, then another girl
- **(ii)** 2 girls and a boy (in any order)
- **(iii)** 3 boys
- **(iv)** 3 children of the same gender
- **(v)** at least 2 girls

Working

(a)

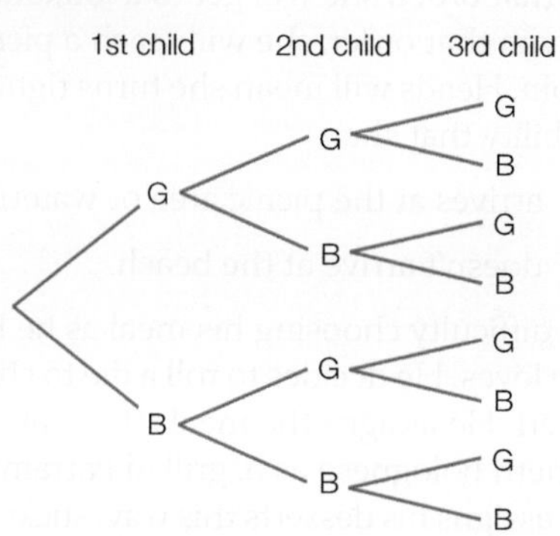

(b) (G, G, G), (G, G, B), (G, B, G), (G, B, B), (B, G, G), (B, G, B), (B, B, G), (B, B, B)

There are 8 equally likely outcomes. These outcomes are equally likely because each child is considered equally likely to be a boy or a girl.

(c) **(i)** P(G, B, G)

$= \frac{1}{8}$

(ii) P(2 girls and 1 boy)

$= \frac{3}{8}$

(iii) P(3 boys)

$= \frac{1}{8}$

(iv) P(3 children of the same gender)

$= \frac{2}{8}$

$= \frac{1}{4}$

(v) P(at least 2 girls)

$= \frac{4}{8}$

$= \frac{1}{2}$

11.1

Answers page 709

11.1 Probability review

Stage 5.2

Navigator

Q1, Q2, Q3, Q4, Q5, Q6, Q7, Q8, Q9, Q10, Q11, Q13, Q14, Q16, Q18, Q20, Q22	Q1, Q2, Q3, Q4, Q5, Q6, Q7, Q8, Q9, Q12, Q13, Q15, Q17, Q18, Q21, Q23	Q1, Q2, Q3, Q4, Q5, Q6, Q7, Q8, Q9, Q12, Q13, Q15, Q17, Q18, Q19, Q21, Q22, Q23

Fluency and understanding

WE1

1 Ben's car has been hard to start on 8 days over the last 4 weeks.

(a) Estimate *P*(Ben's car will be hard to start tomorrow).

(b) Estimate *P*(Ben's car will be easy to start in 3 days' time).

(c) What assumptions have you made to calculate these answers?

WE2

2 Fiona is bushwalking through a national park. She finds that the path keeps dividing and she has to decide whether to take the left path or the path to the right. If she takes 3 left turns, she will arrive at a beach. If she takes 3 right turns, she will reach a waterfall. If she turns left twice and right once, not necessarily in that order, she will get to a lookout. If she turns left once and right twice, not necessarily in that order, she will reach a picnic area. She can't decide where to go so she flips a coin. Heads will mean she turns right and tails will mean that she turns left. Find the probability that she:

(a) gets to the beach

(b) arrives at the picnic area or waterfall

(c) reaches the lookout

(d) doesn't arrive at the beach.

WE3

3 Mario likes to eat at a local restaurant but he has difficulty choosing his meal as he likes 6 main courses equally and there are 2 desserts he loves. He decides to roll a die to choose his main course and flip a coin to decide his dessert. He assigns the meals this way: chicken parmigiana as 1, chilli prawns as 2, spaghetti bolognese as 3, grilled barramundi as 4, pepper steak as 5 and spicy calamari as 6. He assigns his desserts this way: sticky date pudding as H and tiramisu as T. Find the probability that he chooses:

(a) chicken parmigiana and tiramisu

(b) a seafood dish—chilli prawns, grilled barramundi or spicy calamari

(c) pepper steak and a sticky date pudding

(d) grilled barramundi or spaghetti bolognese as a main course with a tiramisu dessert

(e) chilli prawns

(f) sticky date pudding.

WE4

4 A coin is tossed three times.

(a) Represent all the possible outcomes using a tree diagram.

(b) List the sample space.

(c) Find the probability of tossing:

(i) a head, a tail then a head

(ii) a head, a head then a tail

(iii) 3 tails

(iv) all heads or all tails

(v) at least two heads.

5 Nikola and Brad are in a class containing 14 girls and 10 boys. To select a class representative, a person's name is chosen at random from the class roll. Find the probability that the class representative is:

(a) a girl

(b) not a girl

(c) Nikola

(d) Brad

(e) Nikola or Brad

(f) neither Nikola nor Brad.

6 Pedro has two drawers containing loose socks. There are a black, a white and a green sock in each. He is half asleep and grabs one sock, at random, from each drawer.

(a) Represent all the possible outcomes using a tree diagram.

(b) Why is it reasonable to regard these as equally likely outcomes?

(c) Find the probability of Pedro obtaining:

(i) a pair of black socks **(ii)** one green and one white sock

(iii) a matching pair of socks **(iv)** two socks that do not match.

7 Jacob and Ted are 2 members of a 12-member cricket team. The captain is chosen randomly each week. Find the probability that the captain is:

(a) Ted **(b)** not Jacob **(c)** Jacob or Ted **(d)** neither Jacob nor Ted.

8 The local supermarket is flooded and the labels are washed off 12 cans of peaches, 5 cans of apples and 3 cans of beans. The supermarket now has them all for sale for 50 cents a can, but you will not know what is inside each can until it is opened.

(a) If you buy only 1 can the probability of it being apples is:

A $\frac{1}{5}$ **B** $\frac{1}{4}$ **C** $\frac{1}{3}$ **D** $\frac{5}{8}$

(b) If you buy 1 can the probability of it containing fruit (not beans) is:

A $\frac{3}{20}$ **B** $\frac{5}{20}$ **C** $\frac{12}{20}$ **D** $\frac{17}{20}$

(c) If you want to be sure of purchasing at least 1 can of peaches, the minimum number of cans you must purchase is:

A 1 **B** 3 **C** 9 **D** 12

9 A spinner evenly divided and showing the numbers 1, 2, 3, 4 is spun once, and an eight-sided die showing 1 to 8 is rolled once.

(a) The probability that the total of the two numbers is 7 is:

A $\frac{3}{11}$ **B** $\frac{3}{32}$ **C** $\frac{1}{8}$ **D** $\frac{5}{32}$

(b) The probability that the total of the two numbers is 11 is:

A $\frac{1}{32}$ **B** $\frac{1}{16}$ **C** $\frac{1}{8}$ **D** $\frac{3}{32}$

(c) The probability that the total of the two numbers is a multiple of 3 is:

A $\frac{5}{16}$ **B** $\frac{11}{32}$ **C** $\frac{3}{8}$ **D** $\frac{15}{32}$

10 Suresh and Julian regularly play tennis against one another. Over the past 2 years, Suresh has won 16 times and Julian has won 20 times. What is the estimated probability that Suresh will win their next match?

11 Of the 125 students in Year 10, 15 ride to school every day and 30 ride sometimes. A student is chosen at random from the Year 10 students.

(a) What is the probability that the student always rides to school?

A $\frac{3}{25}$ **B** $\frac{9}{25}$ **C** $\frac{17}{25}$ **D** $\frac{22}{25}$

(b) What is the probability that the student never rides to school?

A $\frac{3}{25}$ **B** $\frac{9}{25}$ **C** $\frac{16}{25}$ **D** $\frac{22}{25}$

11.1

12 If the probability of it raining tomorrow is $\frac{2}{9}$, what is the probability of it not raining?

13 In a game of chance, 45 balls numbered from 1 to 45 are placed in a barrel. If one ball is drawn at random, find the probability that its number is:

(a) 2 **(b)** less than 10 **(c)** even

(d) odd **(e)** greater than 50 **(f)** between 10 and 20 inclusive.

14 In the game of Monopoly, players advance clockwise around the 40 squares of the board according to the total of the numbers shown on the roll of two dice.

(a) Use a two-way table to represent all the possible totals obtained from the roll of two dice.

(b) How many possible outcomes are there?

(c) Why is it reasonable to regard these as equally likely outcomes?

(d) If your token is on Whitechapel Road, what is the probability that your next throw will result in your landing on Euston Road?

(e) Suppose you do land on Euston Road. Find the probability of advancing to Bow Street or further on your next throw.

(f) One way of getting out of jail in the game of Monopoly is to roll a 'double' with the two dice (i.e. to roll the dice and obtain the same result on both). Supposing you are in jail, find the probability of getting out of jail with your next roll of the dice.

15 Skye has the phone numbers of 40 friends saved on her phone. If we were trying to estimate the probability of Skye messaging one of these friends, would it be reasonable to assume they were equally likely outcomes? Explain your thoughts in a sentence.

16 Val and Peter have a bag of sweets each. Val's bag of sweets contains 3 licorice blocks, 2 jelly babies and 1 chocolate caramel. Peter has 2 licorice blocks and 2 jelly babies in his bag. Both Val and Peter offer a sweet to Narelle. If Narelle randomly selects a sweet from each bag, find the probability that she obtains:

(a) 2 licorice blocks **(b)** 2 jelly babies **(c)** no licorice blocks

(d) no jelly babies **(e)** 1 licorice block **(f)** at least 1 jelly baby.

17 In a round-robin competition in tenpin bowling, Shelley plays 4 rounds. She estimates that she has an equal chance of winning or losing each round.

(a) Draw a tree diagram showing all the possible outcomes. (W = a win; L = a loss)

(b) List the sample space. How many possible outcomes are there?

(c) Find the probability that Shelley:

(i) wins all 4 rounds **(ii)** wins only 2 of the 4 rounds

(iii) wins at least 3 rounds **(iv)** wins the second round

(v) doesn't win the third round

(vi) doesn't win the first or the fourth round.

18 There are 12 houses in the court where Zoran lives. Of the 20 cars owned between the people who live in the court, 7 are Holdens, 5 are Fords and the rest are other makes. The owner of one of the Holdens moves out and a new person, who owns a car, moves into the court. Is it reasonable to say that the probability this new person owns a Holden is $\frac{7}{20}$? Explain your thoughts in a sentence.

19 Two fair six-sided dice are rolled. One is numbered 1, 1, 2, 2, 3, 3 and the other is numbered 2, 4, 6, 8, 10, 12.

(a) Draw up a two-way table to show the sample space.

(b) Draw up a two-way table to show the sum of the two numbers shown.

(c) Find the following probabilities.

(i) P(sum is even) **(ii)** P(sum is multiple of 5)

(iii) P(sum is greater than 10) **(iv)** P(sum is not less than 11)

(d) Why is it harder to answer the questions in part **(c)** for this two-way table than for a diagram with consecutive numbers on each side?

(e) Draw up another two-way table that shows the product of the two numbers.

(f) Draw up a table that shows the probability for each possible product.

Reasoning, communicating and open-ended

20 Explain the following observations.

(a) In the long run, the observed probability of Aiden wearing a tie to work is $\frac{19}{20}$, but on three consecutive days he is seen not wearing a tie.

(b) Lesley has kept track of the lottery numbers and has noticed that 16 has not occurred for 12 months. Thinking the probability of it being drawn soon must be high, she includes 16 in her numbers for the next 2 months, but it still isn't drawn.

21 You are given a 12-sided die, numbered 1 to 12. Write an event that would have the following probabilities.

(a) $P(A) = \frac{1}{6}$ **(b)** $P(B) = \frac{1}{3}$ **(c)** $P(C) = \frac{5}{12}$

22 The long-term probability of Maria's team beating Abbey's team at netball is $\frac{4}{5}$. List three possible sets of long-term results that would produce this figure.

23 Sharanpreet calculates the probability of a family having at least one boy in a family of three children is $\frac{3}{8}$. What is the correct answer and what mistake has Sharanpreet made?

Hint

Assume that there is an equal probability of having a boy as having a girl.

Problem solving

A chance of freedom

A prisoner is brought before a judge who says: "I have decided to give you a chance of freedom. Here are four red balls and four black balls. Divide the eight balls into two groups, any way you wish, and then place each group in a bag. I will hide the two bags behind my back and you will choose a bag from either my right or my left hand. You can then choose a ball at random from the bag. If it is red you can go free. If it is black your sentence will double."

(a) How can the prisoner divide the balls to give him the best chance of freedom?

(b) What division would give him the worst chance?

Strategy options

- Act it out.
- Test all possible combinations.
- Solve a simpler problem.

11.2 Venn diagrams and two-way tables

Stage 5.2

Need to Know

Venn diagrams

A **Venn diagram** consists of a rectangle with one or more circles inside it. The rectangle represents the **universal set**, which is every element under consideration for the particular situation being examined. On the Venn diagram it is represented by ξ. The circles represent particular events and are represented by capital letters.

There are three situations for which Venn diagrams are really useful. These are when dealing with the words 'and', 'or' and 'not'.

These can be shown on Venn diagrams as follows.

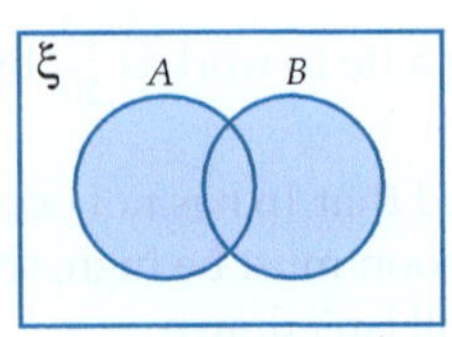

A or *B*

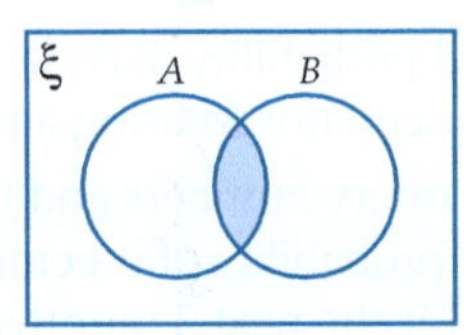

A and *B*

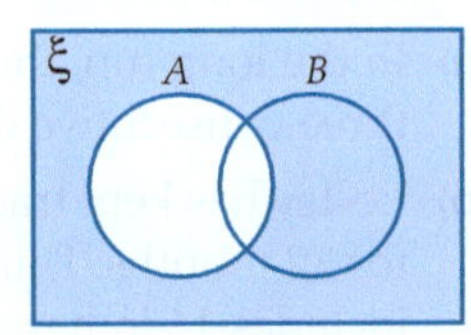

not *A*

Two-way tables

We can use a two-way table to show similar information.

	B	**not B**
A	*A* and *B*	*A* and not *B*
not A	*B* and not *A*	not *A* and not *B*

Together, the sections of the table make up the universal set. In addition to this, no section has any overlap with any other section.

11.2

Worked Example 5

WE 5

A single card is drawn from a set of 25 cards numbered from 1 to 25.

Let A = 'odd numbers' and B = 'multiples of 3'.

(a) Represent this on a Venn diagram.

(b) Use the Venn diagram to find:

- **(i)** $P(A)$
- **(ii)** $P(A \text{ and } B)$
- **(iii)** $P(\text{not } B)$

(c) Draw a two-way table to display the probability of possible events.

Working

(a) Step 1: Draw a Venn diagram with sets A and B.

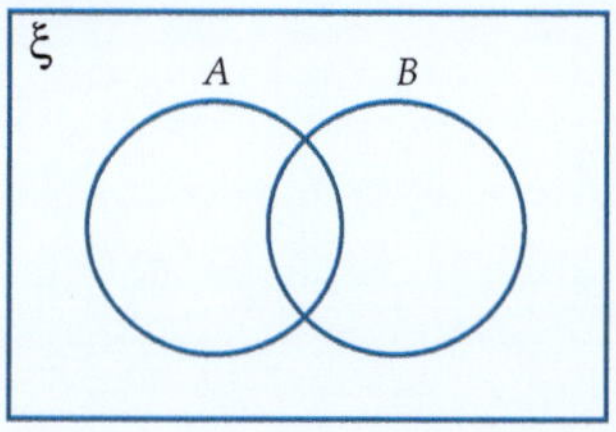

Step 2: Write in the elements common to both sets.

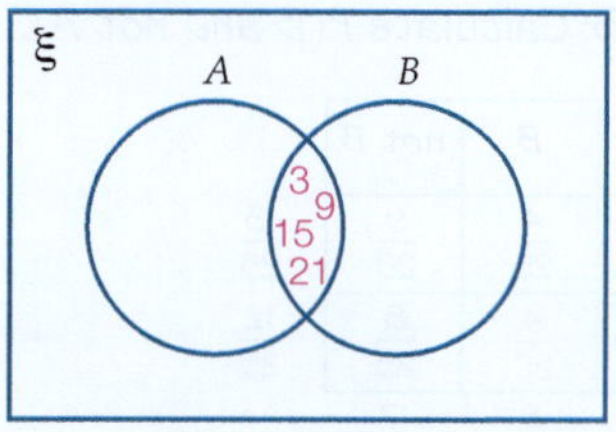

Step 3: Write the remaining elements in either set.

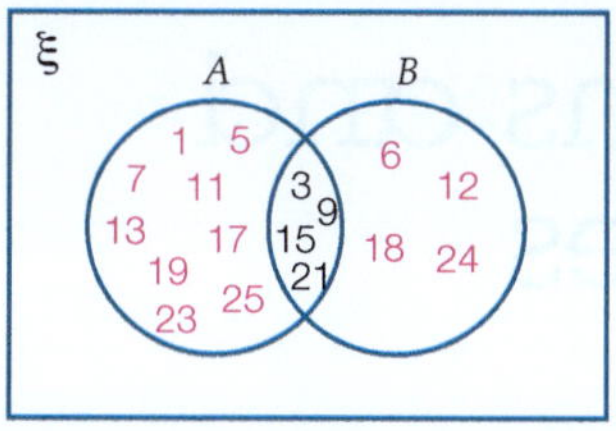

Step 4: Write the remaining elements in the universal set.

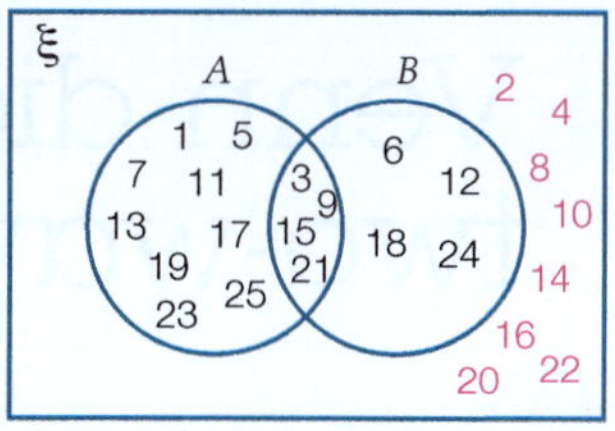

(b) **(i)** $P(A) = \dfrac{13}{25}$

(ii) $P(A \text{ and } B) = \dfrac{4}{25}$

(iii) $P(\text{not } B) = 1 - P(B)$

$$= 1 - \frac{8}{25}$$

$$= \frac{17}{25}$$

11.2

(c) Step 1: Draw a two-way table with the known information.

	B	not B	
A	$\frac{4}{25}$		$\frac{13}{25}$
not A			
		$\frac{17}{25}$	

Step 2: Calculate P(A and not B).

	B	not B	
A	$\frac{4}{25}$	$\frac{9}{25}$	$\frac{13}{25}$
not A			
		$\frac{17}{25}$	

Step 3: Calculate P(not A).

	B	not B	
A	$\frac{4}{25}$	$\frac{9}{25}$	$\frac{13}{25}$
not A			$\frac{12}{25}$
		$\frac{17}{25}$	

Step 4: Calculate P(B) and P(not A and not B).

	B	not B	
A	$\frac{4}{25}$	$\frac{9}{25}$	$\frac{13}{25}$
not A		$\frac{8}{25}$	$\frac{12}{25}$
	$\frac{8}{25}$	$\frac{17}{25}$	

Step 5: Calculate P(B and not A).

	B	not B	
A	$\frac{4}{25}$	$\frac{9}{25}$	$\frac{13}{25}$
not A	$\frac{4}{25}$	$\frac{8}{25}$	$\frac{12}{25}$
	$\frac{8}{25}$	$\frac{17}{25}$	

Answers page 710

11.2 Venn diagrams and two-way tables

Stage 5.2

Navigator

Q1, Q2, Q3, Q5, Q8, Q9, Q11, Q14	Q1, Q2, Q3, Q4, Q5, Q6, Q7, Q9, Q10, Q12, Q13	Q1, Q2, Q3, Q5, Q6, Q7, Q9, Q10, Q12, Q13

Fluency and understanding

WE5

1 In a non-leap year, there are 28 days in February. A date in February is to be chosen to celebrate the 100th birthday of a city.

Let A = 'even numbered dates' and B = 'dates that are multiples of 5'.

(a) Represent this information on a Venn diagram.

(b) Use the Venn diagram to find:

(i) $P(A)$ (ii) $P(A \text{ and } B)$ (iii) $P(\text{not } B)$

(c) Draw a two-way table to display the probability of possible events.

2 Use the two-way table shown at right to find:

(a) $P(A \text{ and } B)$ (b) $P(\text{not } A \text{ and not } B)$

(c) $P(B \text{ and not } A)$ (d) $P(A \text{ and not } B)$

(e) $P(A)$ (f) $P(B)$

	B	not B
A	$\frac{3}{16}$	$\frac{5}{16}$
not A	$\frac{1}{16}$	$\frac{7}{16}$

3 In a class of 20, 8 students learn the piano, 6 take singing lessons and 10 students neither learn the piano nor have singing lessons.

(a) Represent this information in a two-way table.

(b) Use the two-way table to find the number of people who learn piano and take singing lessons.

(c) Use the two-way table to find the probability of:

(i) taking singing lessons but not piano **(ii)** taking piano lessons but not singing.

4 In a group of 30 people, there are 15 who like cats, 20 who like dogs and 6 who dislike both cats and dogs. One person is chosen from the group at random.

(a) Represent this information on a Venn diagram.

(b) Use the Venn diagram to find the probability that the person chosen at random is:

(i) a cat lover only

(ii) a dog lover only

(iii) a person who likes both.

5 Complete these two-way tables.

(a) $P(A) = \frac{7}{16}$

	B	**not B**
A		$\frac{5}{16}$
not A	$\frac{1}{16}$	

(b) $P(D) = 0.7$

	C	**not C**
D	0.2	
not D		0.1

6 Given the following Venn diagram, identify the numbers represented by each of the following.

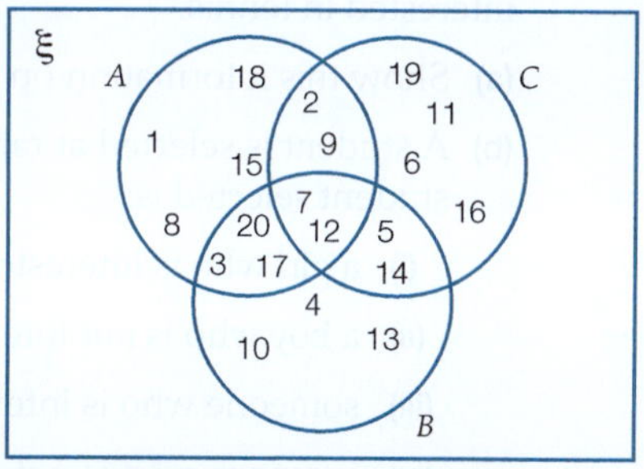

(a) *A* or *B* **(b)** *A* or *C*

(c) *B* or *C* **(d)** *A* and *B*

(e) *A* and *C* **(f)** *B* and *C*

(g) *A* and *B* and *C* **(h)** not *A* and not *B*

7 A coin is tossed, a six-sided die is rolled and a ball is selected from a bag containing 4 red and 3 green balls. *M* is the set of outcomes that contains 'coin shows heads', *N* is the set of outcomes that contains 'a 2 is not rolled' and *P* is the set of outcomes that contains 'a red ball is selected'.

(a) Draw a Venn diagram to show the sample space.

(b) Write down the elements of not *M*, not *N* and not *P*.

(c) Is the probability of H1R the same as H1G? Explain your answer.

8 A poll of 2000 students revealed that 554 respondents only watch TV and do not browse the internet while they watch. 1200 respondents only use the internet and the TV is not on while they're browsing. 10 students claim to neither watch TV nor use the internet.

(a) Draw a two-way table to represent this situation.

(b) What is the probability of a student in this survey watching TV and browsing the internet at the same time?

11.2

Reasoning, communicating and open-ended

9 A survey of 1200 university students revealed that $\frac{3}{8}$ use Twitter, $\frac{3}{4}$ use Facebook, and $\frac{1}{4}$ use both.

(a) Represent the information in a two-way table.

(b) If a student is chosen at random, find the probability that the student:

(i) uses either **(ii)** uses either but not on both.

10 Ms Lauder is creating the timetable for Year 12 students. The request forms have been entered online:

- 163 want to study mathematics, 95 physics, 124 chemistry
- 61 requested maths and physics
- 67 requested maths and chemistry
- 49 requested physics and chemistry
- 27 want to study all three subjects
- 17 want to study none of these subjects.

(a) Draw a Venn diagram to illustrate the situation.

(b) How many students requested only mathematics?

(c) How many students requested only physics?

(d) How many students requested only chemistry?

(e) What was the total number of requests?

(f) What is the probability of a Year 12 student at Ms Lauder's school selecting mathematics as one of their subjects? Will the answer be the same at every school?

11 In a scout troop there are 15 boys and 12 girls. 10 of the boys and 8 of the girls say they are interested in tennis.

(a) Show this information on a Venn diagram.

(b) A student is selected at random. Use the diagram to find the probability that the student selected is:

(i) a girl who is interested in tennis

(ii) a boy who is not interested in tennis

(iii) someone who is interested in tennis

(iv) someone who is not interested in tennis.

12 Human blood can contain up to three different antigens: A, B and Rh. The following table outlines the antigens present in each blood type. (The actual situation is slightly more complicated for the O group.) The values given in the table refer to Australia; countries with different ethnic make-up would have different values.

Blood group	A	B	Rh	% in Australia
A+	Yes	No	Yes	31
A–	Yes	No	No	7
B+	No	Yes	Yes	8
B–	No	Yes	No	2
AB+	Yes	Yes	Yes	2
AB–	Yes	Yes	No	1
O+	No	No	Yes	40
O–	No	No	No	9

The antigens present in the blood of patients at a clinic were recorded for 1 week:

- 39 patients had the A antigen
- 2 had the A and B antigens
- 12 had the B antigen
- 9 had the B and Rh antigens
- 67 had the Rh antigen
- 30 had the A and Rh antigens
- 2 had all three antigens
- 5 had no antigens.

(a) Draw a Venn diagram to show this situation.

(b) How many patients were there?

(c) What was the *P*(exactly one antigen)?

(d) What was the *P*(exactly two antigens)?

(e) Use the data to calculate the probability of each of the eight blood groups.

(f) Compare the answers from part **(e)** with those expected from long-term data as shown in the table.

13 From a class of 25 students, 14 like watching the news on TV and 16 said they use the internet to find out the news. Mert says there must be something wrong with the statistics because there are 30 different opinions from only 25 students. Sasha says Mert hasn't taken into account students who do both. Neither of them is correct. What is the correct interpretation of the statistics and what mistakes have Mert and Sasha made?

14 What is wrong with the following question?

In a survey of 200 people with a driving licence, 18% have never received a speeding fine, 44 people have received a parking ticket but have never received a speeding fine, and 78% have received both a speeding fine and a parking ticket. How many people have received a speeding fine but have never received a parking ticket?

11.3 Mutually exclusive events

Stage 5.2

Need to Know

Mutually exclusive events

If two events cannot occur at the same time they are called **mutually exclusive** events. In a Venn diagram, mutually exclusive sets do not intersect. They are called **disjoint sets**.

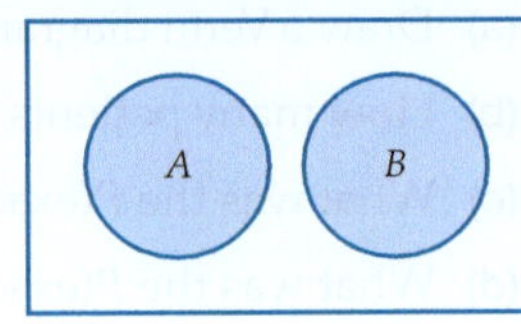

For example:

The spinner at right is spun. Since the spinner is divided into four equal pieces, each number has an equal chance of occurring.

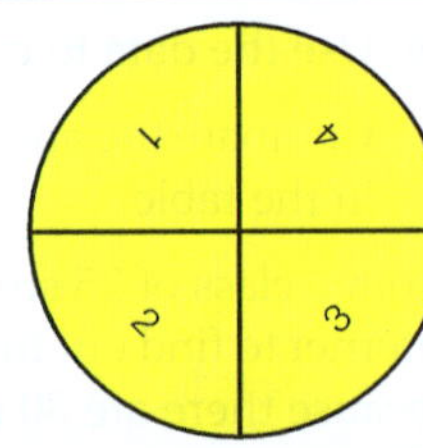

Consider two events.

A = obtaining an even number

= obtaining a 2 or a 4

B = obtaining a 3

These events are said to be mutually exclusive as it is impossible for both of them to occur at the same time.

$P(A \text{ and } B) = 0$

Addition law for mutually exclusive events

If A and B are mutually exclusive events (i.e. they cannot occur at the same time), then:

$P(A \text{ or } B) = P(A) + P(B)$

Worked Example 6

WE6

A six-sided die is rolled. What is the probability that the number obtained is:

(a) a 2 or a 5?

(b) neither odd nor a 4?

Working

(a)
$$\begin{aligned} &P(2 \text{ or } 5) \\ &= P(2) + P(5) \\ &= \frac{1}{6} + \frac{1}{6} \\ &= \frac{2}{6} \\ &= \frac{1}{3} \end{aligned}$$

(b)
$$\begin{aligned} &P(\text{neither odd nor a } 4) \\ &= P[\text{not}(\text{odd or } 4)] \\ &= 1 - P(\text{odd or } 4) \\ &= 1 - [P(\text{odd}) + P(4)] \\ &= 1 - \left(\frac{1}{2} + \frac{1}{6}\right) \\ &= 1 - \frac{2}{3} \\ &= \frac{1}{3} \end{aligned}$$

11.3 Mutually exclusive events

Answers page 711

Stage 5.2

Navigator

Q1, Q2, Q3, Q4, Q5, Q6 Column 1, Q7, Q8, Q10	Q1, Q2, Q3, Q4, Q5, Q6 Column 2, Q7, Q10	Q1, Q2, Q3, Q4, Q5, Q6 Column 2, Q7, Q8, Q9, Q10

Fluency and understanding

WE6

1 A fair eight-sided die numbered 1 to 8 is rolled. What is the probability that the number obtained is:

(a) a 3 or a 6 (b) neither 5 nor even?

2 A spinner is coloured as shown.

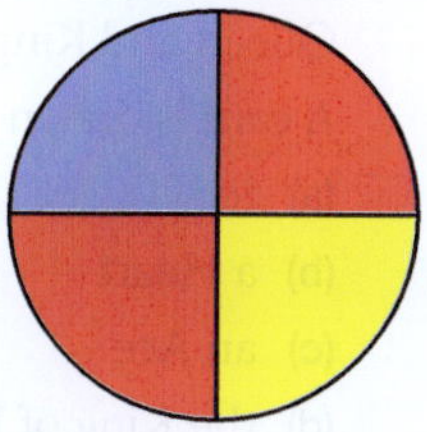

(a) What is the probability of spinning blue?

(b) What is the probability of spinning red?

(c) What is the probability of spinning blue or red?

(d) What is the probability of spinning blue or yellow?

3 Are any of the following pairs of events mutually exclusive?

(a) Obtaining a head on the toss of a coin and a tail on the toss of a coin.

(b) Choosing an odd-numbered horse in a race and horse number 7 in another race.

(c) Winning first prize in a raffle with a blue ticket and first prize in a raffle with a yellow ticket.

4 (a) A basket contains 3 red, 2 white and 5 black scarves. The probability that a scarf drawn at random will be red or black is:

A $\frac{1}{5}$ B $\frac{2}{3}$ C $\frac{7}{10}$ D $\frac{4}{5}$

(b) A jar contains 6 red lollies, 3 green lollies and 2 yellow lollies. The probability that a lolly drawn at random will be green or yellow is:

A $\frac{5}{11}$ B $\frac{6}{11}$ C $\frac{3}{11}$ D $\frac{5}{6}$

(c) A drawer contains 6 blue socks, 4 black socks and 2 brown socks. The probability that a sock drawn at random will *not* be black is:

A $\frac{1}{4}$ B $\frac{1}{3}$ C $\frac{1}{2}$ D $\frac{2}{3}$

5 These probabilities apply to different modes of transport students use to travel to and from a particular school. The modes are all mutually exclusive.

$P(\text{car}) = \frac{7}{20}$ $P(\text{train}) = \frac{9}{20}$

$P(\text{bus}) = \frac{1}{20}$ $P(\text{walking}) = \frac{3}{20}$

A student is asked how he travels to school. Find the probability that:

(a) he travels by car or train

(b) he walks or catches a bus

(c) he does not walk nor catches a bus

(d) he does not travel by car nor by bus

(e) he travels by car, train or bus.

11.3

6 Twenty tokens numbered in order from 1 to 20 are placed in a bag and one is randomly selected. Find the probability that the number on the token is:

(a) 19 or 20
(b) a multiple of 3
(c) a multiple of 3 or a multiple of 7
(d) even or the number 15
(e) less than 10 or greater than 18
(f) neither 1 nor 2
(g) neither odd nor a multiple of 4
(h) 5, 6 or 7.

7 A standard pack of playing cards contains 52 cards, half red and half black. The red cards are divided into two suits: Hearts and Diamonds. The black cards are divided into two suits: Clubs and Spades. Each suit is made up of the cards Ace, 2, 3, 4, 5, 6, 7, 8, 9, 10, Jack, Queen and King.

A card is drawn at random from the pack. Find the probability that the card is:

(a) red
(b) a Heart
(c) an Ace
(d) the King of Hearts
(e) red or the Ace of Spades
(f) a Diamond or the 4 of Hearts
(g) the 5 of Clubs or the Queen of Spades
(h) a King or an Ace
(i) any King or any Ace or the 10 of Hearts
(j) the Jack of Diamonds, any 2, or the 7 of Clubs
(k) neither a 6 nor a 9 of any suit
(l) neither a Spade nor red.

Reasoning, communicating and open-ended

8 A normal six-sided die is rolled. The following events have been defined:

A = rolling an even number | B = rolling a 2 or a 6
C = rolling a 1 or a 3 | D = rolling a multiple of 3

(a) List all possible pairs of events (order does not matter).
(b) Are any of these pairs mutually exclusive?
(c) Find the probability of each event.
(d) Find the probability that either event occurs for each of the pairs listed in part **(a)**.
(e) Find the probability that both of the events occur for each of the pairs listed in part **(a)**.
(f) Find the sum of the probabilities of each event in all the pairs listed in part **(a)**.
(g) Compare your answers to parts **(d)**, **(e)** and **(f)**. What do you find? Write an equation to summarise your findings.

9 A class conducted a traffic survey. The types of vehicles they recorded were:

sedan	station wagon	4WD	hatchback
truck	van	bus	2 door

(a) What difficulties can you see arising from the data collection?
(b) How could these difficulties have been avoided?

10 Consider a fair 20-sided die numbered 1 to 20. Write down at least eight pairs of mutually exclusive events related to this die.

Half-time 11

Stage 5.2 Q1–5

1 **(a)** Complete the following two-way table which relates to the number of surveyed students who own either, or both, an iPhone or an iPad. Ex. 11.2

	iPhone	**not iPhone**	
iPad	60		75
not iPad			
	210		400

(b) Now use the completed two-way table to state the probability of a randomly chosen student owning:

(i) an iPhone but not an iPad **(ii)** an iPad but not an iPhone.

2 Ten balls numbered from 1 to 10 are placed in a bag and one is drawn at random. Find the probability that the number on the ball is: Ex. 11.1

(a) 1 or 2

(b) an even number

(c) a 6 or an odd number

(d) neither an odd number nor a multiple of 4.

3 The probability of it raining on a given day of the weekend is 0.3. What is the probability that it rains on both Saturday and Sunday? Ex. 11.1

4 In a chess tournament, each contestant must play three others. Suzanne estimates that she has an equal chance of winning or losing each game. Ex. 11.1

(a) Draw a tree diagram and list the sample space for the possible outcomes of Suzanne's three games.

(b) How many possible outcomes are there?

(c) Find the probability of Suzanne:

(i) winning all three games **(ii)** winning only one game

(iii) winning the first game **(iv)** not winning the third game

(v) winning at least two games **(vi)** not winning either of the first two games.

5 A fair eight-sided die is numbered 1 to 8. The die is rolled once. The following events are defined: Ex. 11.3

A = 'rolling an even number'

B = 'rolling a prime number'

C = 'rolling a multiple of 5'

(a) Given that order does not matter, list the pairs of events that are possible.

(b) Find the probability of each individual event.

(c) State which, if any, of the pairs of events are mutually exclusive.

11.4 Probability tree diagrams

Stage 5.2

Need to Know

Tree diagrams

Consider the case of flipping a fair coin three times. There are eight equally likely outcomes: HHH, HHT, HTH, HTT, THH, THT, TTH, TTT.

So $P(\text{HHH}) = P(\text{HHT}) = P(\text{HTH}) = P(\text{HTT}) = P(\text{THH}) = P(\text{THT}) = P(\text{TTH}) = P(\text{TTT}) = \frac{1}{8}$

and

$P(\text{HHH}) + P(\text{HHT}) + P(\text{HTH}) + P(\text{HTT}) + P(\text{THH}) + P(\text{THT}) + P(\text{TTH}) + P(\text{TTT}) = 1.$

These outcomes can be represented with a probability tree diagram.

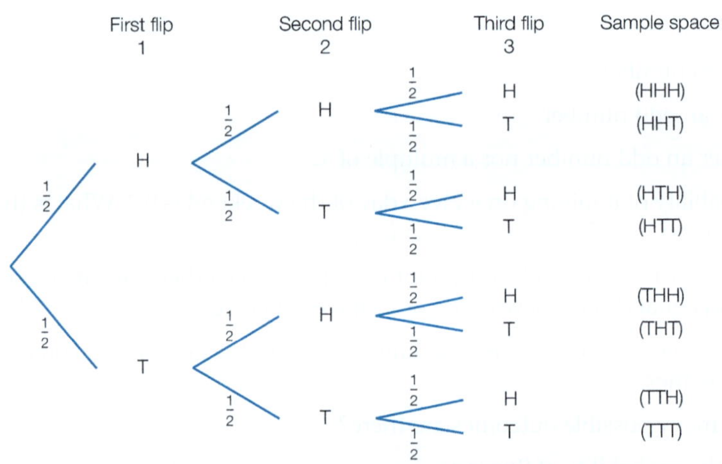

Calculating probabilities using a probability tree diagram

Probabilities can be calculated by multiplying the probabilities along the branches.

A **probability tree diagram** can also be used for events that do not have equally likely outcomes.

Consider a person playing a game three times with $P(\text{winning}) = \frac{3}{5}$ and $P(\text{losing}) = \frac{2}{5}$.

There are only two outcomes as $P(\text{winning}) + P(\text{losing}) = 1$.

A probability tree diagram can represent this information.

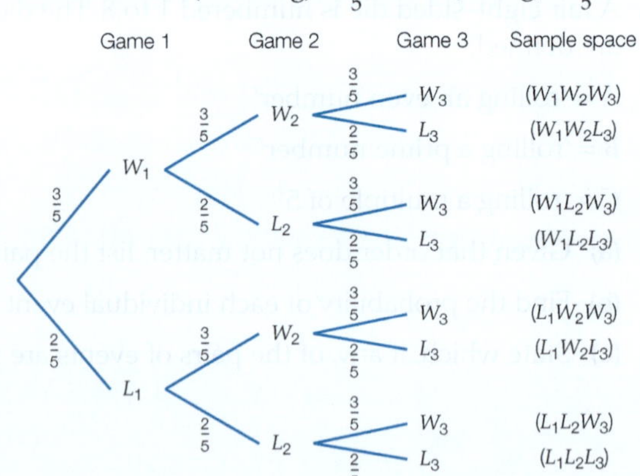

Now consider the probability of winning two out of three games. There are three possible outcomes $W_1W_2L_3$, $W_1L_2W_3$ and $L_1W_2W_3$ and the outcomes are not equally likely.

$$P(W_1W_2L_3) = \frac{3}{5} \times \frac{3}{5} \times \frac{2}{5}$$
$$= \frac{18}{125}$$

Likewise, $P(W_1L_2W_3) = P(L_1W_2W_3) = \frac{18}{125}$.

As these events are mutually exclusive, these probabilities can be added.

$$P(\text{winning 2 out of 3 games}) = P(W_1W_2L_3) + P(W_1L_2W_3) + P(L_1W_2W_3)$$

$$= \frac{18}{125} + \frac{18}{125} + \frac{18}{125}$$

$$= \frac{54}{125} \text{ or } 0.432$$

Worked Example 7

WE 7

At supermarkets, cartons of eggs are inspected for breakages by the checkout assistant before payment is made. The probability that a carton of eggs will be rejected is found to be 0.1. Two cartons of eggs are brought to a checkout.

(a) Copy and complete the tree diagram on the right. Evaluate the probability of each outcome.

Let R = carton of eggs is rejected and A = carton of eggs is not rejected (is accepted).

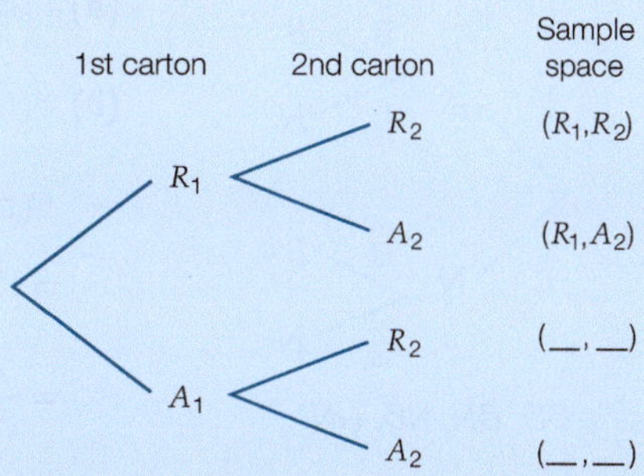

(b) Find the probability that:

(i) both cartons will be rejected

(ii) only one carton will be rejected

(iii) neither carton will be rejected

(iv) at least one carton will be rejected.

Working

(a) $P(R) = 0.1$, so $P(A) = 0.9$

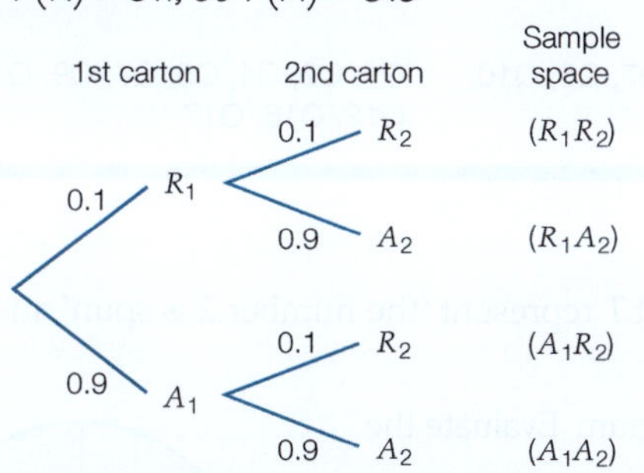

$P(R_1R_2) = 0.1 \times 0.1$
$= 0.01$

$P(R_1A_2) = 0.1 \times 0.9$
$= 0.09$

$P(A_1R_2) = 0.9 \times 0.1$
$= 0.09$

$P(A_1A_2) = 0.9 \times 0.9$
$= 0.81$

(b) **(i)** $P(\text{both cartons will be rejected})$
$= P(R_1R_2)$
$= 0.01$

(ii) $P(\text{only one carton will be rejected})$
$= P(R_1A_2 \text{ or } A_1R_2)$
$= P(R_1A_2) + P(A_1R_2)$
$= 0.09 + 0.09$
$= 0.18$

(iii) $P(\text{neither carton will be rejected})$
$= P(A_1A_2)$
$= 0.81$

(iv) $P(\text{at least one carton will be rejected})$
$= P(\text{one carton rejected or both cartons rejected})$
$= P(\text{one carton rejected}) + P(\text{both cartons rejected})$
$= 0.18 + 0.01$
$= 0.19$

11.4

Worked Example 8

WE8

The lucky dip at the school fete contains 16 different gifts, 10 for the 4–7 age group and 6 for the 8–11 age group. Unfortunately, the 16 gifts have been mixed together. As you cannot tell what is inside without opening the gift, the lucky dips are offered as a special two for the price of one. A 5-year-old takes his two gifts without replacement. Let *B* represent a gift is chosen in the 4–7 age group and *N* represent a gift in the 4–7 age group is not chosen.

(a) Find *P*(both gifts are in the 4–7 age group).

(b) Find *P*(the 5-year-old chooses at least one suitable gift).

Working

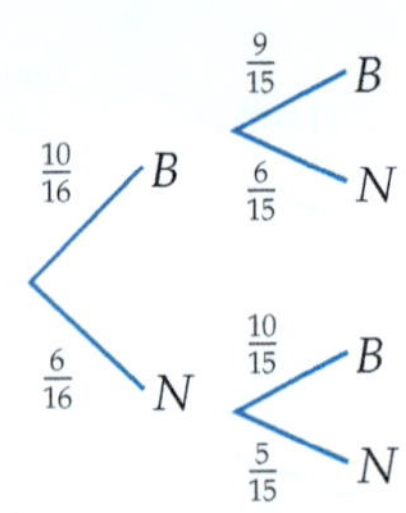

BB, BN, NB, NN

(a) $P(BB) = \frac{10}{16} \times \frac{9}{15} = \frac{3}{8}$

(b) $P(NN) = \frac{6}{16} \times \frac{5}{15} = \frac{1}{8}$

P(the 5-year-old chooses at least one suitable gift)

$= 1 - \frac{1}{8}$

$= \frac{7}{8}$

Answers page 712

11.4 Probability tree diagrams

Stage 5.2

Navigator

Q1, Q2, Q3, Q4, Q6, Q7, Q9, Q10, Q12, Q14, Q17	Q1, Q2, Q3, Q4, Q7, Q8, Q10, Q13, Q15, Q16	Q1, Q2, Q4, Q5, Q7, Q9, Q11, Q12, Q16, Q17

Fluency and understanding

WE7

1 A spinner numbered 1 to 3 is spun twice. Let *T* represent 'the number 2 is spun' and *N* represent 'the number 2 is not spun'.

(a) Copy and complete the following tree diagram. Evaluate the probability of each outcome.

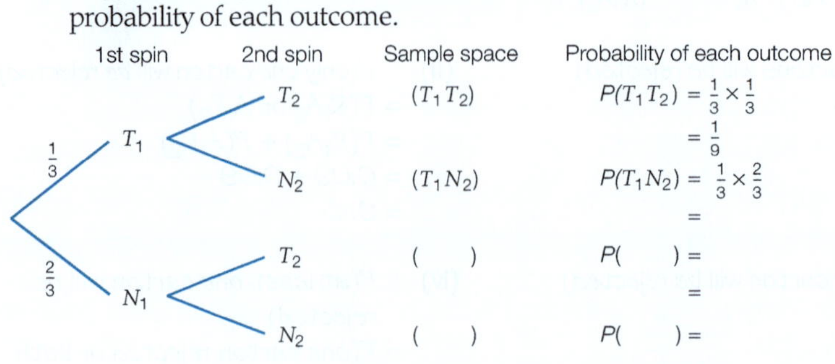

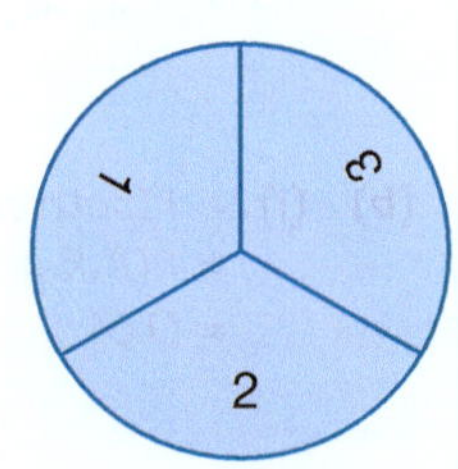

(b) Find the probability of:

(i) spinning 2 twice

(ii) spinning 2 once only

(iii) *not* spinning a 2

(iv) spinning 2 once at least.

11.4

WE 8

2 Two balls are drawn without replacement from a bag containing 4 green marbles and 7 blue marbles. Draw a tree diagram to represent the situation and find:

(a) P(both marbles being blue) (b) P(getting at least one blue marble).

3 The spinner described in Question **1** is now spun three times.

(a) Draw a probability tree diagram for this situation.

(b) What is the probability of spinning a 2 three times?

(c) What is the probability of *not* obtaining a 2 on any of the three spins?

4 A biased coin is tossed twice. Let H represent 'obtaining a head with the coin' and T represent 'obtaining a tail with the coin'. The following probability tree diagram can be drawn.

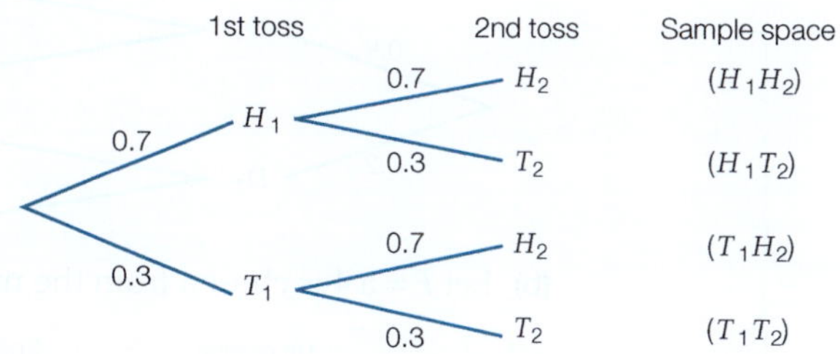

(a) The probability of obtaining two heads is:

A $0.7 + 0.7$ B 0.7×0.7 C 0.25 D 0.7×0.3

(b) The probability of obtaining two tails is:

A 0.6 B 0.3 C 0.21 D 0.09

(c) The probability of obtaining a tail first, and then a head is:

A 0.25 B 0.21 C 0.49 D 1.0

(d) The probability of obtaining only one tail is:

A 0.21 B 0.3 C 0.42 D 0.63

(e) The probability of obtaining at least one tail is:

A 0.75 B 0.51 C 0.42 D 0.39

5 Two cards are drawn, without replacement, from a set of 16 cards numbered from 1 to 16. Let M = the event a multiple of four is drawn and N = a multiple of four is not drawn.

(a) Draw a tree diagram and assign a probability to each branch.

(b) List the sample space.

(c) Find P(both cards are a multiple of 4).

(d) Find P(only one card is a multiple of 4).

(e) Find P(neither card is a multiple of 4).

6 For afternoon tea, patrons of a cafe can select from tea, coffee or hot chocolate, and either scones or carrot cake. The probability of a person selecting tea is 0.45, whereas the probability of a person selecting coffee is 0.35. The probability of a person selecting scones is 0.6.

(a) What is the probability of a person selecting hot chocolate?

(b) What is the probability of a person selecting carrot cake?

(c) Draw a probability tree diagram to represent this situation.

(d) Find the probability of a person:

(i) selecting tea and scones

(ii) selecting hot chocolate and carrot cake

(iii) selecting tea and scones or hot chocolate and carrot cake

(iv) *not* selecting coffee and scones.

11.4

7 For each of the following:

(i) copy and complete the probability tree diagram

(ii) write down the probability of each possible outcome

(iii) find the sum of the probabilities of all outcomes.

(a) Let G = scores a goal and D = does not score a goal.

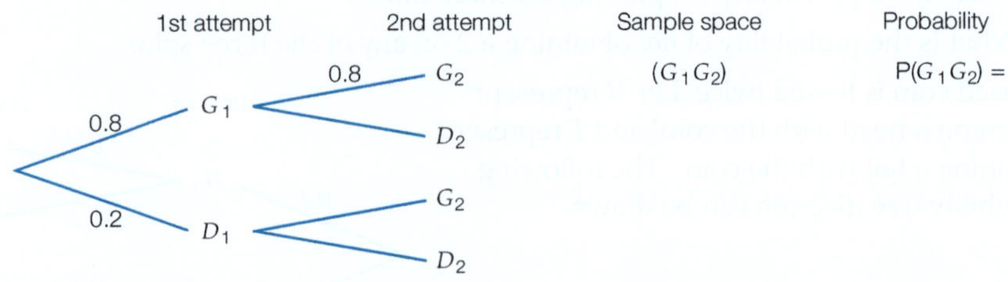

(b) Let F = a 4 is chosen from the numbers 1 to 6 and N = a 4 is not chosen.

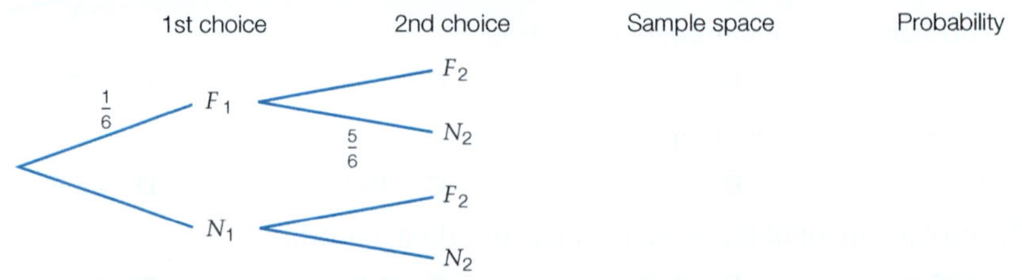

(c) H = Henry is chosen from a group of boys.

A = Alia is chosen from a group of girls.

(d) Let B = student travels to school by bus and C = student does not travel by bus.

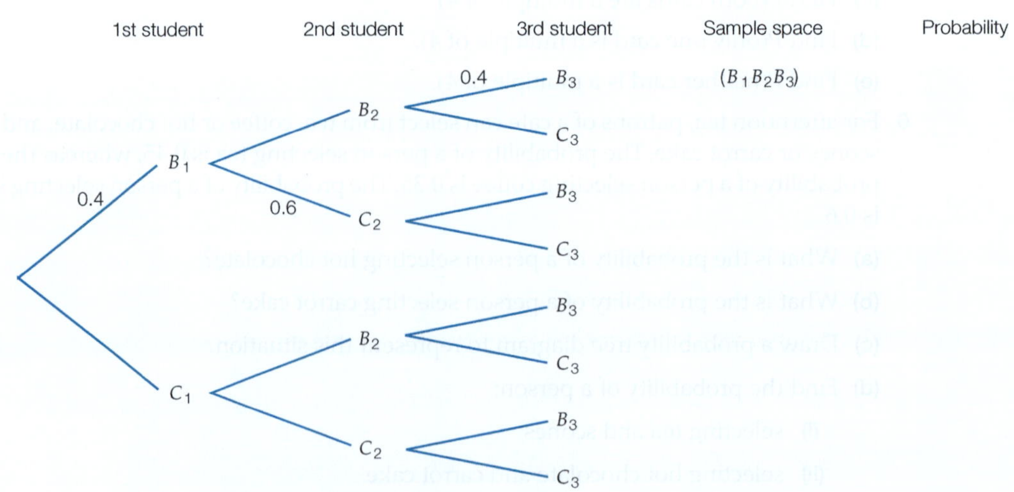

8 In a game of darts, each player has a turn to throw three darts one at a time. Seline estimates her chances of scoring a bullseye with any one throw of a dart to be 0.2. Find the probability that:

(a) Seline scores three bullseyes in her turn

(b) she scores two bullseyes only

(c) she scores at least one bullseye.

9 Two fair six-sided dice are rolled. Let T represent the event 'a 3 is rolled'.

(a) What would not T represent?

(b) Draw a probability tree diagram to illustrate this situation.

(c) Find the probability of:

(i) rolling two 3s (ii) not rolling any 3s (iii) rolling one 3 only

(iv) rolling at least one 3 (v) rolling no more than one 3.

10 A card is selected from a standard pack of 52 playing cards and then replaced. Another card is selected. Find the probability that:

(a) both cards are Aces

(b) the first card is a Club and the second card is red

(c) both cards are of the same colour

(d) one card is red and the other is black.

11 Wesley thinks his lucky number is 5. At the races, he decides to bet on horse number 5 in the first three events. In race number 1, there is a field of 12 horses. In race number 2, there is a field of 9 horses. In race number 3, there is a field of 10 horses.

(a) If each horse in an event has an equal chance of winning, draw a probability tree diagram to represent Wesley's chances of winning or losing on each race.

(b) Find the probability that:

(i) Wesley wins all three racing events

(ii) only one of his chosen horses wins

(iii) at least one of his chosen horses wins.

Reasoning, communicating and open-ended

12 A mathematics quick quiz consists of five multiple-choice questions, each with five choices for the solution. Draw up a table to show the probability of obtaining each of the totals of correct answers from 0 to 5 if you randomly guess the answers.

13 A fair six-sided die is rolled until a 6 is obtained.

(a) Copy and complete the following table, which shows the probability that this will take the indicated number of rolls. Round your answers to three decimal places.

Number of rolls	1	2	3	4	5	6
Probability						

(b) Add the probabilities you found in part (a). What sum do you get?

(c) Explain what this answer means.

14 A party game for 29 people starts by assigning everyone a number from 1 to 29. In the first round, 7 numbers are drawn at random and those people eliminated. In the second round 6 people are eliminated, the third round eliminates 5 people, then 4, 3, 2 and finally 1, to leave the winner.

(a) Find *P*(being eliminated in round 1).

(b) Assuming you are not eliminated in the first round, find *P*(being eliminated in round 2).

(c) Assuming you are not eliminated in the previous rounds, find the probability of being eliminated in each round.

(d) Add the results for the probability of being eliminated in each round. Explain the result.

15 A set of volleyball is won when one side reaches 25 points. If we knew nothing about the ability of the teams, we would assign the chance of winning as 0.5 each. If, however, one side was leading 15–0, we may rethink the original assignment of 0.5 each. Estimate the new probability for each team to win the set. Justify your answer.

16 Draw up a probability tree diagram that finishes with at least 9 branches where one of the final branches has a probability of $\frac{1}{9}$ and another has a probability of $\frac{1}{8}$. Explain what your tree diagram represents.

17 The estimated probability of a 25-year-old male playing cricket is 0.2 and of a 25-year-old male playing golf is 0.18. Armapreet calculates that the probability of a man playing cricket or golf must be 0.38. Is she right? If not, what mistake has she made?

Problem solving

The origins of probability theory

The origins of modern probability theory involve a problem known as the 'problem of the points' or 'division of the stakes'. It involves a two-person game where each has made an equal contribution to the prize pot and the agreement is that the first person to get to a certain number of points will be the winner and collect the entire prize pot. The problem arises when the game is interrupted, and cannot be concluded. How should the prize pot be divided?

Let's say the winner is the first to 10 points.

1 How would you divide the pot if the score was 7 to 5?

2 One of the early solutions suggested that the pot should be divided in proportion to the number of games won. Using this solution, the pot above would be divided $\frac{7}{12}:\frac{5}{12}$. What do you think of this solution? What would happen if the score was 1 to 0? Pascal and Fermat thought the crucial issue was the possible ways the game could have continued rather than what had already happened.

3 Think of a game where the score is 8 to 7. Write all the ways the game could continue and use this to decide how to divide the pot.

4 What if the score was 2 to 5? What is the problem with this sort of score?

5 What do you think would be the easiest scores to deal with if the game was interrupted?

Strategy options

- Act it out.
- Draw a diagram.

Extension

Pascal and Fermat are considered to have solved this problem. Conduct some research to discover how they resolved the issue.

Conditional statements

11.5

Stage 5.2

Need to Know

Identifying conditional statements

When we are dealing with a **conditional statement** we are given information that has restricted the universal set. Words such as 'given', 'knowing that' and 'if … then' are indications of conditional statements.

Consider rolling two fair six-sided dice, one green and one red. All outcomes are listed in the array shown. There are 36 equally likely outcomes. They are equally likely as each die is assumed to be fair.

		Green die					
		1	2	3	4	5	6
Red die	1	(1, 1)	(2, 1)	(3, 1)	(4, 1)	(5, 1)	(6, 1)
	2	(1, 2)	(2, 2)	(3, 2)	(4, 2)	(5, 2)	(6, 2)
	3	(1, 3)	(2, 3)	(3, 3)	(4, 3)	(5, 3)	(6, 3)
	4	(1, 4)	(2, 4)	(3, 4)	(4, 4)	(5, 4)	(6, 4)
	5	(1, 5)	(2, 5)	(3, 5)	(4, 5)	(5, 5)	(6, 5)
	6	(1, 6)	(2, 6)	(3, 6)	(4, 6)	(5, 6)	(6, 6)

So, $P(1, 1) = \frac{1}{36}$.

If the number on the green die is known to be 1, then $P(1, 1)$ would be $\frac{1}{6}$ because only the first column of results is possible from our array. The universal set has been restricted in this case, so effectively only the outcomes that have a 1 on the green die are considered.

$P(1, 1, \text{given the green die is } 1) = \frac{1}{6}$.

The word 'given' implies a conditional statement. The information after the word 'given' is used to restrict the universal set. The number of outcomes in the restricted set becomes the denominator for the probability fraction.

The order of the statements is important.

For example:

		Green die	
		3	6
Red die	1	(3, 1)	(6, 1)
	2	(3, 2)	(6, 2)
	3	(3, 3)	(6, 3)
	4	(3, 4)	(6, 4)
	5	(3, 5)	(6, 5)
	6	(3, 6)	(6, 6)

		Green die					
		1	2	3	4	5	6
Red die	2	(1, 2)	(2, 2)	(3, 2)	(4, 2)	(5, 2)	(6, 2)
	4	(1, 4)	(2, 4)	(3, 4)	(4, 4)	(5, 4)	(6, 4)
	6	(1, 6)	(2, 6)	(3, 6)	(4, 6)	(5, 6)	(6, 6)

Using the arrays above,

$P(\text{red die is even given green die is a multiple of } 3) = \frac{1}{2}$, whereas

$P(\text{green die is a multiple of three given the red die is even}) = \frac{1}{3}$.

Venn diagrams and tree diagrams can also be used to find conditional probability.

11.5

Worked Example 9

WE9

A 16-sided die is rolled, with faces numbered 1–16. The following events are defined.

A = 'numbers less than 7' and B = 'multiples of 4'

(a) Draw a Venn diagram to show this information.

(b) Calculate P(a number less than 7 given it is a multiple of 4).

(c) Calculate P(a multiple of 4 given it is a number less than 7).

Working

(a)

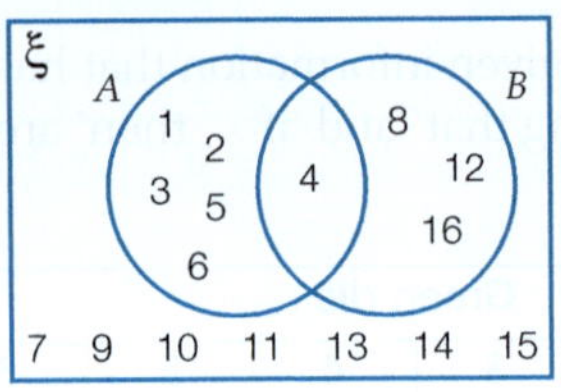

(b) The denominator is the number of elements in B (a multiple of 4): 4.

The numerator is the number of elements in B and in A (a number less than 7): 1.

P(a number less than 7 given it is a multiple of 4) = $\frac{1}{4}$.

(c) The denominator is the number of elements in A (a number less than 7): 6.

The numerator is the number of elements in A and in B (a multiple of 4): 1.

P(a multiple of 4 given it is a number less than 7) = $\frac{1}{6}$.

Worked Example 10

WE10

In a particular electorate, 55% of the voting population is female. It is also known that 60% of male voters support the Progressive Party, whereas only 45% of female voters support this party. Find the probability that a randomly selected voter from that electorate is male if you know that they support the Progressive Party. Round your answer to four decimal places.

Working

F = 'female'

M = 'male'

P = 'Progressive Party supporter'

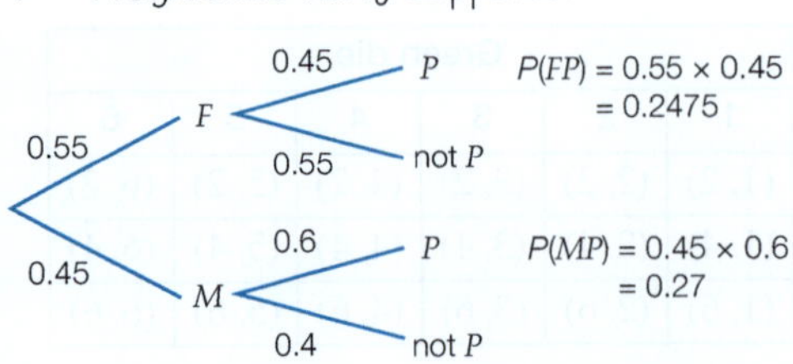

P(support Progressive Party)

$= P(FP) + P(MP)$

$= 0.2475 + 0.27$

$= 0.5175$

P(male and support Progressive Party)

$= P(MP)$

$= 0.27$

P(male given they support Progressive Party)

$= P(M$ given $P)$

$= \frac{0.27}{0.5175}$

$= 0.5217$

11.5 Conditional statements

Answers page 714

Stage 5.2

Navigator

Q1, Q2, Q3, Q4, Q5, Q6, Q9 (a)–(c), Q11	Q1, Q2, Q4, Q5, Q7, Q9 (a)–(c), Q11	Q1, Q2, Q4, Q5, Q7, Q8, Q9, Q10, Q11

Fluency and understanding

1 **(a)** A 16-sided die is rolled, with faces numbered 1–16. The following events are defined. WE9

A = 'numbers less than 9' and B = 'multiples of 3'

(i) Draw a Venn diagram to show this information.

(ii) Calculate P(number less than 9 given it is a multiple of 3).

(iii) Calculate P(multiple of 3 given it is a number less than 9).

(b) A 16-sided die, with faces numbered 1–16, is rolled and the following events are defined.

A = 'numbers greater than 5' and B = 'prime numbers'

(i) Draw a Venn diagram to show this information.

(ii) Calculate P(number greater than 5 given it is a prime number).

(iii) Calculate P(prime number given it is a number greater than 5).

2 In a particular electorate, 65% of the voting population are pensioners. It is also known that 55% of non-pensioners support the Progressive Party, whereas only 40% of pensioners support this party. Find the probability that a randomly selected voter from that electorate is a pensioner if you know that they support the Progressive Party. Round your answer to four decimal places. WE10

3 Use the information in the Venn diagram to find:

(i) $P(A$ given $B)$ **(ii)** $P(B$ given $A)$

(a)

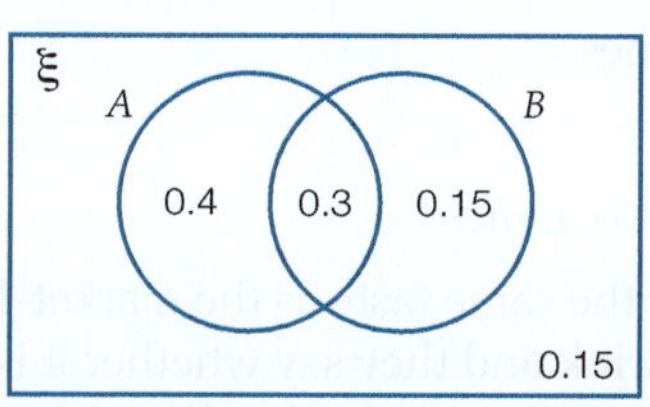

(b)

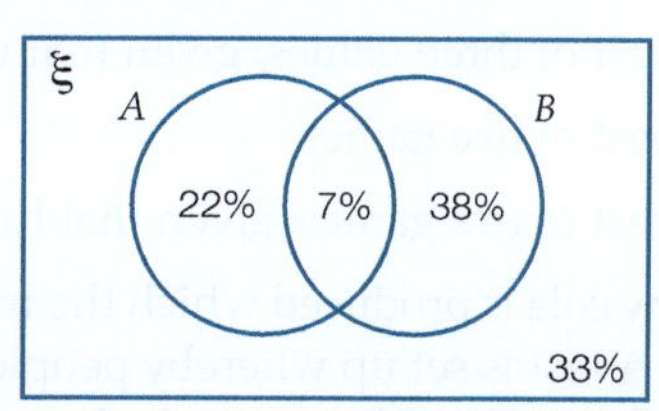

4 A fair (unbiased) coin is flipped three times.

(a) The probability that the third flip results in a head, given the previous two flips were both heads, is:

A $\frac{1}{8}$ **B** $\frac{1}{4}$ **C** $\frac{1}{2}$ **D** 1

(b) The probability that the third flip results in a head, given at least two of the three flips are heads, is:

A $\frac{3}{8}$ **B** $\frac{1}{2}$ **C** $\frac{2}{3}$ **D** $\frac{3}{4}$

5 You have a reduced pack of playing cards consisting only of the 26 red cards. A card is selected at random from the pack. Find the probability that the card is:

(a) an Ace given it is a Heart

(b) a Heart given it is a picture card (J, Q, K)

(c) the five of Diamonds given it is a Diamond

(d) the five of Diamonds given it is a five.

11.5

6 A new disease is known to infect 2.5% of the population. A test for the disease has been developed, but it is not completely accurate. It correctly identifies the disease in infected patients 95% of the time, but also (incorrectly) identifies the disease in 7% of the uninfected population.

(a) Draw a tree diagram to show this information.

(b) Find *P*(have disease given test positive), rounding your answer to four decimal places.

(c) Find *P*(don't have disease given test negative), rounding your answer to four decimal places.

7 A sportswear company obtains its shoes from two suppliers, *A* and *B*. It is known that 5% of the shoes from supplier *A* are defective, whereas the figure is 9% for supplier B. Artemis obtains 65% of its shoes from supplier *A*. Where necessary, round your answers to four decimal places.

(a) Draw a tree diagram to show this information.

(b) Find *P*(*A* given the shoe is defective).

(c) Find *P*(*B* given the shoe is not defective).

8 A university student is studying both Economics and Asian Studies. The probability that she will pass Economics is 0.95 and that she will pass Asian Studies is 0.8. There is a 0.75 chance she will pass both. Find:

(a) *P*(passes Economics given passes Asian Studies)

(b) *P*(passes Asian Studies given passes Economics)

(c) *P*(passes Economics given passes only one of the subjects)

Reasoning, communicating and open-ended

9 Abbud plays a lot of table-tennis. He has a 70% chance of winning any particular game. Find the probability that Abbud wins a match decided by the:

(a) best of three games

(b) best of three games, given that he won the first game

(c) best of three games, given that it lasts three games

(d) best of five games

(e) best of five games, given that he won the first two games.

10 A new cola is produced which the maker claims has the same taste as the market-leading cola. A test is set up whereby people are offered a drink and they say whether it is the market leader or the new cola. The new cola is correctly identified 40% of the time and the market leader is correctly identified 75% of the time. If a person has an equal chance of being given either cola and they say the cup contains the market leader, what is the probability that this is true?

11 A tired teacher wrote this question:

Given $P(A) = 0.6$ and $P(B) = 0.4$, find *P*(*A* given *B*).

(a) What is wrong with the question?

(b) Rewrite the question so it is not wrong, and then find the answer.

Dependent and independent events

11.6

Stage 5.2

Need to Know

Independent events

Two events A and B are **independent events** if the result of one event has no effect on the result of the other event.

When A and B are independent events:

$$P(A \text{ and } B) = P(A) \times P(B)$$

For example:

Consider the situation where a fair coin is tossed and a spinner that is equally divided into three sections labelled 1, 2 and 3 is spun. The outcomes are equally likely as the coin is assumed to be fair and the spinner is equally divided.

$P(\text{H}) = \frac{1}{2}$ $\quad P(1) = \frac{1}{3}$ $\quad P(\text{H}, 1) = \frac{1}{6}$

$$\begin{aligned} P(\text{H}) \times P(1) &= \frac{1}{2} \times \frac{1}{3} \\ &= \frac{1}{6} \\ &= P(\text{H}, 1) \end{aligned}$$

$P(\text{H}, 1)$ is the product of $P(\text{H})$ and $P(1)$ because the events are independent.

Sampling with replacement is a common example of an independent event.

Dependent events

Dependent events are influenced by the outcome of another event; sampling without replacement is an example of a dependent event.

For example:

A bag contains 5 red and 5 green marbles. Two marbles are selected without replacement. Find the probability that both are red.

$P(\text{Red 1st draw}) = \frac{5}{10}$

$P(\text{Red 2nd draw}) = \frac{4}{9}$

$$\begin{aligned} P(\text{Red, Red}) &= \frac{5}{10} \times \frac{4}{9} \\ &= \frac{2}{9} \end{aligned}$$

11.6

Worked Example 11

WE11

Two representatives, one girl and one boy, are chosen at random from a group of 6 girls and 4 boys. If Linda and Adam are in this group, what is the probability that both Linda and Adam will be chosen?

Working

Let L = Linda chosen
and A = Adam chosen

L and A are independent events.

$P(L \text{ and } A) = P(L) \times P(A)$

$P(L) = \frac{1}{6}$

$P(A) = \frac{1}{4}$

$$P(L \text{ and } A) = \frac{1}{6} \times \frac{1}{4} = \frac{1}{24}$$

The probability that both Linda and Adam will be chosen is $\frac{1}{24}$ (i.e. 1 chance in 24).

Worked Example 12

WE12

Two cards are drawn out of a 52-card deck of playing cards without replacement.

(a) Find the P(drawing 2 Kings).

(b) Find the P(both cards are from the same suit).

Working

(a) $P(\text{King on 1st draw}) = \frac{4}{52} = \frac{1}{13}$

$P(\text{King on 2nd draw}) = \frac{3}{51} = \frac{1}{17}$

$P(\text{both cards being a King}) = \frac{1}{13} \times \frac{1}{17} = \frac{1}{221}$

(b) $P(\text{Club, Club}) = \frac{13}{52} \times \frac{12}{51} = \frac{1}{17}$

Similarly,
$P(\text{Spade, Spade})$
$= P(\text{Heart, Heart})$
$= P(\text{Diamond, Diamond})$
$= \frac{1}{17}$

$P(\text{both cards are from the same suit})$
$= \frac{1}{17} + \frac{1}{17} + \frac{1}{17} + \frac{1}{17}$
$= \frac{4}{17}$

11.6 Dependent and independent events

Answers page 714

Stage 5.2

Navigator

Q1, Q2, Q3, Q4, Q5, Q6, Q7, Q8, Q9, Q15, Q18, Q19	Q1, Q2, Q3, Q5, Q7, Q8, Q10, Q12, Q13, Q14, Q15, Q17, Q18	Q1, Q2, Q4, Q6, Q7, Q9, Q10, Q11, Q13, Q14, Q16, Q17, Q19

Fluency and understanding

1 Katrina finds that on average her train is late 2 mornings out of 5, whereas Frank's bus is late 1 morning in every 4. Find the probability that on any given day both Katrina and Frank are late. WE11

2 A box containing 40 apples, 10 of each of 4 different varieties, has 5 rotten ones. If 2 apples are taken out without replacement, calculate: WE12

(a) P(2 rotten apples) (b) P(2 apples of the same variety)

3 Phillip has 2 pairs of blue and 1 pair of black jeans in his wardrobe. In a drawer, he has 1 red, 3 black and 5 white T-shirts. If he randomly chooses a pair of jeans and a T-shirt to wear, find the probability that he chooses:

(a) a white T-shirt and blue jeans

(b) a black T-shirt and black jeans

(c) black jeans and a T-shirt that is not black

(d) blue jeans and a T-shirt that is not red.

4 In a group of 20 people, 2 have red hair, 6 are blonde and the rest have dark hair. Of the group, 15 people are right-handed. If a person is chosen at random from the group, find the probability that the person:

(a) has red hair and is right-handed

(b) is blonde and is left-handed

(c) is right-handed and does not have red hair.

5 Label each of these pairs of events as either independent (I) or dependent (D).

(a) Tossing a coin and rolling a die.

(b) Drawing two cards from a pack without replacing the first before drawing the second.

(c) Drawing two cards from a pack replacing the first before drawing the second.

(d) Raining in two cities 100 km apart on the same day.

(e) Hours of study and test result.

(f) The number of hours of practice a learner driver does before getting their license and the chance of having an accident in the first month of having their license.

6 An urn contains 3 red beads, 4 blue beads and 8 green beads. If two beads are drawn without replacement, find:

(a) P(RR) (b) P(BB) (c) P(GG) (d) P(RB)

(e) P(BG) (f) P(GB) (g) P(same colour) (h) P(different colours)

11.6

7 (a) If a 20-cent coin is tossed in the air twice, the probability of obtaining two heads is:

A $\frac{1}{4}$ B $\frac{1}{3}$ C $\frac{1}{2}$ D $\frac{3}{4}$

(b) If a six-sided die is rolled twice, the probability of obtaining a double 6 is:

A $\frac{1}{3}$ B $\frac{1}{6}$ C $\frac{1}{12}$ D $\frac{1}{36}$

(c) If a 20-cent coin is tossed and a six-sided die is rolled, the probability of getting a head and a multiple of 3 is:

A $\frac{1}{6}$ B $\frac{1}{3}$ C $\frac{1}{4}$ D $\frac{1}{2}$

8 Joshua and Tania are competing in different sections of a chess tournament. Joshua estimates that his chance of winning against his opponent is 3 in 5, whereas Tania estimates that her chance of winning against her opponent is 5 in 7. Find the probability that:

(a) both Joshua and Tania win their chess game

(b) they both lose their chess game

(c) Joshua loses his game but Tania wins hers.

9 A drawer contains 6 black socks, 8 red socks and 4 white socks. If two socks are drawn without replacement, find:

(a) *P*(2 black socks)

(b) *P*(2 red socks)

(c) *P*(2 white socks)

(d) *P*(a pair of socks of matching colour)

(e) *P*(two odd socks)

(f) *P*(one black sock)

(g) *P*(at least one black sock)

(h) *P*(no white socks)

10 A spinner numbered 1 to 8 is spun twice. Find the probability that:

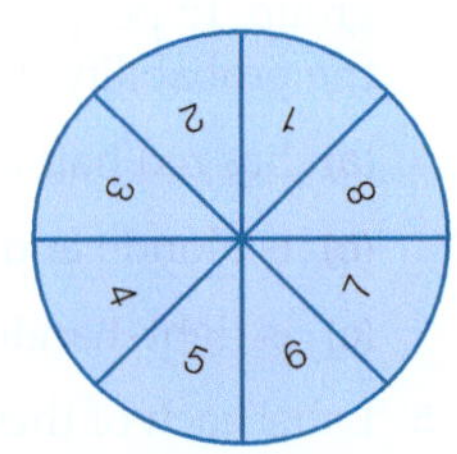

(a) a 5 is obtained on the first spin

(b) a 5, then a 7, is spun

(c) a 5 is obtained on the first spin and a number less than 4 is obtained on the second

(d) an even number is obtained on the first spin and an odd number is obtained on the second.

11 If the probability that person A will be alive in 20 years is 0.7 and the probability that person B will be alive in 20 years is 0.5, find:

(a) the probability that they will both be alive in 20 years

(b) the probability that only one is alive in 20 years.

12 Natalie and her team play in a weekly squash competition. The probability that her team wins in any given week is 0.7. The outcome is not affected by any previous wins or losses. Find the probability that:

(a) Natalie's team does not win in any given week

(b) her team wins 2 weeks running

(c) her team loses 3 weeks running

(d) her team loses 4 weeks running.

13 A bag contains 6 white marbles, 5 black marbles and 4 green marbles. In each draw, a marble is drawn from the bag and not replaced. In three draws, find:

(a) *P*(WBG in that order)

(b) *P*(WBG in any order).

Reasoning, communicating and open-ended

14 Two normal unbiased six-sided dice are rolled. These events are defined:

A = rolling a 6 with the first die B = rolling a 5 with the second die

C = rolling a total of 8 D = rolling a total greater than 10

(a) Write the sample space for this experiment.

(b) Find the probability for each event.

(c) How many pairs of events are possible when order is not important? List them.

(d) How can you check which pairs are independent?

(e) Which, if any, of these pairs of events are independent?

15 You know that two events, A and B, are independent and that $P(A) = 0.3$ and $P(A \text{ and } B) = 0.15$.

(a) Find $P(B)$. **(b)** Now find $P(A \text{ or } B)$. Explain how you obtained this value.

16 A secondary school has 900 students, of whom 475 have a part-time job and 870 study mathematics. There are 470 mathematics students who have part-time jobs. If a student is picked at random, show that the events 'picking a student with a part-time job' and 'picking a maths student' are not independent.

17 For two events, A and B, we are given $P(A) = 0.35$ and $P(B) = 0.45$. Find $P(A \text{ and } B)$ if:

(a) A and B are mutually exclusive

(b) A and B are independent.

18 You have an unbiased ten-sided die, numbered 1 to 10. Write at least five questions related to independent events that deal with this die. Include the answers for your questions.

19 **(a)** Give two examples of pairs of dependent events.

(b) Give two examples of pairs of independent events.

Problem solving

Winning the lottery

In a particular raffle you purchase a ticket that has a unique six-digit number from 000001 to 999999 printed on it. The winning number is drawn at random and the ticket with that number printed on it wins the first prize. In this case, the prize is \$100 000. Any ticket that has the last five of the winning digits in order in the last five positions in their number wins second prize (\$20 000). Any ticket with the last four digits in order in the last four positions wins third prize (\$10 000), and any ticket with the last three digits in order in the last three positions wins fourth prize (\$2000). Each ticket can claim only one prize.

1 What is the probability that you win:

(a) \$100 000? **(b)** \$20 000?

(c) \$10 000? **(d)** \$2000?

2 What is the total payout on the lottery, assuming every prize is claimed?

3 How much do the tickets have to sell for so that the organisers make a profit?

Strategy options

- Break problem into manageable parts.
- Have I seen a similar problem?

Chapter review 11

D.I.Y. Summary

Key Words

Stage 5.2

addition law for mutually exclusive events
complementary events
conditional statement
dependent events
disjoint sets
events
independent events
long-run proportion
mutually exclusive
probability
probability tree diagram
sample space
tree diagram
two-way table
universal set
Venn diagram

Stage 5.2 Q1–8

Copy and complete the following using the words and phrases from the list, where appropriate, to write a summary for this chapter. A word or phrase may be used more than once.

1 Every element under consideration in a particular question is called the ___________.

2 If two events, A and B, are ___________, then $P(A) \times P(B) = P(A \text{ and } B)$.

3 The events 'having brown hair' and 'not having brown hair' are an example of ___________.

4 When we put the probabilities on each of the branches of a tree diagram we then have a ___________.

5 Events that cannot occur at the same time are said to be ___________.

6 A ___________ or a ___________ can be used to help find the sample space.

7 Spinning a spinner that has been equally divided into four pieces labelled 1, 2, 3 and 4, then calculating $P(1) + P(2) + P(3) + P(4) = 1$ is an example of the ___________.

8 Words and phrases such as 'given', 'knowing that' and 'if … then' are indications we are dealing with a ___________.

Stage 5.2 Q1–18

Fluency and understanding

1 A spinner numbered 1 to 5 is spun. Find: Ex 11.1

(a) P(spinning a 2)
(b) P(not spinning a 2)
(c) P(spinning a number greater than 2)
(d) P(spinning an even number)
(e) P(spinning an odd number)
(f) P(spinning a number less than 5).

2 A box contains 15 blocks. Seven are red, five are white and the rest are blue. A block is chosen at random from the box. Find the probability that the block chosen is: Ex 11.1

(a) blue
(b) not red
(c) red or blue.

3 A fair six-sided die is rolled. What is the probability of obtaining an even number or a 5? Ex 11.1

4 A spinner numbered 1 to 4 is spun twice.

Ex 11.1

(a) Represent all the possible outcomes using a two-way table.

(b) Find:

(i) P(spinning a pair of 3s) (ii) P(spinning only one 3)

(iii) P(not spinning any 3s) (iv) P(the sum of the two numbers obtained is 5).

5 A 50-cent coin is tossed twice. Find the probability of obtaining two tails.

Ex 11.1

6 This two-way table shows the number of elements in each of three sets.

Ex 11.2

	A and B	A and not B	not A and B	not A and not B	
C	1	4	3	2	10
not C	2	1	2	3	8
	3	5	5	5	18

Write down the number of elements in:

(a) the universal set (b) A

(c) A or B (d) B and C

(e) not B (f) not A and not B and not C.

7 A cutlery rack contains 6 knives, 5 forks and 2 spoons.

Ex 11.3

(a) The probability that a piece of cutlery drawn at random will be a knife or a spoon is:

A $\frac{2}{13}$ B $\frac{2}{5}$ C $\frac{6}{13}$ D $\frac{8}{13}$

(b) The probability that a piece of cutlery drawn at random will not be a knife is:

A $\frac{6}{13}$ B $\frac{7}{13}$ C $\frac{6}{7}$ D $\frac{7}{6}$

8 Two cards are drawn, without replacement, from a set of 20 cards numbered 1 to 20.

Ex 11.4

Let R = 'a prime number is drawn' and N = 'a prime number is not drawn'

(a) Draw a tree diagram to show this information.

(b) Find P(both cards show a prime number).

(c) Find P(no cards show a prime number).

(d) Find P(at least one card shows a prime number).

9 Use the information in each Venn diagram to find the following probabilities. Where necessary, give your answers correct to four decimal places.

Ex 11.5

(i) $P(A$ given $B)$ (ii) $P(B$ given $A)$

(a)

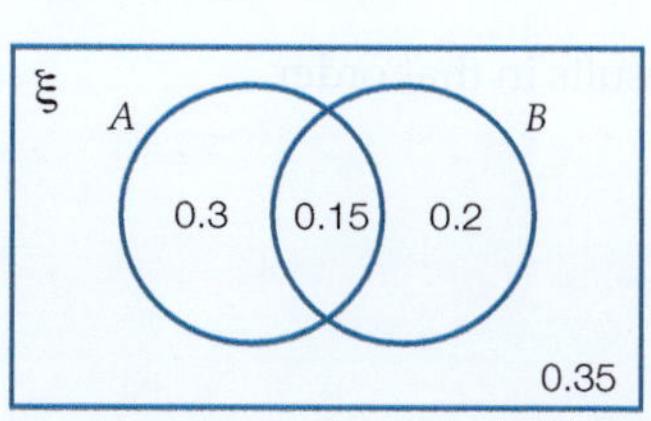

(b)

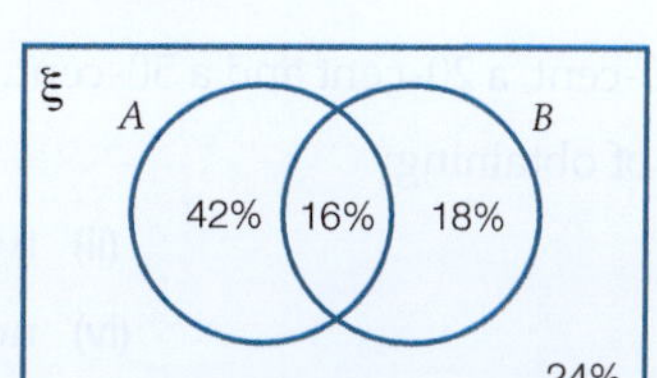

10 John and Petula are in the same class at school and both have trouble getting to school on time. John is late 2 days out of 9, whereas Petula is late 3 days out of 11. What is the probability that they will both be late on any particular day?

Ex 11.6

11 A fair seven-sided spinner, numbered 1 to 7, is spun. Find the probability of:

(a) obtaining a 5
(b) obtaining an even number
(c) obtaining a number less than 4
(d) not obtaining a 6
(e) obtaining a multiple of 3
(f) not obtaining a prime number.

12 In clay pigeon shooting, a clay block in the shape of a pigeon is used as a target for shooting. Jackie estimates her chance of hitting the moving clay target with one shot to be 0.75.

(a) What is the probability of her not hitting the target with one shot?

(b) Draw a probability tree diagram to represent the possible outcomes of shooting at three clay targets.

(c) Find the probability that Jackie hits:

(i) all three clay targets
(ii) only one clay target
(iii) at least one clay target.

13 In a survey of 60 people, it was found that 40 people liked lemonade, whereas 30 people liked cola. Two people said they disliked both.

(a) Present this information in a two-way table.

(b) If one person is chosen at random, find the probability that this person:

(i) likes lemonade
(ii) likes lemonade only
(iii) likes cola
(iv) likes cola only
(v) likes both lemonade and cola
(vi) does not like lemonade
(vii) likes lemonade or cola
(viii) likes neither cola nor lemonade.

14 These probabilities apply to eye colour for students at a particular school. The colours are all mutually exclusive.

$P(\text{green}) = \frac{3}{16}$ $P(\text{blue}) = \frac{1}{4}$ $P(\text{hazel}) = \frac{3}{8}$ $P(\text{brown}) = \frac{3}{16}$

A student is chosen at random. What is the probability that he or she has:

(a) blue or green eyes
(b) neither blue nor brown eyes
(c) blue, hazel or brown eyes
(d) neither hazel nor blue eyes?

15 Nerida and Ngaio play each other at badminton every week. The probability that Nerida wins any particular game is 0.65. Find, rounded to three decimal places where necessary, the probability that:

(a) Ngaio wins in any particular week

(b) Nerida wins for 3 weeks in a row

(c) Ngaio wins for 4 weeks in a row

(d) over a 6-week period the winning sequence is Nerida, Ngaio, Nerida, Ngaio, Nerida, Ngaio.

16 You toss three coins: a 10-cent, a 20-cent and a 50-cent. You note the results in that order.

(a) Find the probability of obtaining:

(i) three heads
(ii) two heads
(iii) one head
(iv) no heads.

(b) What do the answers in part **(a)** add up to? Is this an expected result? Explain your reasoning.

(c) Without doing any further calculations, explain how you can find the probability of each number of tails.

(d) Find the probability of getting:

(i) HHT **(ii)** HTH **(iii)** THH.

(e) Write a statement about what you have found in part **(d)**.

Reasoning and communicating

17 Justine and Kim are entered for a tennis tournament that begins with eight players. The draw profile looks like the following (because they are so good, Justine and Kim are never scheduled to play until the final).

Justine and Kim are very confident players and they think they have a probability of $\frac{2}{3}$ of beating any other player, except for each other. If they play each other, Kim has a $\frac{3}{5}$ chance of winning. Find the following probabilities.

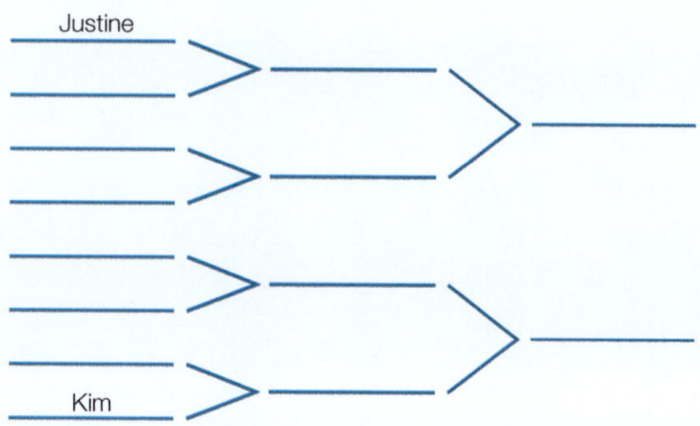

(a) The two girls play each other in the final.

(b) Kim wins the tournament (not necessarily against Justine).

(c) Justine wins the tournament (not necessarily against Kim).

18 A weather forecast predicting the chance of rain in the next 24 hours in some capital cities is recorded below.

City	Brisbane	Sydney	Canberra	Melbourne	Hobart	Adelaide	Perth	Darwin
Chance of rain	0.15	0.4	0.3	0.1	0.2	0.4	0.4	0.1

Assuming the events to be independent, find:

(a) P(rain in Melbourne and Brisbane)

(b) P(rain in Sydney and Perth)

(c) P(rain in Adelaide, Hobart and Perth)

(d) P(rain in Sydney and Canberra).

Long-term statistics are kept and found that the probability of rain in both Sydney and Canberra on the same day is actually 0.2.

(e) Explain the difference between your answer to **(d)** and the long-term probability.

SECTION 4

Statistics

Statistics

30

Objectives: This chapter:

- introduces the three common averages – mean, median and mode – and shows how to calculate them
- explains what is meant by the variance and standard deviation and shows how to calculate them

30.1 Introduction

In the previous chapter we outlined how data gathered from a range of sources can be represented. In order to get the most out of such data and be able to interpret it sensibly and reliably, techniques have been developed for its analysis. **Statistics** is the name given to this science. One important statistical quantity is an 'average', the purpose of which is to represent a whole set of data by a single number that in some way represents the set. This chapter explains three types of average and shows how to calculate them. Finally the 'variance' and 'standard deviation' are introduced. These are numbers that describe how widely the data is spread.

30.2 Averages: the mean, median and mode

When presented with a large amount of data it is often useful to ask 'is there a single number that typifies the data?' For example, following an examination taken by a large number of candidates, examiners may be dealing with hundreds of examination scripts. The marks on these will vary from those with very poor marks to those with very high or even top marks. In order to judge whether the group of students as a whole found the examination difficult, the examiner may be asked to provide a single value that gives a measure of how well the students have performed. Such a value is called an **average**. In statistics there are three important averages: the arithmetic mean, the median and the mode.

The arithmetic mean

The **arithmetic mean**, or simply the **mean**, of a set of values is found by adding up all the values and dividing the result by the total number of values in the set. It is given by the following formula:

Key point

$$\text{mean} = \frac{\text{sum of the values}}{\text{total number of values}}$$

WORKED EXAMPLES

30.1 Eight students sit a mathematics test and their marks out of 10 are 4, 6, 6, 7, 7, 7, 8 and 10. Find the mean mark.

Solution The sum of the marks is $4+6+6+7+7+7+8+10=55$. The total number of values is 8. Therefore,

$$\text{mean} = \frac{\text{sum of the values}}{\text{total number of values}} = \frac{55}{8} = 6.875$$

The examiner can quote 6.875 out of 10 as the 'average mark' of the group of students.

30.2 In a hospital a patient's body temperature is recorded every hour for six hours. Find the mean temperature over the six-hour period if the six temperatures, in °C, were

36.5 36.8 36.9 36.9 36.9 37.0

Solution To find the mean temperature the sum of all six values is found and the result is divided by 6. That is,

$$\text{mean} = \frac{36.5+36.8+36.9+36.9+36.9+37.0}{6}$$

$$= \frac{221.0}{6} = 36.83\ ^\circ\text{C}$$

In more advanced work we make use of a formula for the mean that requires knowledge of some special notation. Suppose we have n values and we call these x_1, x_2, x_3 and so on up to x_n. The mean of these values is

INTRODUCTION TO MATHEMATICAL THINKING
A publication from Pearson Custom Publishing

given the symbol $\bar{x}$, pronounced 'x bar'. To calculate the mean we must add up these values and divide by n, that is

$$\text{mean} = \bar{x} = \frac{x_1 + x_2 + x_3 + \cdots + x_n}{n}$$

A notation is often used to shorten this formula. In mathematics, the Greek letter sigma, written $\sum$, stands for a sum. The sum $x_1 + x_2$ is written

$$\sum_{i=1}^{2} x_i$$

and the sum $x_1 + x_2 + x_3 + x_4$ is written

$$\sum_{i=1}^{4} x_i$$

Note that i runs through all integer values from 1 to n.

where the values below and above the sigma sign give the first and last values in the sum. Similarly $x_1 + x_2 + x_3 + \cdots + x_n$ is written $\sum_{i=1}^{n} x_i$. Using this notation, the formula for the mean can be written in the following way:

Key point

$$\text{mean} = \bar{x} = \frac{\sum_{i=1}^{n} x_i}{n}$$

There is a further treatment of sigma notation in Chapter 12, §12.5.

WORKED EXAMPLES

30.3 Express the following in sigma notation:

(a) $x_1 + x_2 + x_3 + \cdots + x_8 + x_9$ (b) $x_{10} + x_{11} + \cdots + x_{100}$

Solution (a) $x_1 + x_2 + x_3 + \cdots + x_8 + x_9 = \sum_{i=1}^{9} x_i$

(b) $x_{10} + x_{11} + \cdots + x_{100} = \sum_{i=10}^{100} x_i$

30.4 Find the mean of the values $x_1 = 5$, $x_2 = 7$, $x_3 = 13$, $x_4 = 21$ and $x_5 = 29$.

Solution The number of values equals 5, so we let $n = 5$. The sum of the values is

$$\sum_{i=1}^{5} x_i = x_1 + x_2 + x_3 + x_4 + x_5 = 5 + 7 + 13 + 21 + 29 = 75$$

The mean is

$$\text{mean} = \bar{x} = \frac{\sum_{i=1}^{5} x_i}{n} = \frac{75}{5} = 15$$

When the data is presented in the form of a frequency distribution the mean is found by first multiplying each data value by its frequency. The results are added. This is equivalent to adding up all the data values. The mean is found by dividing this sum by the sum of all the frequencies. Note that the sum of the frequencies is equal to the total number of values. Consider the following example.

WORKED EXAMPLE

30.5 Thirty-eight students sit a mathematics test and their marks out of 10 are shown in Table 30.1. Find the mean mark.

Solution Each data value, in this case the mark, is multiplied by its frequency, and the results are added. This is equivalent to adding up all the 38 individual marks. This is shown in Table 30.2.

Table 30.1
Marks of 38 students in a test

Mark, m	**Frequency, f**
0	0
1	0
2	1
3	0
4	1
5	7
6	16
7	8
8	3
9	1
10	1
Total	38

Table 30.2
Marks of 38 students multiplied by frequency

Mark, m	**Frequency, f**	**$m \times f$**
0	0	0
1	0	0
2	1	2
3	0	0
4	1	4
5	7	35
6	16	96
7	8	56
8	3	24
9	1	9
10	1	10
Totals	38	236

Note that the sum of all the frequencies is equal to the number of students taking the test. The number 236 is equal to the sum of all the individual marks. The mean is found by dividing this sum by the sum of all the frequencies:

$$\text{mean} = \frac{236}{38} = 6.21$$

The mean mark is 6.21 out of 10.

Using the sigma notation the formula for the mean mark of a frequency distribution with N classes, where the frequency of value x_i is f_i, becomes

Key point

$$\text{mean} = \bar{x} = \frac{\sum_{i=1}^{N} f_i \times x_i}{\sum_{i=1}^{N} f_i}$$

Note that $\sum_{i=1}^{N} f_i = n$; that is, the sum of all the frequencies equals the total number of values.

When the data is in the form of a grouped distribution the class midpoint is used to calculate the mean. Consider the following example.

WORKED EXAMPLE

30.6 The heights, to the nearest centimetre, of 100 students are given in Table 30.3. Find the mean height.

Solution Because the actual heights of students in each class are not known we use the midpoint of the class as an estimate. The midpoint of the class 164–165 is 164.5. Other midpoints and the calculation of the mean are shown in Table 30.4.

Then

$$\text{mean} = \bar{x} = \frac{\sum_{i=1}^{N} f_i \times x_i}{\sum_{i=1}^{N} f_i} = \frac{17150}{100} = 171.5$$

The mean height is 171.5 cm.

Table 30.3
Heights of 100 students

Height (cm)	Frequency
164–165	4
166–167	8
168–169	10
170–171	27
172–173	30
174–175	10
176–177	6
178–179	5
Total	100

Table 30.4
Heights of 100 students with midpoints multiplied by frequency

Height (cm)	Frequency f_i	Midpoint x_i	$f_i \times x_i$
164–165	4	164.5	658.0
166–167	8	166.5	1332.0
168–169	10	168.5	1685.0
170–171	27	170.5	4603.5
172–173	30	172.5	5175.0
174–175	10	174.5	1745.0
176–177	6	176.5	1059.0
178–179	5	178.5	892.5
Total	100		17150.0

The median

A second average that also typifies a set of data is the **median**.

Key point

The **median** of a set of numbers is found by listing the numbers in ascending order and then selecting the value that lies halfway along the list.

WORKED EXAMPLE

30.7 Find the median of the numbers

1 2 6 7 9 11 11 11 14

Solution The set of numbers is already given in order. The number halfway along the list is 9, because there are four numbers before it and four numbers after it in the list. Hence the median is 9.

When there is an even number of values, the median is found by taking the mean of the two middle values.

WORKED EXAMPLE

30.8 Find the median of the following salaries: £24,000, £12,000, £16,000, £22,000, £10,000 and £25,000.

Solution The numbers are first arranged in order as £10,000, £12,000, £16,000, £22,000, £24,000 and £25,000. Because there is an even number of values there are two middle figures: £16,000 and £22,000. The mean of these is

$$\frac{16{,}000 + 22{,}000}{2} = 19{,}000$$

The median salary is therefore £19,000.

The mode

A third average is the **mode**.

Key point

The **mode** of a set of values is that value that occurs most often.

WORKED EXAMPLE

30.9 Find the mode of the set of numbers

1 1 4 4 5 6 8 8 8 9

Solution The number that occurs most often is 8, which occurs three times. Therefore 8 is the mode. Usually a mode is quoted when we want to represent the most popular value in a set.

Sometimes a set of data may have more than one mode.

WORKED EXAMPLE

30.10 Find the mode of the set of numbers

20 20 21 21 21 48 48 49 49 49

Solution In this example there is no single value that occurs most frequently. The number 21 occurs three times, but so does the number 49. This set has two modes. The data is said to be **bimodal**.

Self-assessment questions 30.2

1. State the three different types of average commonly used in statistical calculations.
2. In an annual report, an employer of a small firm claims that the median salary for the workforce is £18,500. However, over discussions in the canteen it is apparent that no worker earns this amount. Explain how this might have arisen.

Exercise 30.2

MyMathLab Global

1. Find the mean, median and mode of the following set of values: 2, 3, 5, 5, 5, 5, 8, 8, 9.
2. Find the mean of the set of numbers 1, 1, 1, 1, 1, 1, 256. Explain why the mean does not represent the data adequately. Which average might it have been more appropriate to use?
3. The marks of seven students in a test were 45%, 83%, 99%, 65%, 68%, 72% and 66%. Find the mean mark and the median mark.
4. Write out fully each of the following expressions:

   (a) $\sum_{i=1}^{4} x_i$ (b) $\sum_{i=1}^{7} x_i$

   (c) $\sum_{i=1}^{3} (x_i - 3)^2$ (d) $\sum_{i=1}^{4} (2 - x_i)^2$
5. Write the following concisely using sigma notation:

   (a) $x_3 + x_4 + x_5 + x_6$

   (b) $(x_1 - 1) + (x_2 - 1) + (x_3 - 1)$

6. Calculate the mean, median and mode of the following numbers: 1.00, 1.15, 1.25, 1.38, 1.39 and 1.40.

7. The prices of the eight executive homes advertised by a local estate agent are

   £290,000 £375,000 £325,000 £299,950
   £319,950 £327,500 £299,500 £329,500

   Find the mean price of these houses.

8. The data in the table gives the radius of 20 ball bearings. Find the class midpoints and hence calculate the average radius.

Radius (mm)	Frequency
20.56–20.58	3
20.59–20.61	6
20.62–20.64	8
20.65–20.67	3

30.3 The variance and standard deviation

Suppose we consider the test marks of two groups of three students. Suppose that the marks out of 10 for the first group are

4 7 and 10

while those of the second group are

7 7 and 7

It is easy to calculate the mean mark of each group: the first group has mean mark

$$\frac{4+7+10}{3}=\frac{21}{3}=7$$

The second group has mean mark

$$\frac{7+7+7}{3}=\frac{21}{3}=7$$

We see that both groups have the same mean mark even though the marks in the first group are widely spread, whereas the marks in the second group are all the same. If the teacher quotes just the mean mark of each group this gives no information about how widely the marks are spread. The **variance** and **standard deviation** are important and widely used statistical quantities that contain this information. Most calculators are pre-programmed to calculate these quantities. Once you understand the processes involved, check to see if your calculator can be used to find the variance and standard deviation of a set of data.

Suppose we have n values x_1, x_2, x_3 up to x_n. Their mean is $\bar{x}$ given by

$$\bar{x} = \frac{\sum_{i=1}^{n} x_i}{n}$$

The variance is found from the following formula:

Key point

$$\text{variance} = \frac{\sum_{i=1}^{n} (x_i - \bar{x})^2}{n}$$

If you study this carefully you will see that to calculate the variance we must:

- calculate the mean value $\bar{x}$
- subtract the mean from each value in turn, that is find $x_i - \bar{x}$
- square each answer to get $(x_i - \bar{x})^2$
- add up all these squared quantities to get $\sum_{i=1}^{n} (x_i - \bar{x})^2$
- divide the result by n to get

$$\frac{\sum_{i=1}^{n} (x_i - \bar{x})^2}{n}$$

which is the variance

The standard deviation is found by taking the square root of the variance:

Key point

$$\text{standard deviation} = \sqrt{\frac{\sum_{i=1}^{n} (x_i - \bar{x})^2}{n}}$$

Let us calculate the variance and standard deviation of each of the two sets of marks 4, 7, 10 and 7, 7, 7. We have already noted that the mean of each set is 7. Each stage of the calculation of the variance of the set 4, 7, 10 is shown in Table 30.5. The mean is subtracted from each number, and the results are squared and then added. Note that when a negative number is squared the result is positive. The squares are added to give 18. In this

Table 30.5

x_i	$x_i - \bar{x}$	$(x_i - \bar{x})^2$
4	$4 - 7 = -3$	$(-3)^2 = 9$
7	$7 - 7 = 0$	$0^2 = 0$
10	$10 - 7 = 3$	$3^2 = 9$
Total		18

example, the number of values equals 3. So, taking $n = 3$,

$$\text{variance} = \frac{\sum_{i=1}^{n}(x_i - \bar{x})^2}{n} = \frac{18}{3} = 6$$

The standard deviation is the square root of the variance, that is $\sqrt{6} = 2.449$. Similarly, calculation of the variance and standard deviation of the set 7, 7, 7 is shown in Table 30.6. Again, n equals 3. So

$$\text{variance} = \frac{\sum_{i=1}^{n}(x_i - \bar{x})^2}{n} = \frac{0}{3} = 0$$

Table 30.6

x_i	$x_i - \bar{x}$	$(x_i - \bar{x})^2$
7	$7 - 7 = 0$	$0^2 = 0$
7	$7 - 7 = 0$	$0^2 = 0$
7	$7 - 7 = 0$	$0^2 = 0$
Total		0

The standard deviation is the square root of the variance and so it also equals zero. The fact that the standard deviation is zero reflects that there is no spread of values. By comparison it is easy to check that the standard deviation of the set 6, 7 and 8, which also has mean 7, is equal to 0.816. The fact that the set 4, 7 and 10 has standard deviation 2.449 shows that the values are more widely spread than in the set 6, 7 and 8.

WORKED EXAMPLE

30.11 Find the variance and standard deviation of 10, 15.8, 19.2 and 8.7.

Solution First the mean is found:

$$\bar{x} = \frac{10 + 15.8 + 19.2 + 8.7}{4} = \frac{53.7}{4} = 13.425$$

The calculation to find the variance is given in Table 30.7.

$$\text{variance} = \frac{\sum_{i=1}^{n}(x_i - \bar{x})^2}{n} = \frac{73.049}{4} = 18.262$$

The standard deviation is the square root of the variance:

$\sqrt{18.262} = 4.273$

Table 30.7

x_i	$x_i - \bar{x}$	$(x_i - \bar{x})^2$
10	$10 - 13.425 = -3.425$	$(-3.425)^2 = 11.731$
15.8	$15.8 - 13.425 = 2.375$	$2.375^2 = 5.641$
19.2	$19.2 - 13.425 = 5.775$	$5.775^2 = 33.351$
8.7	$8.7 - 13.425 = -4.725$	$(-4.725)^2 = 22.326$
Total		73.049

When dealing with a grouped frequency distribution with N classes the formula for the variance becomes

Key point

$$\text{variance} = \frac{\sum_{i=1}^{N} f_i(x_i - \bar{x})^2}{\sum_{i=1}^{N} f_i}$$

As before, the standard deviation is the square root of the variance.

WORKED EXAMPLE

30.12 In a period of 30 consecutive days in July the temperature in °C was recorded as follows:

18	19	20	23	24	24	21	18	17	16
16	17	17	17	18	19	20	20	22	23
24	24	25	23	21	21	20	19	19	18

(a) Produce a grouped frequency distribution showing data grouped from 16 to 17 °C, 18 to 19 °C and so on.
(b) Find the mean temperature of the grouped data.
(c) Find the standard deviation of the grouped data.

Solution (a) The data is grouped using a tally chart as in Table 30.8.

Table 30.8

Temperature range (°C)	Tally	Frequency
16–17	~~////~~ /	6
18–19	~~////~~ ///	8
20–21	~~////~~ //	7
22–23	////	4
24–25	~~////~~	5
Total		30

(b) When calculating the mean temperature from the grouped data we do not know the actual temperatures. We do know the frequency of each class. The best we can do is use the midpoint of each class as an estimate of the values in that class. The class midpoints and the calculation to determine the mean are shown in Table 30.9. The mean is then

$$\bar{x} = \frac{\sum_{i=1}^{N} f_i \times x_i}{\sum_{i=1}^{N} f_i} = \frac{603}{30} = 20.1\ ^\circ\text{C}$$

Table 30.9

Temperature range (°C)	Frequency f_i	Class midpoint x_i	$f_i \times x_i$
16–17	6	16.5	99.0
18–19	8	18.5	148.0
20–21	7	20.5	143.5
22–23	4	22.5	90.0
24–25	5	24.5	122.5
Total	30		603

(c) To find the variance and hence the standard deviation we must subtract the mean, 20.1, from each value, square and then add the results. Finally this sum is divided by 30. The complete calculation is shown in Table 30.10.

Finally the variance is given by

Table 30.10

Temperature range (°C)	Frequency f_i	Class midpoint x_i	$x_i - \bar{x}$	$(x_i - \bar{x})^2$	$f_i(x_i - \bar{x})^2$
16–17	6	16.5	−3.60	12.96	77.76
18–19	8	18.5	−1.60	2.56	20.48
20–21	7	20.5	0.40	0.16	1.12
22–23	4	22.5	2.40	5.76	23.04
24–25	5	24.5	4.40	19.36	96.80
Total	30				219.20

$$\text{variance} = \frac{\sum_{i=1}^{N} f_i(x_i - \bar{x})^2}{\sum_{i=1}^{N} f_i} = \frac{219.20}{30} = 7.307$$

The standard deviation is the square root of the variance, that is $\sqrt{7.307} = 2.703$.

Self-assessment questions 30.3

1. Why is an average often an insufficient way of describing a set of values?
2. Describe the stages involved in calculating the variance and standard deviation of a set of values $x_1, x_2, \ldots, x_n$.

Exercise 30.3

MyMathLab Global

1. Find the standard deviation of each of the following sets of numbers and comment upon your answers:
   (a) 20, 20, 20, 20 (b) 16, 17, 23, 24
   (c) 0, 20, 20, 40
2. Calculate the variance and standard deviation of the following set of numbers: 1, 2, 3, 3, 3, 4, 7, 8, 9, 10.
3. The examination results of two students, Jane and Tony, are shown in the table. Find the mean and standard deviation of each of them. Comment upon the results.

	Jane	Tony
Maths	50	29
Physics	42	41
Chemistry	69	60
French	34	48
Spanish	62	80

4. The marks of 50 students in a mathematics examination are given as

45	50	62	62	68	47	45	44	48
73	62	63	62	67	80	45	41	40
23	55	21	83	67	49	48	48	62
63	79	58	71	37	32	58	54	50
62	66	68	62	81	92	62	45	49
71	72	70	49	51				

   (a) Produce a grouped frequency distribution using classes 25–29, 30–34, 35–39 and so on.
   (b) The modal class is the class with the highest frequency. State the modal class.
   (c) Calculate the mean, variance and standard deviation of the grouped frequency distribution.

Challenge Exercises 30

1. Show that

$$\sum (x_i - \bar{x})^2 = \sum x_i^2 - n\bar{x}^2$$

2. The data set, X, comprises n items, $x_1, x_2, x_3, \ldots x_n$ with mean $\bar{x}$ and standard deviation σ_x. Each data point, x_i, is linearly transformed to produce a new data point, y_i with

$$y_i = a + bx_i \qquad i = 1, 2, 3, \ldots, n$$

where a and b are constants.

(a) If $\bar{y}$ is the mean of the new data set show that

$$\bar{y} = a + b\bar{x}$$

(b) If σ_y is the standard deviation of the new data set show that

$$\sigma_y = b\sigma_x$$

Test and assignment exercises 30

1. Calculate the mean, median and mode of the following numbers: 2.7, 2.8, 3.1, 3.1, 3.1, 3.4, 3.8, 4.1.

2. Calculate the mean of
   (a) the first 10 whole numbers,

   $$1, 2, \ldots, 10$$

   (b) the first 12 whole numbers,

   $$1, 2, \ldots, 12$$

   (c) the first n whole numbers,

   $$1, 2, \ldots, n$$

3. A manufacturer of breakfast cereals uses two machines that pack 250 g packets of cereal automatically. In order to check the average weight, sample packets are taken and weighed to the nearest gram. The results of checking eight packets from each of the two machines are given in the table.

Machine								
1	250	250	251	257	253	250	268	259
2	249	251	250	251	252	258	254	250

   (a) Find the mean weight of each sample of eight packets.
   (b) Find the standard deviation of each sample.
   (c) Comment upon the performance of the two machines.

4. What is the median of the five numbers 18, 28, 39, 42, 43? If 50 is added as the sixth number what would the median become?

5. Express the following sums in sigma notation:

   (a) $y_1 + y_2 + y_3 + \cdots + y_7 + y_8$
   (b) $y_1^2 + y_2^2 + y_3^2 + \cdots + y_7^2 + y_8^2$
   (c) $(y_1 - \bar{y})^2 + (y_2 - \bar{y})^2 + \cdots + (y_7 - \bar{y})^2 + (y_8 - \bar{y})^2$

6. The ages of people attending a theatre performance were recorded as follows:

Age (years)	Frequency
0–9	0
10–19	6
20–29	24
30–39	42
40–49	59
50–59	53
60–69	17
70–79	8
80–89	3

(a) Calculate the mean age of the audience.
(b) Calculate the standard deviation of the distribution.

29 Tables and charts

Objectives: This chapter:

- explains the distinction between discrete and continuous data
- shows how raw data can be organised using a tally chart
- explains what is meant by a frequency distribution and a relative frequency distribution
- shows how data can be represented in the form of bar charts, pie charts, pictograms and histograms

29.1 Introduction to data

In the modern world, information from a wide range of sources is gathered, presented and interpreted. Most newspapers, television news programmes and documentaries contain vast numbers of facts and figures concerning almost every aspect of life, including environmental issues, the lengths of hospital waiting lists, crime statistics and economic performance indicators. Information such as this, which is gathered by carrying out surveys and doing research, is known as **data**.

Sometimes data must take on a value from a specific set of numbers, and no other values are possible. For example, the number of children in a family must be 0, 1, 2, 3, 4 and so on, and no intermediate values are allowed. It is impossible to have 3.32 children, say. Such data is said to be **discrete**. Other examples of discrete data include:

- the number of car thefts in a city in one week – this must be 0, 1, 2, 3 and so on. You cannot have three-and-a-half thefts.
- shoe sizes – these can be ..., $3\frac{1}{2}$, 4, $4\frac{1}{2}$, 5, $5\frac{1}{2}$, 6, $6\frac{1}{2}$, 7 and so on. You cannot have a shoe size of 9.82.

Sometimes data can take on *any* value within a specified range. Such data is called **continuous**. The volume of liquid in a 1 litre jug can take any value from 0 to 1 litre. The lifespan of an electric light bulb could be any non-negative value.

Frequently data is presented in the form of tables and charts, the intention being to make the information readily understandable. When data is first collected, and before it is processed in any way, it is known as **raw data**. For example, in a test the mathematics marks out of 10 of a group of 30 students are

7	5	5	8	9	10	9	10	7	8	6	3	5	9	6
10	8	8	7	8	6	7	8	8	10	9	4	5	9	8

This is raw data. To try to make sense of this data it is helpful to find out how many students scored each particular mark. This can be done by means of a **tally chart**. To produce this we write down in a column all the possible marks, in order. We then go through the raw data and indicate the occurrence of each mark with a vertical line or tally like /. Every time a fifth tally is recorded this is shown by striking through the previous four, as in ~~////~~. This makes counting up the tallies particularly easy. The tallies for all the marks are shown in Table 29.1. The number of occurrences of each mark is called its **frequency**, and the frequencies can be found from the number of tallies. We see that the tally chart is a useful way of organising the raw data into a form that will enable us to answer questions and obtain useful information about it.

Table 29.1
Tally chart for mathematics marks

Mark	Tally	Frequency
0		0
1		0
2		0
3	/	1
4	/	1
5	////	4
6	///	3
7	////	4
8	~~////~~ ///	8
9	~~////~~	5
10	////	4
Total		30

Nowadays much of the tedium of producing tally charts for very large amounts of data can be avoided using computer programs available specifically for this purpose.

Self-assessment questions 29.1

1. Explain the distinction between discrete data and continuous data. Give a new example of each.
2. Explain the purpose of a tally chart.

Exercise 29.1

MyMathLab Global

1. State whether each of the following is an example of discrete data or continuous data:
   (a) the number of matches found in a matchbox
   (b) the percentage of sulphur dioxide found in an air sample taken above a city
   (c) the number of peas in a packet of frozen peas
   (d) the radius of a ball bearing
   (e) the weight of a baby
   (f) the number of students in a particular type of accommodation
   (g) the weight of a soil sample
   (h) the temperature of a patient in hospital
   (i) the number of telephone calls received by an answering machine during one day
2. Twenty-five people were asked to state their year of birth. The information given was

   1975 1975 1974 1974 1975 1976 1976 1974 1972 1973 1973 1970 1975
   1975 1973 1970 1975 1974 1975 1974 1970 1973 1974 1974 1975

   Produce a tally chart to organise this data and state the frequency of occurrence of each year of birth.

29.2 Frequency tables and distributions

Data is often presented in a **table**. For example, in a recent survey, 1000 students were asked to state the type of accommodation in which they lived during term time. The data gathered is given in Table 29.2. As we saw from the work in the previous section, the number of occurrences of each entry in the table is known as its **frequency**. So, for example, the frequency of students in private rented accommodation is 250. The table summarises the various frequencies and is known as a **frequency distribution**.

A **relative frequency distribution** is found by expressing each frequency as a proportion of the total frequency.

Table 29.2
Frequency distribution

Type of accommodation	Number of students
Hall of residence	675
Parental home	50
Private rented	250
Other	25
Total	1000

Key point

The relative frequency is found by dividing a frequency by the total frequency.

In this example, because the total frequency is 1000, the relative frequency is easy to calculate. The number of students living at their parental home is 50 out of a total of 1000. Therefore the relative frequency of this group of students is $\frac{50}{1000}$ or 0.050. Table 29.3 shows the relative frequency distribution of student accommodation. Note that the relative frequencies always sum to 1.

Table 29.3
Relative frequency distribution

Type of accommodation	Number of students	Relative frequency
Hall of residence	675	0.675
Parental home	50	0.050
Private rented	250	0.250
Other	25	0.025
Total	1000	1.000

It is sometimes helpful to express the frequencies as percentages. This is done by multiplying each relative frequency by 100, which gives the results shown in Table 29.4. We see from this that 5% of students questioned live at their parental home.

Table 29.4
Frequencies expressed as percentages

Type of accommodation	Number of students	Relative frequency × 100
Hall of residence	675	67.5%
Parental home	50	5.0%
Private rented	250	25.0%
Other	25	2.5%
Total	1000	100%

Key point

A frequency can be expressed as a percentage by multiplying its relative frequency by 100.

WORKED EXAMPLE

29.1

In 1988 there were 4320 areas throughout the world that were designated as national parks. Table 29.5 is a frequency distribution showing how these parks are distributed in different parts of the world.

(a) Express this data as a relative frequency distribution and also in terms of percentages.

(b) What percentage of the world's national parks are in Africa?

Table 29.5
The number of national parks throughout the world

Region	Number of national parks
Africa	486
N. America	587
S. America	315
Former Soviet Union	168
Asia	960
Europe	1032
Oceania	767
Antarctica	5
Total	4320

Solution

Table 29.6
The number of national parks throughout the world

Region	Number	Relative frequency	Rel. freq. × 100
Africa	486	0.1125	11.25%
N. America	587	0.1359	13.59%
S. America	315	0.0729	7.29%
Former Soviet Union	168	0.0389	3.89%
Asia	960	0.2222	22.22%
Europe	1032	0.2389	23.89%
Oceania	767	0.1775	17.75%
Antarctica	5	0.0012	0.12%
Total	4320	1.000	100%

(a) The relative frequencies are found by expressing each frequency as a proportion of the total. For example, Europe has 1032 national parks out of a total of 4320, and so its relative frequency is $\frac{1032}{4320} = 0.2389$. Figures for the other regions have been calculated and are shown in Table 29.6. To express a relative frequency as a percentage it is multiplied by 100. Thus we find that Europe has 23.89% of the world's national parks. The percentages for the other regions are also shown in Table 29.6.

(b) We note from Table 29.6 that 11.25% of the world's national parks are in Africa.

When dealing with large amounts of data it is usual to group the data into classes. For example, in an environmental experiment the weights in grams of 30 soil samples were measured using a balance. We could list all 30 weights here but such a long list of data is cumbersome. Instead, we have already grouped the data into several weight ranges or **classes**. Table 29.7 shows the number of samples in each class. Such a table is called a **grouped frequency distribution**.

Table 29.7
Grouped frequency distribution

Weight range (grams)	Number of samples in that range
100–109	5
110–119	8
120–129	11
130–139	5
140–149	1
Total	30

A disadvantage of grouping the data in this way is that information about individual weights is lost. However, this disadvantage is usually outweighed by having a more compact set of data that is easier to work with. There are five samples that lie in the first class, eight samples in the second and so on. For the first class, the numbers 100 and 109 are called **class limits**, the smaller number being the **lower class limit** and the larger being the **upper class limit**. Theoretically, the class 100–109 includes all weights from 99.5 g up to but not including 109.5 g. These numbers are called the **class boundaries**. In practice the class boundaries are found by adding the upper class limit of one class to the lower limit of the next class and dividing by 2. The **class width** is the difference between the larger class boundary and the smaller. The class 100–109 has width $109.5 - 99.5 = 10$. We do not know the actual values of the weights in each class. Should we require an estimate, the best we can do is use the value in the middle of the class, known as the class **midpoint**. The midpoint of the first class is 104.5, the midpoint of the second class is 114.5 and so on. The midpoint can be found by adding half the class width to the lower class boundary.

WORKED EXAMPLE

29.2

The systolic blood pressure in millimetres of mercury of 20 workers was recorded to the nearest millimetre. The data collected was as follows:

121 123 124 129 130 119 129 124 119 121 122 124
124 128 129 136 120 119 121 136

(a) Group this data into classes 115–119, 120–124, and so on.
(b) State the class limits of the first three classes.
(c) State the class boundaries of the first three classes.
(d) What is the class width of the third class?
(e) What is the midpoint of the third class?

Solution

(a) To group the data a tally chart is used as shown in Table 29.8.

Table 29.8
Tally chart for blood pressures

Pressure (mm mercury)	Tally	Frequency
115–119	///	3
120–124	~~////~~ ~~////~~	10
125–129	////	4
130–134	/	1
135–139	//	2
Total		20

(b) The class limits of the first class are 115 and 119; those of the second class are 120 and 124; those of the third are 125 and 129.

(c) Theoretically, the class 115–119 will contain all blood pressures between 114.5 and 119.5 and so the class boundaries of the first class are 114.5 and 119.5. The class boundaries of the second class are 119.5 and 124.5, and those of the third class are 124.5 and 129.5.

(d) The upper class boundary of the third class is 129.5. The lower class boundary is 124.5. The difference between these values gives the class width, that is $129.5 - 124.5 = 5$.

(e) The midpoint of the third class is found by adding half the class width to the lower class boundary, that is $2.5 + 124.5 = 127$.

Self-assessment questions 29.2

1. Why is it often useful to present data in the form of a *grouped* frequency distribution?
2. Explain the distinction between class boundaries and class limits.
3. How is the class width calculated?
4. How is the class midpoint calculated?

Exercise 29.2

MyMathLab Global

1. The age of each patient over the age of 14 visiting a doctor's practice in one day was recorded as follows:

18	18	76	15	72	45	48	62	21
27	45	43	28	19	17	37	35	34
23	25	46	56	32	18	23	34	32
56	29	43						

   (a) Use a tally chart to produce a grouped frequency distribution with age groupings 15–19, 20–24, 25–29, and so on.
   (b) What is the relative frequency of patients in the age group 70–74?
   (c) What is the relative frequency of patients aged 70 and over?
   (d) Express the number of patients in the age group 25–29 as a percentage.

2. Consider the following table, which shows the lifetimes, to the nearest hour, of 100 energy saving light bulbs.

   (a) If a light bulb has a life of 5499.4 hours, into which class will it be put?
   (b) If a light bulb has a life of 5499.8 hours, into which class will it be put?
   (c) State the class boundaries of each class.
   (d) Find the class width.
   (e) State the class midpoint of each class.

Lifetime (hours)	Frequency
5000–5499	6
5500–5999	32
6000–6499	58
6500–6999	4
Total	100

3. The percentages of 60 students in an Information Technology test are given as follows:

45	92	81	76	51	46	82	65	61
19	62	58	72	65	66	97	61	57
63	93	61	46	47	61	56	45	39
47	55	58	81	71	52	38	53	59
82	92	87	86	51	29	19	79	55
18	53	29	87	87	77	85	67	89
17	29	86	57	59	57			

   Form a frequency distribution by drawing up a tally chart using classes 0–9, 10–19 and so on.

4. The data in the table opposite gives the radius of 20 ball bearings. State the class boundaries of the second class and find the class width.

Radius (mm)	Frequency
20.56–20.58	3
20.59–20.61	6
20.62–20.64	8
20.65–20.67	3

29.3 Bar charts, pie charts, pictograms and histograms

In a **bar chart** information is represented by rectangles or bars. The bars may be drawn horizontally or vertically. The length of each bar corresponds to a frequency.

The data given earlier in Table 29.2 concerning student accommodation is presented in the form of a horizontal bar chart in Figure 29.1. Note that the scale on the horizontal axis must be uniform, that is it must be evenly spaced, and that the type of accommodation is clearly identified on each bar. The bar chart must be given a title to explain the information that is being presented.

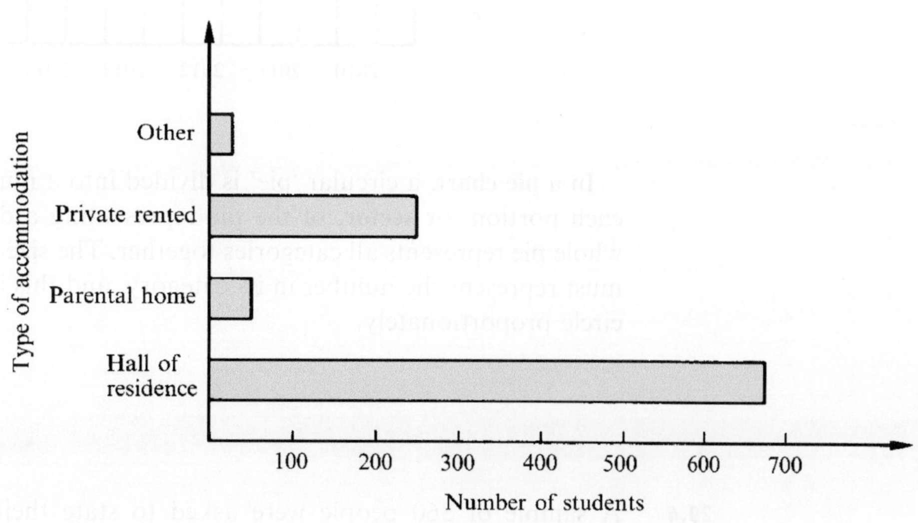

Figure 29.1 Horizontal bar chart showing student accommodation

WORKED EXAMPLE

29.3 A reproduction antique furniture manufacturer has been producing pine wardrobes for the past six years. The number of wardrobes sold each year is given in Table 29.9. Produce a vertical bar chart to illustrate this information.

Table 29.9
Wardrobes sold

Year	Number sold
2010	15
2011	16
2012	20
2013	20
2014	24
2015	30

Solution The information is presented in the form of a vertical bar chart in Figure 29.2.

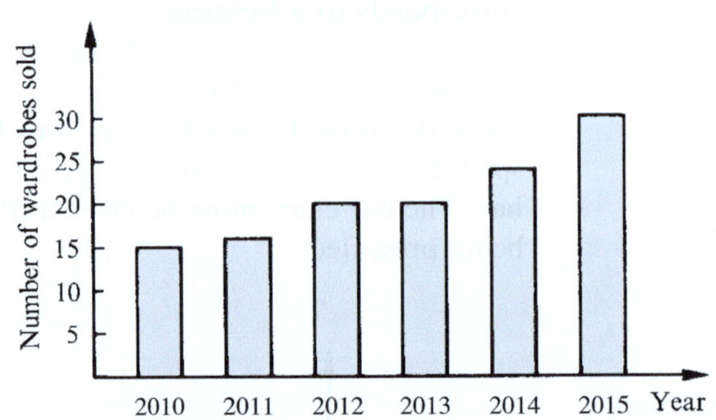

Figure 29.2
Vertical bar chart showing the number of wardrobes sold each year

In a **pie chart**, a circular 'pie' is divided into a number of portions, with each portion, or **sector**, of the pie representing a different category. The whole pie represents all categories together. The size of a particular portion must represent the number in its category, and this is done by dividing the circle proportionately.

WORKED EXAMPLE

29.4 A sample of 360 people were asked to state their favourite flavour of potato crisp: 75 preferred 'plain' crisps, 120 preferred 'cheese and onion',

90 preferred 'salt and vinegar', and the remaining 75 preferred 'beef'. Draw a pie chart to present this data.

Solution

All categories together, that is all 360 people asked, make up the whole pie. Because there are 360° in a circle this makes the pie chart particularly easy to draw. A sector of angle 75° will represent 'plain crisps', a sector of 120° will represent 'cheese and onion' and so on. The pie chart is shown in Figure 29.3.

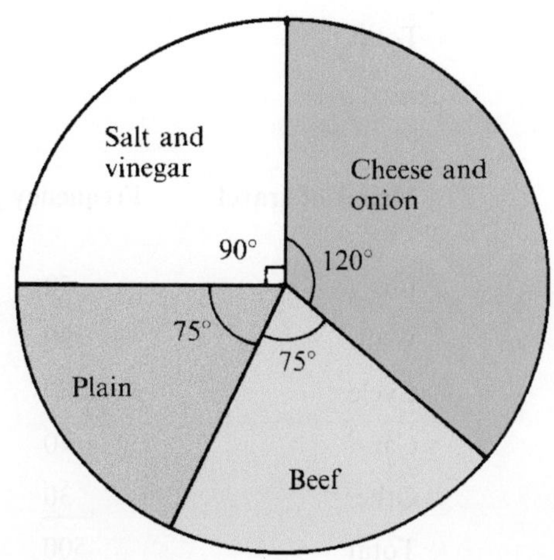

Figure 29.3 Pie chart showing preferred flavour of crisps of 360 people

---

The pie chart in the previous worked example was simple to draw because the total number of all categories was 360, the same as the number of degrees in a circle. Let us now see how to deal with a situation where this is not the case. First each relative frequency is found and then this is multiplied by 360. The result is the angle, in degrees, of the corresponding sector.

Key point

The angle, in degrees, of each sector in a pie chart is found by multiplying the corresponding relative frequency by 360. This gives the angle of each sector.

WORKED EXAMPLE

29.5 Five hundred students were asked to state how they usually travelled to and from their place of study. The results are given in Table 29.10. Present this information in a pie chart.

Solution

To find the angle of each sector of the pie chart we first find the corresponding relative frequencies. Recall that this is done by dividing each frequency by the total, 500. Then each relative frequency is multiplied by 360. The results are shown in Table 29.11. The table gives

Table 29.10
Means of travel

Means of travel	Frequency f
Bus	50
Walk	180
Cycle	200
Car	40
Other	30
Total	500

Table 29.11
Means of travel

Means of travel	Frequency f	Rel. freq. $f/500$	Rel. freq. $\times$ 360
Bus	50	$\frac{50}{500}$	$\frac{50}{500} \times 360 = 36.0°$
Walk	180	$\frac{180}{500}$	$\frac{180}{500} \times 360 = 129.6°$
Cycle	200	$\frac{200}{500}$	$\frac{200}{500} \times 360 = 144.0°$
Car	40	$\frac{40}{500}$	$\frac{40}{500} \times 360 = 28.8°$
Other	30	$\frac{30}{500}$	$\frac{30}{500} \times 360 = 21.6°$
Total	500		360°

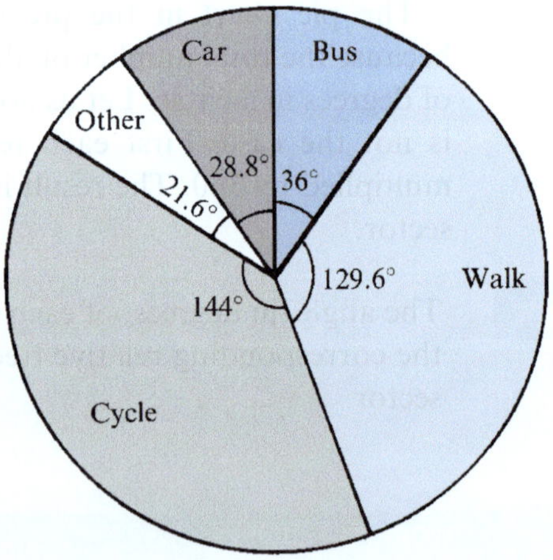

Figure 29.4
Pie chart showing how students travel to work

the required angles. The pie chart can then be drawn. This is shown in Figure 29.4.

A **pictogram** uses a picture to represent data in an eye-catching way. There are many ways to produce pictograms but generally the number of objects drawn represents the number of items in a particular category. This is a common form of representation popular on television when the viewer sees the information for only a few seconds.

WORKED EXAMPLE

29.6 The number of hours of sunshine in each of the four months January, April, July and October in the holiday resort of Torquay is given in Table 29.12. Show this information using a pictogram. Use one 'sun' to represent 30 hours of sunshine.

Table 29.12
Pictogram showing hours of sunshine.

Month	Number of hours
January	120
April	270
July	240
October	180

Solution The pictogram is shown in Figure 29.5.

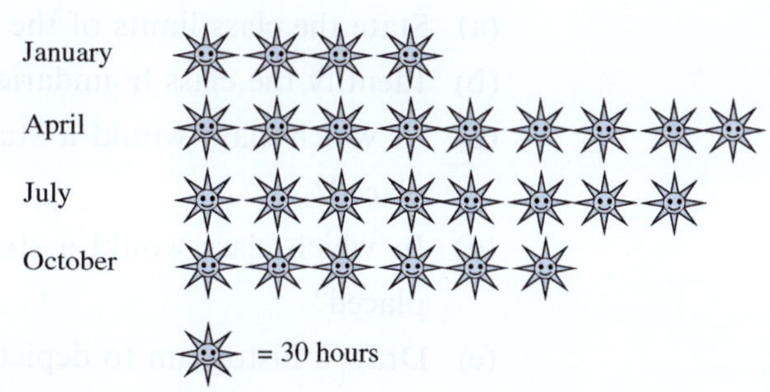

Figure 29.5
Pictogram showing hours of sunshine in each of four months.

A **histogram** is often used to illustrate data that is continuous or has been grouped. It is similar to a vertical bar chart in that it is drawn by constructing vertical rectangles. However, in a histogram it is the *area* of each rectangle, and not its length, that is proportional to the frequency of each class. When all class widths are the same this point is not important and you can assume that the length of a rectangle represents the frequency. Class boundaries and not class limits must be used on the horizontal axis to distinguish one class from the next. Consider the following example.

WORKED EXAMPLE

29.7 The heights, to the nearest centimetre, of 100 students are given in Table 29.13.

Table 29.13
Height of 100 students

Height (cm)	Frequency
164–165	4
166–167	8
168–169	10
170–171	27
172–173	30
174–175	10
176–177	6
178–179	5
Total	100

(a) State the class limits of the second class.

(b) Identify the class boundaries of the first class and the last class.

(c) In which class would a student whose actual height is 167.4 cm be placed?

(d) In which class would a student whose actual height is 167.5 cm be placed?

(e) Draw a histogram to depict this data.

Solution

(a) The lower class limit of the second class is 166 cm. The upper class limit is 167 cm.

(b) The class boundaries of the first class are 163.5 cm and 165.5 cm. The class boundaries of the last class are 177.5 cm and 179.5 cm.

(c) A student with height 167.4 cm will have his height recorded to the nearest centimetre as 167. He will be placed in the class 166–167.

(d) A student with height 167.5 cm will have her height recorded to the nearest centimetre as 168. She will be placed in the class 168–169.

(e) The histogram is shown in Figure 29.6. Note that the class boundaries are used on the horizontal axis to distinguish one class from the next. The area of each rectangle is proportional to the frequency of each class.

Figure 29.6
Histogram showing the distribution of heights of 100 students

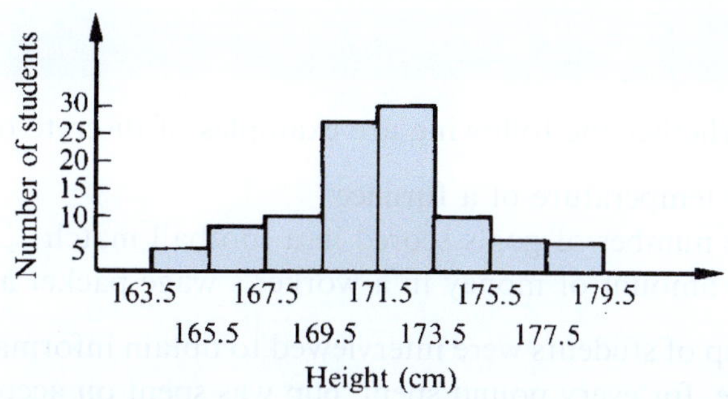

Self-assessment questions 29.3

1. There is an important distinction between a histogram and a vertical bar chart. Can you explain this?
2. Give one advantage and one disadvantage of using a pictogram to represent data.
3. On a histogram the class boundaries are used on the horizontal axis to distinguish one class from the next. Why do you think class boundaries as opposed to class limits are used?

Exercise 29.3

MyMathLab

1. A local council wants to produce a leaflet explaining how the council tax is spent. For every one pound spent, 40p went to Education, 15p went to the Police, 15p went to Cleansing and Refuse, 20p went to Highways and the remaining 10p went to a variety of other services. Show this information (a) using a pie chart, and (b) using a pictogram.
2. The composition of the atmosphere of the Earth is 78% nitrogen, 21% oxygen and 1% other gases. Draw a pie chart to show this information.
3. An insurance company keeps records on the distribution of the amount of small claims for theft it receives. Amounts are given to the nearest pound. This information is given in the following table.

Amount (£)	Frequency
0–99	5
100–199	8
200–299	16
300–399	24
400–499	26

(a) State the class boundaries.
(b) In which class would a claim for £199.75 be placed?
(c) Draw a histogram to show this information.

Test and assignment exercises 29

1. State whether the following are examples of discrete or continuous data:

   (a) the temperature of a furnace
   (b) the number of goals scored in a football match
   (c) the amount of money in a worker's wage packet at the end of the week

2. A group of students were interviewed to obtain information on their termly expenditure. On average, for every pound spent, 60p was spent on accommodation, fuel etc., 25p was spent on food, 10p was spent on entertainment and the remainder on a variety of other items. Draw a pie chart to show this expenditure.

3. Forty lecturers were asked to state which newspaper they bought most regularly. The results were: *Guardian*, 15; *Independent*, 8; *The Times*, 2; *Daily Express*, 7; *Daily Mail*, 2; None, 6. Produce a pie chart to show this data.

4. The percentage of total body weight formed by various parts of the body of a newborn baby is as follows: muscles, 26%; skeleton, 17%; skin, 20%; heart, lungs etc., 11%; remainder, 26%. Depict this information (a) on a pie chart, (b) using a bar chart.

5. The following marks were obtained by 27 students in a test. Draw up a frequency distribution by means of a tally chart.

   5 4 6 2 9 1 10 10 5 5 6 5 5 9 8 7 9 6 7 6 8 7 6 9 8 8 7

6. The number of cars sold by a major manufacturer of luxury cars in the last 10 years is given in the table below. Show this data on a horizontal bar chart.

Year	**Number**
2006	2800
2007	1500
2008	1425
2009	1300
2010	1560
2011	1870
2012	2100
2013	2340
2014	2610
2015	2780

7. The table below shows the frequency distribution of the lifetimes in hours of 100 light bulbs tested.

   (a) State the class limits of the first and second classes.

(b) State the class boundaries of all classes.
(c) Determine each class width and midpoint.
(d) Draw a histogram for this frequency distribution.

Lifetime (hours)	Frequency
600–699	13
700–799	25
800–899	32
900–999	10
1000–1099	15
1100–1199	5

8. In each year from 2010 to 2015 the numbers of visitors to a museum were respectively 12000, 13000, 14500, 18000, 18000 and 17500. Show this information in a pictogram.

SECTION 5

Transformations

Transformation and visualisation 10

Recall 10

Prepare for this chapter by attempting the following questions. If you have difficulty with a question, go to Pearson Places and download the Recall Worksheet.

1 Find the size of the unknown angle in each of the following.

(a)

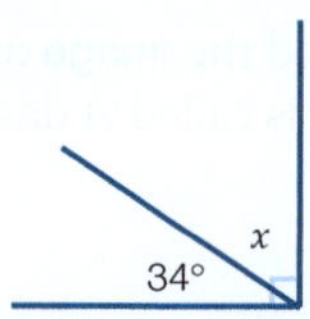

(b)

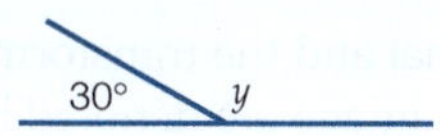

(c)

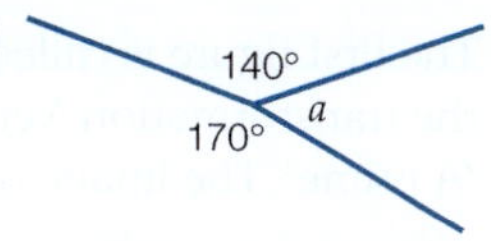

2 Find the size of the unknown angle in each of the following.

(a)

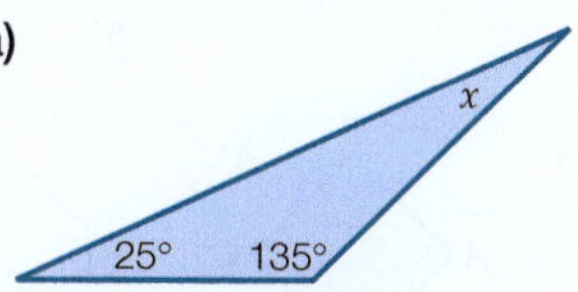

(b)

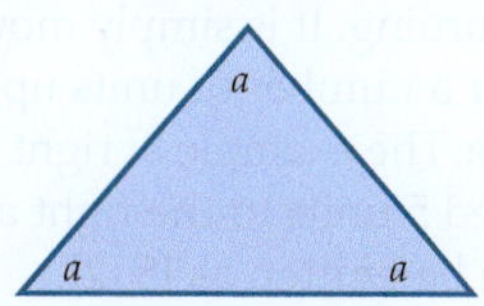

(c)

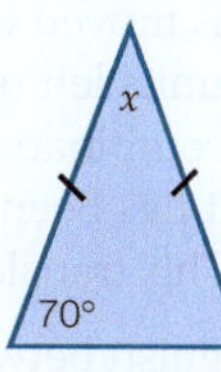

3 Find the size of the unknown angle in each of the following.

(a)

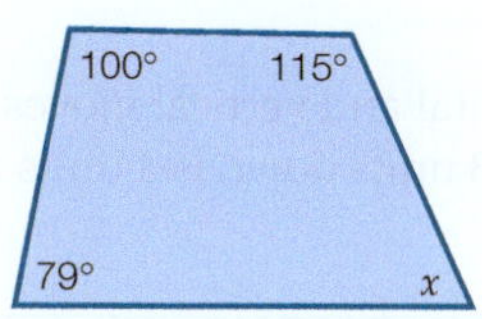

(b)

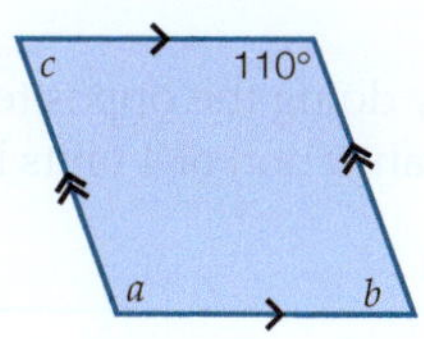

(c)

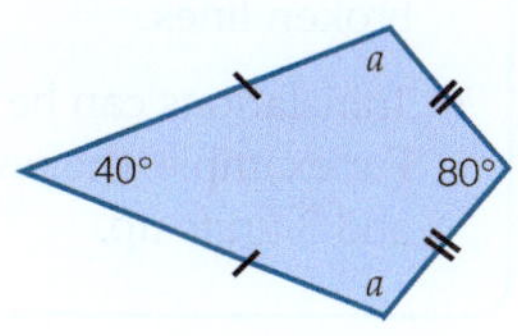

4 Draw an example of:

(a) a rhombus **(b)** a trapezium **(c)** an equilateral triangle

(d) a kite **(e)** an isosceles triangle **(f)** a parallelogram.

Mark all sides that are equal and/or parallel. Use letter names to name your figures.

Syllabus references

Number and Algebra: Linear Relationships

✓ Describe translations, reflections in an axis, and rotations of multiples of 90° on the Cartesian plane using coordinates **(ACMMG181)**

Measurement and Geometry: Volume

✓ Draw different views of prisms and solids formed from combinations of prisms **(ACMMG161)**

Measurement and Geometry: Properties of Geometrical Figures 1

✓ Identify line and rotational symmetries **(ACCMMG181)**

10.1 Translations

Need to Know

Transformations

A **transformation** changes a particular shape (or figure) or set of points according to a defined rule. A transformation may slide, flip, turn or change the shape or size of a figure. The transformations dealt with here are those that move a shape but retain its original shape and size. These transformations are translations, reflections and rotations.

The first figure is called the **original** and the transformed shape is called the **image** of the transformation. Vertices of an image are denoted by a dash; e.g. A' is called 'A dash' or 'A prime'. The image of an original figure $ABCD$ is written $A'B'C'D'$.

Translations

A **translation** occurs when a figure is shifted or slid across the page to change its position.

The image is moved without turning. It is simply moved a number of units left or right, or a number of units up or down, or a combination of both. The example at right shows a figure that has been translated 5 units to the right and 2 units up. This translation can be written as [5 2].

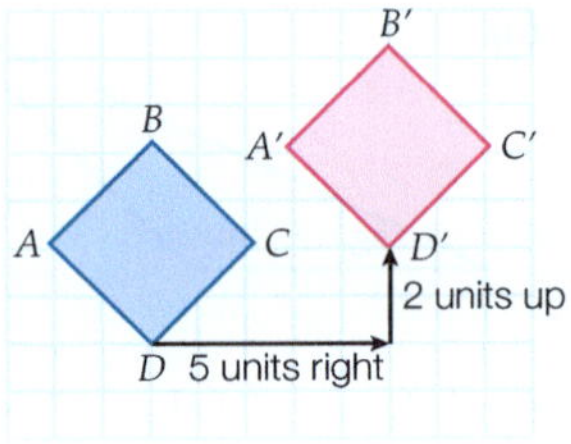

- To distinguish between the original figure and the translated image, you can use a different colour or broken lines.
- Translations can be reversed by doing the opposite horizontal and vertical moves. For example, the reverse of a translation of 4 units left and 3 units down is 4 units right and 3 units up.

Worked Example 1

Describe the translation shown in the diagram.

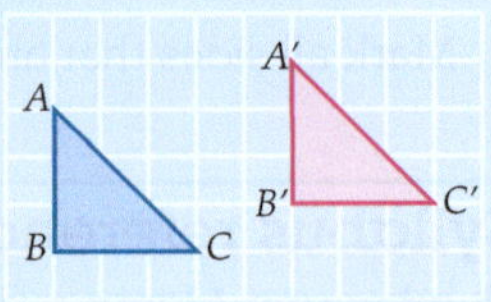

Working

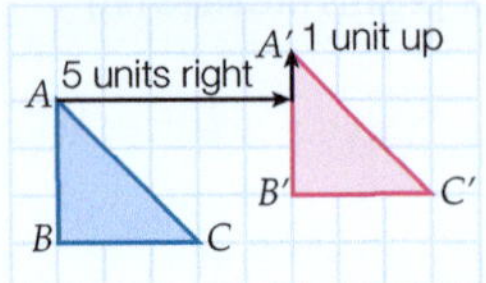

5 units right and 1 unit up or [5 1].

Worked Example 2

Copy the following onto grid paper and draw the resulting image after the translation: 3 units left and 4 units down.

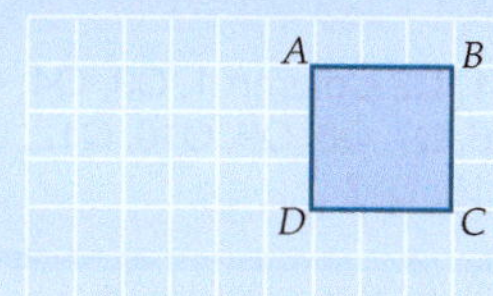

Working

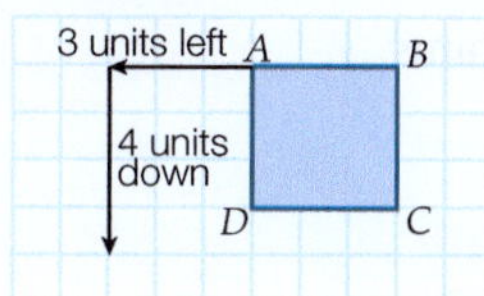

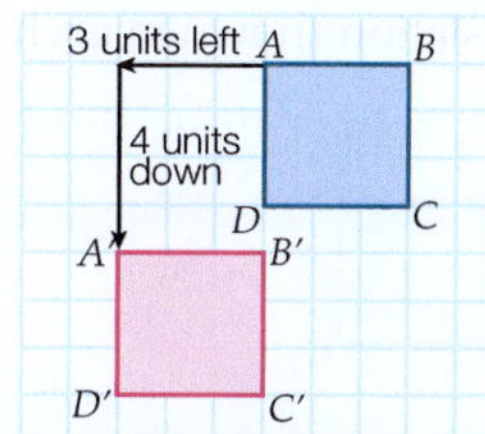

Worked Example 3

(a) Write the coordinates of each vertex on the given shape.

(b) Copy the shape onto grid paper and draw the resulting image after the translation of 2 units to the right and 3 units down.

(c) Write the coordinates of the translated vertices.

(d) Explain how the coordinates of the image could be found without drawing the shape.

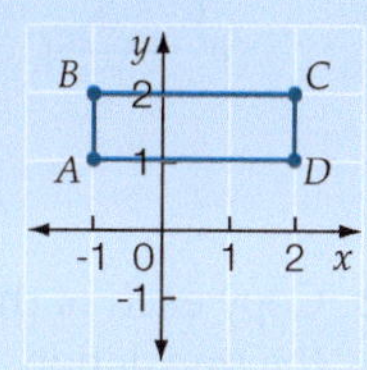

Working

(a) $A = (-1, 1)$, $B = (-1, 2)$, $C = (2, 2)$, $D = (2, 1)$

(b)

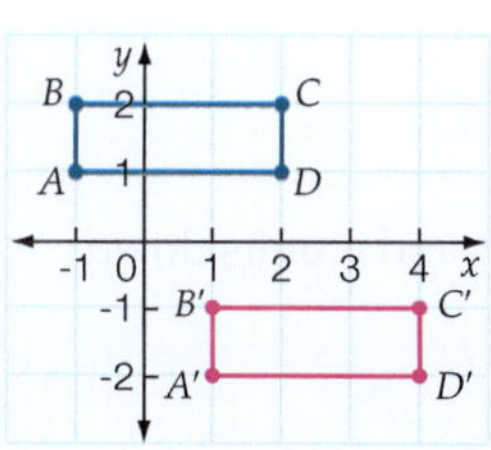

(c) $A' = (1, -2)$, $B' = (1, -1)$, $C' = (4, -1)$, $D' = (4, -2)$

(d) Adding 2 to the *x*-coordinate of the original and subtracting 3 from the *y*-coordinate of the original will give the image coordinates.

10.1

Answers page 504

10.1 Translations

Navigator

Q1, Q2 Column 1, Q3, Q4, Q5, Q6, Q7, Q8, Q9, Q10, Q12, Q13, Q16, Q17	Q1, Q2 Column 2, Q3, Q4, Q5, Q6, Q7, Q8, Q9, Q10, Q11, Q12, Q13, Q14, Q16, Q17	Q1, Q2 Column 2, Q3, Q4, Q5, Q6, Q7, Q8, Q9, Q10, Q11, Q12, Q13, Q14, Q15, Q16, Q17

Equipment required: Grid paper for Questions **2**, **3**, **9**, **13(b)**, **15(c)**, **16** and **17**

Fluency and understanding

WE1

Hint

Translations can be written in square brackets.

1 Describe the translation shown in each of the following diagrams.

(a)

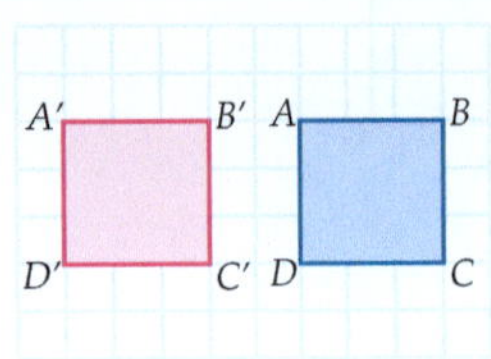

(b)

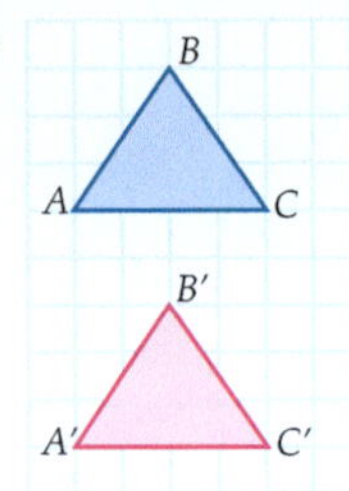

(c)

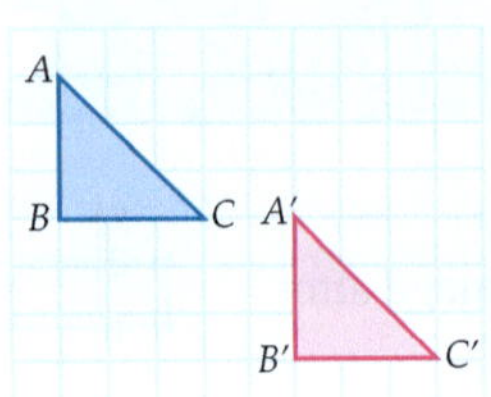

(d)

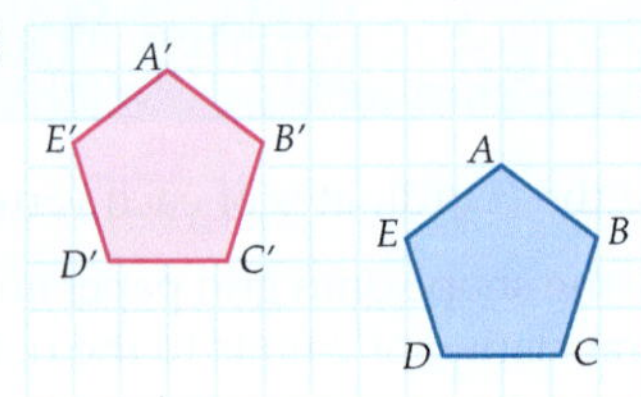

WE2

2 Copy each of the following onto grid paper and draw the resulting image after the translation.

(a) 6 units right and 2 units up

(b) 6 units left and 6 units down

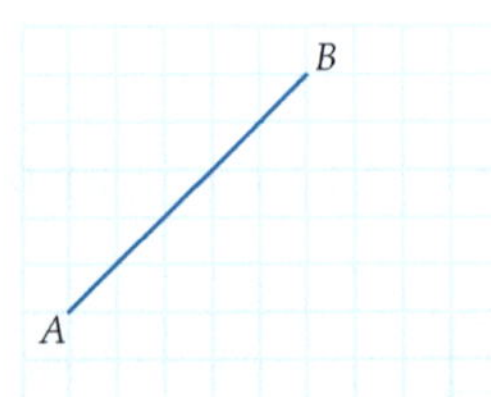

(c) 6 units left and 1 unit down

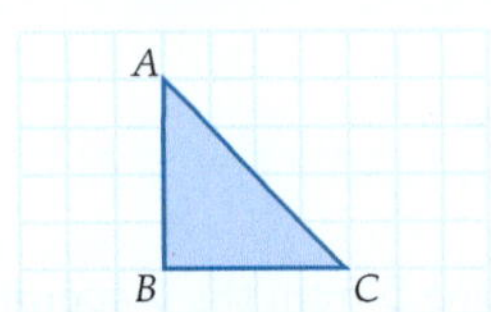

(d) 1 unit right and 6 units down

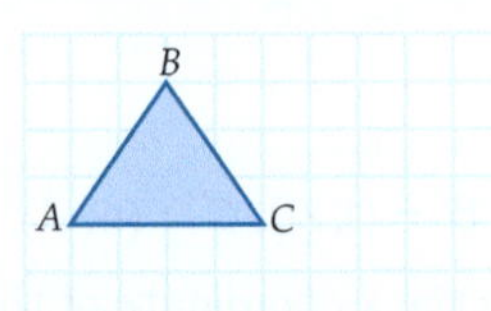

(e) [7 -5]

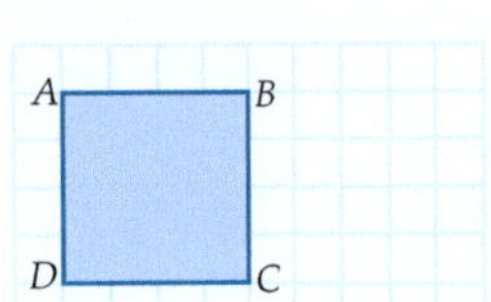

(f) [-6 4]

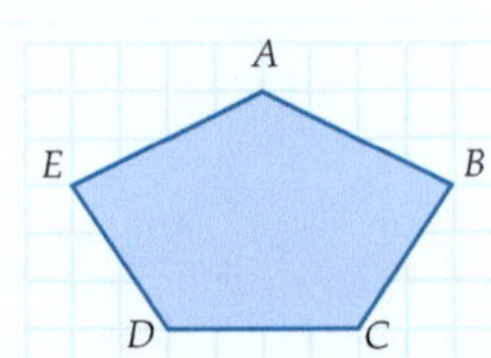

10.1

WE3

3 **(a)** Write the coordinates of each vertex on the given shape.

(b) Copy the shape onto grid paper and draw the resulting image after the translation of 1 unit to the left and 2 units up.

(c) Write the coordinates of the translated vertices.

(d) Explain how the coordinates of the image could be found without drawing the shape.

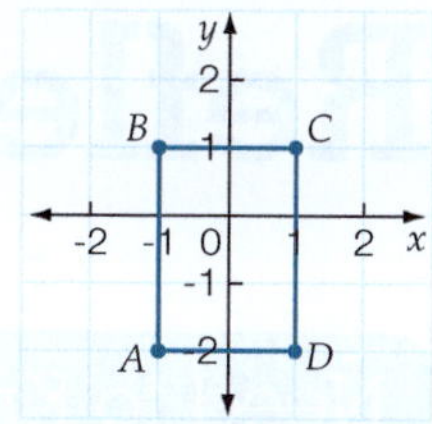

4 How has the hexagon in the diagram been translated?

A 7 units up and 2 units right

B 7 units right and 2 units up

C 7 units left and 2 units down

D 6 units right and 2 units up

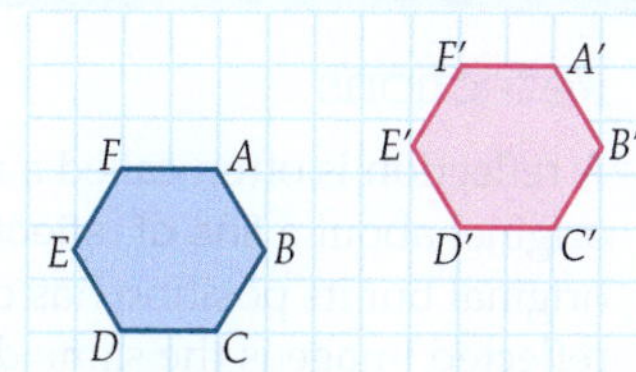

5 What would be the reverse translation of 4 units left and 8 units down?

A 4 units up and 8 units right **B** 4 units left and 8 units up

C 4 units left and 8 units down **D** 4 units right and 8 units up

6 Write the translation [-2 6] using direction instructions (i.e. right, left, up, down).

7 Write the translation of 3 units left and 4 units down using square brackets.

8 Write the reverse translation of 10 units right and 4 units up using square brackets.

9 **(a)** Plot the points (-1, 4), (2, 3) and (0, 2) on a Cartesian plane and label them *A*, *B* and *C*.

(b) Join the points to form triangle *ABC*, then translate the shape 2 units right and 1 unit down, labelling the image with image notation.

(c) Write the coordinates of the translated points.

10 A figure is translated 4 units left, 3 units down, 6 units left and 7 units up. What would be the final position of the image compared to that of the original figure?

11 What single translation would be equivalent to 12 units down, 5 units right, 7 units left, 8 units up and 2 units right?

12 What single translation would be equivalent to a translation of [2 3] followed by a translation of [-2 6]?

Reasoning, communicating and open-ended

13 **(a)** Without plotting points, write the coordinates of the image after the line joining *A*(4, 2) and *B*(6, -1) is translated 2 units left and 2 units up.

(b) Plot the line *AB* and its image *A′B′* on a Cartesian plane to check your answer to part **(a)**.

14 If a particular translation was 2 units right, what would be the single translation that would be equivalent to repeating 2 units right a hundred times?

15 The following diagram is of a translated image and a single point of the original figure.

(a) What translation has occurred to the original figure?

(b) What is the reverse translation?

(c) Use this to reverse the translation and draw the original figure.

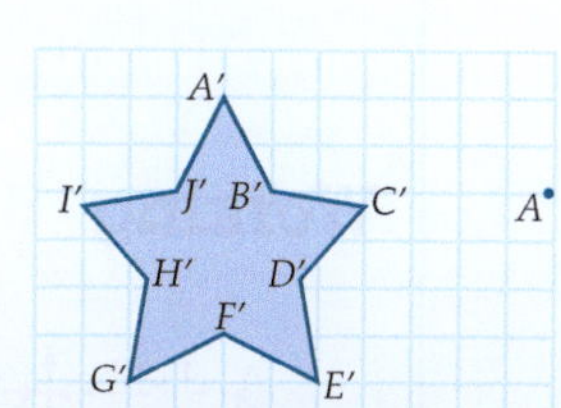

16 Plot three points on a Cartesian plane and write their coordinates. Decide upon a particular translation and use this translation to translate your shape. Draw the image after the translation and write down the coordinates of the translated points using image notation.

17 On grid paper, draw a figure that has 4 vertices. Then, translate that figure in any direction and identify the reverse translation required to get the image back to the original figure.

10.2 Reflections

Need to Know

Reflections

A reflection is often called a mirror image. It is a reversed or 'flipped over' version of the original about a **line of reflection**. A reflected image has the same shape and size as the original but its position has changed and its orientation is reversed. Each point on a reflected image is the same distance from the line of reflection as the original, but on the opposite side. The distance used is the **perpendicular distance**, which is measured by a line that makes a right angle with the line of reflection.

- A **reflection** always flips the original, so the order in which the vertices are labelled in the image will be opposite to that in the original figure.
- A reflection is as far from one side of the line of reflection as the original figure is from the other side of it.
- When reflected, the properties of the shape will remain the same; for example, the size of the side lengths or angles will stay the same.

Reflecting a figure

1 Measure the perpendicular distance from each vertex on the original to the line of reflection.

2 Use this distance between the vertices and the line of reflection to plot the points on the opposite side of the line of reflection.

3 Join the points to produce the reflected image of the original figure.

4 Label the image vertices with image notation.

Worked Example 4

WE4

Copy this figure onto grid paper and draw the resulting image when the triangle is reflected in the line of reflection shown.

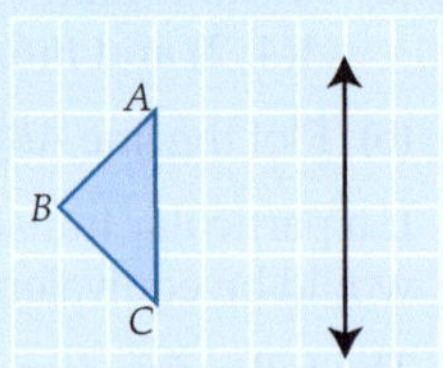

Working

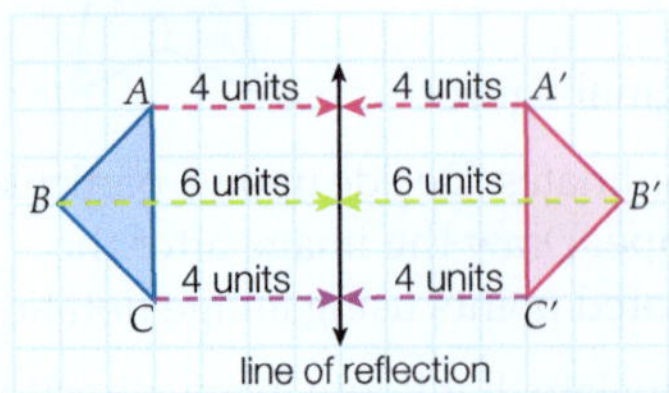

Hint

A point on the mirror line or line of reflection doesn't move when you reflect it.

Worked Example 5

WE5

(a) (i) Write down the coordinates of each vertex on the given shape.

(ii) Copy the shape onto grid paper and draw the resulting image after a reflection in the x-axis, labelling each vertex with image notation.

(iii) Write the coordinates of the reflected vertices.

(iv) Explain how the coordinates of the image could be found without drawing the shape.

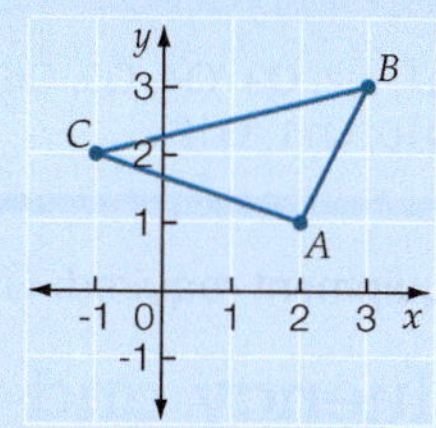

(b) (i) Write down the coordinates of each vertex on the given shape.

(ii) Copy the shape onto grid paper and draw the resulting image after a reflection in the y-axis, labelling each vertex with image notation.

(iii) Write the coordinates of the reflected vertices.

(iv) Explain how the coordinates of the image could be found without drawing the shape.

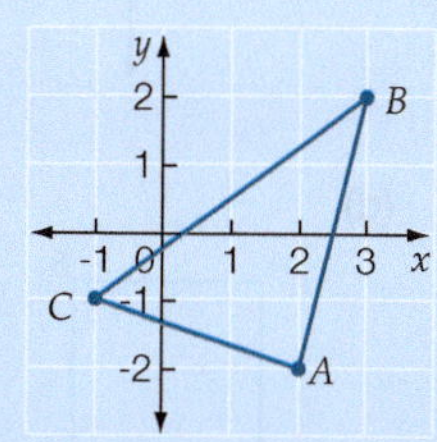

Working

(a) (i) $A = (2, 1)$, $B = (3, 3)$, $C = (-1, 2)$

(ii)

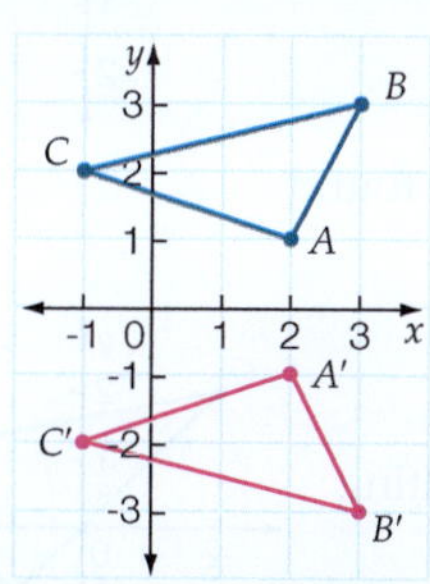

(iii) $A' = (2, -1)$, $B' = (3, -3)$, $C' = (-1, -2)$

(iv) The x-coordinate of the image stays the same as that of the original. The y-coordinate of the image is the negative of that of the original.

(b) (i) $A = (2, -2)$, $B = (3, 2)$, $C = (-1, -1)$

(ii)

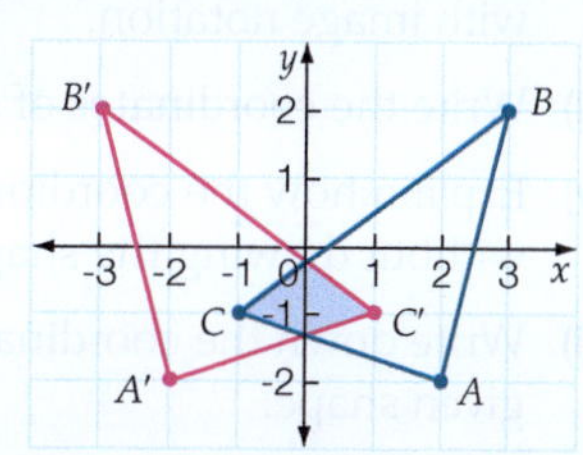

(iii) $A' = (-2, -2)$, $B' = (-3, 2)$, $C' = (1, -1)$

(iv) The x-coordinate of the image is the negative of that of the original. The y-coordinate of the image stays the same as that of the original.

Answers page 505

10.2 Reflections

Navigator

Q1, Q2, Q3, Q4, Q5, Q6, Q8, Q10, Q11, Q12	Q1, Q2, Q3, Q4, Q5, Q6, Q7, Q8, Q9, Q10, Q13	Q1, Q2, Q3, Q4, Q5, Q6, Q7, Q8, Q9, Q10, Q13

Equipment required: Grid paper for Questions **1–3**, **5–8** and **10–13**

Fluency and understanding

WE4

1 Copy the following figures onto grid paper and draw the resulting image when each is reflected in the line of reflection shown.

(a)

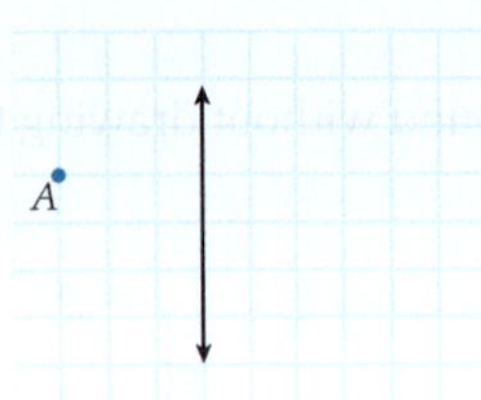

(b)

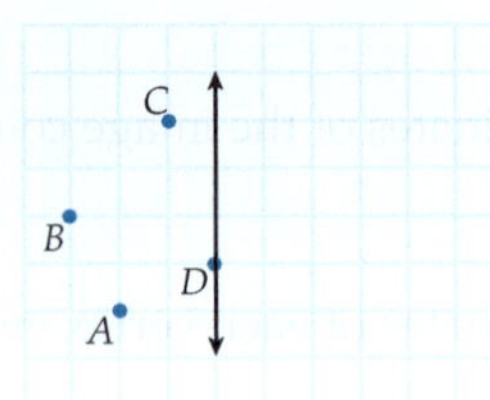

(c)

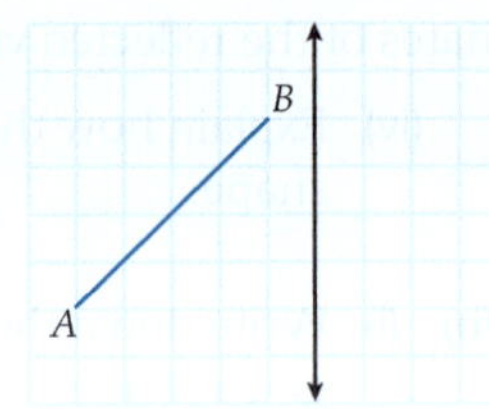

(d)

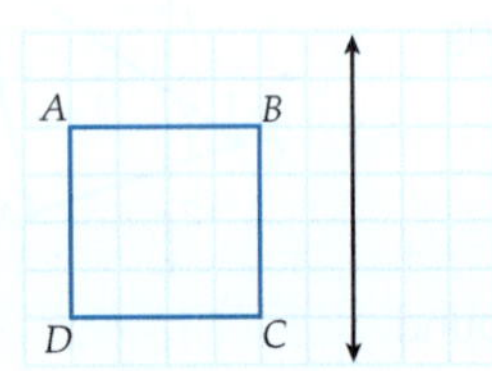

(e)

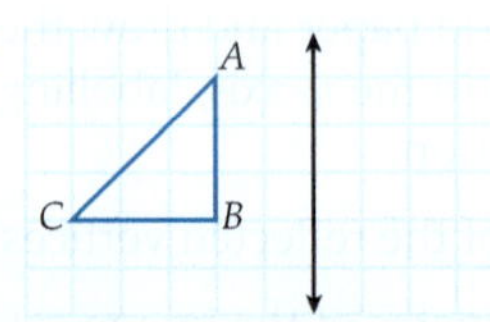

(f)

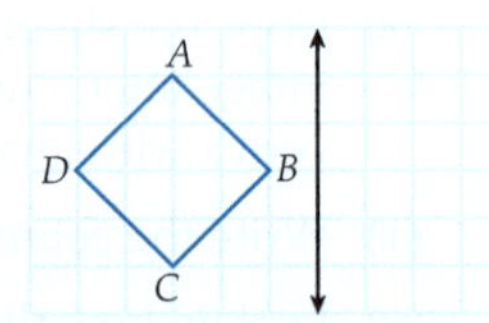

WE5

2 (a) **(i)** Write down the coordinates of each vertex on the given shape.

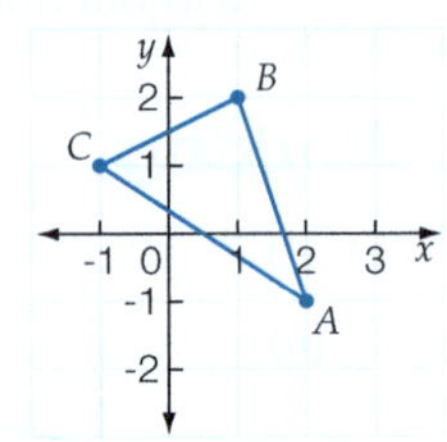

(ii) Copy the shape onto grid paper and draw the resulting image after a reflection in the *x*-axis, labelling each vertex with image notation.

(iii) Write the coordinates of the reflected vertices.

(iv) Explain how the coordinates of the image could be found without drawing the shape.

(b) **(i)** Write down the coordinates of each vertex on the given shape.

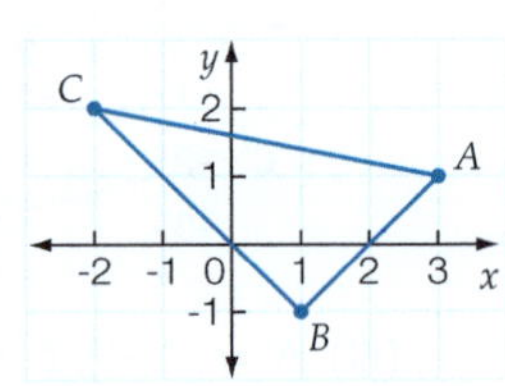

(ii) Copy the shape onto grid paper and draw the resulting image after a reflection in the *y*-axis, labelling each vertex with image notation.

(iii) Write the coordinates of the reflected vertices.

(iv) Explain how the coordinates of the image could be found without drawing the shape.

3 Copy each of the following figures onto grid paper, then draw the reflection of each in the line of reflection shown.

(a)
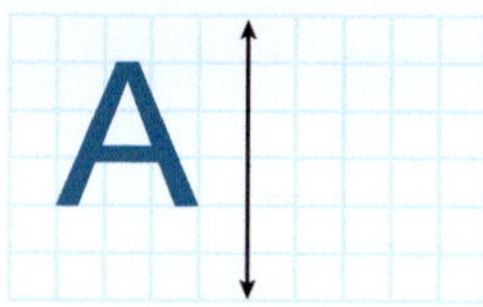

(b)
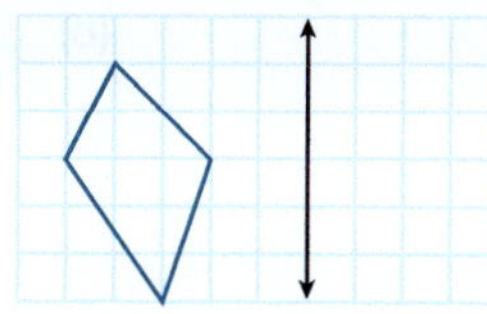

(c)
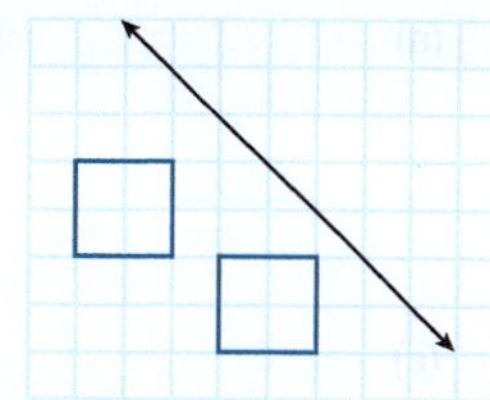

(d)
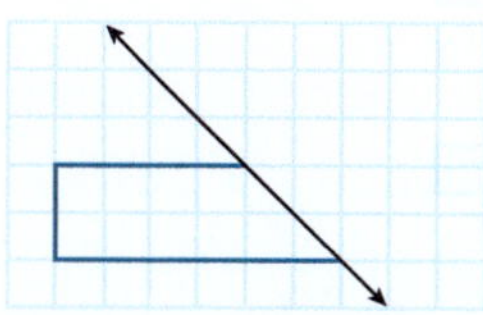

(e)
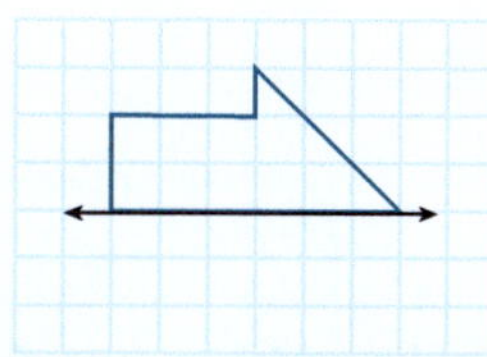

(f)
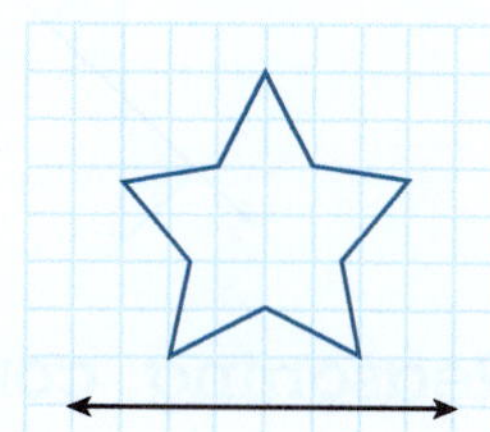

4 Which of the following options is the correct reflection of the capital F in the vertical line of reflection?

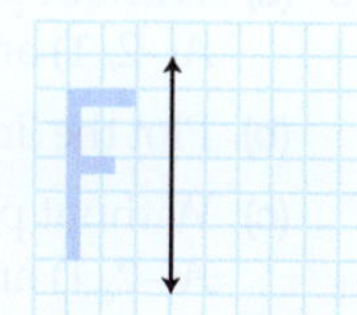

A
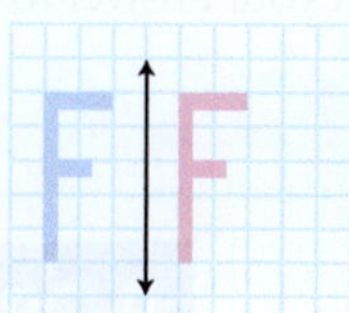

B
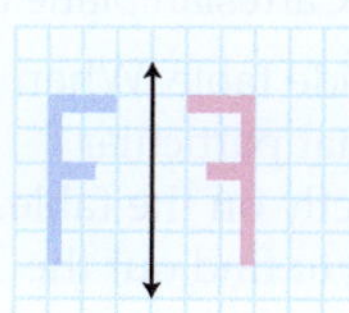

C

D
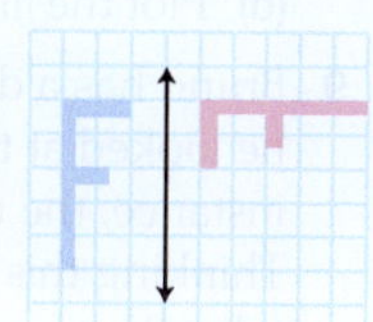

5 **(a)** Plot the points (-1, 5), (3, 1) and (2, -1) on a Cartesian plane and label them A, B and C.

(b) Join the points to form triangle ABC, then reflect the shape first in the x-axis and then in the y-axis.

(c) Write the coordinates of the points after the second reflection.

6 Copy each of the following onto grid paper and indicate where the line of reflection should be placed to produce the given image.

(a)
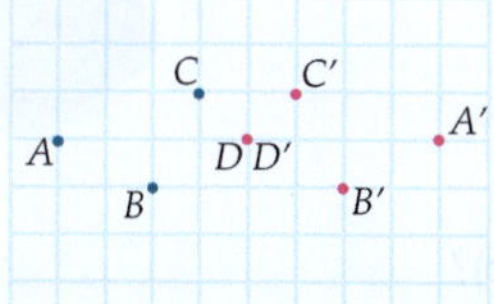

(b)

(c)

(d)
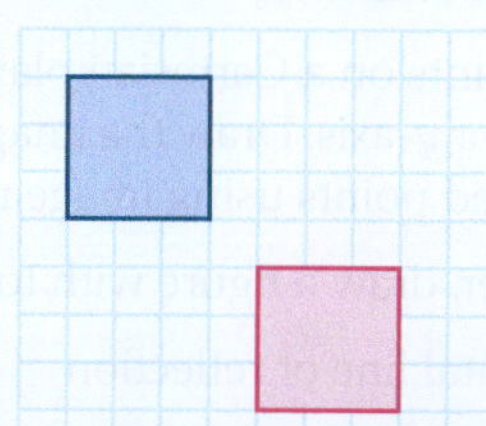

7 Copy each of the following onto grid paper and draw the reflected images. The image may be on both sides of the line of reflection.

(a)
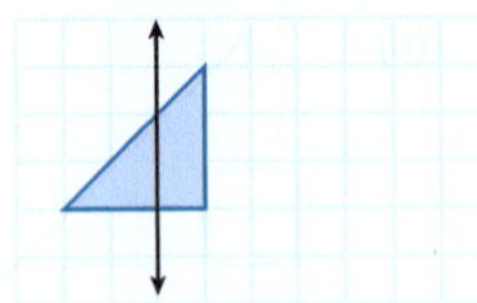

(b)
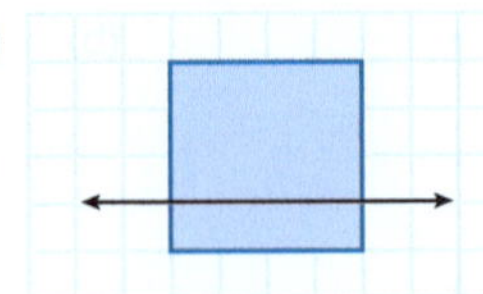

(c)
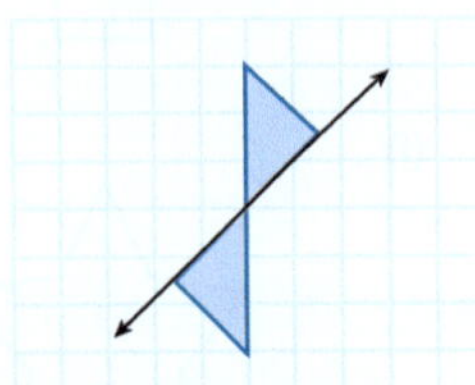

(d)
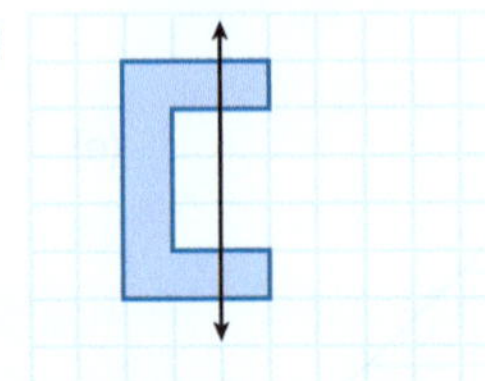

Reasoning, communicating and open-ended

8 **(a)** Without plotting points, write the coordinates of the image after the line joining $A(-2, 1)$ and $B(4, -3)$ is reflected in the x-axis.

(b) Plot the line AB and its image $A'B'$ on a Cartesian plane to check your answer to part **(a)**.

(c) Without plotting points, write the coordinates of the image after the line joining $A(-2, 1)$ and $B(4, -3)$ is reflected in the y-axis.

(d) Plot the line AB and its image $A'B'$ on a Cartesian plane to check your answer to part **(c)**.

9 Bruno has a digital clock on his glass bedside table. When he looked at the time, he noticed that at that particular instance, the time 8:30 was reflected perfectly on the table. Thinking this to be a little odd, he decided to find out the following.

(a) List the number of minutes past the hour when the minute part of the time reflects the correct time (include 00).

(b) In a 12-hour mode, list the hours when the hour part of the time reflects the correct time.

(c) In a 12-hour period, use your answers to parts **(a)** and **(b)** to find how many times the reflection on the table gives the correct time in a 12-hour mode.

(d) In a 24-hour period, how many times does the reflection in the table give the correct time in a 24-hour mode?

(e) Is your answer to part **(d)** double your answer to part **(c)**? Explain your answer.

10 Plot three points on a Cartesian plane and write their coordinates. Reflect this shape in either the x- or y-axis. Draw the image after the reflection and write down the coordinates of the reflected points using image notation.

11 On grid paper, draw a figure with four vertices. Reflect this figure in:

(a) a horizontal line of reflection

(b) a vertical line of reflection.

12 Repeat Question **11** with a figure with five, six, or seven vertices.

13 On grid paper, draw a figure with four vertices.

(a) Reflect this figure in a horizontal line of reflection and then in a vertical line of reflection.

(b) Reflect the same figure in the vertical line of reflection and then in the horizontal line of reflection.

(c) Comment on the images found in parts **(a)** and **(b)**.

Problem solving

Mirror, mirror on the wall, does it matter where I stand at all?

Equipment required: 1 brain, grid paper

The diagram below consists of two blue triangles and two vertical lines of reflection.

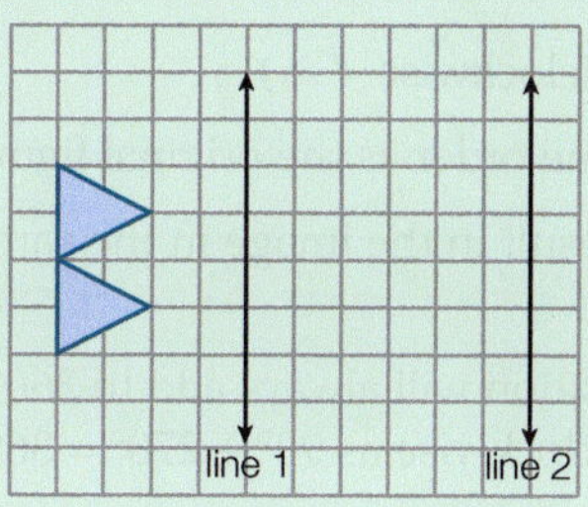

To begin, reflect the two triangles in the line of reflection labelled 'line 1' and then in the line of reflection labelled 'line 2' and note the final location of the image. Next, switch the order in which the two triangles are reflected so that they are reflected in line 2 first, and then in line 1, and note the final location of the image.

Does the order in which these triangles are reflected in either line affect the final location and orientation of the final image? What did you notice about the final image? Can this transformation be described another way?

To justify your decision, experiment with different situations, such as drawing the two triangles originally on the right-hand-side of line 2 or in-between the lines. You can also alter the position of the lines of reflection. For example, line 1 and line 2 could be two horizontal lines or one could be horizontal and the other vertical.

Strategy options

- Draw a diagram.
- Test all possible combinations.

10.3 Rotations

Need to Know

Rotations

A figure is also transformed when it is **rotated** (turned) around a fixed point. This transformation is called a **rotation**. The fixed point is called the **centre of rotation** and is labelled with the letter *O*.

A centre of rotation may be inside or outside the original shape. This makes no difference to the method used to **rotate** the figure. The distance of each point from the centre of rotation remains the same after rotation.

When rotating any figure, these three things must be known:

- the location of the **centre of rotation**
- the size of the angle of rotation
- the direction of the rotation (clockwise ↷ or anticlockwise ↶).

The vertices of a rotated image are labelled in the same order as the original figures.

Rotations of 360°, 720° and other multiples of 360° result in the image in the same position as the original figure.

Two rotations that result in an image in the same position will always add to 360°. For example, 90° clockwise is equivalent to 270° anticlockwise as 90° + 270° = 360°.

A rotation in an anticlockwise direction is a positive rotation, e.g. 90°.

A rotation in a clockwise direction is a negative rotation, e.g. -90°.

Rotating a figure

1. Select a key point, such as a vertex, and join this point to the centre of rotation with a straight line.
2. Place the centre of a protractor on the centre of rotation, with the chosen vertex on the baseline.
3. Measure the given angle of rotation in the given direction from this vertex.
4. Plot the resulting point the same distance from the centre of rotation as the original point.
5. Repeat the process with the other vertices and connect the resulting points to produce the rotated image of the original figure.

Hint

A clockwise rotation of 180° is the same as an anticlockwise rotation of 180°.

10.3

Worked Example 6

WE 6

Copy this figure onto grid paper and draw the resulting image after the given rotation. Use a protractor to help you. Rotate 90° in an anticlockwise direction (90°) about *O*.

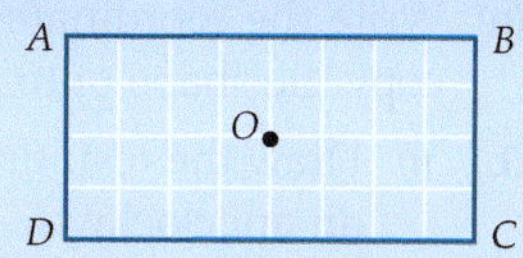

Working

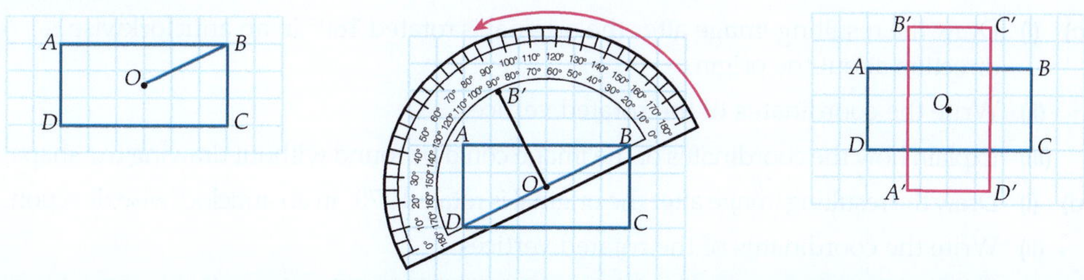

Worked Example 7

WE 7

Copy the following figure onto grid paper and draw the resulting image after the given rotation. Use a protractor to help you. Rotate 180° in a clockwise direction (-180°) about *O*.

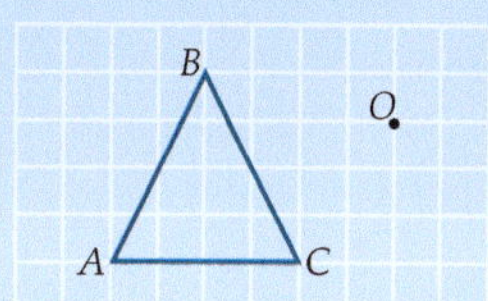

Working

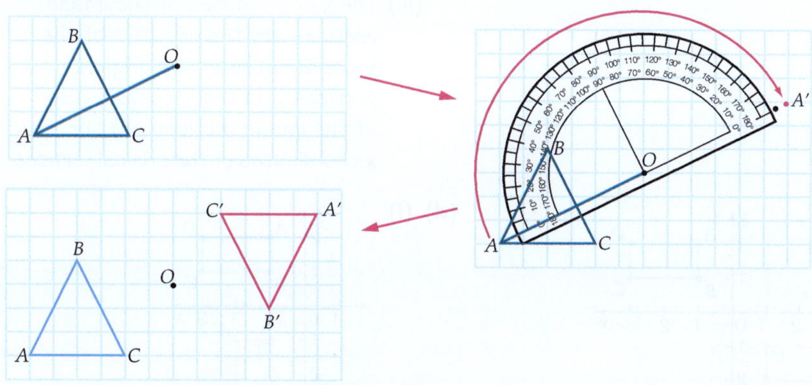

Worked Example 8

WE8

(a) Write the coordinates of each vertex on the given shape and copy the shape onto grid paper.

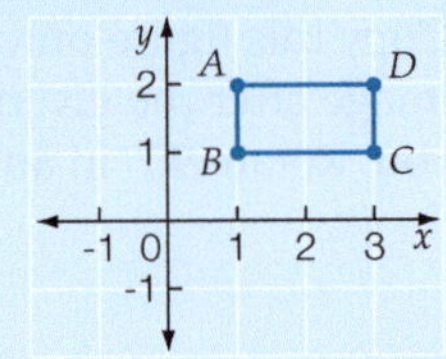

(b) **(i)** Draw the resulting image after the original is rotated 90° in an anticlockwise direction about the origin.

(ii) Write the coordinates of the rotated vertices.

(iii) Explain how the coordinates of the image could be found without drawing the shape.

(c) **(i)** Draw the resulting image after the original is rotated 180° in an anticlockwise direction about the origin.

(ii) Write the coordinates of the rotated vertices.

(iii) Explain how the coordinates of the image could be found without drawing the shape.

(d) **(i)** Draw the resulting image after the original is rotated 270° in an anticlockwise direction.

(ii) Write the coordinates of the rotated vertices.

(iii) Explain how the coordinates of the image could be found without drawing the shape.

Working

(a) $A = (1, 2)$, $B = (1, 1)$, $C = (3, 1)$, $D = (3, 2)$

(b) **(i)**

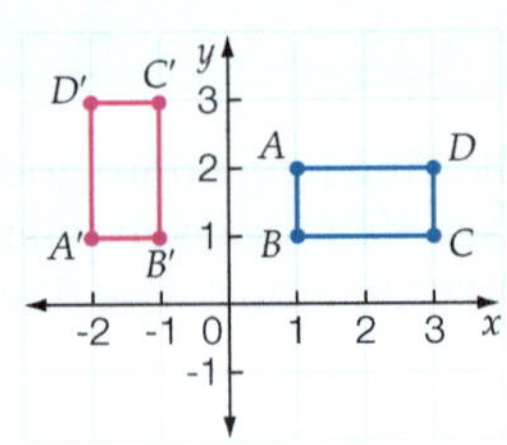

(ii) $A' = (-2, 1)$, $B' = (-1, 1)$, $C' = (-1, 3)$, $D' = (-2, 3)$

(iii) The x-coordinates of the image vertices are the negative of the y-coordinates of the original vertices and the y-coordinates of the image vertices are the x-coordinates of the original vertices.

(c) **(i)**

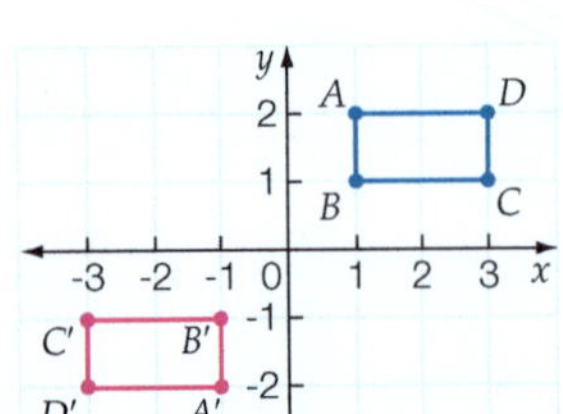

(ii) $A' = (-1, -2)$, $B' = (-1, -1)$, $C' = (-3, -1)$, $D' = (-3, -2)$

(iii) All the coordinates of the image vertices are the negative of the original coordinates of the vertices.

(d) **(i)**

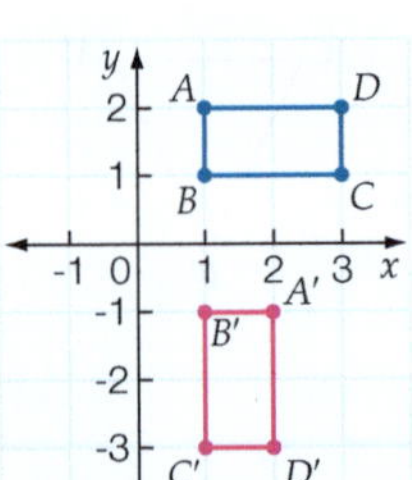

(ii) $A' = (2, -1)$, $B' = (1, -1)$, $C' = (1, -3)$, $D' = (2, -3)$

(iii) The x-coordinates of the image vertices are the y-coordinates of the original vertices and the y-coordinates of the image vertices are the negative of the x-coordinates of the original vertices.

10.3 Rotations

Answers page 508

Navigator

Q1, Q2, Q3, Q4, Q5, Q6, Q7, Q8, Q9, Q10 (a), Q11, Q14, Q15	Q1, Q2, Q3, Q4, Q5, Q6, Q7, Q8, Q9, Q10, Q11, Q12 Column 1, Q14, Q15	Q1, Q2, Q3, Q4, Q5, Q6, Q7, Q9, Q10, Q11, Q12 Column 2, Q13, Q14, Q15, Q16

Equipment required: Protractor and grid paper for Questions **1–3**, **6**, **9 (b)**, **10**, **14** and **16**

Fluency and understanding

1 Copy each of the following figures onto grid paper and draw the resulting image after the given rotation. Use a protractor to help you. WE6

(a) Rotate 180° in an anticlockwise direction (180°) about *O*.

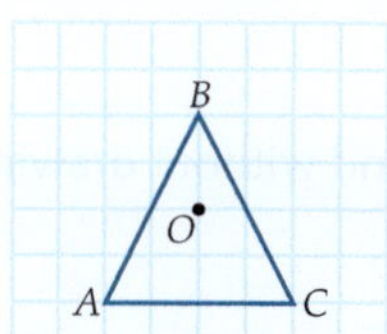

(b) Rotate 270° in an anticlockwise direction (270°) about *O*.

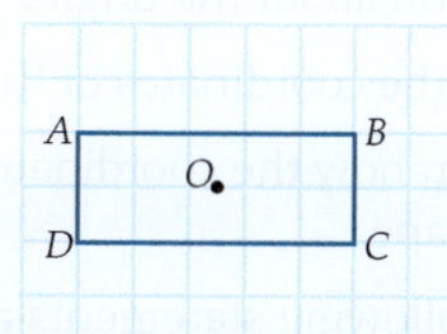

(c) Rotate 180° in a clockwise direction (-180°) about *O*.

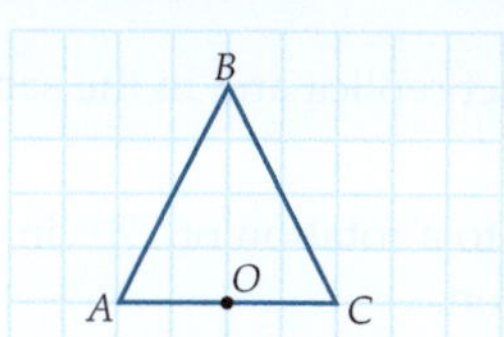

(d) Rotate 90° in a clockwise direction (-90°) about *O*.

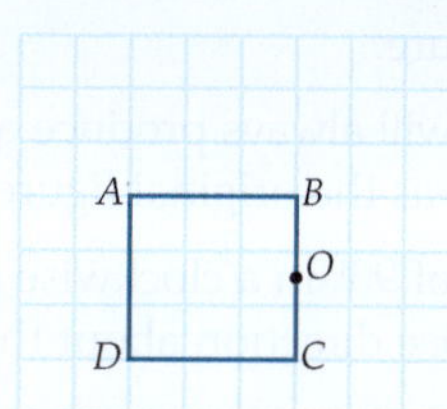

2 Copy each of the following onto grid paper and draw the resulting image after the given rotation. Use a protractor to help you. WE7

(a) Rotate each of the points 90° in a clockwise direction (-90°) about *O*.

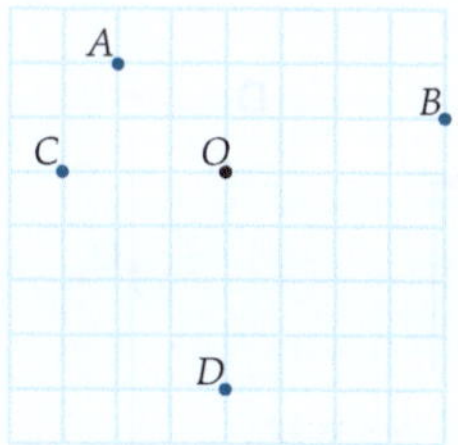

(b) Rotate 270° in an anticlockwise direction (270°) about *O*.

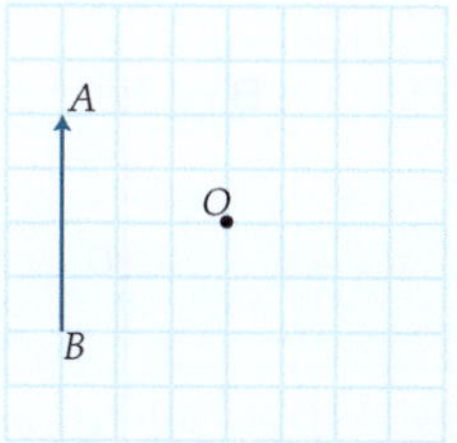

(c) Rotate 180° in a clockwise direction (-180°) about *O*.

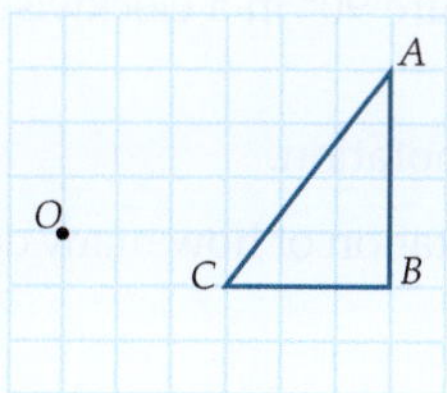

(d) Rotate 90° in an anticlockwise direction (90°) about *O*.

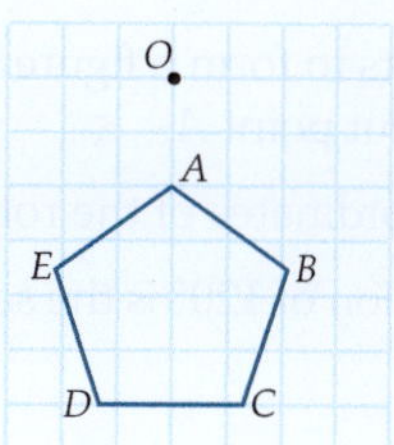

10.3

WE8

3 **(a)** Write the coordinates of each vertex on the given shape and copy the shape onto grid paper.

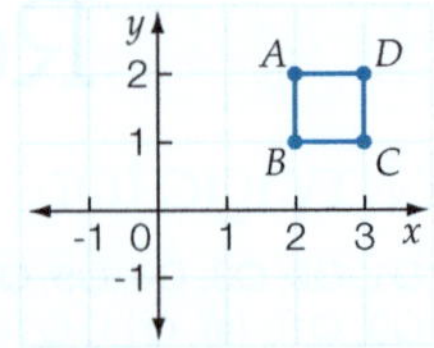

(b) **(i)** Draw the resulting image after the original is rotated 90° in an anticlockwise direction about the origin.

(ii) Write the coordinates of the rotated vertices.

(iii) Explain how the coordinates of the image could be found without drawing the shape.

(c) **(i)** Draw the resulting image after the original is rotated 180° in an anticlockwise direction about the origin.

(ii) Write the coordinates of the rotated vertices.

(iii) Explain how the coordinates of the image could be found without drawing the shape.

(d) **(i)** Draw the resulting image after the original is rotated 270° in an anticlockwise direction about the origin.

(ii) Write the coordinates of the rotated vertices.

(iii) Explain how the coordinates of the image could be found without drawing the shape.

4 Which of the following statements about rotation is incorrect?

A Rotation can be in either a clockwise or an anticlockwise direction.

B A rotation of 360° will result in the image located in the same position as the original figure.

C A rotation will always produce an image that is a perfect replica and in the same orientation as the original figure.

D A rotation of 90° in a clockwise direction is equivalent to a rotation of 270° in an anticlockwise direction about the same centre of rotation.

5 This arrow turns about the centre of rotation *O*. What does the shape look like after a turn of 180° anticlockwise?

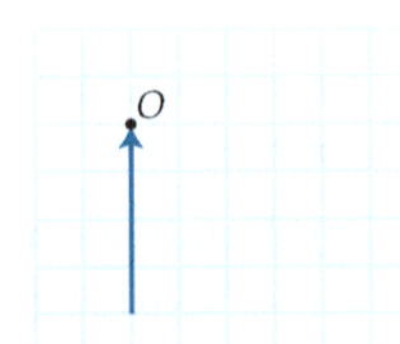

A

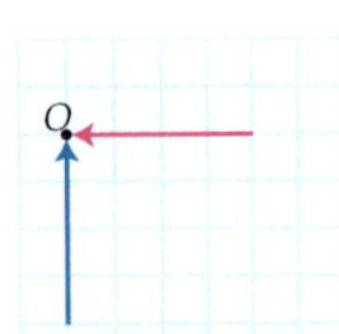

B

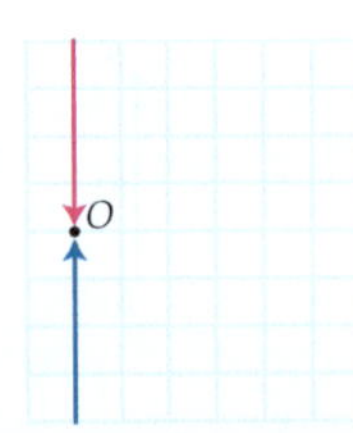

C

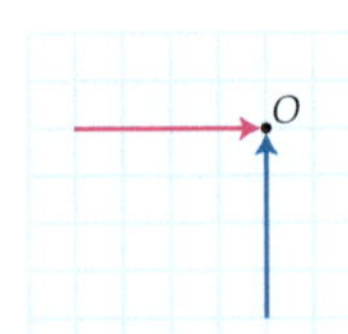

D

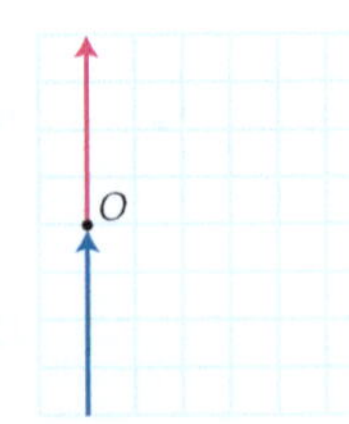

6 **(a)** Plot the points (1, 4), (2, 4), (2, 2) and (1, 2) on a Cartesian plane and label them *A*, *B*, *C* and *D*.

(b) Join the points to form a figure *ABCD*, then rotate the figure 90° in a clockwise direction about point *A*.

(c) Write the coordinates of the rotated points using image notation.

7 A clockwise rotation of 120° is the same as an anticlockwise rotation of how many degrees?

8 Find the rotation (the size of the rotation and the direction) that has taken place to produce the following images. Identify more than one rotation that will achieve the same result.

(a)

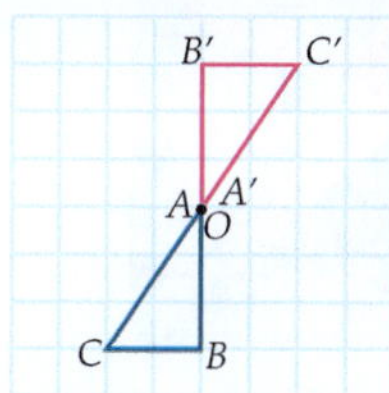

(b)

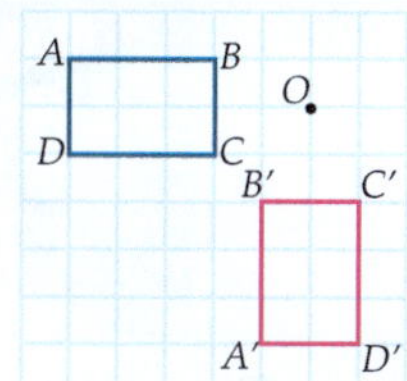

(c)

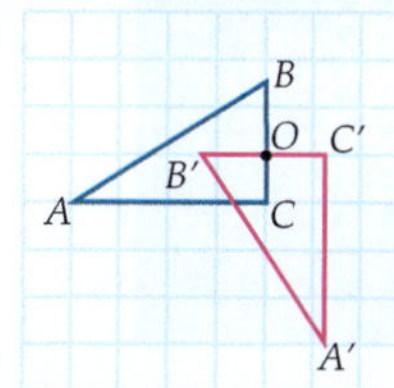

(d)

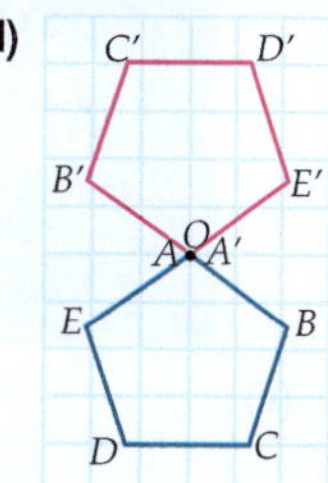

Reasoning, communicating and open-ended

9 (a) Without plotting points, write the coordinates of the image after the line joining $A(4, 2)$ and $B(3, -2)$ is rotated 180° clockwise about the origin.

(b) Plot the line AB and its image $A'B'$ on a Cartesian plane to check your answer to part (a).

10 Copy each of the following figures onto grid paper and draw the resulting image after the given rotation. Use a protractor to help you.

(a) Rotate 90° in an anticlockwise direction about O.

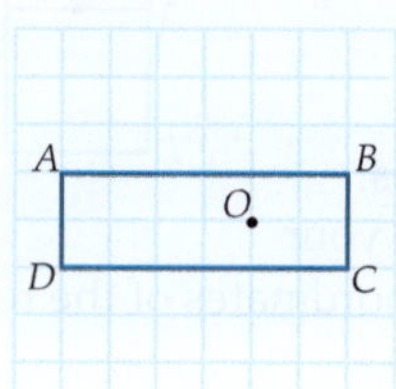

(b) Rotate 90° in a clockwise direction about O.

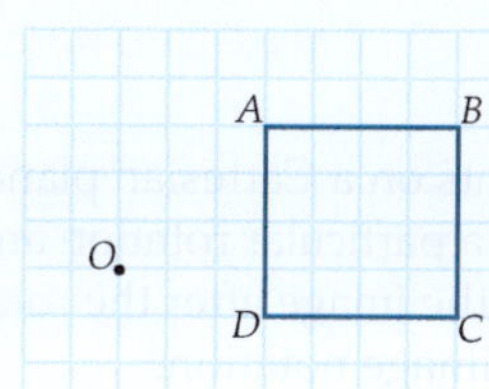

11 The following are the capital letters of the alphabet.

A B C D E F G H I J K L M
N O P Q R S T U V W X Y Z

(a) Which letters of the alphabet are able to be rotated about a centre of rotation in either direction and at an angle of less than 360° to produce an identical image in the same orientation as the original letter?

(b) Four of these letters have something else in common with each other. Can you identify these four letters? Describe the similarity you have noticed.

12 Estimate the smallest angle through which the following figures need to be rotated so the resulting image will match the original figure.

(a)

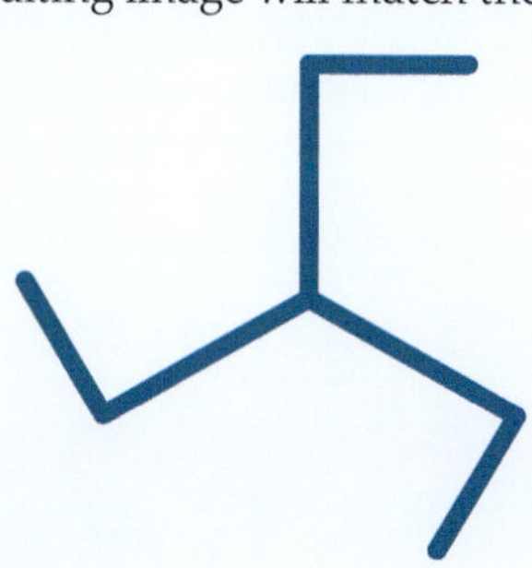

(b)

10.3

(c)

(d)

(e)

(f)

13 Is the following transformation a rotation of 180° around point O, a reflection along the line AB or both? Provide an explanation.

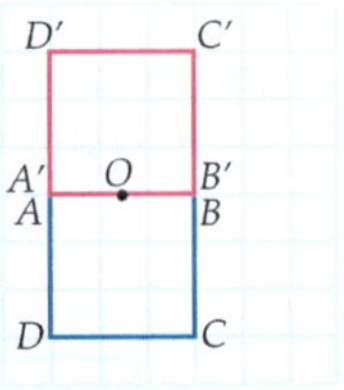

14 Plot four points on a Cartesian plane and write their coordinates. Decide upon a particular rotation and use this rotation to rotate your shape. Draw the image after the rotation and write down the coordinates of the rotated points using image notation.

15 Explain whether you agree or disagree with the following statement.

'A rotation of 180° in either direction and about any point of rotation can be equivalent to reflecting the same figure in a line of reflection through that point.'

16 On grid paper, draw two figures, a regular and an irregular polygon with the same number of sides. Decide upon a centre of rotation for each polygon that is on a vertex of that polygon.

(a) Rotate each of the figures by 90° about the centre of rotation in a clockwise direction and draw the image.

(b) Repeat this four times.

(c) Rotate the figures 45° about the centre of rotation.

(d) Repeat this eight times.

(e) Comment on the final images in parts **(b)** and **(d)**.

Combined transformations

10.4

Need to Know

- Translations, reflections and rotations can be combined into a sequence of transformations.
- A side length can act as the line of reflection during a combined transformation.
- Different colours can be used to distinguish between several transformations.
- The properties of the three different types of transformations still hold true for combined transformations.
- After each transformation, the vertices are labelled with an additional dash. For example, the vertex A''' has been transformed by three different transformations.
- Label each image after each transformation.

Worked Example 9

WE 9

Copy the following figure onto grid paper and draw the resulting image after the combined transformation has been carried out. Use a protractor to help you. Label each vertex appropriately after each transformation.

A translation of 3 units right and 2 units up, followed by a rotation of 180° in a clockwise direction about the vertex A'.

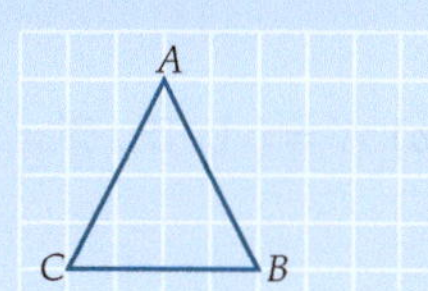

Working

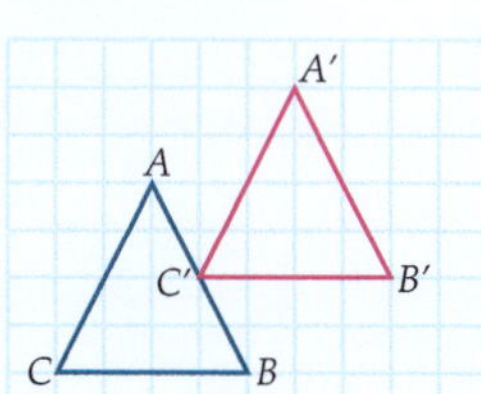

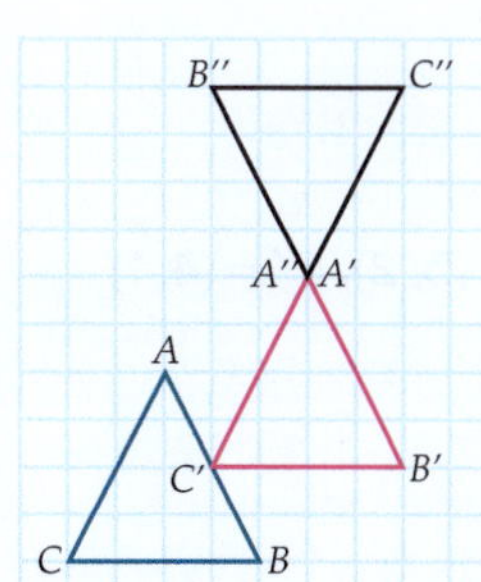

INTRODUCTION TO MATHEMATICAL THINKING
A publication from Pearson Custom Publishing

Worked Example 10

WE10

For this figure:

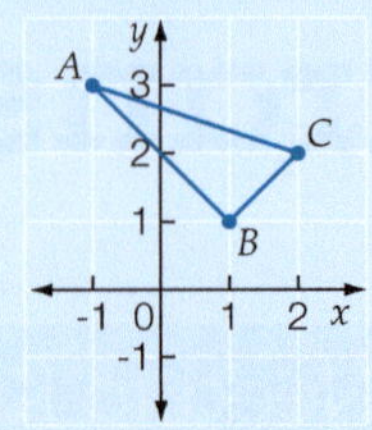

(a) write the coordinates of each vertex on the given shape and copy the shape onto grid paper

(b) reflect the shape in the x-axis

(c) write the coordinates of the reflected vertices

(d) translate this image 2 units right and 3 units up

(e) write the coordinates of the translated vertices

(f) rotate this translated image 90° in an anticlockwise direction about the origin

(g) write the coordinates of the vertices of the final image.

Working

(a) $A = (-1, 3)$, $B = (1, 1)$, $C = (2, 2)$

(b)

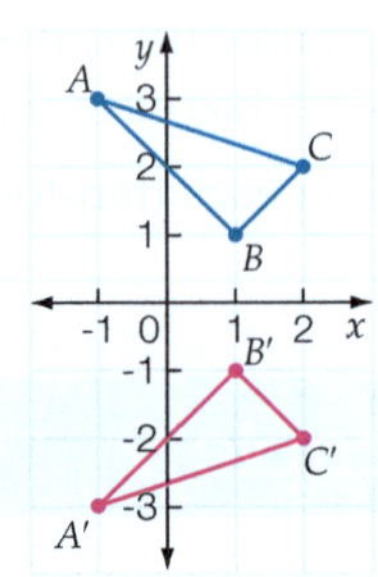

(c) $A' = (-1, -3)$, $B' = (1, -1)$, $C' = (2, -2)$

(d)

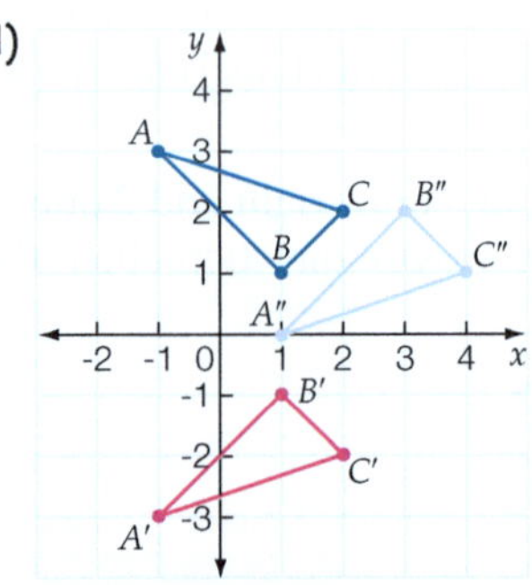

(e) $A'' = (1, 0)$, $B'' = (3, 2)$, $C'' = (4, 1)$

(f)

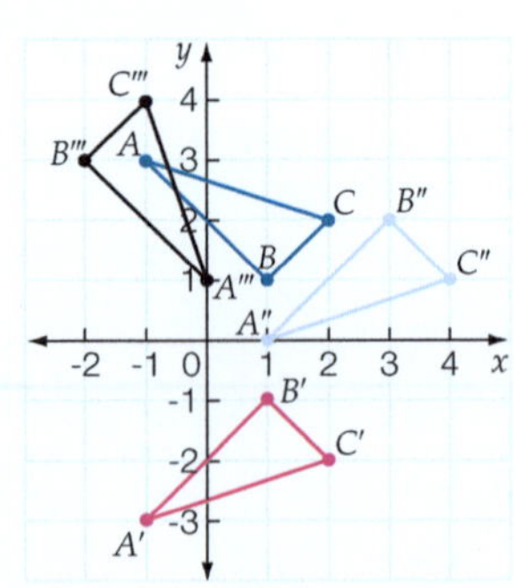

(g) $A''' = (0, 1)$, $B''' = (-2, 3)$, $C''' = (-1, 4)$

10.4 Combined transformations

Answers page 509

Navigator

Q1, Q2, Q3 Column 1, Q4, Q5, Q6, Q7, Q8 Column 1, Q9 Column 1, Q10, Q11, Q13, Q15	Q1, Q2, Q3 Column 2, Q4, Q5, Q6, Q7, Q8 Column 2, Q9 Column 2, Q10, Q11, Q12, Q13, Q15	Q1, Q2, Q3 Column 2, Q4, Q5, Q6, Q7, Q8 Column 2, Q9 Column 2, Q10, Q11, Q12, Q13, Q14, Q15

Equipment required: Protractor and grid paper for Questions **1–3**, **6**, **10 (b)** and **13–15**

Fluency and understanding

1 Copy each of the following figures onto grid paper and draw the resulting image after the combined transformation has been carried out. Use a protractor to help you. Label each vertex appropriately after each transformation. WE 9

(a) A translation of 3 units left and 1 unit down, followed by a reflection in the line $B'C'$.

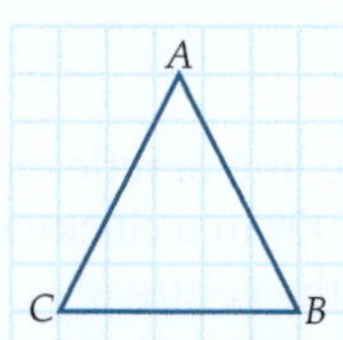

(b) A translation of 5 units right and 2 units down, followed by a reflection in the line $D'C'$.

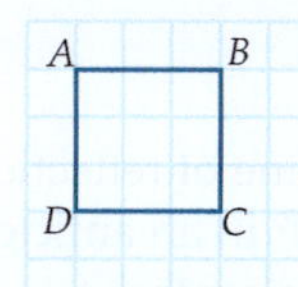

(c) A translation of 5 units right, followed by a rotation of 90° about B' in a clockwise direction.

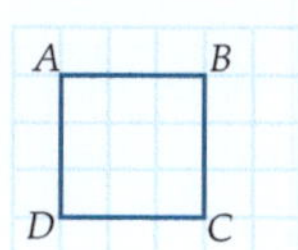

(d) A translation of 3 units left and 4 units down, followed by a rotation of 180° about C' in a clockwise direction.

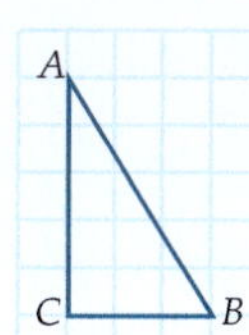

2 For the figure given opposite: WE 10

(a) write the coordinates of each vertex on the given shape and copy the shape onto grid paper

(b) reflect the shape in the x-axis

(c) write the coordinates of the reflected vertices

(d) translate this image 5 units left and 3 units up

(e) write the coordinates of the translated vertices

(f) rotate this translated image 180° in a clockwise direction about the origin

(g) write the coordinates of the vertices of the final image.

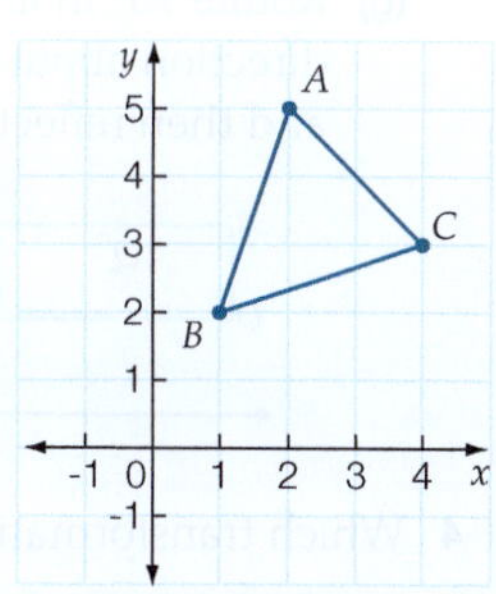

10.4

3 Copy each of the following figures onto grid paper and draw the resulting image after the combined transformation has been carried out. Use a protractor to help you. Label each of the vertices appropriately after each transformation (for **(e)–(h)**, point O is a point on or inside the figure and is reflected with the shape).

(a) Reflect along the line CB and translate 6 units right and 2 units up.

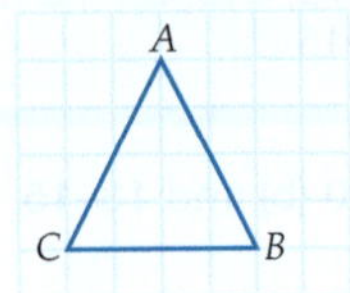

(b) Reflect along the line BC and translate 1 unit left and 3 units down.

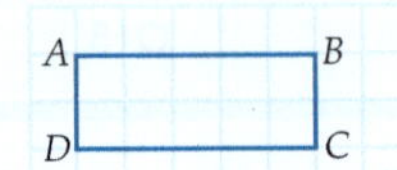

(c) Rotate 90° in a clockwise direction about vertex A and translate 8 units down.

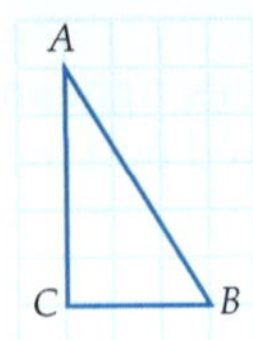

(d) Rotate 180° in an anticlockwise direction about vertex D and move 2 units left and 6 units up.

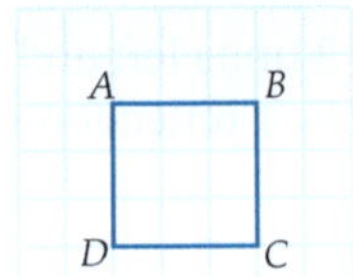

(e) Reflect along the line of reflection and then rotate 90° in an anticlockwise direction about the centre of rotation O'.

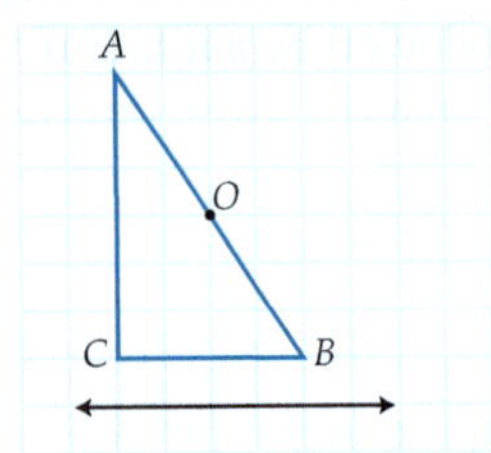

(f) Reflect along the line of reflection and then rotate 90° in a clockwise direction about the centre of rotation O'.

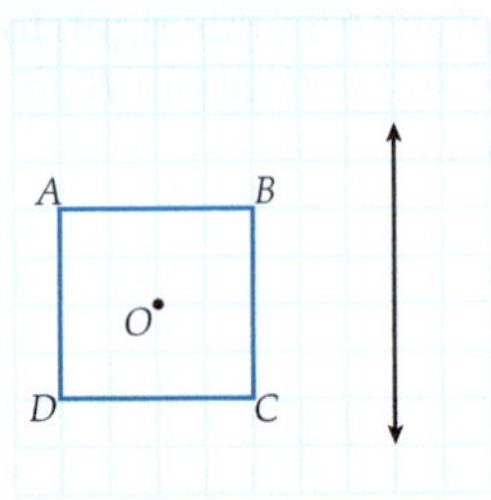

(g) Rotate 90° in an anticlockwise direction about the centre of rotation O and then reflect in the line of reflection.

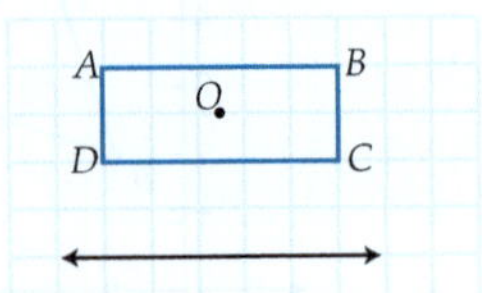

(h) Rotate 180° in a clockwise direction about the centre of rotation O and then reflect in the line of reflection.

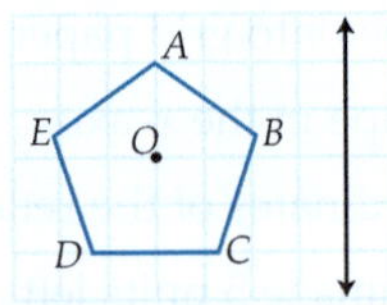

4 Which transformation reverses the order in which the vertices are labelled?

A rotation and translation
B reflection
C translation
D rotation

5 Which transformation can result in an image in the same position and orientation as the original figure?

A rotation and translation
B translation
C reflection
D rotation

10.4

6 **(a)** Plot the points (-1, 2), (2, 3) and (3, -1) on a Cartesian plane and label them *A*, *B* and *C*.

(b) Join the points to form a triangle *ABC*, then translate the shape 3 units right and 1 unit down.

(c) Now, rotate the figure 90° in a clockwise direction about the origin.

(d) Write the coordinates of the vertices of the final image using image notation.

7 For the combined transformations in Question **1**, identify the necessary combined transformation that would move the image back to the position and orientation of the original figure.

8 Describe the transformations that have taken place to move the original figure in blue to the position in pink, then to its final position in black. List the relevant names of each transformation.

(a)

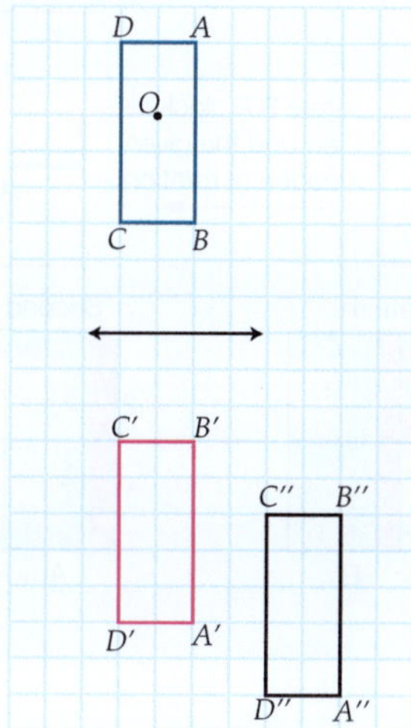

(b)

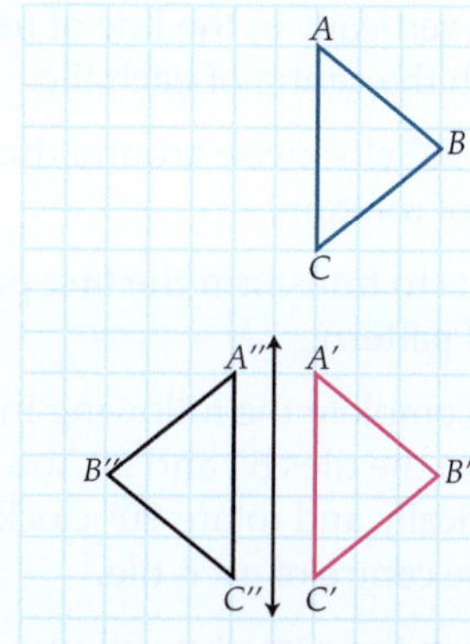

(c)

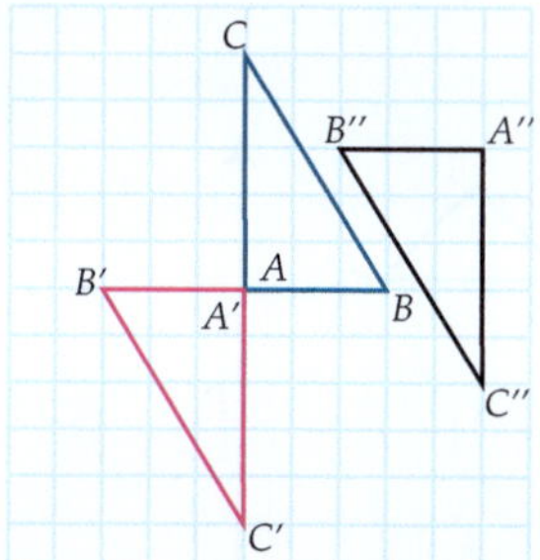

(d)

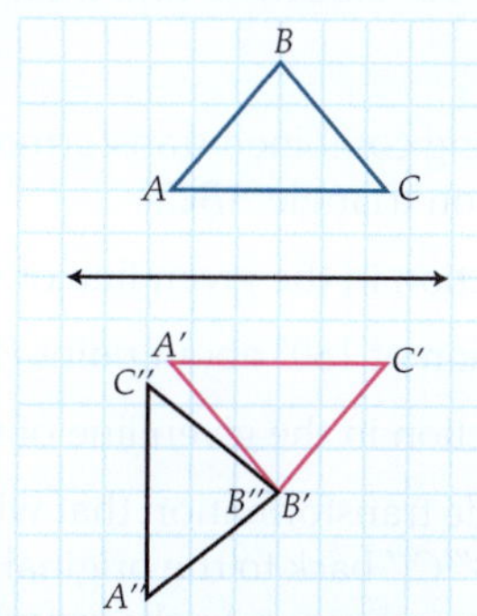

(e)

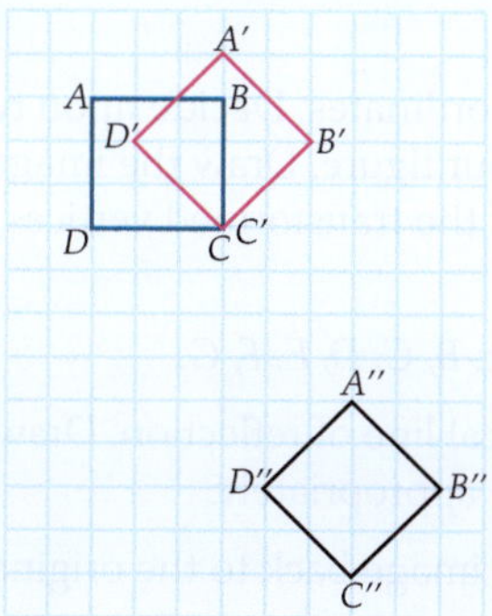

(f)

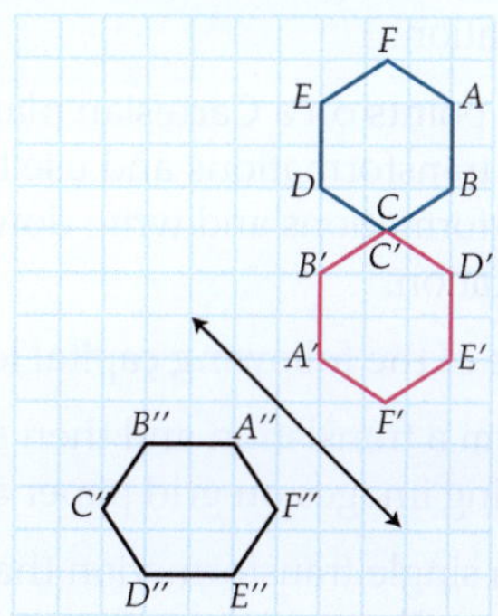

10.4

9 For the combined transformations stated in each part of Question **8**, identify the transformations required to move the final image back to the original position and orientation of the figure.

Reasoning, communicating and open-ended

10 **(a)** Without plotting points, write the coordinates of the image after the line joining $A(-2, 3)$ and $B(4, -1)$ is translated 1 unit right and 5 units down, then reflected in the y-axis.

(b) Plot the line AB, the first image $A'B'$ and the final image $A''B''$ on a Cartesian plane to check your answer to part **(a)**.

11 Sarah has written a computer program to transform pictures of tiles. There are only two instructions in her program.

1 Reflect vertically in the line of reflection through the centre of each tile.

2 Rotate 90° clockwise around the given centre of rotation.

Sarah wants to transform the first pattern to the second pattern.

Copy and complete the following instructions to transform the tiles B1 and B2. You must only reflect vertically and rotate 90° clockwise through the centre of each tile.

A1 – Tile is in the correct position.

A2 – Reflect vertically, and then rotate 90° clockwise.

B1 – Rotate 90° clockwise and then …

B2 – …

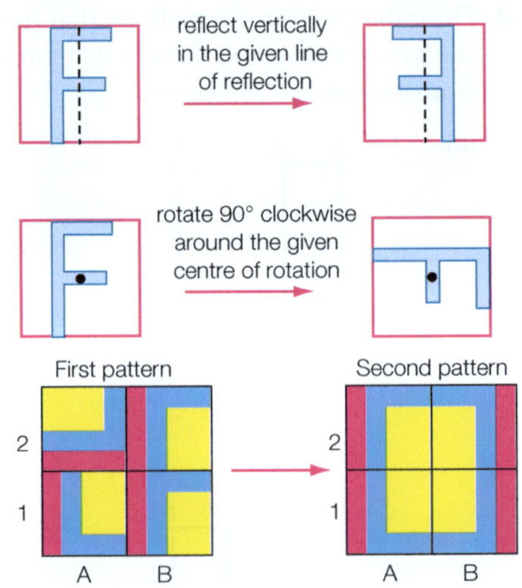

12 The following combined transformation is performed on triangle ABC.

- A reflection in the given line of reflection.
- A rotation of 180° about point P.
- A reflection in the given line of reflection.

Find a single transformation that will move the image $A'''B'''C'''$ back to the original position of the figure. Draw a diagram to show each required transformation.

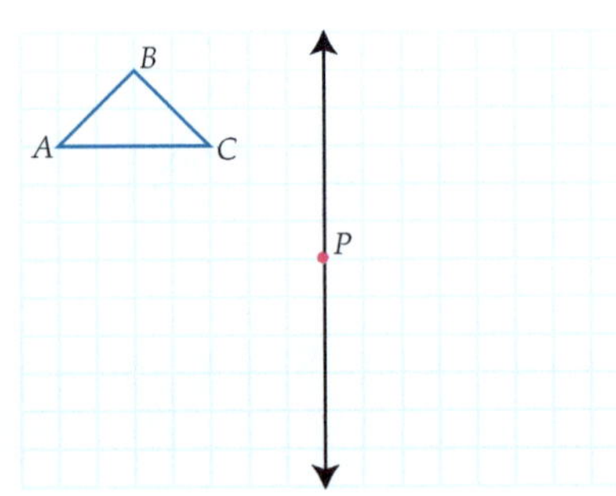

13 Plot three points on a Cartesian plane and write their coordinates. Decide upon two particular transformations and use them to transform your figure. Draw the image after both transformations and write down the coordinates of the transformed vertices using image notation.

14 Select one of the following capital letters of the alphabet: *B, C, D, E, F, G*.

(a) Perform a translation and then a reflection in a vertical line of reflection. Draw the resulting images on grid paper and label each image appropriately.

(b) Find a single transformation that will move the final image back to the original position. Show this transformation on your diagram.

(c) Use your answer for part **(b)** to find a single transformation that would produce the same result as the two transformations in part **(a)**.

15 A shape is rotated through 180° about a point O and then its image is reflected in a horizontal line of reflection passing through O.

(a) Choose any shape and, on grid paper, carry out the transformations described above.

(b) Find a single transformation that would have the same results as the two transformations described above.

Problem solving

Conway's army

Equipment required: 1–2 brains, grid paper, approximately 40 small counters

Conway's army is a one-person puzzle that was created by mathematician John Horton Conway in 1961. It is a variant of the peg solitaire game and uses a checkerboard of any dimension. The board is divided by a horizontal line; above it are empty cells and below it are a random number of game pieces, which are referred to as the 'soldiers'. Examples of different boards and arrangements of soldiers are as follows:

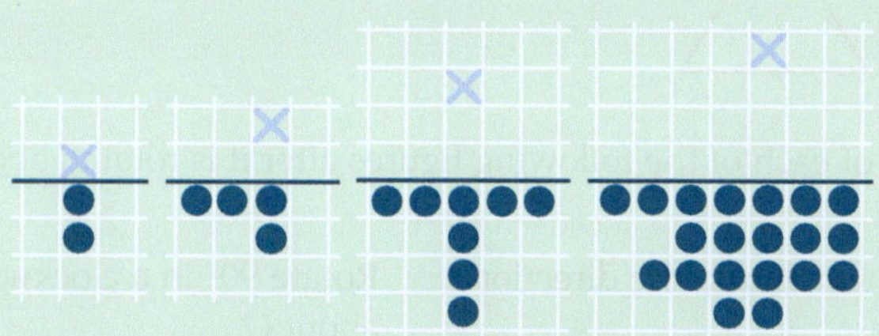

How to play:

Play proceeds by jumping horizontally or vertically (not diagonally) over other pieces onto an empty space, where jumped-over pieces are then removed.

How to win:

The goal is to place a soldier as far above the horizontal line as possible. The cross in each of the diagrams at left is the maximum distance that a soldier can be placed from the horizontal line for the given number of original soldiers.

Strategy options

- Act it out.
- Test all possible combinations.

Questions

1 Create your own checkerboards with the horizontal line drawn. Begin by investigating how far four counters can be moved beyond the horizontal line. Try at least three different arrangements of the four counters. Is it possible to move a soldier past the second row?

2 Continue to investigate with 4, 5, 6, 7, 8, 9, 10, 15 and 20 counters to see how far past the horizontal line the counters can be moved.

3 Use your observations thus far to predict if it is possible to move a soldier to the fifth row above the horizontal line. Explain your answers.

Half-time 10

Ex 10.2

1 Reflect each of the following figures in the line of reflection shown.

(a)

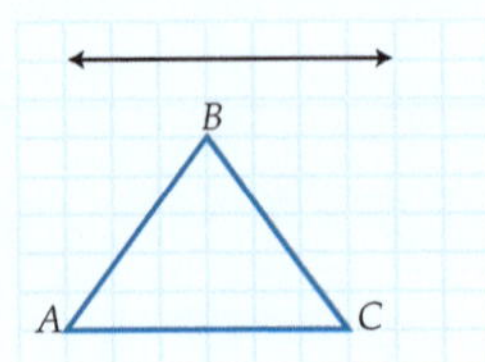

(b)

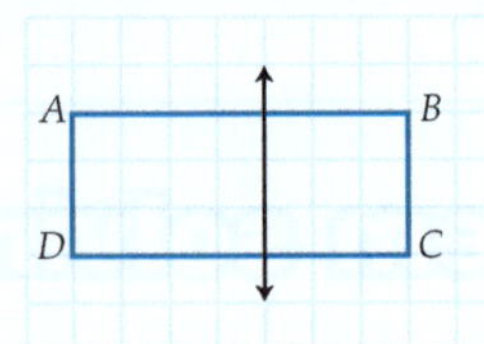

Ex 10.1

2 Describe the translation shown in each of the following diagrams.

(a)

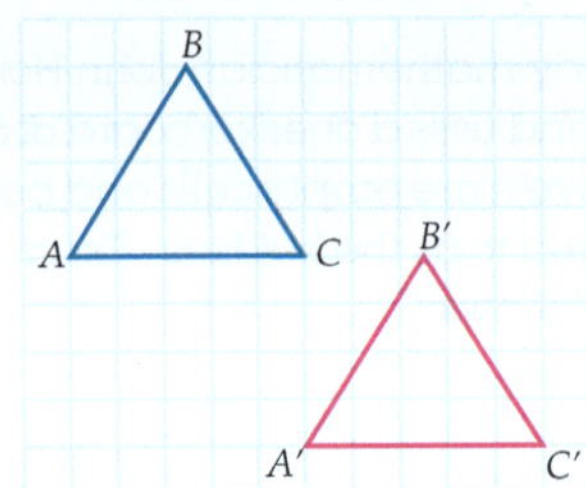

(b)

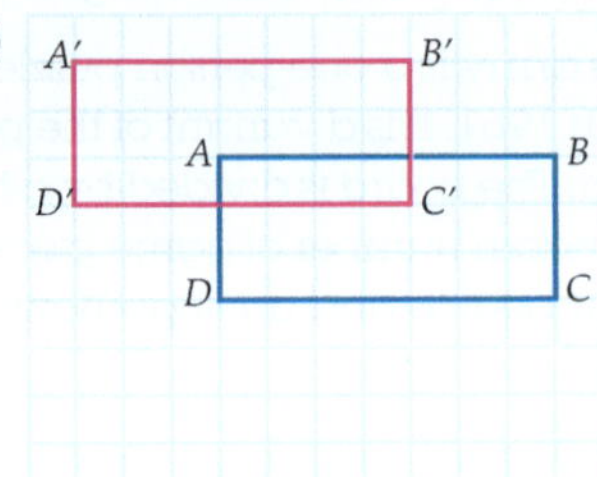

Ex 10.3

3 Show the image of each of the following figures after it is rotated about the given centre of rotation.

Rotate 180° in an anticlockwise direction about *O*.

(a)

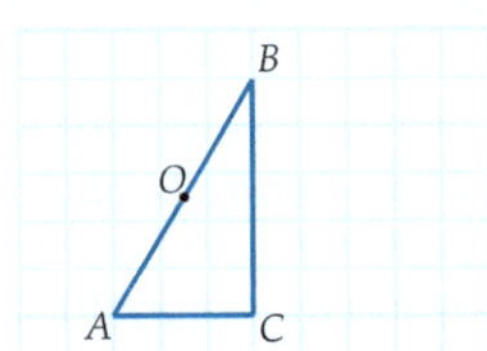

Rotate 90° in a clockwise direction about *O*.

(b)

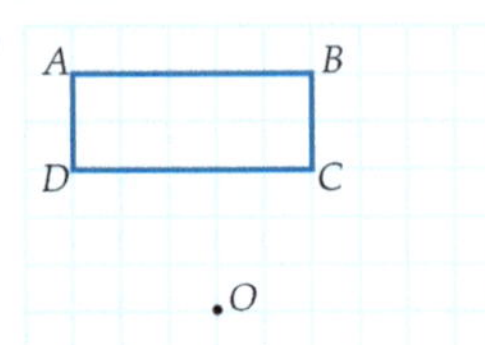

Ex 10.4

4 Reflect the following shape in the line of reflection shown, then translate it 2 units right and 1 unit down.

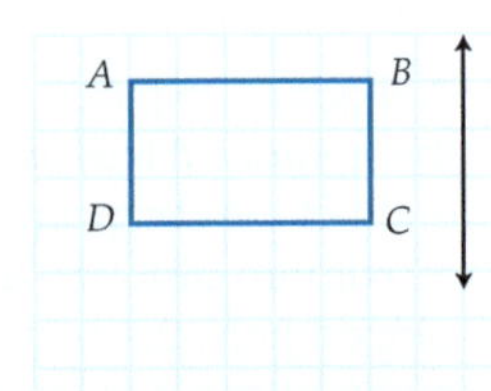

Ex 10.4

5 **(a)** Plot the points (1, 2), (3, 4) and (5, -2) on a Cartesian plane and join them to form a triangle.

(b) Draw the image after a translation of 6 units left and 1 unit up.

(c) Write the coordinates of the vertices of the translated points.

(d) Draw the image after the image drawn in part **(b)** is reflected in the x-axis.

(e) Write the coordinates of the vertices of this new image.

(f) Finally, rotate the image drawn in part **(d)** 180° clockwise about the origin.

(g) Write the coordinates of your final image.

Symmetry

10.5

Need to Know

Reflectional symmetry

Reflectional symmetry or mirror symmetry occurs when a line can be drawn through an object to divide it into two identical but reversed parts. Each part is the mirror image of the other. This line is called the **axis of symmetry**.

- An axis of symmetry is a mirror line that can be drawn through a shape to divide it into two identical but reversed halves.
- Some objects have multiple axes of reflectional symmetry.
- The **order of reflectional symmetry** is the number of axes of reflectional symmetry that can be drawn through a shape. Draw as many axes of symmetry as possible so that the object is divided into two identical but reversed halves. Count the number of axes of symmetry. This gives us the order of reflectional symmetry.
- An **asymmetrical** object has no **reflectional symmetry**.

Rotational symmetry

If a figure can be turned less than 360° about a point so that the image is identical to the original, that figure is said to have **rotational symmetry**. The point about which the figure is rotated is called the centre of rotation.

- The **order of rotational symmetry** is the number of times in a complete 360° rotation around the centre of rotation that an identical image of the original would be produced. Rotate the original around this point until an image identical to the original is reproduced. Repeat this process until the figure has rotated through 360°. Count the number of times in a complete 360° rotation that an identical image would be produced. This is the order of rotational symmetry.
- Dividing the order of symmetry into 360° gives the angle of rotation.

Worked Example 11

WE11

Copy each of the following figures and determine **(i)** whether it has reflectional symmetry by showing every possible axis of symmetry and **(ii)** the order of reflectional symmetry, if present.

(a)

(b)

Working

(a) (i)

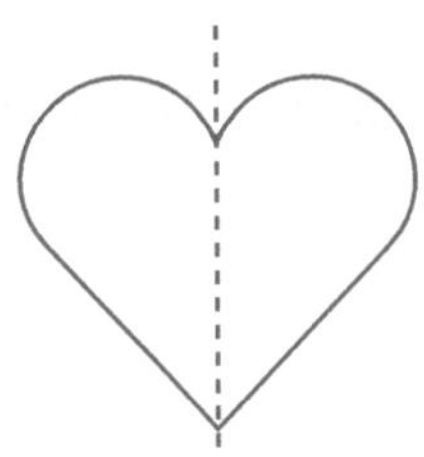

(ii) Order of reflectional symmetry = 1.

(b) (i)

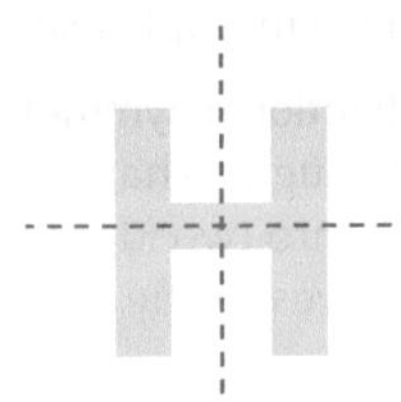

(ii) Order of reflectional symmetry = 2.

Worked Example 12

WE12

Determine **(i)** which of the following figures has rotational symmetry, **(ii)** the order of rotational symmetry and **(iii)** the size of the angle of rotation required to produce an identical image.

(a)

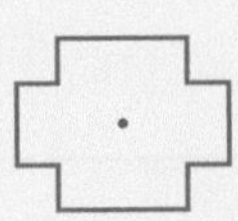

(b)

Working

(a) (i) Yes, it has rotational symmetry.

(ii) Order of rotational symmetry = 2.

(iii) $\frac{360}{2}$

$= 180°$

(b) (i) Yes, it has rotational symmetry.

(ii) Order of rotational symmetry = 5

(iii) $\frac{360}{5}$

$= 72°$

10.5 Symmetry

Answers page 513

Navigator

Q1, Q2, Q3, Q4, Q5, Q6, Q7, Q10, Q11, Q12, Q13, Q15	Q1, Q2, Q3, Q4, Q5, Q6, Q7, Q10, Q11, Q12, Q13, Q14, Q16	Q1, Q2, Q3, Q7, Q8, Q9, Q10, Q11, Q12, Q13, Q14, Q15, Q16, Q17

Equipment required: Grid paper for Question **12**

Fluency and understanding

1 Copy each of the following figures and determine **(i)** whether it has reflectional symmetry by showing every possible axis of symmetry and **(ii)** the order of reflectional symmetry, if present. WE11

(a) 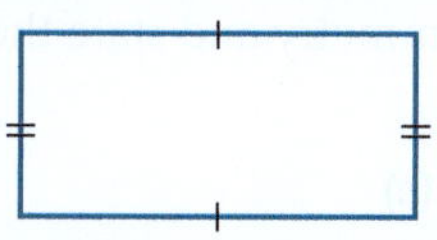

(b) 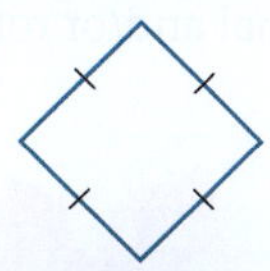

(c) 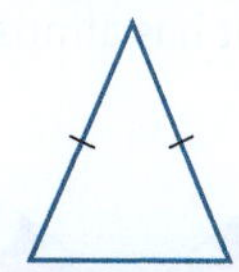

(d) 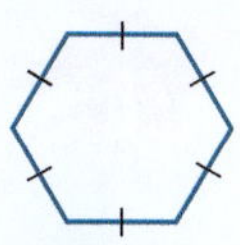

(e) **W**

(f) **J**

(g) **F**

(h) **X**

(i) 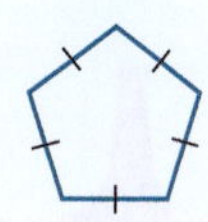

2 Determine **(i)** which of the figures from Question **1** has rotational symmetry, **(ii)** the order of rotational symmetry and **(iii)** the size of the angle of rotation required to produce an identical image. WE12

3 How many axes of symmetry do the following objects have?

(a) 

(b) 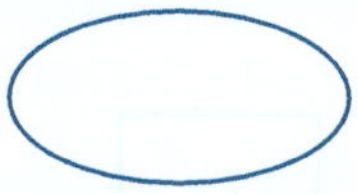

(c) 

For Questions **4**, **5** and **6** refer to the letters of the word M A T H S.

4 Which letter does not have an axis of symmetry?

A M **B** A **C** H **D** S

5 How many of the letters have one axis of symmetry?

A 5 **B** 2 **C** 3 **D** 4

6 Which letter has two axes of symmetry?

A H **B** M **C** A **D** S

7 Copy the lower case letters of the alphabet into your book.

a b c d e f g h i j k l m
n o p q r s t u v w x y z

(a) Which lower case letters are asymmetrical?

(b) Which lower case letters have one axis of symmetry?

(c) Which lower case letters have two or more axes of symmetry?

8 Copy the upper case letters of the alphabet into your book.

A B C D E F G H I J K L M
N O P Q R S T U V W X Y Z

(a) Which upper case letters are asymmetrical?

(b) Which upper case letters have one axis of symmetry?

(c) Which upper case letters have two or more axes of symmetry?

9 Do any of the capital letters from Question **8** have rotational symmetry? If so, which ones are they?

Reasoning, communicating and open-ended

10 Symmetry is present everywhere in nature. For each of the following images, determine whether it has almost perfect reflectional and/or rotational symmetry and state the order of each.

(a)

(b)

(c)

(d)

(e)

(f)

11 For each of the following shapes, state whether or not the dotted line is an axis of symmetry.

(a)
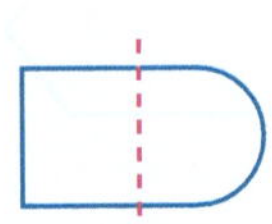

(b)
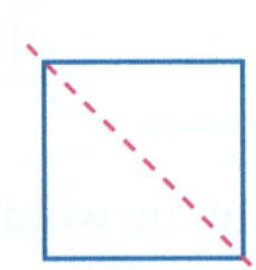

(c)
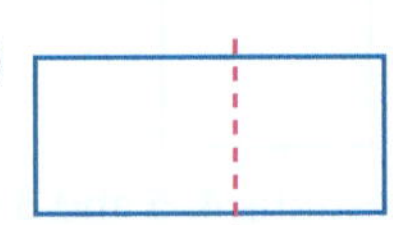

(d)

(e)
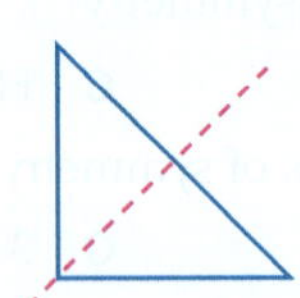

(f)
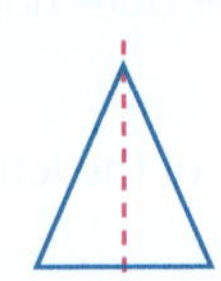

12 For each of the following diagrams, use the dotted line as an axis of symmetry to complete the diagram using grid paper.

(a)

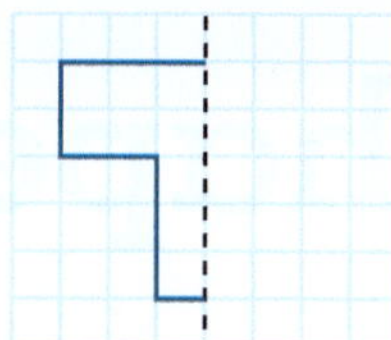

(b)

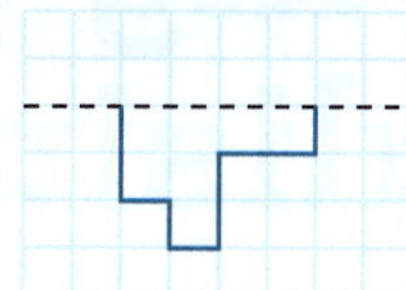

(c)

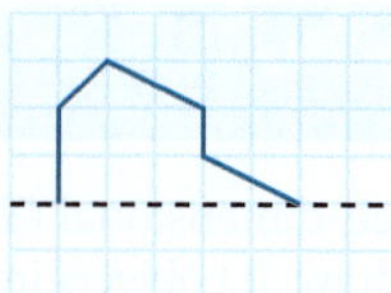

(d)

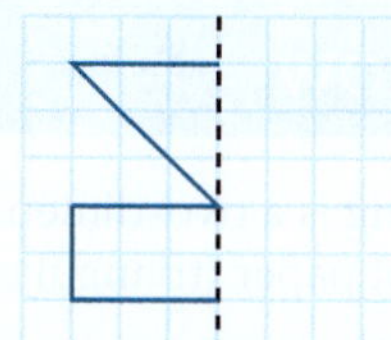

13 Which two quadrilaterals have exactly two lines of reflectional symmetry?

14 Which regular polygon has exactly three lines of reflectional symmetry?

15 WOW is a symmetrical word. Locate the axis of symmetry and find three more words that are also symmetrical.

16 The dates of some years have reflectional and/or rotational symmetry. An example of a year that has rotational symmetry is I96I and a year that has reflectional symmetry is I88I.

Identify an example of a year that has:

(a) rotational symmetry

(b) reflectional symmetry

(c) both reflectional and rotational symmetry.

17 Describe when the following statement is true and when it is false.

'This parallelogram has reflectional symmetry.'

Problem solving

Symmetry challenge

How many symmetrical designs of shaded squares can you find in a blank 3 by 3 square grid?

Hint: Before you shade any squares, can you identify the lines of symmetry present in this 3 by 3 grid shown at left? What symmetries are possible if you shade one square only?

Here are three examples of possible symmetrical designs. Note the position of the line of symmetry.

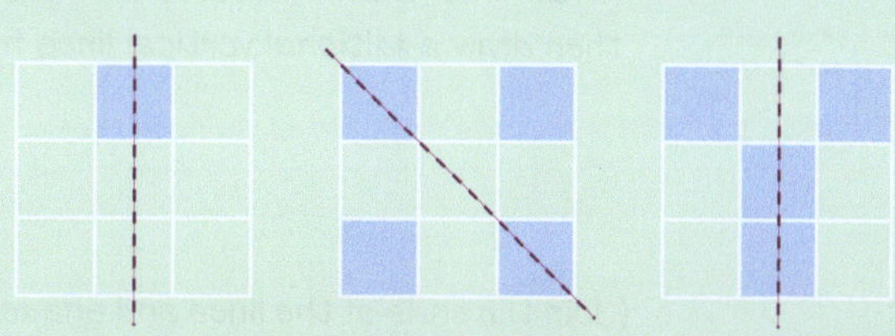

Don't include any rotations;

e.g. 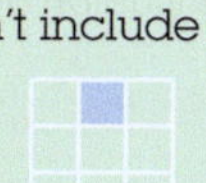 is considered the same as 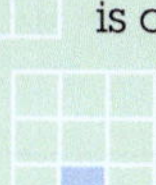.

Strategy options

- Draw a diagram.
- Look for a pattern.
- Break problem into manageable parts.

10.6 Drawing and visualising 3D shapes

Need to Know

An **isometric** drawing is a two-dimensional (2D) drawing of a three-dimensional (3D) object. Isometric dot paper (triangular dot paper) is used to help draw 3D objects in 2D.

It is recommended that isometric dot paper be held in this orientation:

Worked Example 13

WE 13

Use isometric dot paper to copy this shape.

Working

(Draw the top face and vertical lines from the corners of the face.)

(Draw lines from the bottom of the verticals you have just drawn, and then draw additional vertical lines from the ends of these lines.)

(Join the ends of the lines and shade in faces to create the illusion of 3D.)

10.6 Drawing and visualising 3D shapes

Answers page 514

Navigator

Q1, Q2 (a)–(c), Q3, Q4, Q5, Q6 (a)–(c), Q8, Q9 (a), Q10, Q11, Q13	Q1, Q2 (b), (d), (f), Q4, Q5, Q6 (c)–(e), Q7, Q8, Q9 (b), Q10, Q11, Q14	Q1, Q2 (d)–(f), Q4, Q6 (d)–(f), Q7, Q8, Q9 (c), Q10, Q11, Q12, Q14

Equipment required: Isometric dot paper for Questions **1**, **5**, **8**, **9** and **12–14**

Fluency and understanding

1 Use isometric dot paper to copy the following shapes.

WE 13

Hint

Is your isometric paper up the right way?

(a)

(b)

(c)

2 How many cubes would be required to build these solids? (Assume that there are no cubes missing at the back of the solids where you cannot see.)

(a)

(b)

(c)

(d)

(e)

(f)

3 The number of cubes required to build this solid is:

A 7 **B** 8

C 9 **D** 10

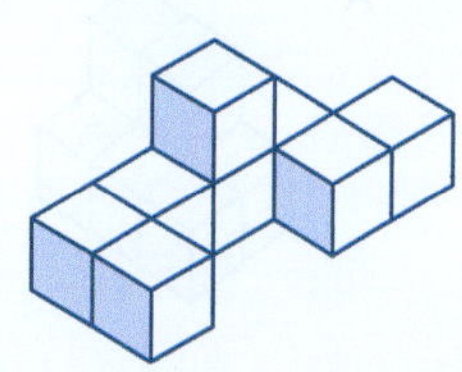

4 Which of the shapes below is *not* a rotation of the one shown?

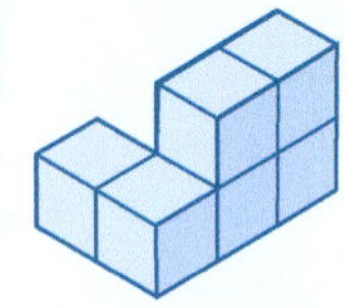

A

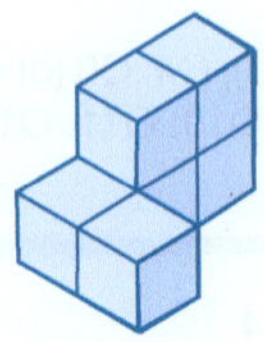

B

C

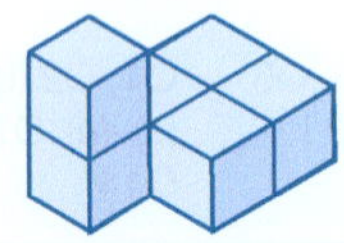

D

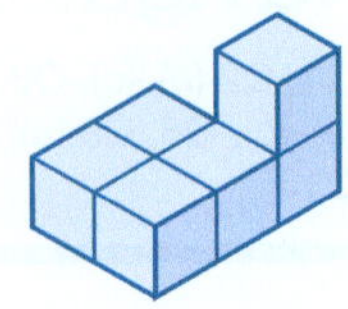

5 Using isometric dot paper, draw a 3D sketch of a Rubik's® cube. (A Rubik's cube is made of 27 cubes stacked three layers high, three layers wide and three layers deep.)

6 How many cubes are in each solid?

(a)

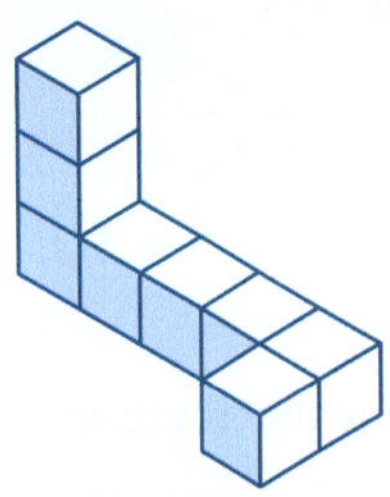

(b)

(c)

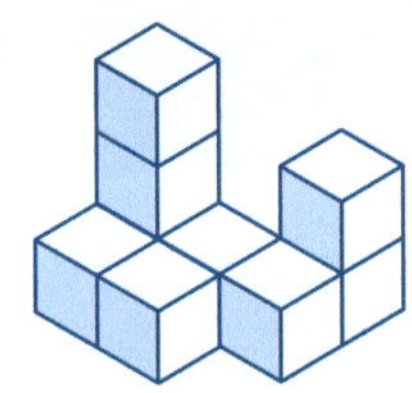

(d)

(e)

(f)

7 Which of the following shapes could result from joining the two shapes below?

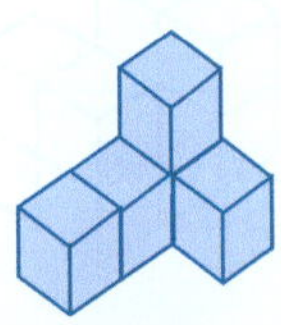

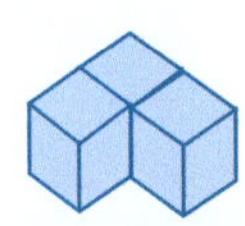

A

B

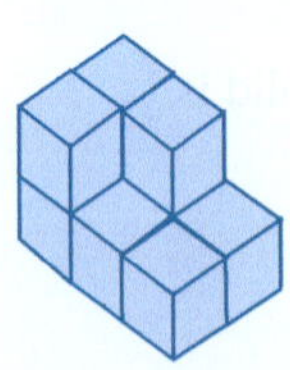

C

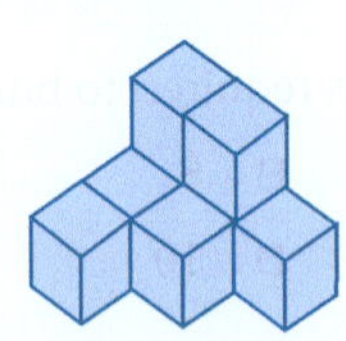

D

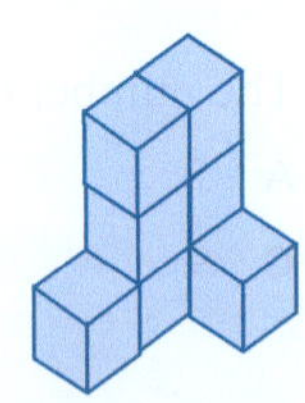

10.6

Reasoning, communicating and open-ended

8 Using this aerial photograph of the Pentagon in Washington, DC:

(a) draw a top view of the Pentagon

(b) draw a side view of the Pentagon.

(c) Does it matter which side you choose? Explain.

9 Draw each solid after the pink shaded blocks have been removed.

(a) 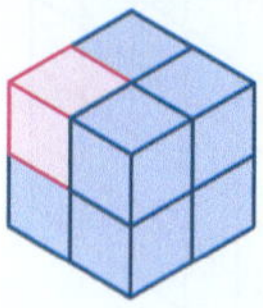

(b) 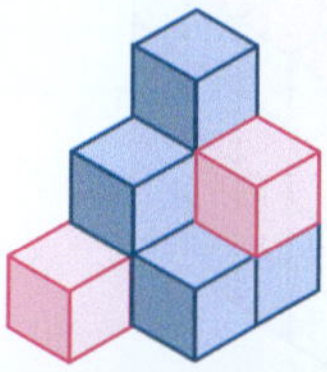

(c) 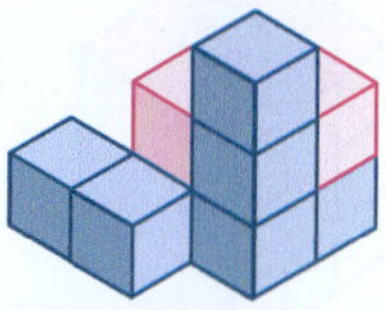

10 This shape is known as the impossible rectangle.

(a) Why is this shape known as the *impossible* rectangle? (You may like to try to make it with blocks.)

(b) Copy the impossible rectangle onto isometric dot paper.

(c) Draw a similar rectangle that is possible. Colour your drawing using three colours.

11 Draw the top view of this house.

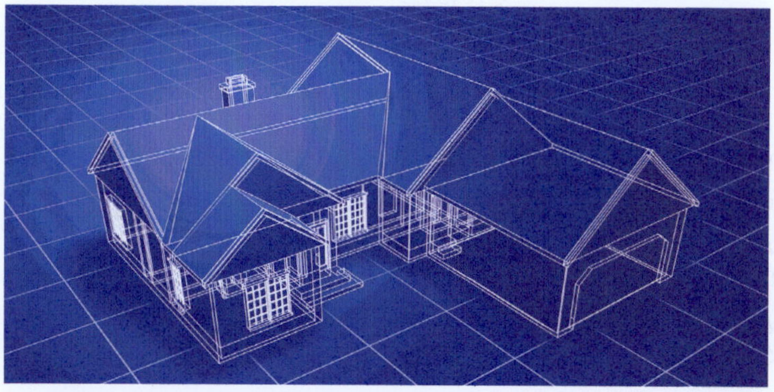

12 A sculptor is designing a sculpture based on an arrangement of 24 cubes. Each cube has a width of 50 cm and the maximum height of the sculpture is to be 2.5 m. The sculptor wants to have a maximum of six cubes at the front and a maximum of four cubes in depth. Each cube must have at least one whole face touching another cube. Use isometric dot paper to design some sculptures that meet all these requirements.

13 Using isometric dot paper, draw three different solids that can be built from 10 cubes that have whole faces touching.

14 Draw three shapes on isometric dot paper that, when joined together, form a $3 \times 3 \times 3$ cube.

Problem solving

What comes next?

Look at the pattern below.

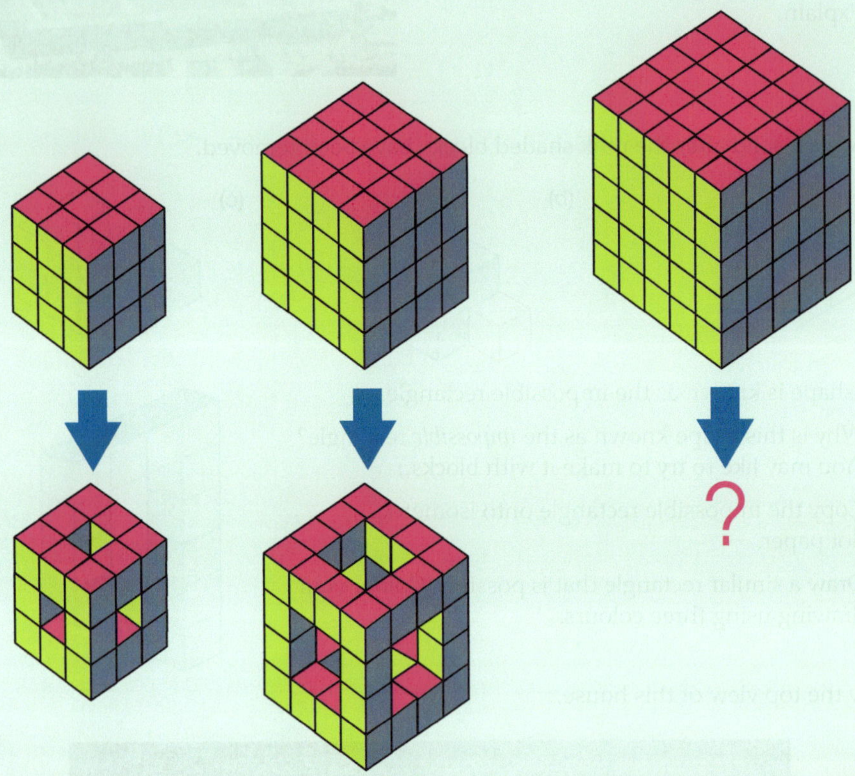

A solid 3 × 3 × 3 cube is made using 27 blocks. When one block is removed from the centre of each of the six faces and one block from the centre of the cube, a frame of the same cube is made using 20 cubes.

- Draw the next shape in the pattern.
- How many cubes are needed to make the frame of a 5 × 5 × 5 cube?
- How many cubes are needed to make the frame of a 6 × 6 × 6 cube?

Strategy options

- Make a table.
- Look for a pattern.

Plan views and elevations

10.7

Need to Know

Three-dimensional objects can be viewed from various positions or views.

Objects looked at from the top, front and side can be drawn in two dimensions, and are known as **plan views**, or orthogonal projections.

Drawings of houses and buildings viewed from a side are known as **elevations**.

Elevations may be referred to as front, rear and side elevations or by compass directions, such as east elevation.

Drawing elevations

- Highlight the face being viewed.
- Draw squares to make the shape you would see if you were looking at the face.

Worked Example 14

WE14

For the following solid, draw the:

(a) top view

(b) front view

(c) side view.

Working

(a)

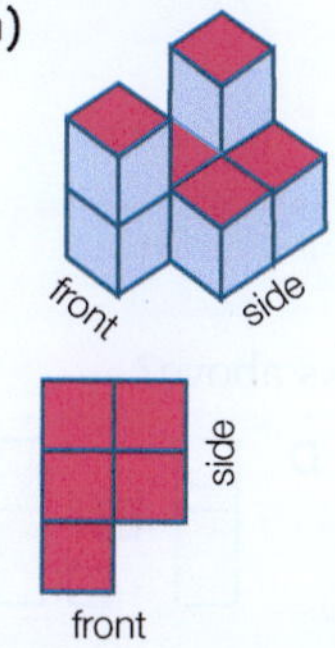

(b)

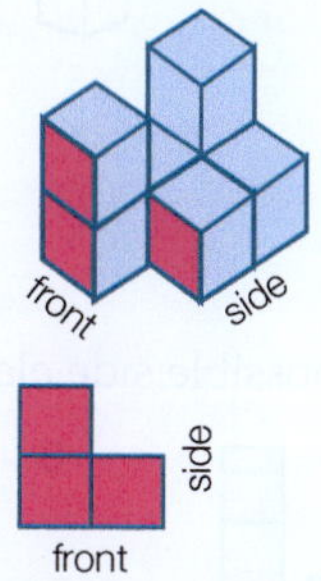

(c)

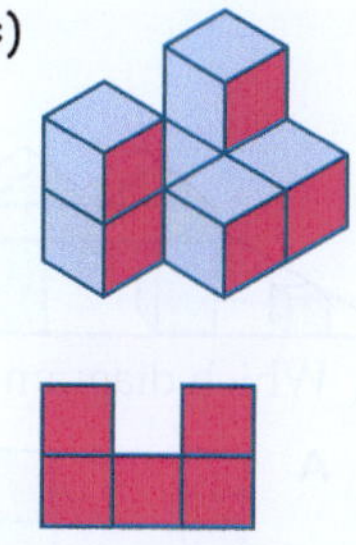

10.7

Answers page 514

10.7 Plan views and elevations

Navigator

Q1, Q2, Q3, Q5, Q6, Q7, Q9, Q10	Q1, Q2, Q3, Q4, Q5, Q6, Q7, Q9, Q10	Q1, Q2, Q3, Q4, Q5, Q6, Q7, Q8, Q9, Q10

Equipment required: Isometric dot paper for Questions **5** and **7–10**

Fluency and understanding

WE14

1 For each of the following solids, draw the:

(i) top view **(ii)** front view **(iii)** side view.

(a)

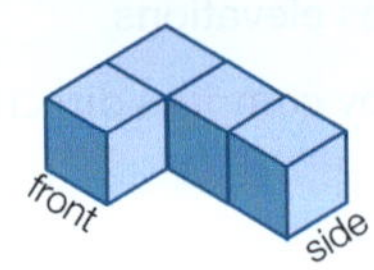

(b)

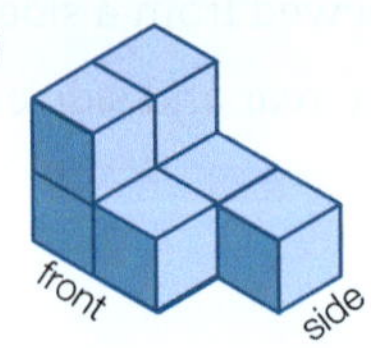

2 The front view of the solid shown has been shaded. The top view of this solid looks like:

A

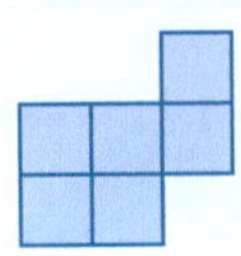

B

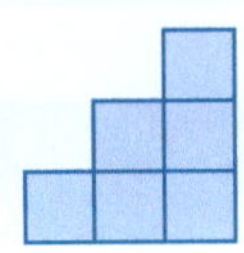

C

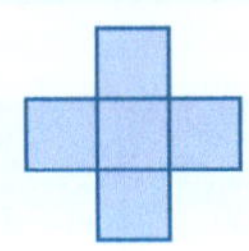

D

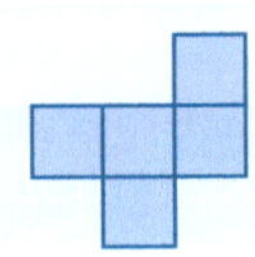

3 The following structures are made with coloured blocks. Draw the front elevation of each structure by imagining you are standing looking from the position of the arrow, then move 90° to the right and draw the side elevation.

(a)

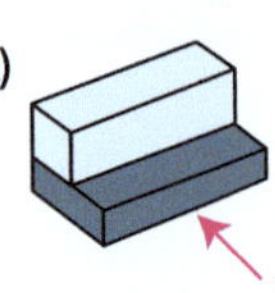

(b)

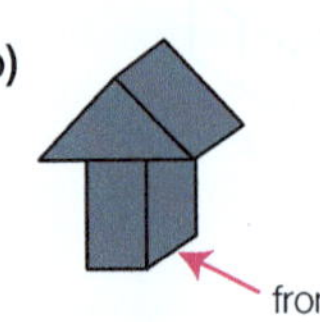

(c)

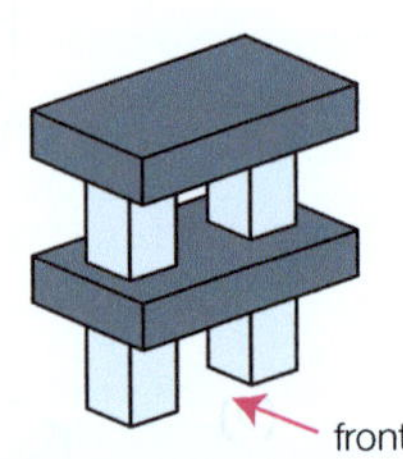

(d)

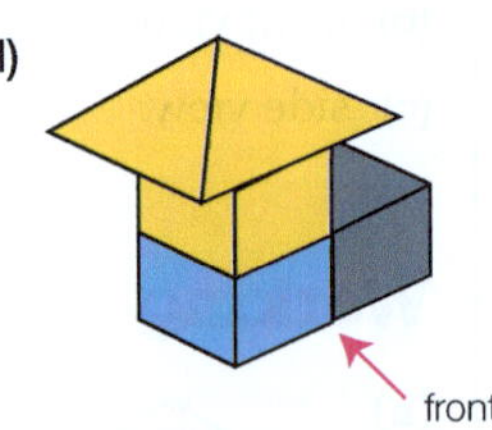

4

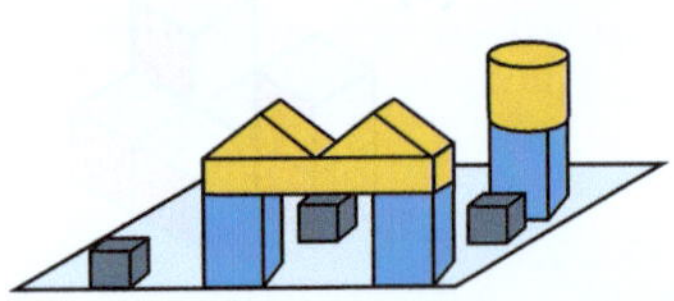

(a) Which diagram below shows a possible side elevation of the blocks above?

A

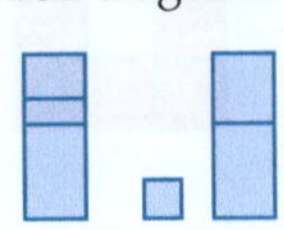

B

C

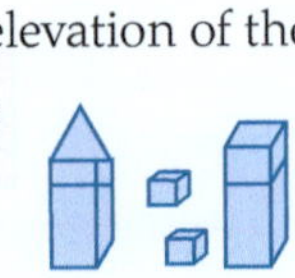

D

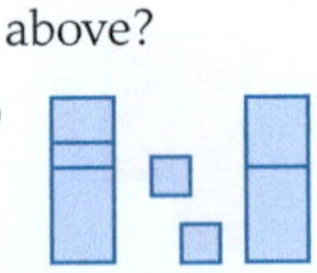

(b) Which elevation of the block structure above is shown opposite?

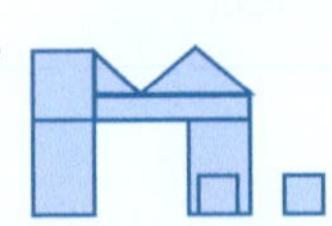

A front **B** rear

C side **D** top

5 For each set of the plan views below, draw the 3D solid on isometric dot paper.

top view | front view | side view

(a)

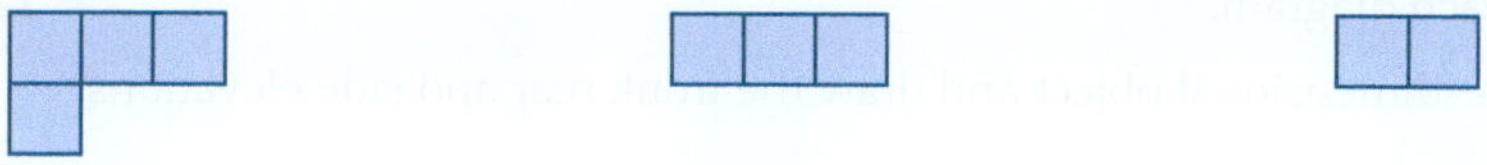

(b)

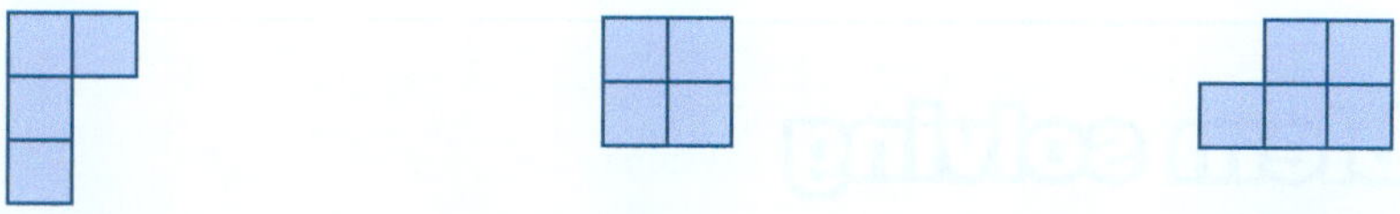

(c)

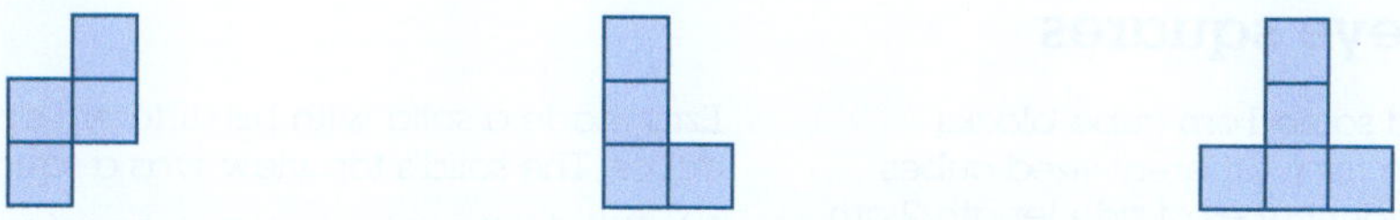

6 A *mat plan* shows the top view of a shape and the number of cubes in each position.

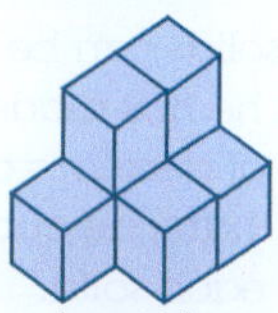

isometric drawing

2	1
2	1
1	

mat plan

Draw the mat plan for each of the shapes below.

(a)

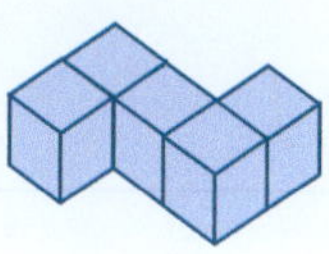

(b)

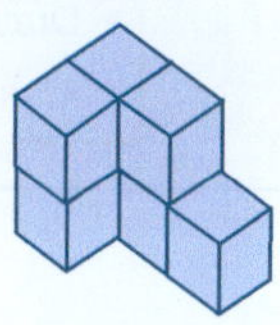

(c)

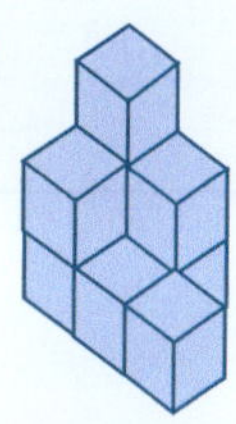

Reasoning, communicating and open-ended

7 Draw each of the following mat plans as 3D shapes on isometric dot paper.

(a)

1	1
1	

(b)

3	2
1	

(c)

3	3	3
2	2	
1		

8 (a) Copy the following diagrams onto isometric dot paper.

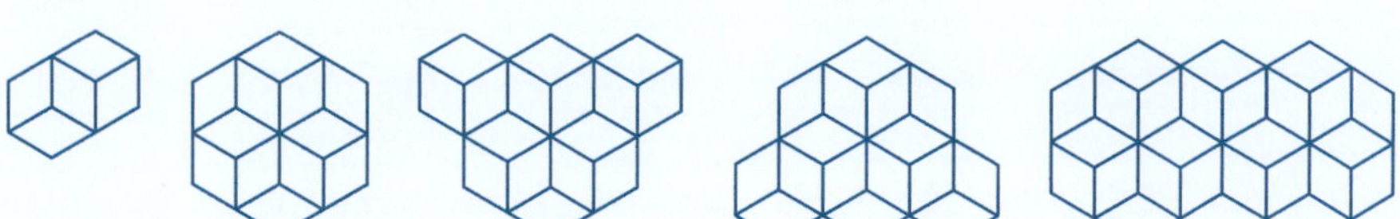

(b) On your drawings, use three colours to shade the different faces of the cubes.

(c) How many cubes are in each of the diagrams?

(d) There is more than one way to colour the diagrams. If the cubes were coloured differently, would each diagram still look like it has the same number of cubes?

(e) Draw the top view of each shape.

10.7

9 A shape made with small cubes looks like this from above.

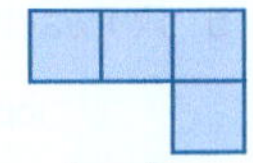
top view

Draw three possible 3D drawings on isometric dot paper, stating the number of cubes in each diagram.

10 Select a three-dimensional object and draw the front, rear and side elevations.

Problem solving

Bird's-eye squares

Ezra glued some 1 cm cube blocks together to form different-sized cubes. He made four cubes of side length 2 cm, four cubes of side length 3 cm and had four 1 cm cubes left over.

Ezra made a solid with his different-sized cubes. The solid's top view was a square.

(a) What is the minimum number of cubes Ezra would need to build his solid?

(b) How many different solids can be made with the cubes he has made that have a top view of a square and are made with more than one cube? Draw the top view of each solid.

Strategy options

- Draw a diagram.
- Make a model.

Challenge 10

1 The equilateral triangle OAB has been transformed into the equilateral triangle $OA'B'$.

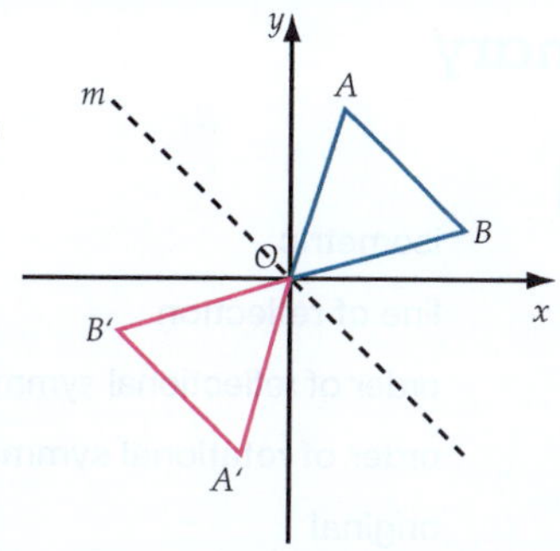

(a) Which single transformation changes OAB into $OA'B'$?

(b) Which two different transformations, applied one after the other, would change OAB into $OA'B'$?

(c) OAB is reflected in the line m. What is the name of the new position of A?

2 Jemima looks in a mirror and sees the clockface behind her. The hands appear to show 2:45. What will the time appear to be in 15 minutes?

3 A horizontal line CD is copied three times, and each copy appears on top of the original. Using the terms 'rotation' and 'translation', describe how each copy must be moved in order to create a square.

4 If the shape 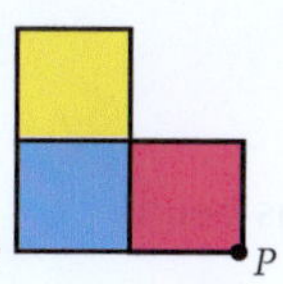 is rotated 90° clockwise about the point P, the resulting figure could be:

A

B

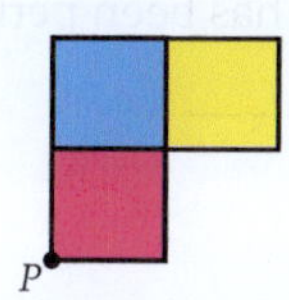

C

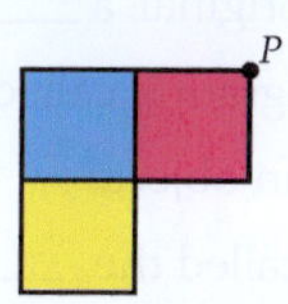

D

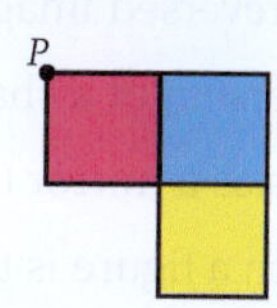

5 To be decoded, a message has to be read by looking in a mirror. Decode the following message: 'TAƎB OT ᗡЯAH ƧI ƎƆƆƎ ᗡƎ⅃IOB A'. To make sense of the message, what did you have to do?

6 A line is drawn on the 8 cm by 3 cm rectangle opposite to make a triangle and a trapezium.

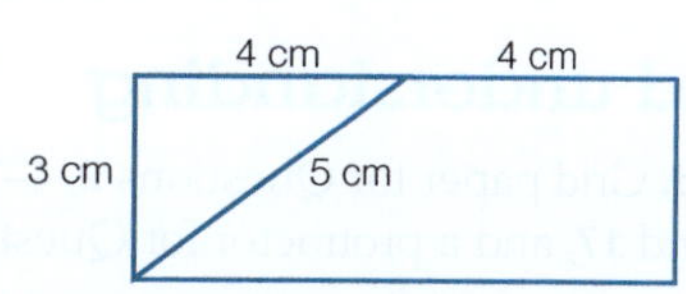

What transformation(s) need to be done to the triangle to create each of the following shapes?

(a)

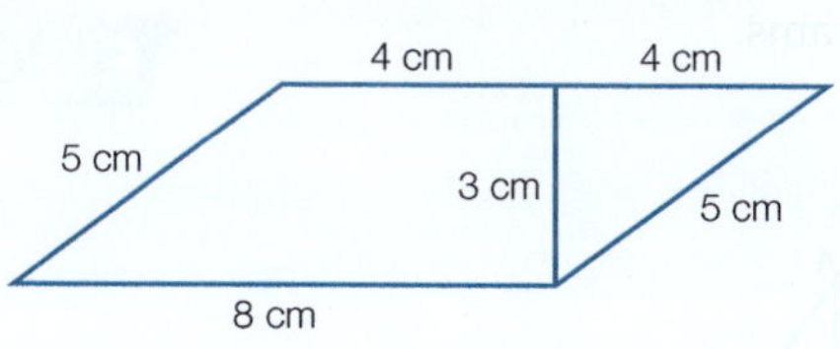

(b)

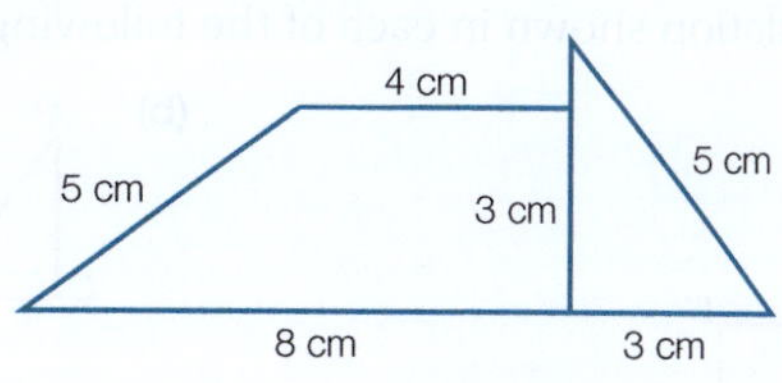

(c)

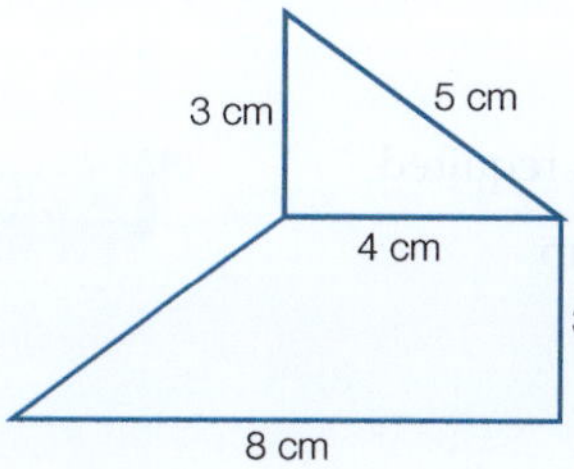

(d)

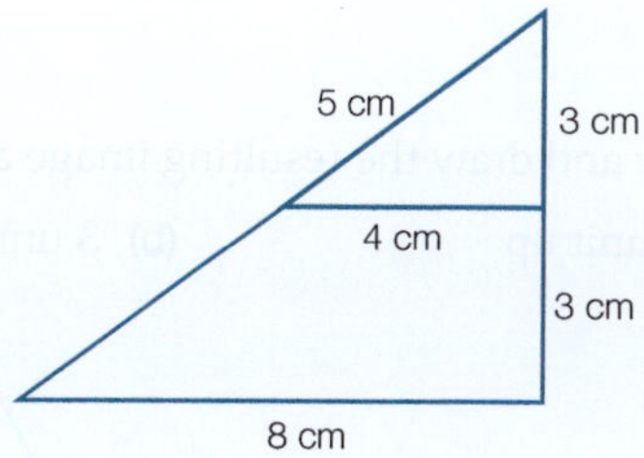

Chapter review 10

D.I.Y. Summary

Key Words

asymmetrical	isometric	perpendicular distance	rotation
axis of symmetry	line of reflection	plan views	rotational symmetry
centre of rotation	order of reflectional symmetry	reflection	transformation
elevations	order of rotational symmetry	reflectional symmetry	translation
image	original	rotate	

Copy and complete the following using the words and phrases from this list, where appropriate, to write a summary for this chapter. A word or phrase may be used more than once.

1 Reflections, rotations and translations are three types of __________s.

2 When an object is turned about a point, a __________ has been performed.

3 When a figure is moved in a horizontal and vertical direction, a __________ has been performed.

4 When a figure is a reversed image of the original, a __________ has been performed.

5 The transformed version of a shape or diagram is called an __________.

6 A __________ acts as a mirror to reflect an object.

7 A point about which a figure is turned is called the __________.

8 The __________ indicates how many times the image is identical to the original in a complete 360° rotation.

9 A 2D representation of a 3D solid can be drawn on __________ paper.

Fluency and understanding

Equipment required: Grid paper for Questions **2**, **4–7**, **9**, **23** and **28**, isometric dot paper for Questions **14**, **16** and **17**, and a protractor for Questions **7**, **9**, **23** and **28**

1 Describe the translation shown in each of the following diagrams. **Ex 10.1**

(a)

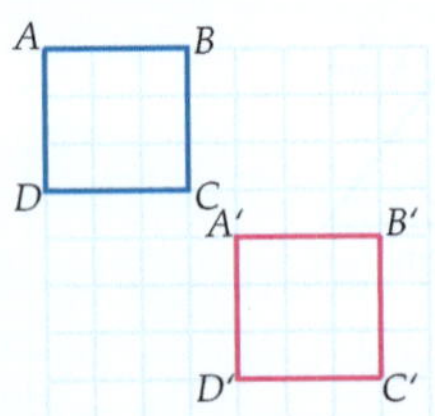

(b)

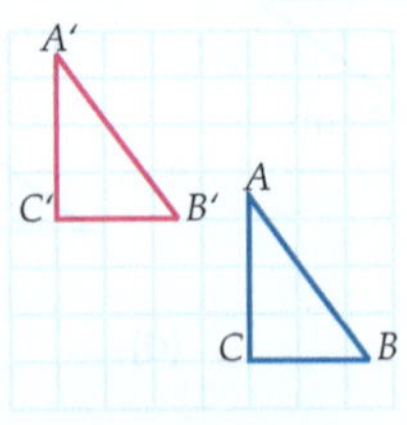

2 Copy each figure below and draw the resulting image after the translation required. **Ex 10.1**

(a) 3 units right and 1 unit up

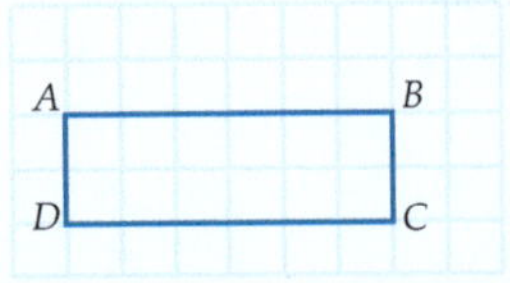

(b) 3 units left and 3 units up

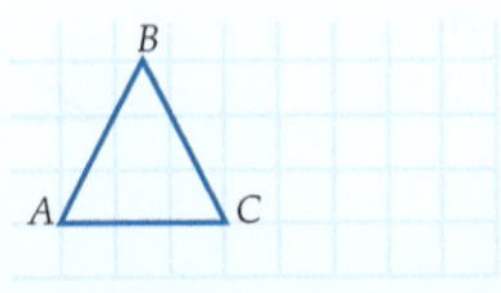

3 What would be the reverse translation of 7 units right and 4 units down? Ex 10.1

A 7 units up and 4 units right
B 7 units left and 4 units up
C 7 units left and 4 units down
D 4 units left and 7 units up

4 For the figure in the diagram opposite: Ex 10.1

(a) write the coordinates of the vertices
(b) copy the figure and translate it 4 units left and 3 units down, marking the image vertices with image notation
(c) write the coordinates of the vertices of the image.

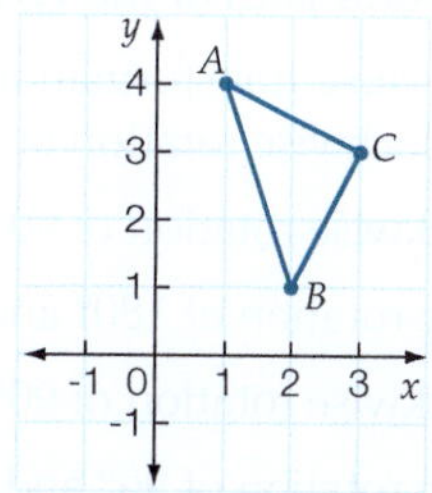

5 Copy each of the following figures onto grid paper and draw the resulting image when the figure is reflected in the line of reflection shown. Ex 10.2

(a)

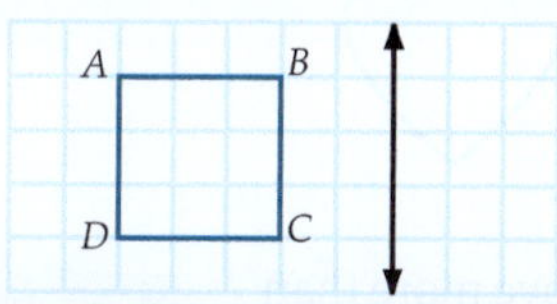

(b)

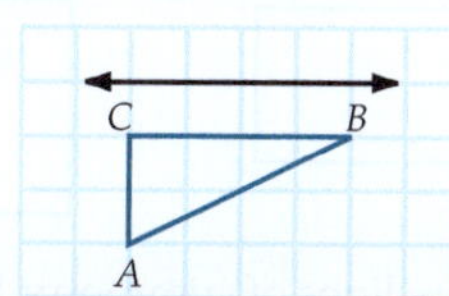

6 For the figure in the diagram opposite: Ex 10.2

(a) write the coordinates of the vertices
(b) copy the figure and reflect it in the x-axis, marking the image vertices with image notation
(c) write the coordinates of the vertices of this image
(d) copy and reflect the original figure in the y-axis, marking the image vertices with image notation
(e) write the coordinates of the vertices of this image.

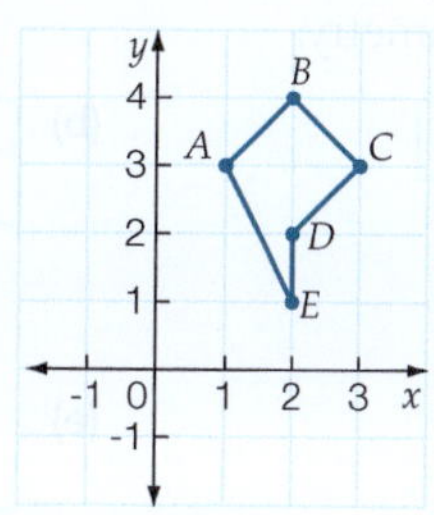

7 Copy each of the following figures onto grid paper and draw the resulting image after the given rotation. Use a protractor to help you. Ex 10.3

(a) Rotate 90° in a clockwise direction about O.

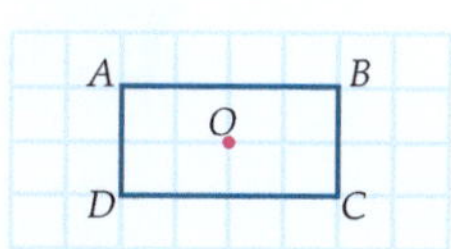

(b) Rotate 180° in an anticlockwise direction about O.

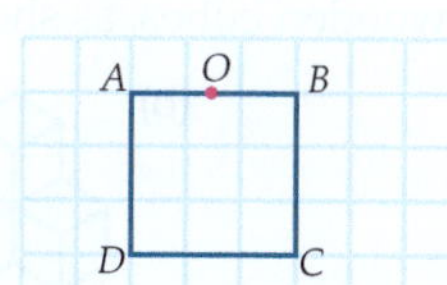

8 A clockwise rotation of 90° is equivalent to an anticlockwise rotation of how many degrees? Ex 10.3

A 180°
B 360°
C 90°
D 270°

9 For the figure in this diagram: Ex 10.3

(a) write the coordinates of the vertices
(b) copy the figure and rotate it 90° in an anticlockwise direction about the origin, marking the image vertices with image notation
(c) write the coordinates of the vertices of this image
(d) copy and rotate the original figure 180° in an anticlockwise direction about the origin, marking the image vertices with image notation

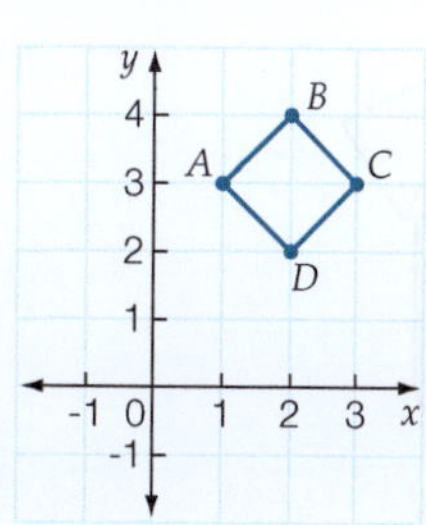

(e) write the coordinates of the vertices of this image

(f) copy and rotate the original figure 270° in an anticlockwise direction about the origin, marking the image vertices with image notation

(g) write the coordinates of the vertices of this image.

10 What are the reverse transformations for a translation of 5 units right and 8 units down followed by a clockwise rotation of 90°? **Ex 10.4**

A An anticlockwise rotation of 90° and a translation of 5 units right and 8 units down

B A clockwise rotation of 180° and a translation of 5 units right and 8 units down

C An anticlockwise rotation of 90° and a translation of 5 units left and 8 units up

D A clockwise rotation of 90° and a translation of 5 units left and 8 units down

11 In which of the following shapes is the dotted line an axis of symmetry? **Ex 10.5**

A
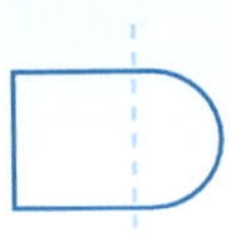
B
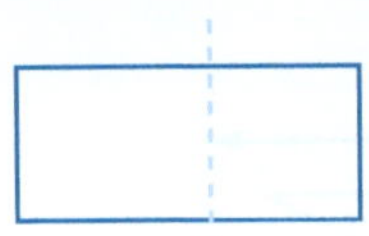
C
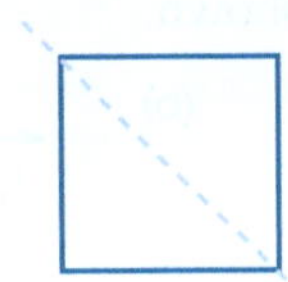
D

12 Copy the following shapes and draw lines of symmetry. Some shapes may have more than one line of symmetry. **Ex 10.5**

(a)
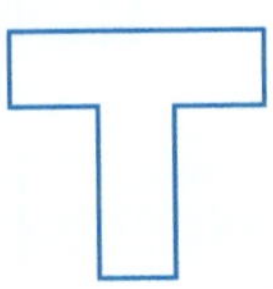
(b)

(c)
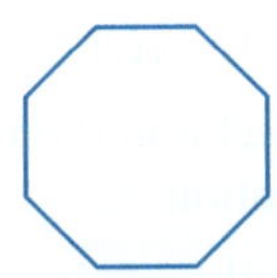
(d)

(e)

(f)

13 Identify the shapes from Question **12** that have rotational symmetry. **Ex 10.5**

14 Two solids have been made using wooden cubes, as shown. **Ex 10.6**

(a)

(b)
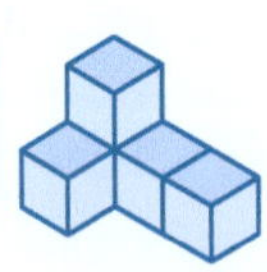

Use isometric dot paper to copy the shapes.

15 For each of the three-dimensional structures below, draw the front and side elevations. **Ex 10.7**

(a)

(b)
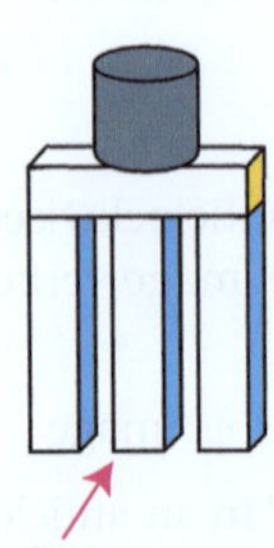

16 Use isometric dot paper to draw the shape shown.

17 Use the shape above to draw:

(a) a front view **(b)** a top view **(c)** a side view.

18 What single translation would be equivalent to 6 units down, 7 units right, 2 units up, 8 units left, 4 units up and 5 units right?

19 **(a)** Using the diagram below, reflect the star in the vertical and then the horizontal lines of reflection.

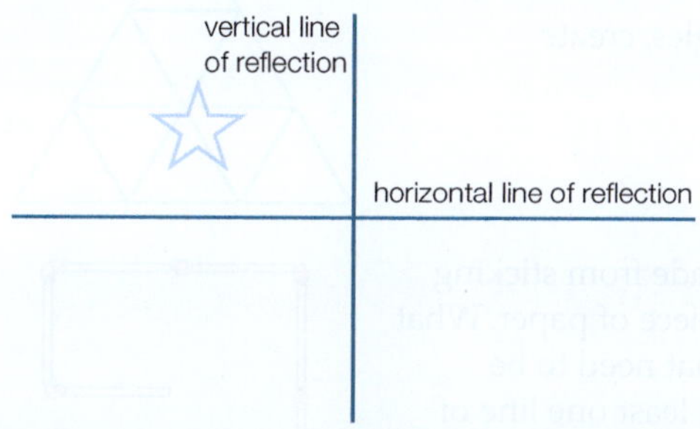

(b) Using the same starting position, reflect the star in the horizontal and then the vertical lines of reflection. Does it result in the same outcome as in **(a)**?

20 What rotation is equivalent to reflecting an object along horizontal, then vertical, lines of reflection?

21 How many letters of the word MATHEMATICS do not have reflectional or rotational symmetry?

22 The plan views of a three-dimensional shape made with cubes are shown. Draw the solid on isometric dot paper.

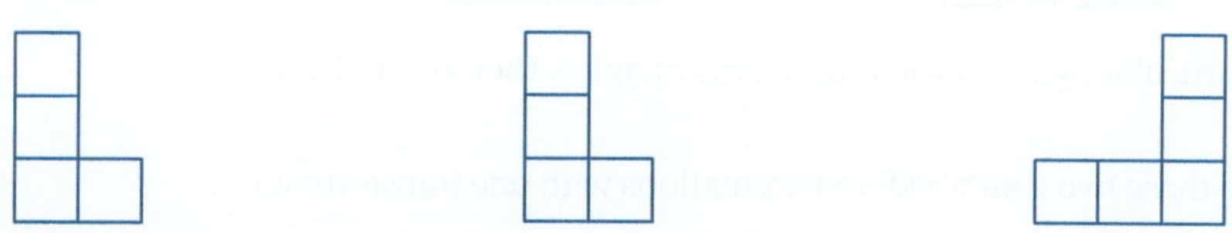

23 On grid paper:

(a) draw the line joining the points $A(-1, 3)$ and $B(2, 5)$

(b) translate this line 3 units right and 2 units down

(c) reflect this image in the x-axis to form a new image

(d) rotate this image 90° in a clockwise direction (about the origin) to form a final image

(e) write the coordinates of the final image.

Reasoning and communicating

24 The following 3 by 3 squares comprise identical symbols, but each one has been reflected horizontally, vertically or along the diagonal and is missing three shapes. Your task is to draw the missing three shapes.

(a) 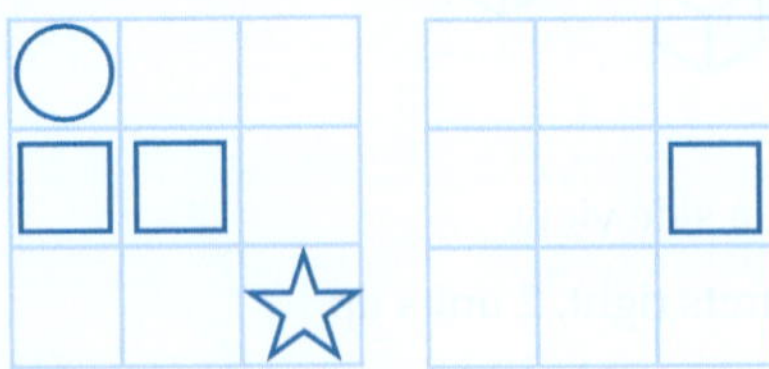

(b) 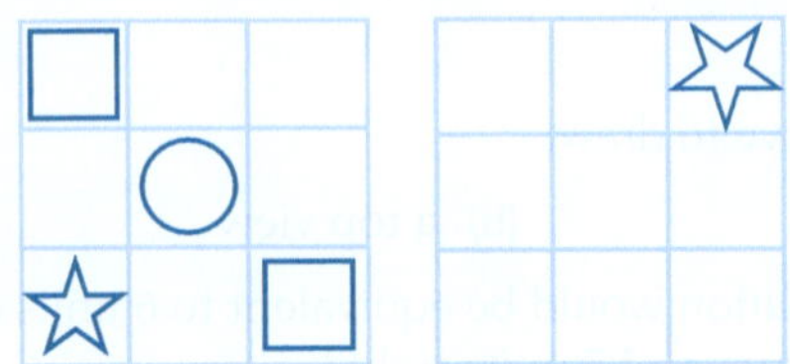

25 The following equilateral triangle comprises nine smaller equilateral triangles.

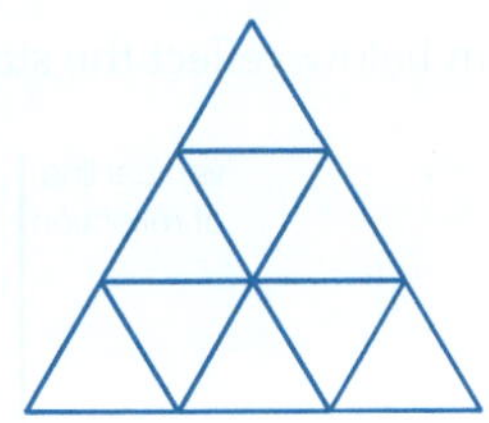

By shading the smaller equilateral triangles, create three different designs that have:

(a) one line of symmetry

(b) three lines of symmetry.

26 The diagram at right shows a pattern made from sticking seven matchsticks of equal length to a piece of paper. What is the smallest number of matchsticks that need to be added so that the resulting image has at least one line of reflectional symmetry?

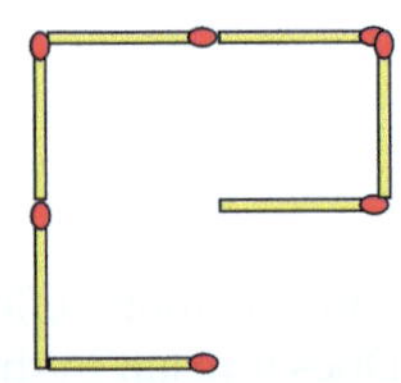

27 A solid is built using cubed blocks. The front, side and top views of the solid are shown. What is the minimum number of blocks this solid can have?

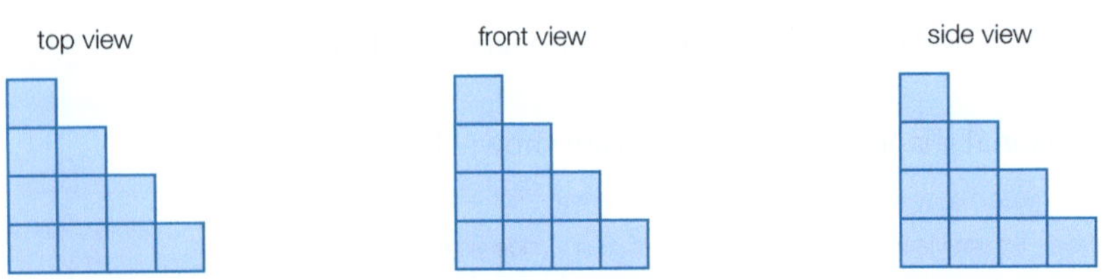

28 Matt is asked to reflect a particular figure in the y-axis. This image is then rotated 180° in a clockwise direction.

Matt believes he can replace these two combined transformations with one transformation.

Choose a shape of your own and perform the two transformations on it that Matt was required to perform. Is he correct? Can one transformation replace the two transformations? If so, write the required transformation.

NAPLAN practice 10

Numeracy: Non-calculator

1 The translation shown is:

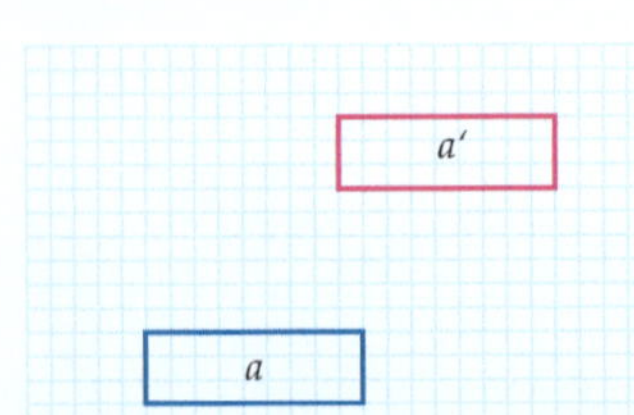

A 6 units right, 8 units up

B 8 units right, 9 units up

C 8 units down, 6 units left

D 6 units down, 8 units left

2 The rotation needed to transform the original shape (blue) to the new rotated image (pink) is:

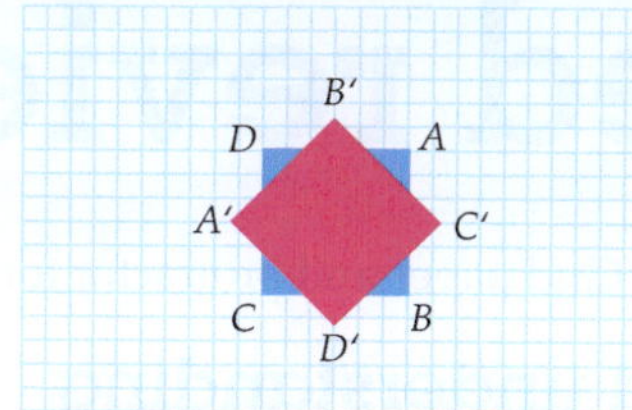

A 45° clockwise

B 135° anticlockwise

C 90° clockwise

D 90° anticlockwise

3 How many of the six faces of a die have fewer than three lines of symmetry?

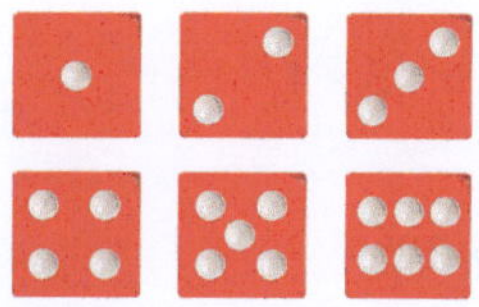

A 1 **B** 2

C 3 **D** 4

4 The top view of a shape is shown at right.

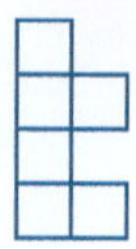

Which solid below has this top view?

A

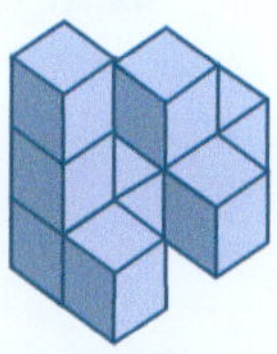

B

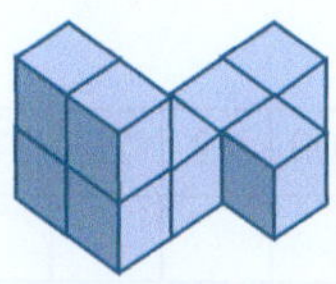

C

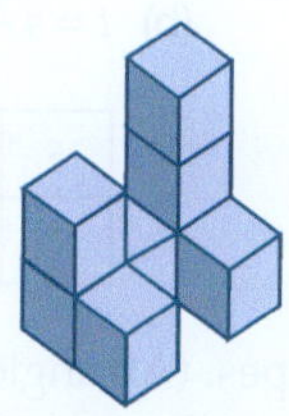

D

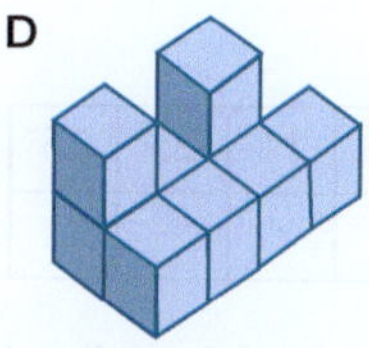

Numeracy: Calculator allowed

5 Mia has these S-shaped tiles.
They are blue on one side and red on the other side.
Mia wants to make a pattern with all the tiles blue side up.
Which one of the patterns can Mia not make?

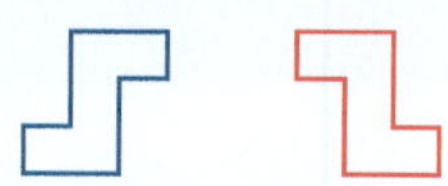

A

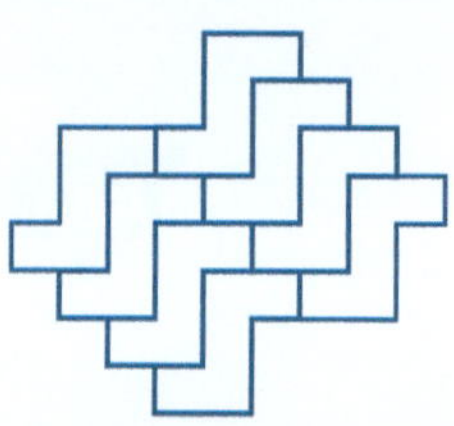

B

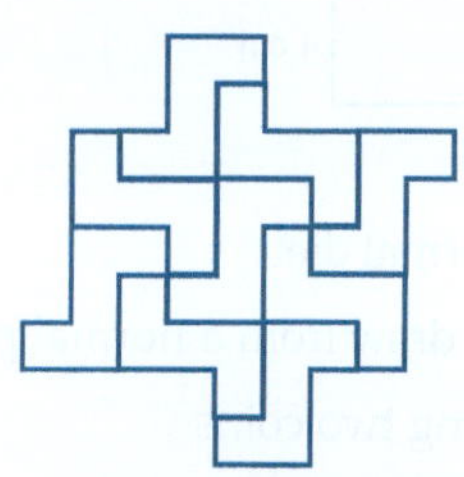

C

D

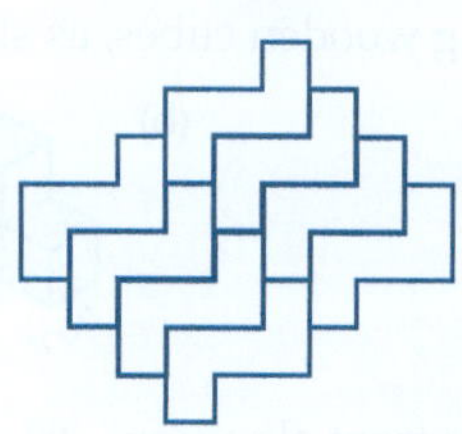

6 Which letter in the word SAME does not have an axis of reflectional symmetry?

A S **B** A **C** M **D** E

7 Janita glued small wooden cubes together to form the solid shown.
She then painted the solid on all surfaces.

How many wooden cubes are painted on just one face?

A 0 **B** 2 **C** 4 **D** 8

Mixed review E

Equipment required: Grid paper and protractor for Questions **11**, **17** and **18**

Fluency and understanding

1 Simplify: **Ex. 1.5**

(a) $44 - 8 \times 4$ (b) $8 \times 6 \div 3 \times 4$ (c) $28 \div 4 + 3 \times 6$

2 Simplify: **Ex. 1.2**

(a) $2^4 - 3^2$ (b) $3^3 \times 10^2$ (c) $10^5 - 10^3$

3 Calculate: **Ex. 4.4–4.6**

(a) $15.006 + 2.45 + 0.059$ (b) 0.2×0.036 (c) $0.597 \div 0.3$

4 Use the following rules to complete the tables. **Ex. 5.3**

(a) $n = 3m$

m	4	0	11	25	15
n					

(b) $t = s - 12$

s	18	100	12	22	312
t					

5 Find the area of each of the following shapes. (All angles are right angles.) **Ex. 6.5**

(a)

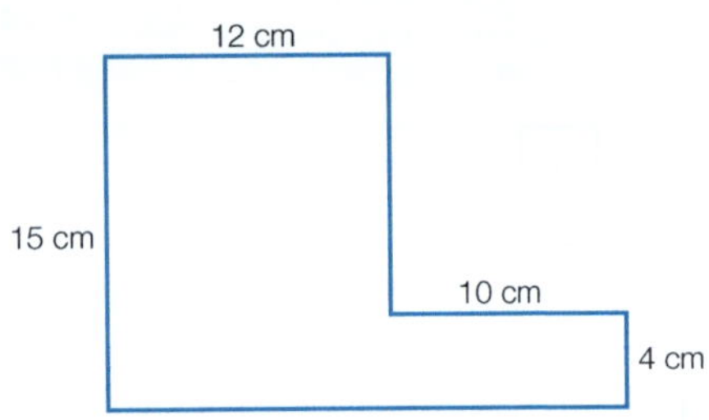

(b)

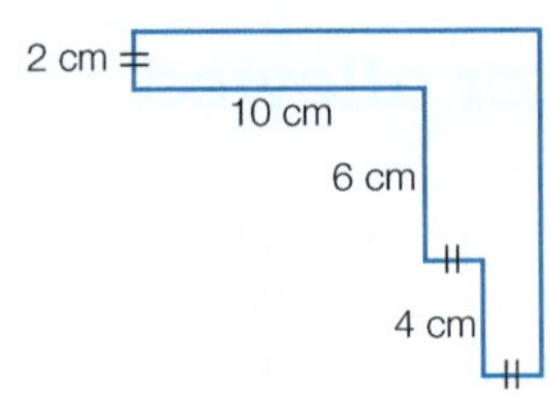

6 Find the following probabilities. **Ex. 9.7**

(a) getting a 6 on one roll of a normal die

(b) getting a red king on a single draw from a normal pack of cards

(c) getting two heads when tossing two coins

7 Two shapes have been made using wooden cubes, as shown. **Ex. 10.7**

(a)

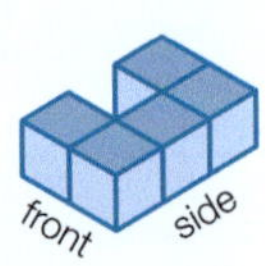

(b)

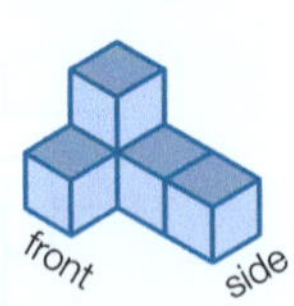

For each shape, draw: (i) the front elevation (ii) the top elevation.

8 Solve each of the following equations using the balance method. Check your answer by substitution. **Ex. 7.4**

(a) $3(x - 1) = 6$ (b) $\frac{x}{4} + 1 = 7$ (c) $2(x + 3) = 12$

9 Find the supplementary angles for each of the following. **Ex. 8.3**

(a) $35°$ (b) $127°$ (c) $90°$

10 The following numbers represent the heights of students in a class, to the nearest centimetre.

148, 163, 158, 162, 151, 153, 170, 167, 165, 162, 172, 173, 157, 162, 158, 155, 163, 159, 160, 161

Draw the frequency table showing the information. You will need to group the results. Use 145–149, 150–154, 155–159 etc.

Ex. 9.1

11 Show the image of each of the following figures after each of the following combined transformations. Use a protractor to help you.

Ex. 10.4

Reflect in the horizontal line of reflection and translate 4 units left and 3 units down.

(a)

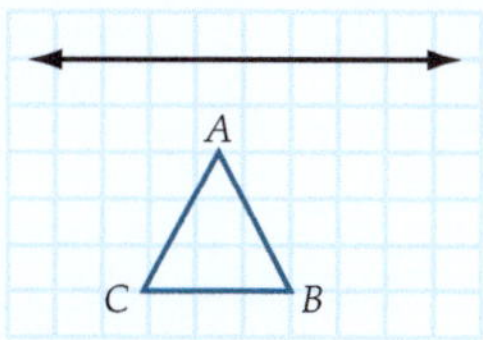

Reflect in the line *BC* and rotate 90° in a clockwise direction about point *C′*.

(b)

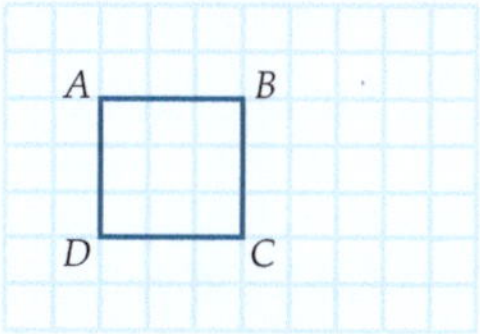

12 For each of the following, identify the number of axes of symmetry.

(a) isosceles triangle **(b)** square **(c)** rhombus

Ex. 10.5

13 Find the mean, mode, median and range of the following data, noting any possible outliers.

23, 34, 45, 26, 35, 42, 6, 34, 22, 36, 21, 29, 32, 34, 41, 39, 25, 27, 31, 30

Ex. 9.2

14 A figure is translated 7 units left, 4 units down, 8 units right and 3 units up. What would be the final position of the image compared to that of the original figure?

15 **(a)** Find the prime factors of 24.

(b) Find the prime factors of 30.

(c) Use the prime factors to find the highest common factor of 24 and 30.

(d) Use the prime factors to find the lowest common multiple of 24 and 30.

16 **(a)** What is the size of each angle in an equilateral triangle?

(b) A parallelogram has one angle that measures 126°. What size is each of the three remaining angles?

17 Which capital letters of the alphabet that have both reflectional and rotational symmetry also have reflectional and rotational symmetry in lower case form?

Reasoning and communicating

18 A point (1, 1) is translated 2 units right and 1 unit up. It is then reflected in the y-axis and then reflected in the x-axis. It is then translated 2 units right and 1 unit up. Finally, it is rotated 180° in a clockwise direction about the origin. What are the coordinates of the final image?

19 **(a)** Find a data set with at least six different values that has a median of 6 and a mean of 8.

(b) Describe the process you followed to find these values.

(c) Why might it be easier to use an odd number of values in a question of this type?

20 A game is played in which three different coins are being tossed. It costs 50 cents to play this game. You win $1 (plus your original 50 cents) if more than one of the coins shows heads, otherwise you lose your money.

(a) Write down the set of results that are possible.

(b) Is the game fair? Explain your reasoning.